CURRENT TRENDS AND CHALLENGES IN RFID

Edited by **Cornel Turcu**

Current Trends and Challenges in RFID
Edited by Cornel Turcu

Published by InTech
Janeza Trdine 9, 51000 Rijeka, Croatia

Publishing Process Manager Davor Vidic
Technical Editor Teodora Smiljanic
Cover Designer Jan Hyrat

First published July, 2011
Printed in Croatia

A free online edition of this book is available at www.intechopen.com
Additional hard copies can be obtained from orders@intechweb.org

Current Trends and Challenges in RFID, Edited by Cornel Turcu
p. cm.
ISBN 978-953-307-356-9

For EU product safety concerns:
IN TECH d.o.o., Prolaz Marije Krucifikse Kozulić 3, 51000 Rijeka, Croatia,
info@intechopen.com or visit our website at intechopen.com.

Meet the editor

Prof. Cornel Turcu was born in 1966 in Adjud, Romania. He received the B.Sc. and Ph.D. degrees in automatic systems, from the University of Iasi, Romania, in 1991, and 1999, respectively. He also holds a degree in Informatics (M.Sc.) from the University of Suceava, Romania. Since 1991, he has been with the Faculty of Electrical Engineering and Computer Science, University of Suceava (USV), where he is a full professor of System Theory and Intelligent Systems and also holds a joint appointment as head of Programmes Management Department. At USV he is also a supervisor for Ph.D. and M.S. theses. He has published over 70 research papers and 4 books. His research interests include intelligent systems, RFID systems and automatic control system design.

Contents

Preface

Radio-frequency identification (RFID) is a technology that uses communication through radio waves to transfer data between a reader and an electronic tag attached to an entity for the purpose of identification, tracking and surveillance. Unlike other identification technologies such as barcodes, RFID technology offers several key benefits such as no line-of-sight necessity, robustness, speed, bidirectional communication, reliability in tough environments, bulk detection, superior data capabilities, etc. Because of this, RFID has become particularly successful for a wide area of applications where traditional identification technologies are inadequate for recent demands: inventory tracking, supply chain management, automated manufacturing, healthcare, etc. As the RFID technology is being spread and applied to real world system, RFID systems have received considerable attention from researchers, engineers and industry personnel. As a result of years of research, a lot of literature has been published on the design and use of the RFID systems, covering a wide range of topics: hardware and software, protocols and algorithms, applications, etc.

This book presents some of the most recent research results of RFID users interested in exchanging ideas on the present development issues of and future trends in RFID technology. It consists in a collection of 24 chapters distributed in 5 parts: RF/RFID Backgrounds, Antennas/Tags, Readers, Protocols and Algorithms, and finally, Case studies/Applications.

The book starts with some background chapters related to Radio Frequency (*Chapter 1*), main RF structures (*Chapter 2*) and RF CMOS (*Chapter 3*). Also, this section contains a chapter that deals with structural design of a CMOS voltage regulator for an implanted device (*Chapter 4*).

The second section of the book focuses on antennas and tags. First, some perspectives and technical considerations of microstrip antennas for multi-band RFID reader are presented (*Chapter 5*). Also, the high gain dual-band antennas and limitations have been described. *Chapter 6* includes low-cost solution for RFID tags in terms of design and manufacture considering that applying the traditional printing technologies to produce the antennas will lower the cost of the antenna part. *Chapter 7* deals with conductive adhesives such as the ultralow cost RFID tag antenna material and includes results which are based on the screen printing method, which is very representative at the stage of lab prototyping. *Chapter 8* is a true experimental research in nature and aims to investigate the process consistency and accuracy of printing RFID tag antennas via the screening printing method with a conductive ink, silver-based (Ag) ink, on PET, PVC, and Wet Strength paper. *Chapter 9* presents a novel CPW-fed-slot antenna on artificial magnetic conductor (AMC) combination prototype suitable to be used in 5.8 GHz RFID tags on metallic objects. The last chapter of this section (*Chapter 10*) proposes a guideline for the design of a new kind of RFID tag to be used in each step of the pharmaceutical supply chain. It describes the main features of the pharmaceutical scenario, mainly focusing on item-level tracing systems and RFID devices' performance.

The third section of the book is dedicated to RFID readers. In *Chapter 11* the authors present a demodulation structure suitable for a reader baseband receiver in a passive RFID environment. *Chapter 12* introduces a new reader obtained by adding HDX functionality to an existing FDX reader, together with some design issues that influence reader performance.

After the chapters focusing on readers design, the following chapters present certain aspects related to protocols and algorithms. In *Chapter 13* the authors propose a new scalable authentication protocol for low-cost RFID systems, for which features are presented, both from the tag's and reader's perspective. *Chapter 14* focuses on an RFID model for simulating framed slotted ALOHA based anti-collision protocol for multi-tag identification. *Chapter 15* describes the implementation of direct sequence code division multiple access channel access methods for the UHF-RFID uplink. *Chapter 16* illustrates an unconditionally secure lightweight RFID authentication protocol with untraceability. *Chapter 17* deals with the application of Monte Carlo method for determining the interrogation zone in anti-collision Radio Frequency. In *Chapter 18*, in order to mitigate the influence of the NLOS propagation, the authors propose an iterative delay compensation algorithm based on NEWTON algorithm which improves the accuracy of positioning items using the DCF and shift vector compensation algorithm. Finally, in *Chapter 19*, an efficient spatial range query method is designed for compensating the lost spatial relationship by the linear mapping mechanisms. The experiments conducted on real data sets demonstrate that the proposed approach is efficient and scalable.

The fifth section of the book includes 5 chapters that describe several RFID applications and studies. *Chapter 20* presents some studies on secure RFID authentication and access control, while *Chapter 21* shows an overview of attacks on the HF physical layer of contactless and RFID systems. *Chapter 22* proposes an effective tag movement direction detection method. *Chapter 23* presents a distance measurement and position estimation application in order to evaluate a WSN system. Finally, in *Chapter 24*, cross-layer design is presented as an attractive tool to optimize RFID platforms. The proposed framework for design of RFID platforms can be potentially used for a wide variety of PHY and MAC algorithms under a cross-layer philosophy.

By presenting design issues related to each component of an RFID system, this book reaches its goal, that of being a collection of actual research results and challenges in RFID domain. It completes a collection of RFID books published by Intech, a collection that is a valuable tool for engineers, researchers and industry personnel, either those that are already familiar with RFID or new to this field.

Cornel Turcu
University of Suceava
Romania

Part 1

RF/RFID Backgrounds

1

Radio Frequency Background

Tales Cleber Pimenta, Paulo C. Crepaldi and Luis H. C. Ferreira
Universidade Federal de Itajuba
Brazil

1. Introduction

Design considerations for the traditional low frequency circuits and the RF circuits are quite different. In low frequency design, the maximum signal transfer occurs when the source presents low impedance while the load presents high impedance. A typical example is a buffer, where the input impedance is high and the output impedance is low. As long as that requirement is fulfilled, the designer is capable of choosing arbitrary levels of impedance that best suits the circuit requirements or applications.

Therefore this chapter aims to provide background on impedance matching between source and load, with or without a transmission line. The analysis can be conducted by using Smith Charts and S-Parameters, which are also presented in this chapter. The analysis in this chapter is oriented to RFID applications whereas other books provide general analysis.

During RF design, the impedances should be matched for maximum signal transfer. Additionally, when the circuits are connected using transmission lines, they should match also the standard values of the transmission lines. At very low frequencies, transmission lines can be thought as just a wire. Nevertheless, at high frequencies, the signal wavelength is comparable to or smaller than the length of the transmissions line and power can be seen as traveling waves. As a matter of fact, even a conductor can be thought as a transmission line in a high frequency circuit.

Most RF equipments and coaxial cables use the standard impedances of 50 or 75 Ω. The value of 75 Ω is used, as an example, in cable TV equipment, since this value provides the minimum losses, as it is desired in transmitting the signal over long distances. In fact, the value of impedance for minimum loss should be 77 Ω, but it was rounded to 75 Ω by convenience.

The value of 50 Ω corresponds to a reasonable compromise, the average, between the minimum loss of a 77 Ω and the maximum power handling capability given of 30 Ω.

2. Transmission line

Fig. 1 shows the lumped component model of a real (lossy) transmission line. The segment indicated corresponds to an infinitesimal segment of the transmission line. The characteristic impedance Z_0 of this line can be found to be [1]:

$$Z_0 = \sqrt{\frac{Z}{Y}} = \sqrt{\frac{R + j\omega L}{G + j\omega C}} \quad (1)$$

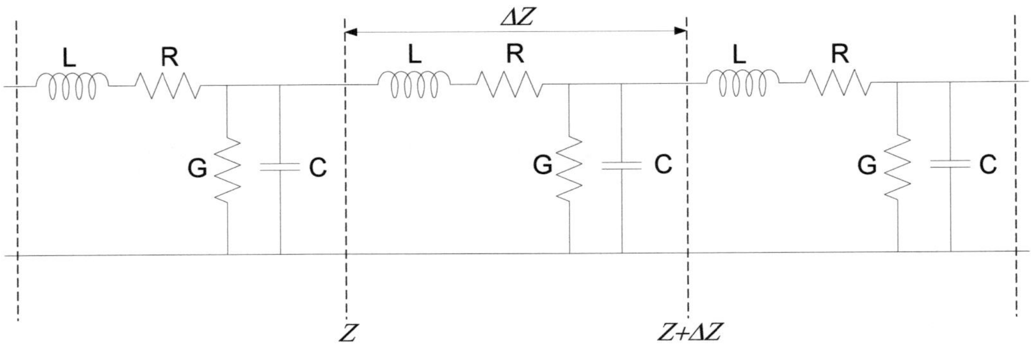

Fig. 1. Lumped component model of a transmission line.

As can be observed, the characteristic impedance Z_0 is dependent on the frequency. Nevertheless, if the resistive terms R and G are negligible, the expression of the characteristic impedance Z_0 can be simplified to:

$$Z_0 = \sqrt{\frac{L}{C}} \tag{2}$$

If the value of RC is equal to GL, expression (1) yields the same value of expression (2). In other words, choosing the L/R time constant of the series impedance equals to the C/G time constant, a lossy line will behave as a lossless line, so that its characteristic impedance will be independent of the frequency[1].

2.1 Reflection coefficient

If a transmission line is terminated by an impedance Z_0, then a signal traveling down the line with a ratio of voltage to current equal to Z_0 will maintain its ratio upon encountering the load and there will be no reflections. On the other hand, when the load is different of Z_0, then it imposes its own particular ratio of voltage to current, and the only way to reconcile the conflict is by having some of the signal reflected back towards the source. In order to distinguish the incident and the reflected signals, the subscripts i and r, respectively, will be used.
The incident signal is given by:

$$Z_0 = \frac{E_i}{I_i} \tag{3}$$

At the load end, the mismatch in impedances gives rise to a reflected signal. Since the system is still linear, the total voltage at any point in the system is the sum of incident and reflected voltages. The net current is superposition of incident and reflected currents. However, since the currents are traveling in opposite directions, the net current is the difference between them. Therefore, the load impedance is given by:

$$Z_L = \frac{E_i + E_r}{I_i - I_r} \tag{4}$$

Expression (4) can be rewritten to express Z_L as function of Z_0 as:

$$Z_L = \frac{E_i + E_r}{I_i - I_r} = \frac{E_i}{I_i}\left[\frac{1 + E_r / E_i}{1 - E_r / E_i}\right] = Z_0\left[\frac{1 + E_r / E_i}{1 - E_r / E_i}\right] \quad (5)$$

The ratio of reflected to incident quantities at the load end of the line is called Reflection Coefficient Γ_L. Therefore, expression (5) can be rewritten as:

$$Z_L = \frac{E_i + E_r}{I_i - I_r} = Z_0\left[\frac{1 + \Gamma_L}{1 - \Gamma_L}\right] \quad (6)$$

Solving for Γ_L yields:

$$\Gamma_L = \frac{Z_L - Z_0}{Z_L + Z_0} \quad (7)$$

As can be observed from expression (7), if the impedances of the load and the line are equal, there will be no reflection. If the line is terminated in either a short or an open circuit, the reflection will be maximum, with a magnitude of 1 [1].

Therefore, if a transmission line is terminated by its characteristic impedance there will be no reflection since all the transmitted power is absorbed by the load and the energy flows in just one direction.

When the line is terminated by a short circuit a reflected wave is sent back to the source since the short can not sustain any voltage, and therefore dissipates zero power. The incident and the reflected voltage waves are of the same magnitude. They are 180º out of phase at the load and they travel in opposite directions.

If the line is terminated by an open circuit a reflected wave is sent back to the source since the open can not sustain any current, and therefore dissipates zero power. The incident and the reflected current waves are of the same magnitude and travel in opposite directions. The current waves are 180º out of phase at the load, but the incident and reflected voltage waves are in phase [1].

If the line is terminated by an impedance different of the short, open and characteristic impedance, part of the signal will be absorbed by the load and part will be reflected back. The amount of reflected signal is given by expression (7).

3. Smith chart

The reflection coefficient Γ_L of expression (7) was obtained from expression (6). By the same way, solving for Z_L in expression (7) yields Γ_L, thus forming a mapping of one complex number into another. The relationship between these two complex numbers forms a bilinear transformation, which means that knowing one is equivalent to knowing the other.

Since Z_L can have any value and $|\Gamma_L|$ cannot exceed unity for passive loads, it is therefore much more convenient plotting Γ_L than plotting Z_L.

The reflection coefficient can become even more convenient by normalizing it to Z_0, as:

$$\Gamma = \frac{\frac{Z_L}{Z_0} - 1}{\frac{Z_L}{Z_0} + 1} = \frac{Z - 1}{Z + 1} \quad (8)$$

On the same way, normalizing (6) results in:

$$Z = \frac{1+\Gamma}{1-\Gamma} \tag{9}$$

Considering the normalized real and imaginary parts of both Γ and Z then:

$$Z = R + jX = \frac{1+\Gamma}{1-\Gamma} = \frac{1+\Gamma_r + j\Gamma_i}{1-\Gamma_r - j\Gamma_i} \tag{10}$$

After some algebraic manipulation (using conjugate), the real and imaginary parts are of Z are:

$$R = \frac{1-\Gamma_r^2-\Gamma_i^2}{1+\Gamma_r^2-2\Gamma_r+\Gamma_i^2} \tag{11}$$

$$X = \frac{2\Gamma_i}{1+\Gamma_r^2-2\Gamma_r+\Gamma_i^2} \tag{12}$$

Expression (11) can be manipulated as:

$$\begin{aligned}
&R + R\Gamma_r^2 - 2R\Gamma_r + R\Gamma_i^2 = 1-\Gamma_r^2-\Gamma_i^2\\
&R\Gamma_r^2 - 2R\Gamma_r + R\Gamma_i^2 + \Gamma_r^2 + \Gamma_i^2 = 1-R\\
&R\Gamma_r^2 - 2R\Gamma_r + R\Gamma_i^2 + \Gamma_r^2 + \Gamma_i^2 = (1-R)\frac{(1+R)}{(1+R)}\\
&R\Gamma_r^2 - 2R\Gamma_r + R\Gamma_i^2 + \Gamma_r^2 + \Gamma_i^2 = \frac{1}{(1+R)} - \frac{R^2}{(1+R)}\\
&\Gamma_r^2(1+R) - 2R\Gamma_r + \frac{R^2}{(1+R)} + \Gamma_i^2(1+R) = \frac{1}{(1+R)}\\
&\Gamma_r^2 - 2\Gamma_r\frac{R}{(R+1)} + \frac{R^2}{(1+R)^2} + \Gamma_i^2 = \frac{1}{(1+R)^2}\\
&\left(\Gamma_r - \frac{R}{R+1}\right)^2 + \Gamma_i^2 = \left(\frac{1}{1+R}\right)^2
\end{aligned} \tag{13}$$

Similarly, expression (12) into:

$$(\Gamma_r - 1)^2 + \left(\Gamma_i - \frac{1}{X}\right)^2 = \frac{1}{X^2} \tag{14}$$

When the two parametric equations (13) and (14) are drawn on a complex coordinate, they build the Smithchart. Equation (13) forms resistance circles, and equation (14) generates reactance circles, as shown in Fig. 2 and Fig. 3, respectively. The resulting Smithchart is illustrated in Fig. 4.

As can be verified from expression (13), the imaginary axis in the Z-plane (resistance equals 0) is mapped as a unity circles into Γ-plane. The other lines of constant resistance in the Z-plane are also mapped as circles, but of different diameter in the Γ-plane. Nevertheless, they are all tangent at the point $\Gamma=1$, as shown in Fig. 2. The larger the value of the resistance, the smaller becomes the circle [1, 2].

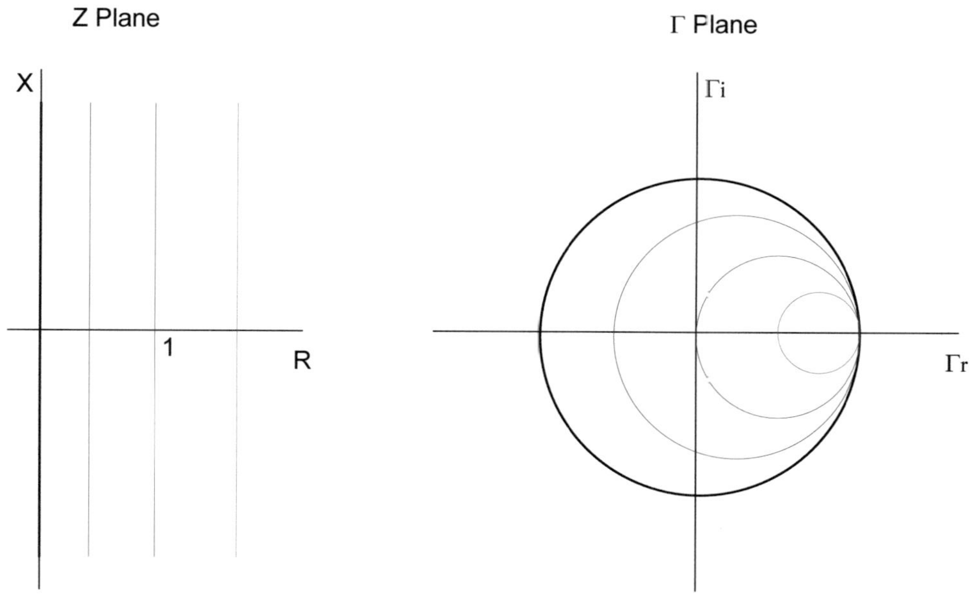

Fig. 2. Mapping of Z-plane into Γ-plane for constant resistances.

As can be verified from expression (14), lines of constant reactance are perpendicular to lines of constant resistance in the Z-plane. This same orthogonality is preserved in the Γ-plane by having arcs perpendicular to the constant resistance lines, as indicated in Fig. 3.

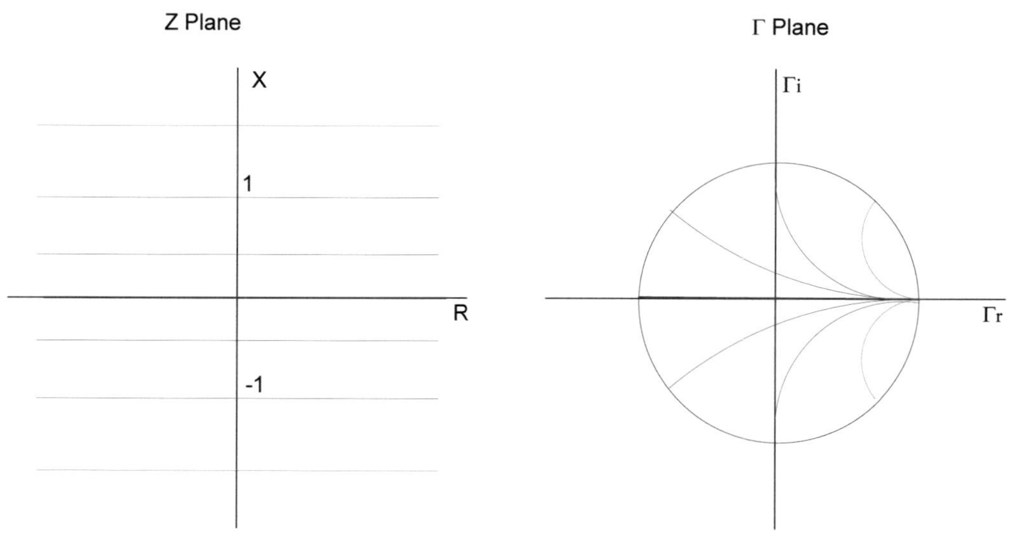

Fig. 3. Mapping of Z-plane into Γ-plane for constant reactances.

The Smith, as shown in Fig. 4, is just the plotting of both constant resistance and constant reactance, but without the presence of the Γ axes. The center of the Smith chart corresponds to zero reflection (Z_L equals Z_0)[1].

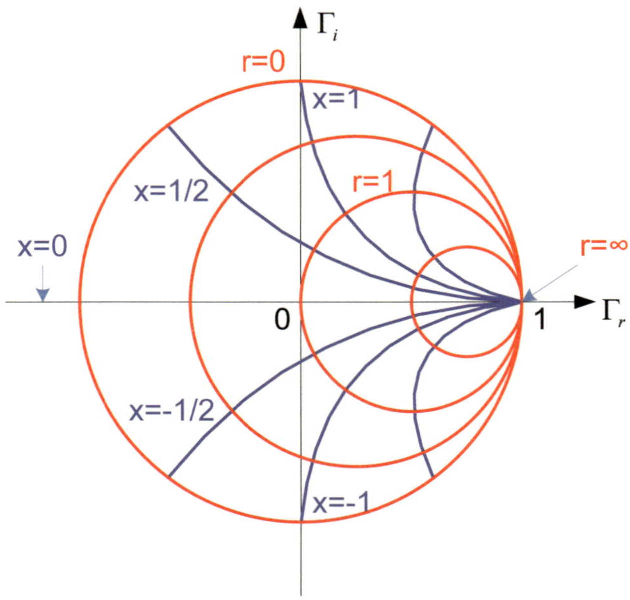

Fig. 4. Smith chart.

The upper half of the Smith chart corresponds to the upper half part of the Z-plane, and therefore presents inductive loads. On the same way, the bottom half of the Smith chart corresponds to the bottom half part of the Z-plane, thus representing capacitive loads. Obviously, the Re axis of the Smith chart represents purely resistive loads.
Although the Smith chart presents many interesting and useful properties, they will no be presented here due to the focus of this material.

2.1 Admittance chart

The Smithchart is built by considering impedance (resistance and reactance). Once the Smithchart is developed, a similar approach can be used for admittance analysis. The concept that admittance is the inverse of impedance is very important for parallel circuit synthesis. Adding new elements in series can be resolved easily by adding the impedance values. However, summing elements in parallel can be cumbersome in terms of impedance. Thus, admittance is often considered for parallel elements.
By definition, admittance is expressed as:

$$Y = \frac{1}{Z} = G + jB \tag{15}$$

where G is conductance and B is susceptance of the element. The reflection coefficient Γ can be expressed in terms of normalized admittance as:

$$\Gamma = \Gamma_r + j\Gamma_i = \frac{Z_L - Z_0}{Z_L + Z_0} = \frac{\frac{1}{Y_L} - \frac{1}{Y_0}}{\frac{1}{Y_L} + \frac{1}{Y_0}} = \frac{Y_0 - Y_L}{Y_0 + Y_L} = \frac{1-y}{1+y} = \frac{1-g-jb}{1+g+jb} \quad (16)$$

The admittance reflection coefficient $\Gamma(y)$ is given by definition as:

$$\Gamma_{(Y)} = \frac{Y_L - Y_0}{Y_L + Y_0} \quad (17)$$

As can be observed, the value of the admittance reflection coefficient $\Gamma(y)$ is equal to $-\Gamma$, the reflection coefficient in terms of impedance. Therefore, once Γ is known, $\Gamma(y)$ can be located a point at the same distance from the origin. Rotating the admittance point by 180° around the origin achieves the same result.
The admittance Smithchart can be obtained using the same procedure used to construct the impedance Smithchart. The normalized real and imaginary parts of Γ and Y can be given as:

$$Y = G + jB = \frac{1 - \Gamma_r^2 - \Gamma_i^2 - j2\Gamma_i}{1 + \Gamma_r^2 + 2\Gamma_i + \Gamma_i^2} \quad (18)$$

After some algebraic manipulation (using conjugate), the real and imaginary parts are of Y are:

$$G = \frac{1 - \Gamma_r^2 - \Gamma_i^2}{1 + \Gamma_r^2 + 2\Gamma_r - \Gamma_i^2} \quad (19)$$

$$B = \frac{-2\Gamma_i}{1 + \Gamma_r^2 + 2\Gamma_r + \Gamma_i^2} \quad (20)$$

Using the same procedure presented for expressions (13) and (14), then the parametric equations of the admittance Smithchart are:

$$\left(\Gamma_r + \frac{G}{G+1}\right)^2 + \Gamma_i^2 = \left(\frac{1}{1+G}\right)^2 \quad (21)$$

$$(\Gamma_r + 1)^2 + \left(\Gamma_i + \frac{1}{B}\right)^2 = \frac{1}{B^2} \quad (22)$$

When the two parametric equations (21) and (22) are drawn on a complex coordinate, they build the Admittance Smithchart. Equation (21) forms resistance circles, and equation (22) generates reactance circles, as shown in Fig. 5.

3. S Parameters

At low frequencies, linear systems can be analyzed by means of voltages and currents applied to its ports. The two port circuit shown in Fig. 5 could be analyzed from its impedance (Z-parameters), admittance (Y-parameters), or a mixture of them, which could be hybrid (H-parameters) and inverse-hybrid (G-parameters).

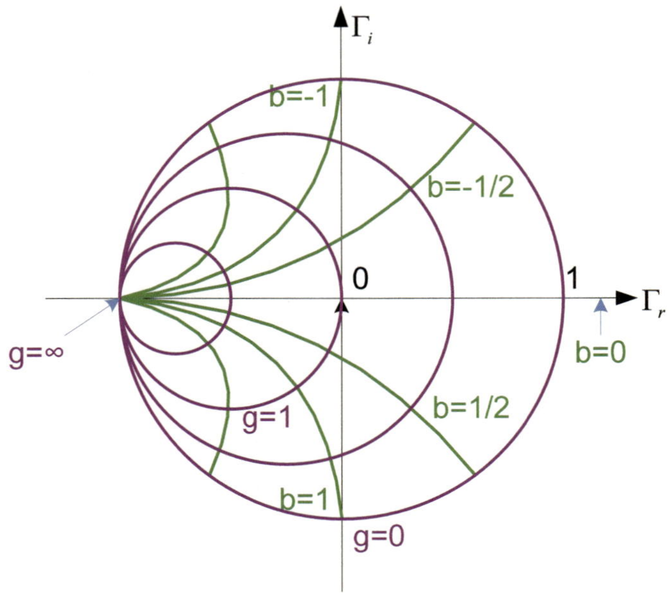

Fig. 5. Admittance Smithchart.

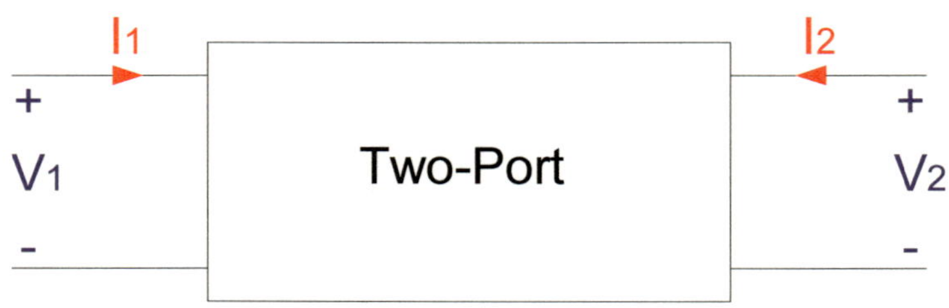

Fig. 6. Two port circuit representation.

As an example, the circuit of Fig. 5 could be analyzed using H parameters whose equations are:

$$\begin{aligned} V_1 &= h_{11}I_1 + h_{12}V_2 \\ I_2 &= h_{21}I_1 + h_{22}V_2 \end{aligned} \tag{16}$$

As can be observed from the equations, the value of h_{11} (port 1 impedance) can be obtained directly from the relationship of V_1 and I_1 when V_2 is set to zero. A voltage source is set to zero by shortening its terminals. The value of h_{21} (current gain from port 1 to port 2) is obtained from the relationship of I_1 and I_2 also when V_2 is set to zero.

By the same way, h_{12} (voltage gain from port 2 to port 1) can be obtained from the relationship of V_1 and V_2 when I_1 is set to zero. A current source is set to zero by opening its

terminals. Finally, h_{21} (port 2 conductance) is obtained from the relationship of V_2 and I_2 when I_1 is set to zero.
Shorting or opening terminals is feasible at low frequencies but virtually impossible at high frequencies, particularly over a broad range of frequencies. Additionally, RF circuits are very sensitive to impedances, and they may oscillate or just quit working when terminated with open or short circuits. Therefore, Z-parameters, Y-parameters, H-parameters and G-parameters are not suitable to high frequency operations.
For high frequency operations, the *scattering* parameters (or just S-parameters) are employed. The input and output variables in S-parameters are based on incident and reflected voltage waves instead of voltages and currents.
The S-parameters takes advantage of the fact that there is no reflection in a line terminated in its characteristic impedance. Therefore, it is necessary a circuit representation for S-parameters, where source and the load terminations are Z_0, as shown in Fig. 6.
The S-parameters equations are:

$$\begin{aligned} b_1 &= s_{11}a_1 + s_{12}a_2 \\ b_2 &= s_{21}a_1 + s_{22}a_2 \end{aligned} \tag{16}$$

where

$$\begin{aligned} a_1 &= E_{i1} / \sqrt{Z_0} \\ a_2 &= E_{i2} / \sqrt{Z_0} \\ b_1 &= E_{r1} / \sqrt{Z_0} \\ b_1 &= E_{r2} / \sqrt{Z_0} \end{aligned} \tag{17}$$

The normalization by $\sqrt{Z_0}$ is very convenient since the square of a and b corresponds to the power of the incident and reflected waves.
S_{11} and S_{21} can be obtained by measuring the incident, the reflected and the transmitted signals at the input when the output is terminated in Z_0. Once the output is terminated by Z_0 there is no reflection. The values of S_{11} and S_{21} are:

$$\begin{aligned} S_{11} &= \frac{b_1}{a_1} = \frac{E_{r1}}{E_{i1}} = \Gamma_1 \\ S_{21} &= \frac{b_2}{a_1} = \frac{E_{r2}}{E_{i1}} \end{aligned} \tag{18}$$

Similarly, S_{12} and S_{22} can be obtained by measuring the incident, the reflected and the transmitted signals at the output when the input is terminated in Z_0. Since the input is terminated by Z_0 there is no reflection. The values of S_{12} and S_{22} are:

$$\begin{aligned} S_{12} &= \frac{b_1}{a_2} = \frac{E_{r1}}{E_{r2}} \\ S_{22} &= \frac{b_2}{a_2} = \frac{E_{r2}}{E_{i2}} = \Gamma_2 \end{aligned} \tag{19}$$

S_{11} corresponds to the input reflection coefficient, S_{21} is the input to output (direct) gain, S_{12} is the reverse transmission gain and S_{22} corresponds to the output reflection coefficient.
The magnitudes of S_{11} and S_{22} are always less than 1, where a value of zero represents a perfect matching (no reflections), and the closer to 1, the higher the reflection.
The magnitudes of S_{21} (transfer characteristic) and S_{12} (reverse) are smaller than 1 for passive circuits but can exceed 1 for active circuits (amplification). A positive value means the input and output signal are in phase and a negative indicates a phase shift.

Fig. 7. Two port circuit representation for S-parameters.

3.1 Measurements of S-Parameters

The circuit topology used to measure S-Parameters is given in Fig. 7.

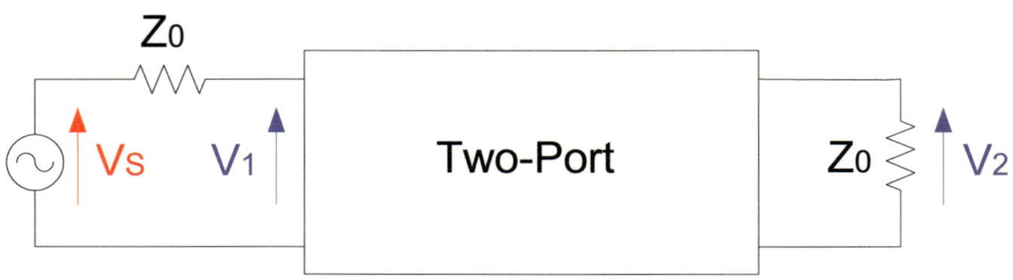

Fig. 8. Circuit topology used to measure S-Parameters.

The input reflection coefficient S_{11} from expression (18) can be modified as:

$$S_{11} = \Gamma_1 = \frac{Z_L - Z_0}{Z_L - Z_0} = 2.\frac{Z_L}{Z_L - Z_0} - 1 \quad (20)$$

This expression, using the concepts of voltage division, corresponds to:

$$S_{11} = \Gamma_1 = 2.\frac{Z_L}{Z_L - Z_0} - 1 = 2.\frac{V_1}{V_S} - 1 \quad (21)$$

Here, Z_L corresponds to the input impedance of the two-port circuit. The value of the input to output gain, S_{21} is given as:

$$S_{21} = 2.\frac{V_2}{V_S} \quad (22)$$

The value of S_{22} and S_{12} can be obtained in a similar way by just exchanging input and output terminals.

3.1 S Parameters in the Smith chart

The center point of the Smith chart corresponds to the point of zero reflection, where Z_L equals Z_0. Plots of S_{11} and S_{22} on the real axis represent ohmic resistors, above the axis indicate inductive load while below the axis indicate capacitive loads.
Plots of S_{12} and S_{21} inside the Smith chart indicate that damping signal between ports, whereas plots outside the chart indicate amplification [1].
As the frequency increases, the S-Parameters plots in the Smith chart move clockwise.
Given the value of S_{11}, the circuit impedance can be found from (6), as:

$$Z_L = Z_0 \left[\frac{1+S_{11}}{1-S_{11}} \right] \tag{23}$$

3.2 Application example

The S_{11} and S_{21} parameters can be obtained as given by expressions (21) and (22), indicated in Fig. 7, and the values are calculated as for S_{11} and S_{21}, respectively. As an application example, consider the cascode amplifier shown in Fig. 8 [2].

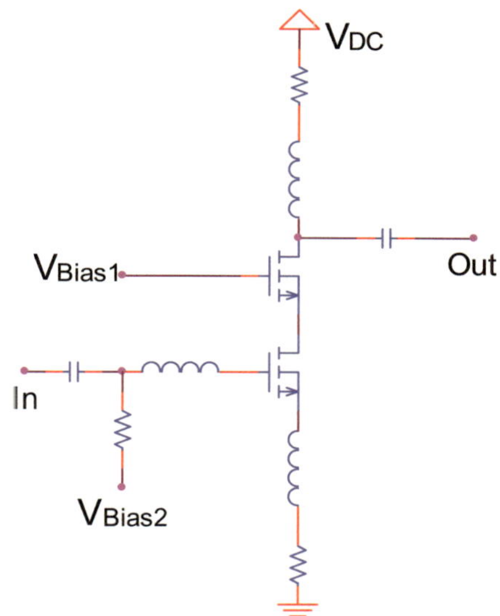

Fig. 9. Cascode amplifier.

Fig. 10 shows its S_{11} parameter, in the Smith chart. The frequency is ranging from 4GHz to 6GHz. As the frequency increases, the plot moves clockwise. At approximately 5GHZ, the circuit presents a pure resistive impedance of approximately 50Ω (it crosses the horizontal axis). The circuit presents a capacitive behavior for frequencies bellow 5HGz and an inductive behavior for higher frequencies [1].
The same parameters could be plotted in a standard dB format, as shown in Fig. 11. The graph of Fig. 10 provides more information and insight than the graph of Fig. 11. The last one provides only the magnitude, whereas the first one provides both the imaginary and real part, so that it is possible to infer a capacitive and/or inductive behavior of the circuit, among other information.

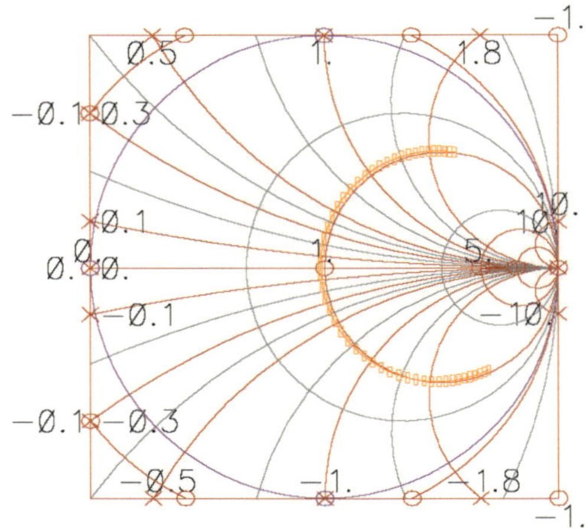

Fig. 10. S_{11} parameter of the circuit from Fig. 8, in a Smith chart.

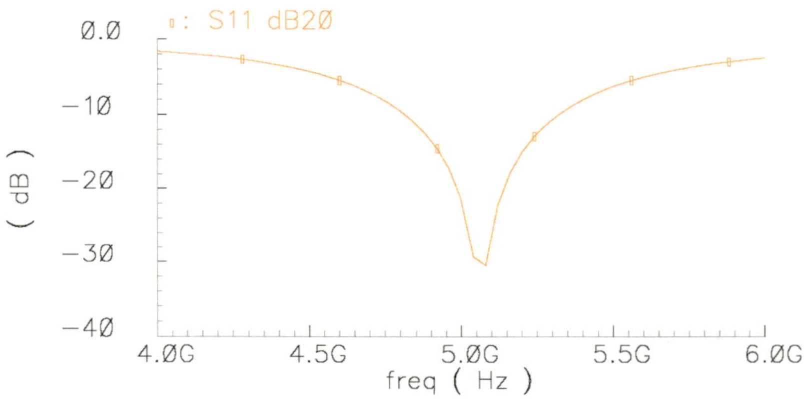

Fig. 11. S_{11} parameter of the circuit from Fig. 8, in a dB chart.

Unfortunately, it is not always possible to analyze S-parameters using Smith chart. One such case is S_{11} that is usually larger than 1 for active circuits. If it is larger than one, it does not fit the Smith chart!

4. Noise figure/factor

In analog circuits at low frequency, the signal-to-noise figure (SNR), defined as the ratio of signal power to the noise power, is an important and very used parameter. As an example, in a radio receiver, it indicates the quality of the demodulated signal [2-5].

Nevertheless, as the signal passes through the RF circuits, the SNR changes. This signal-to-noise degradation along the system is described by the noise factor (F), as:

$$F = \frac{SNR_i}{SNR_o} = \frac{S_i/N_i}{S_o/N_o} \tag{24}$$

where the index ***i*** refers to input and the index ***o*** refers to output.

If a system has no noise, then $SNR_o = SNR_i$ regardless of the gain, and $F=1$. This would be case of a hypothetical noiseless amplifier. Therefore, the noise figure of a real system will be always larger than 1.
Considering A as the system gain, expression (24) can be modified to:

$$F = \frac{S_i/N_i}{S_o/N_o} = \frac{S_i/N_i}{S_i.A/N_o} = \frac{N_o}{A.N_i} \tag{25}$$

which can be seen as the total output noise power over the output noise due to the input source.
The total output noise is the sum of the original noise at the input (which was amplified) and the noise added by the circuit. This can be denoted as:

$$N_o = A.N_i + N_{added} \tag{26}$$

Therefore, expression (18) can be expressed also as:

$$F = \frac{N_o}{A.N_i} = \frac{A.N_i + N_{added}}{A.N_i} = 1 + \frac{N_{added}}{A.N_i} \tag{27}$$

Again, if the circuit adds no noise, F becomes 1.
Another important figure of merit is the noise figure, NF, expressed as:

$$NF = 10.\log_{10} F \tag{28}$$

While the noise factor of a noiseless circuit is 1, the noise figure is 0dB.

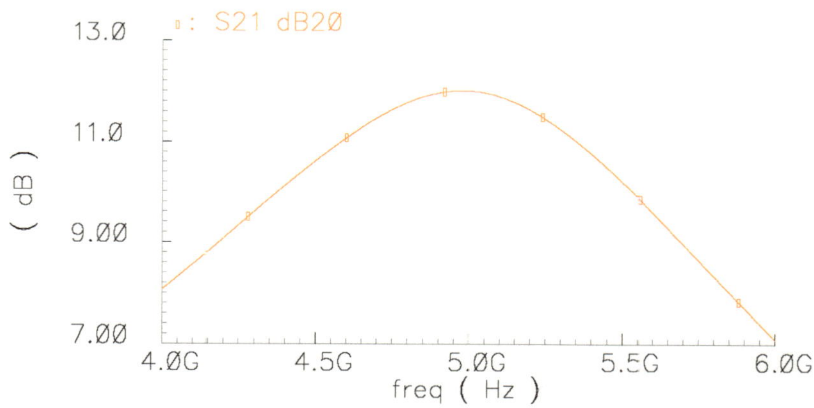

Fig. 12. S_{21} parameter of the circuit Fig. 8, in a dB chart.

4.1 Noise figure of a cascade system

Fig. 12 shows a cascade amplifying system, whose gain of each stage is A_i and the noise added by each stage is $N_{i\text{-}added}$.
The output noise due to the source is:

$$N_{o-source} = N_i.A = N_i.A_1.A_2.A_3 \tag{29}$$

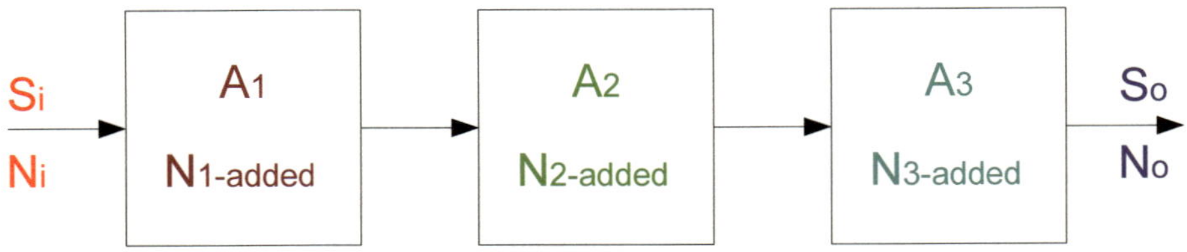

Fig. 13. Cascade amplifying system.

While the total output noise is:

$$N_o = N_i.A_1.A_2.A_3 + N_{1-added}.A_2.A_3 + N_{2-added}.A_3 + N_{3-added} \tag{30}$$

which is the input noise multiplied by the gain of the three stages, plus the noise generated by the first stage and amplified by the stages 2 and 3, plus the noise generated by the second stage and amplified by the last stage, and plus the noise generated by the last stage.
Thus, combining expressions (29) and (30) into expression (25), the noise factor can be found as:

$$F = \frac{N_o}{N_{o-source}} = 1 + \frac{N_{1-added}}{N_i A_1} + \frac{N_{2-added}}{N_i A_1 A_2} + \frac{N_{3-added}}{N_i A_1 A_2 A_3} \tag{31}$$

This expression, with the aid of expression (27), can be re-written as:

$$F = F_1 + \frac{F_2 - 1}{A_1} + \frac{F_3 - 1}{A_1 A_2} \tag{32}$$

As can be observed from expression (32), the noise factor of the first stage is the most relevant to the total noise factor. That is the reason for putting most effort in the first stage in terms of noise minimization, thus requiring low noise amplifiers at the front of the system.

5. Conclusions

The basic knowledge of impedance matching between source and load, either with or without a transmission line is essential to the design of RF circuits. The analysis presented can be conducted by using Smith Charts and S-Parameters. The analysis in this chapter was oriented to RFID applications whereas other books provide general analysis.

6. References

Kurokawa, K. (1965). Power Waves and the Scattering Matrix, *IEEE Trans. Microwave Theory and Tech.*, v.13, March 1965, pp. 194-202, ISSN 0018-9480.

Lee, T. H. (2004). *The Design of CMOS Radio-Frequency Integrated Circuits – 2nd Edition*, Cambridge University Press, ISBN 0521835399.

Rogers, J. & Plett, C. (20030 *Radio Frequency Integrated Circuit Design*, Artech House Inc, ISBN 1607839792.

Coleman, C. (20040 *An Introduction to Radio Frequency Engineering*, Cambridge University Press, ISBN 0521834813.

Razavi, B. (1998) *RF Microelectronics*, Prentice Hall, ISBN 0138875715.

2

Main RF Structures

Tales Cleber Pimenta, Paulo C. Crepaldi, Luis H. C. Ferreira,
Robson L. Moreno and Leonardo B. Zoccal
Universidade Federal de Itajuba
Brazil

1. Introduction

The low noise amplifiers – LNA and the mixers are among the most used structures used in RF integrated circuits. Therefore the goal of this chapter is to present an analysis overview of them as well as the main considerations of their design. Nevertheless, since their interconnections play an important role on performance and noise isolation, this chapter will also describe their AC and DC coupling.

2. Inter-connection

Consider initially a simple common source amplifier stage, with the load impedance Z_L, as given in Fig. 1. Consider also the simplified transistor model as shown in Fig. 2.

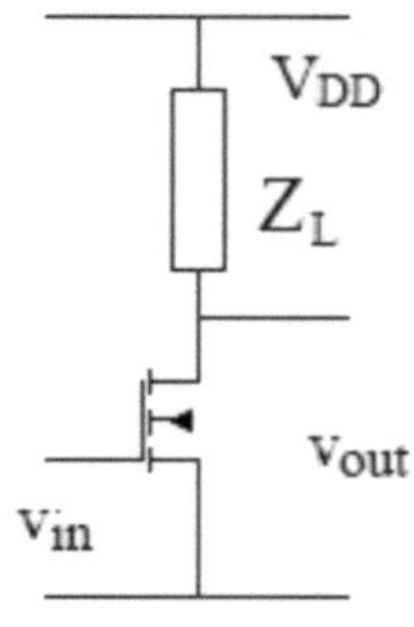

Fig. 1. Simple common source stage.

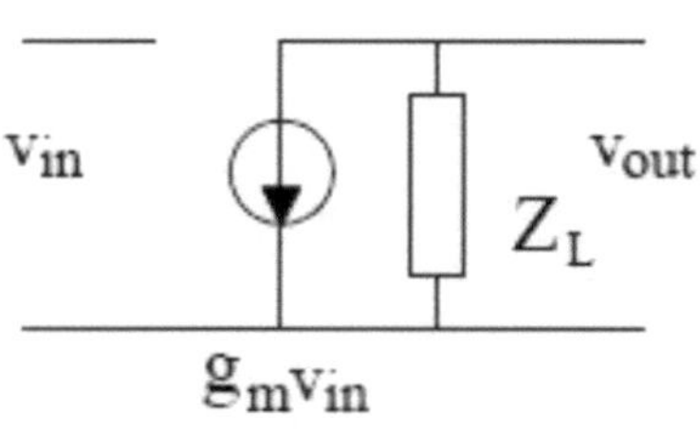

Fig. 2. Simplified circuit model of Fig. 1.

The gain of this stage can be easily calculated as below.

$$A_v = \frac{V_{out}}{V_{in}} = -g_m Z_L \tag{1}$$

Assume this common source stage drives the gate of the following circuit. This next stage needs proper biasing. Usually the DC bias of one stage does not interfere with the bias of another stage, thus the output of the common source stage and the input of the following stage are separated by a DC block capacitor as indicated in Fig. 3. Usually, the biasing resistor R_{bias} is large enough to prevent RF or analog signal from flowing into a bias source.

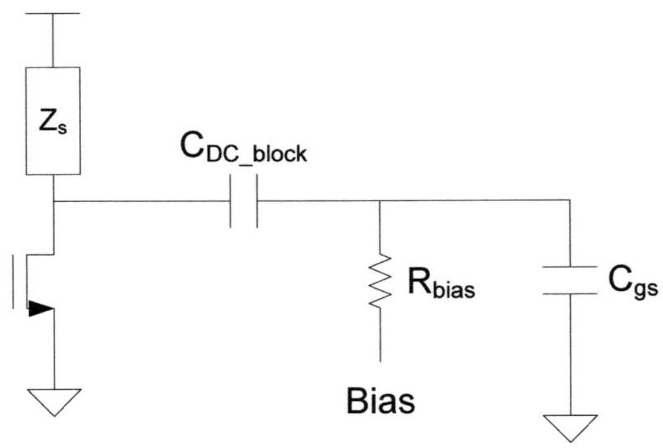

Fig. 3. AC coupling with DC block capacitor.

At high frequency, the effect of the DC block capacitor is negligible, since the DC block capacitor is virtually short. The effect of the large biasing is also negligible since it is connected in parallel with the drain resistor of the first stage. The simplified circuit model is presented in Fig. 4.

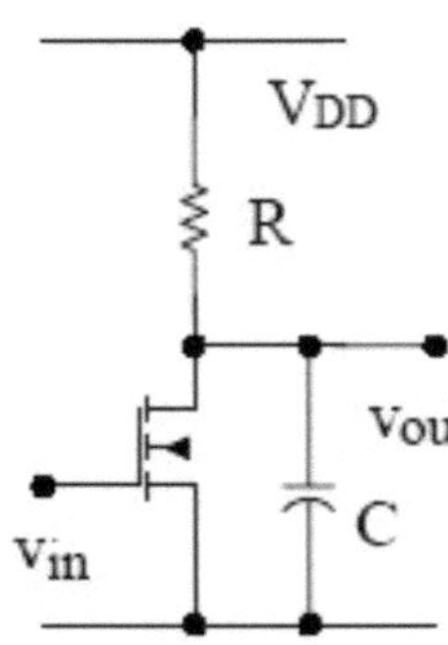

Fig. 4. Common-source stage with RC-load.

When no inductor is used, the only available load is RC-load, thus the load amplifier becomes:

$$Z_L = \frac{R}{1 + RCs} \tag{1}$$

The circuit topology of Fig. 4 is well known and corresponds to a low pass filter configuration. The frequency response of this filter is shown in Fig. 5 and its gain is given by:

$$A_v = -\frac{g_m R}{1 + RC} \tag{2}$$

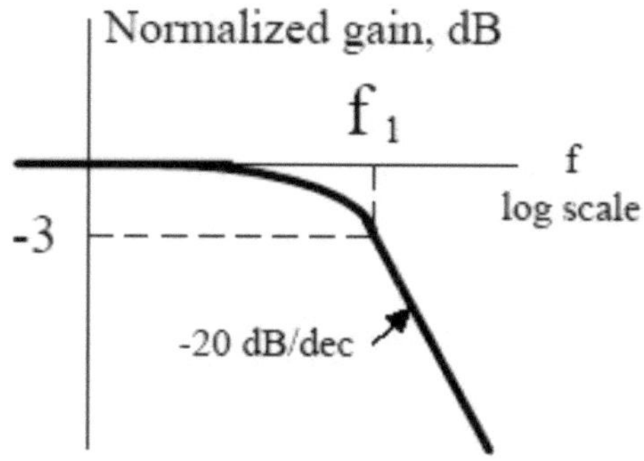

Fig. 5. Frequency response of Fig. 4.

Hence, the DC gain is $-g_m R$ and the bandwidth is $\omega_1 = 1/RC$. The frequency ω_1 is called "uncompensated bandwidth". After frequency ω_1 the gain decreases at the rate of -20dB/dec.
Unfortunately, at low frequency, the effects of DC block capacitor and the bias resistor are more severe, since the equivalent circuit, at low frequency, is a high pass filter.
Therefore, one can expect huge loss of information at very low frequencies for some applications such as Direct-Conversion transceiver, which carries information around DC. Even for DC-free applications, the cutoff frequency should be considered for the high pass filter. Since the 3dB cutoff frequency is defined as 1/RC, one can increase resistance and capacitance. However the capacitor normally occupies more space in integrated circuit than a resistance. Therefore, only the resistance should be increased, only up to few Mega ohms, so that a smaller capacitance can be used.

2.1 DC coupling

As reviewed in previous section, the AC coupling is suitable for RF circuitry, but may present DC blocking problems for baseband analog circuitry. Thus, if information around DC is concerned, one should integrate blocks with DC coupling. The DC coupling consists of combining two blocks so that the DC output voltage level of the previous block is same as the DC input bias voltage of the following block, and thus there is no reason to insert a DC block capacitor between them. The DC coupling is certainly advantageous at low frequency, and since the common-source stage model of Fig. 4 is valid for both low and high frequency, it is also suitable for high frequencies; nevertheless it may restrict the freedom of biasing.
Since the modern integrated technology allows construction of inductors, the designer should know the advantages the inductor can add in the circuit design. This section shows how to enhance the bandwidth using the 'shunt-peaking' technique. It consists of adding an inductor in series with the resistor, as shown in Fig. 6.

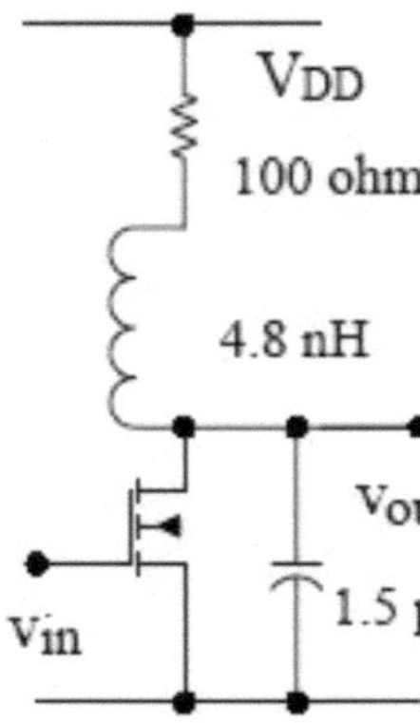

Fig. 6. Common-source stage with RLC-load.

The load impedance for this case becomes:

$$Z_L = (R + Ls) \,||\, \frac{1}{Cs} = \frac{R + Ls}{(R + Ls)Cs + 1} \tag{3}$$

And, substituting this value in (2), one can find that:

$$A_v = -\frac{g_m(R + Ls)}{(R + Ls)Cs + 1} = -\frac{g_m R\left[s(L/R) + 1\right]}{s^2 LC + sRC + 1} \tag{4}$$

Observe that the inductor added a zero, which always increases the bandwidth, and also two poles. These poles can be complex conjugate, and this also can increase bandwidth, yet they introduce peaking, hence the name of the method. On the other side, the difference between the number of finite poles and finite zeros is still one. This means that the asymptotic decrease of gain is the same as in the previous circuit, -20 dB/dec. Thus the inductor allows modifying the gain locally, in the vicinity of the frequency ω_1, and the designer should use this possibility to his/her advantage.
Consider the amplitude of the frequency response for this circuit, given as

$$|A_v(j\omega)| = g_m R \sqrt{\frac{(\omega L/R)^2 + 1}{\left(1 - \omega^2 LC\right)^2 + (\omega RC)^2}} \tag{5}$$

To facilitate subsequent derivations, it is introduced a factor *m*, defined as the ratio of the *RC* and $\tau = L/R$ time constants,

$$m = \frac{RC}{L/R} = \frac{R^2}{L/C} = \frac{R^2}{\rho^2} \tag{6}$$

Here $\rho = \sqrt{L/C}$ is the wave resistance of the load. This allows writing two more useful relationships, namely, $\tau^2 m = \frac{L^2}{R^2}\frac{R^2}{L/C} = LC$ and $\tau m = \frac{L}{R}\frac{R^2}{L/C} = RC$. Using these relationships (5) can be written as:

$$\frac{A_v(j\omega)}{g_m R} = \sqrt{\frac{(\omega\tau)^2 + 1}{\left(1 - \omega^2\tau^2 m\right)^2 + (\omega\tau m)^2}} \tag{7}$$

The right side of (7) is considered the normalized gain.
First, the bandwidth will be maximized without any consideration regarding the behavior to the gain in the bandwidth. The frequency where the right side equals $1/\sqrt{2}$ is denoted as ω_{-3dB}. Considering a new parameter defined as $x = \omega_{-3dB}\,\tau$, then one has the equation:

$$2\left(x^2 + 1\right) = \left(1 - x^2 m\right)^2 + (xm)^2 \tag{8}$$

or

$$x^4 m^2 + x^2\left(m^2 - m - 2\right) - 1 = 0 \tag{9}$$

From this equation one can find that:

$$x^2 m^2 = -\frac{m^2}{2} + m + 1 + \sqrt{\left(-\frac{m^2}{2} + m + 1\right)^2 + m^2} \tag{10}$$

But:

$$x^2 m^2 = \omega_{-3dB}{}^2 \tau^2 m^2 = \omega_{-3dB}{}^2 (RC)^2 = \left(\frac{\omega_{-3dB}}{\omega_1}\right)^2 \tag{11}$$

And maximizing the right side of (10) by proper choice of m one can find the maximum available bandwidth, given as:

$$f_{-3dB}(m) = \sqrt{\left(m^2 - 2m - 2\right)^2 + 4m^2} - m^2 + 2m \tag{12}$$

Differentiating and equating the derivative of (12) to zero, one can obtain:

$$\left(m^2 - 2m - 2\right)(m - 1) + 2m = (m - 1)\sqrt{\left(m^2 - 2m - 2\right)^2 + 4m^2} \tag{13}$$

Squaring both sides of this equation, then:

$$(m - 1)\left(m^2 - 2m - 2\right) = m\left[(m - 1)^2 - 1\right] \tag{14}$$

And from this equation one finally finds that the required value of m is $\sqrt{2}$.
Substituting this value of m in the right side of (10), then:

$$\left(\omega_{-3dB} / \omega_1\right)\Big|_{\max} = \sqrt{\sqrt{2} + 2} = 1.847 \tag{15}$$

Hence the bandwidth is improved nearly two times as shown in Fig. 7. Consider as an example improving the bandwidth from 1 GHz to 1.85 GHz. This is tremendous improvement with the addition of just one inductor.

Unfortunately, however, this choice of m leads to nearly 20% peaking. Indeed, with this choice of m:

$$\left|\frac{A_v(j\omega)}{g_m R}\right|^2 = \frac{x^2+1}{\left(1-x^2\sqrt{2}\right)^2+\left(x\sqrt{2}\right)^2} = \frac{y+1}{2y^2-2\left(\sqrt{2}-1\right)y+1} \tag{16}$$

Where $x=\omega\tau$, and $y = x^2$. Differentiating the right side of (16) and equating the derivative to zero, one obtains that the maximal value of the right side occurs at y obtained from the equation:

$$2y^2+4y-\left(2\sqrt{2}-1\right)=0 \tag{17}$$

The solution of this equation gives y=0.3836, i.e. $x=\sqrt{y}=0.6193=\left(\omega\tau\right)\big|_{peaking}$. Therefore:

$$\omega_{peaking}=\frac{0.693}{\tau}=\frac{0.693m}{RC}=\frac{0.693\sqrt{2}}{RC}=0.98\omega_1\approx\omega_1 \tag{18}$$

And the normalized amplitude frequency response has the value of:

$$\left|\frac{A_v(j\omega_{peaking})}{g_m R}\right|^2=\frac{0.3836+1}{2\cdot 0.3836^2-2\left(\sqrt{2}-1\right)\cdot 0.3836+1}=\left(1.1904\right)^2 \tag{19}$$

This corresponds to a peaking about 1.5dB, as shown in Fig. 7.

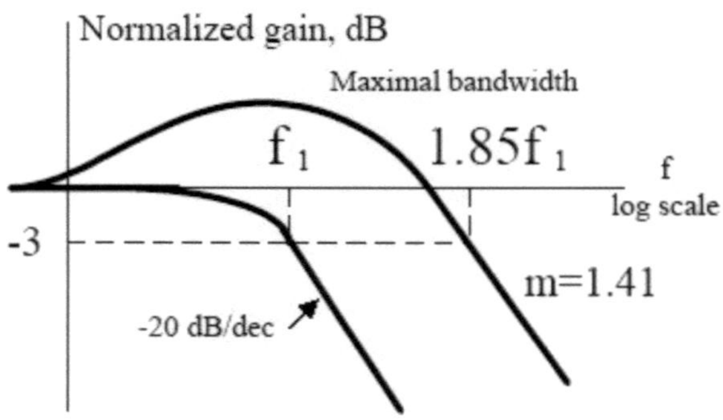

Fig. 7. Frequency enhancement by Fig. 6.

However, there are many applications where the frequency response should be completely free of peaking. Therefore, consider again:

$$\left|\frac{A_v(j\omega)}{g_m R}\right|^2=\frac{x^2+1}{\left(1-x^2m\right)^2+\left(xm\right)^2} \tag{20}$$

Where x=ω, as it was before, and require that the right side does not have any other maximums, except x = 0. The search of maximum leads to:

$$2x\left(1-2x^2m+x^2m^2+x^4m^2\right) \\ =\left(x^2+1\right)\left(-4xm+2x^3m^2+2xm^2\right) \tag{21}$$

One of possible solutions of this equation is $x = 0$. Other solutions can be obtained from the equation:

$$2x^2m+m^2-2m-1=0 \tag{22}$$

One can see that two other solutions will be at $x = 0$ as well, if:

$$m^2-2m-1=0 \tag{23}$$

This gives:

$$m=1+\sqrt{2}=2.414 \tag{24}$$

Direct calculation using (10) shows that this value of m leads to a bandwidth:

$$\omega_{-3dB}/\omega_1=1.707 \tag{25}$$

The corresponding amplitude frequency response is shown in Fig. 8.

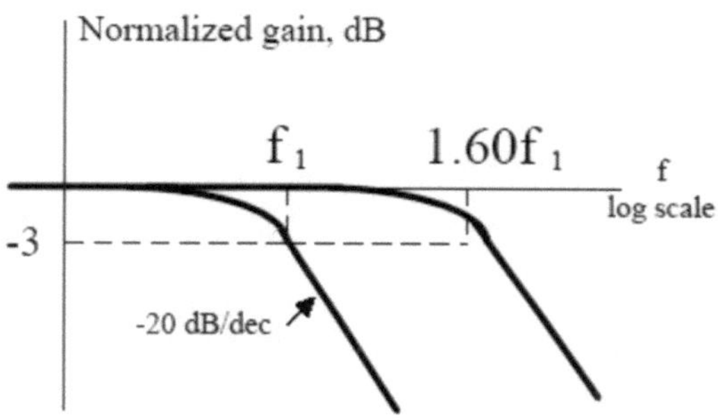

Fig. 8. Maximally flat frequency response.

For this choice of m, both the first and second derivatives of the right side of (20) equal zero at $x = 0$. This amplitude frequency response can be considered as maximally flat. For this reason this choice of m is also very frequently used.

In other situations, there may be a specification on the time response of the amplifier, rather than on frequency response. The amplifier must not only amplify uniformly the various spectral components of the signal over as large a bandwidth as practical, but the phase relationships among its Fourier components must be preserved as well. If all frequencies are delayed by an equal amount of time, then this fixed amount of time delay must represent a linearly increasing amount of phase shift as frequency increases. Phase distortion will be minimized if the deviation from this ideal linear phase shift is minimized. Evidently, then, the delay as the function of frequency must be examined. If this delay is the same for all frequencies, there will be no phase distortion. The delay is defined as

$$T_D(\omega)=-\frac{d\phi}{d\omega} \tag{26}$$

Where ϕ is the phase shift of the amplifier at frequency ω.
Using (4), then:

$$\frac{A_v(j\omega)}{g_m R} = -\frac{1 + j\omega\tau}{1 - \omega^2\tau^2 m + j\omega\tau m} \tag{27}$$

And from this expression, one can find that:

$$\phi(\omega) = -\tan^{-1}\omega\tau + \tan^{-1}\left(\frac{\omega\tau m}{1 - \omega^2\tau^2 m}\right) \tag{28}$$

It is impossible for this amplifier to provide a constant time delay over an infinite bandwidth. It is reasonable to provide, then, with an approximation to a constant delay over some finite bandwidth. A maximally flat time delay will result the number of derivatives of $T_D(\omega)$, whose value is zero at DC, is maximized.
This derivation is rather complicated. Ultimately, however, on may derive the following cubic equation for m as:

$$m^3 - 3m - 1 = 0 \tag{29}$$

whose relevant root is:

$$m = 1 + \left(\frac{3+\sqrt{5}}{2}\right)^{1/3} + \left(\frac{3-\sqrt{5}}{2}\right)^{1/3} \approx 3.104 \tag{30}$$

which is corresponding to a bandwidth improvement factor a little bit less than 1.6.
Since the conditions for maximally flat amplitude frequency response and maximally flat time delay do not coincide, one can compromise. Depending on requirements, there is a range of useful inductance value. A larger L (smaller m) gives the bandwidth extension but poorer phase linearity, whereas a smaller L yields less bandwidth improvement but better phase linearity. All considered cases are summarized in Table 1.

Condition	m=R2C/L	Normalized bandwidth	Normalized peak frequency response
Maximum bandwidth	1.414	1.85	1.19
Maximally flat bandwidth	2.414	1.707	1
Maximally flat time delay	3.104	1.6	1
No shunt peaking	∞	1	1

Table 1. Shunt peaking design summary.

3. Low noise amplifier

Low noise amplifier – LNA is the most critical block in the receiver signal chain, since it determines the overall noise Fig. of the received signal, so that it determines the quality of communication system.
There are several issues on LNA design for UWB applications. First, it must provide wideband impedance matching for both optimal power transfer and noise characteristic. Second, it should be a low power implementation with high power gain. According to the

802.13a specification [1] [2], it is required a power gain of at least 15dB with less than 3dB noise Fig. Since, one of the biggest applications of UWB systems is low-power implementation, the LNA should be able to operate in low supply voltage. The third issue is gain flatness to avoid any signal distortion over such a wide bandwidth.
In terms of wideband impedance matching, the most popular methods are the feedback topology, the distributed impedance matching, the BPF configuration matching network, and the common-gate topology. Nevertheless, each method has advantages and disadvantages, so it is difficult to select one single method for UWB LNA design. For example, feedback topology has good noise and impedance matching performance, but degrades the achievable power gain. The other side, BPF configuration matching is able to achieve high power gain with spurious impedance matching performance in addition to great frequency selection characteristics, while increasing noise Fig. with more passive components used to implement the filter.
This section discussed a unique UWB CMOS LNA, which utilizes both feedback, and BPF configuration method, as presented in [3].

3.1 LNA circuit synthesis

In general, it is very difficult to establish a systematic method for LNA design with satisfying simultaneously low noise factor, impedance matching, and high gain. The major difficulty comes from the fact that the optimal source impedance for optimal noise is different from the matching condition for maximum power delivery. So it is very important to confirm initial design decisions of circuit parameters because two matching conditions are highly related. Also, too simplified circuit model forces trial-and-error strategy for optimizing the circuit. Therefore, accurate circuit evaluation is required to avoid the tedious effort for circuit optimization. Thus, the accurate Miller effect of source degenerative topology with cascode topology, and a methodology to utilize the Miller effect for the input matching network implementation are presented in this section.
The overall LNA schematic, including input and output impedance matching network, is shown in Fig. 9. The LNA looks like a simple conventional narrowband LNA with one gate

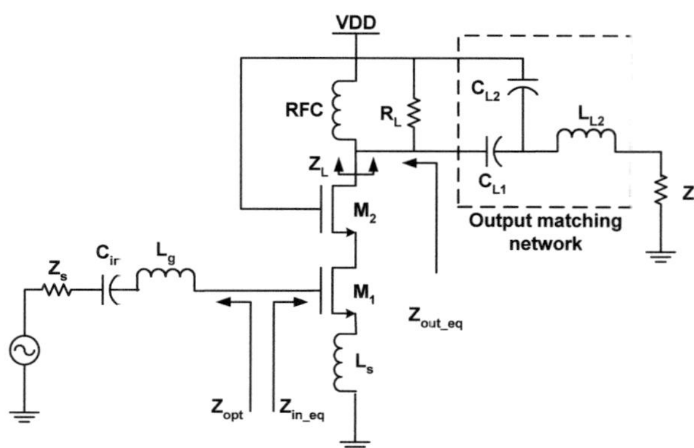

Fig. 9. Overall LNA architecture.

inductor. However, the LNA can achieve wideband input matching by using Miller effect as explained later. Also, the UWB LNA architecture does not make use of a source follower for output matching, but has passive output matching network, which consists of bandpass filter and impedance inverting scheme.

3.2 Transistor sizing and bias condition

Since the size of transistors and their bias condition determine power dissipation, it is often recommended to establish them under a certain power budget. However, the size of transistor versus its bias condition should be evaluated carefully, because they are also related to impedance seen by input gate. Thus, the best choice is to determine the size and bias condition to satisfy both impedance matching and noise matching with limited bias current. In fact, there is no much freedom for this choice technically. According to the MOSFET noise analysis [4], the generator admittance for optimal noise performance is known as (31) and (32).

$$G_{opt} = \alpha\omega C_{gs}\sqrt{\frac{\delta}{5\gamma}\left(1-|c|^2\right)} \tag{31}$$

$$B_{opt} = -\omega C_{gs}\left(1+\alpha|c|\sqrt{\frac{\delta}{5\gamma}}\right) \tag{32}$$

where $\alpha = g_m/g_{d0}$, the parameters δ and γ are given in Chapter 3, and c is defined as the correlation between the drain noise i_{nd} and the gate noise i_{ng} currents, given as:

$$c = \frac{\overline{i_{ng}.i_{nd}^*}}{\sqrt{\overline{i_{ng}^2}.\overline{i_{nd}^2}}} \tag{33}$$

For the sake of simplicity, initially the correlation of noise can be ignored, so that c has to be 0. Therefore, (31) and (32) can be simplified as:

$$R_{opt} \approx \frac{1}{\alpha\omega C_{gs}}\sqrt{\frac{5\gamma}{\delta}} \tag{34}$$

$$X_{opt} \approx \frac{1}{\omega C_{gs}} \tag{35}$$

Furthermore, (35) can be modified to (36) in order to take account of the degenerative inductor at the source-end.

$$X_{opt} \approx \frac{1}{\omega C_{gs}} - \omega L_s \tag{36}$$

Note that expressions (34) and (35) represent real and imaginary terms of impedance, while (31) and (32) presents admittance expressions.

Observe from expression (36) that the imaginary term of the optimal noise generator impedance is inversely proportional to the gate-source capacitance. Since the gate-source capacitance is always positive, than noise matching can be achieved with inductive generator impedance. However, increasing L_s will reduce the gain, but at the same time, the inductive term of generator impedance (L_g) can be decreased. According to the above observation, it is clear that optimal noise condition and maximum power transfer are obtained simultaneously when $Z_{opt} = Z_{in\_eq}^*$, where Z_{in_eq} is the equivalent input impedance seen by input gate of amplifying transistor given as:

$$Z_{in\_eq} = R_{in\_eq} + jX_{in\_eq} = \frac{g_m L_s}{C_{gs}} + j\left(\omega L_s - \frac{1}{\omega C_{gs}}\right) \quad (37)$$

However, it is not easy to make both Z_{opt} and $Z_{in\_eq}^*$ to have same value. Nevertheless, high gain can be achieved if the inequality shown in (38) is satisfied. Obviously, smaller resistive term of input impedance seen by gate-end leads higher gain.

$$R_{in\_eq} \le R_{opt} \le Z_s \quad (38)$$

where Z_s is the source impedance.

Since the reactance term of Z_{opt} and $Z_{in\_eq}^*$ are almost always matched according to (36) and (37), inequality (38) will force Z_{in_eq} to be positioned in outer side of Z_{opt} in Smith chart until the frequency exceeds the desired frequency range.

As mentioned already, the bias condition should be achieved under a limited current, thus I_{DS} is a limited value. For the sake of simple procedure, assumed the g_m and C_{gs} are given as (39) and (40), which ignore overlapped channel length L_{ov}, The initial value of V_{eff} is given by (40).

$$g_m = \mu_n C_{ox} \frac{W}{L} V_{eff} \quad (39)$$

$$C_{gs} = \frac{2}{3} WLC_{ox} \quad (40)$$

$$V_{eff} \le \frac{2Z_s L^2}{3L_s \mu_n} \quad (41)$$

Note that considers minimum channel length L. Once V_{eff} is obtained, then the minimum value of g_m is:

$$g_m \ge \frac{2I_{DS}}{V_{eff\_max}} \quad (42)$$

where V_{eff_max} is the maximum effective voltage.

Assume, roughly, that $\gamma \approx 2$, $\delta \approx 4$ and $\alpha \approx 5$, since $g_{ds} \approx 0.2 g_m$ in active region, so that (34) can be simplified even more as:

$$R_{opt} \approx \frac{1}{\sqrt{10}\,\omega C_{gs}} \quad (43)$$

Finally, the minimum channel width W given in (44), is based on (38), (40) and (43):

$$W \ge \frac{3}{2\sqrt{10}\,\omega Z_s L C_{ox}} \quad (44)$$

Again, minimum channel length is assumed and the results are roughly selected so that they must be optimized later. The obtained Z_{opt} and Z_{in_eq} are shown in Fig. 10 over the frequency range of 100MHz to 20GHz, and one can notice that $Z_{in\_eq}^*$ is almost matched to Z_{opt}. $Z_{in\_eq}^*$

remains positioned in outer circle of Z_{opt} in Smith chart up to *6GHz*, which is higher than the desired frequency range.

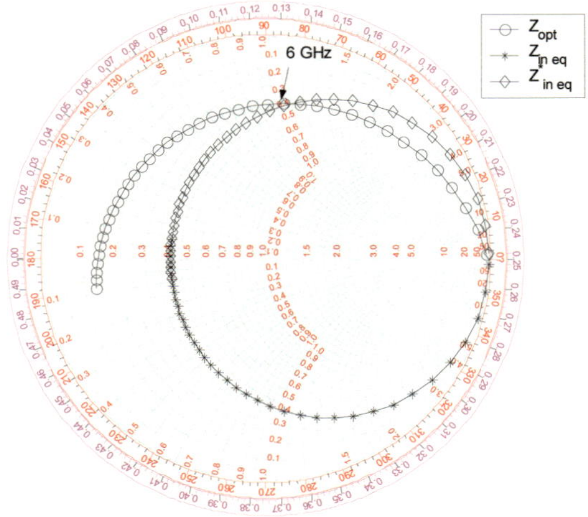

Fig. 10. Zopt, Zin_eq, and Zin_eq*.

The obtained condition so far should be applied to M_1 in Fig. 9.

3.3 Miller effect in cascode topology

The Miller effect implies that the effective capacitance is increased by negative voltage gain between input and output. However, since the input impedance of the cascode device M_2 is capacitive, the voltage gain is high in low frequency and low in high frequency, which implies the effective Miller capacitance will be high in low frequency and low in high frequency. Therefore, it explains that the Miller effect creates not only a single capacitor, but also an inductor in parallel with the Miller capacitor.
The input impedance Z_{Load} of the cascode device M_2 seen at the source of M_2 is described as

$$Z_{Load} = \frac{R_{ds2} + Z_L}{1 + g_{m2} R_{ds2} + sC_{gs2}\left(R_{ds2} + Z_L\right)} \tag{45}$$

where Z_L is the output load connected to drain of M_2, and this is assumed as pure resistor over the frequency of interest, for simplicity.
The load impedance of the cascode device, therefore, can be expressed as R and C parallel circuit as shown in Fig 11, whose values are:

$$C_{Load} = C_{gs2} \tag{46}$$

$$R_{Load} = \frac{R_{ds2} + Z_L}{1 + g_{m2} R_{ds2}} \tag{47}$$

The resistance term of the cascode load is equal to $1/g_{m2}$, when R_{ds2} is infinite. Note that the R_{ds2} is relatively large for low power design due to the relation $R_{ds} = 1/\lambda I_{DS}$, where λ is the depletion length coefficient (channel length modulation), and I_{DS} is the bias DC current, which is small for low power design.

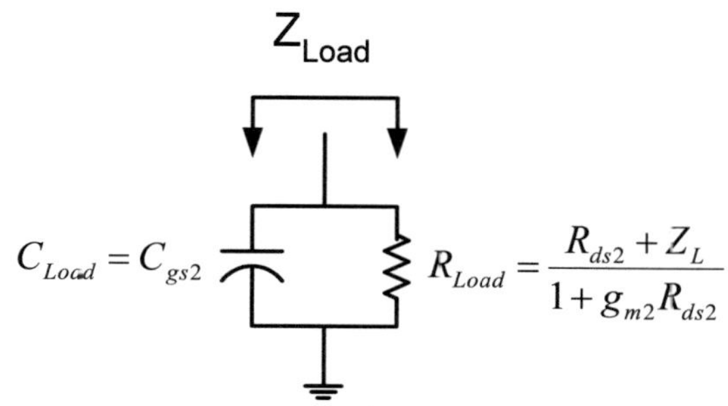

Fig. 11. Input impedance of cascode device M_2.

The effective transconductance for source degenerative topology can be obtained as:

$$G_m = \frac{g_{m1}}{1 + g_{m1} L_s s + C_{gs1} L_s s^2} \tag{48}$$

Thus, the overall open voltage gain A_{vo} is:

$$A_{vo} = -G_m Z_{Load} = \frac{-g_{m1}(R_{ds2} + Z_L)}{\left(1 + g_{m1} L_s s + C_{gs1} L_s s^2\right)\left(1 + g_{m2} R_{ds2} + C_{gs2}(R_{ds2} + Z_L)s\right)} \tag{49}$$

According to the non-flat open voltage gain between gate and drain of M_1, the Miller capacitor is not a simple capacitor anymore, but an RLC combination circuit.
The Miller capacitance C_{mil} is:

$$C_{mil} = (1 - A_{vo})C_{gd1} == \left(1 + \frac{g_{m1}(R_{ds2} + Z_L)}{\left(1 + g_{m1} L_s s + C_{gs1} L_s s^2\right)\left(1 + g_{m2} R_{ds2} + C_{gs2}\left(R_{ds2} + Z_L\right)s\right)}\right) C_{gd1} \tag{50}$$

Finally, the overall Miller impedance caused by the non-flat voltage gain is:

$$Z_{mil} = \frac{1}{sC_{mil}} \approx \frac{s^2 L_s (C_{gs1} g_{m2} + C_{gs2} g_{m1}) + s(C_{gs2} + L_s g_{m1} g_{m2}) + g_{m2}}{s^3 C_{gd1} L_s (C_{gs1} g_{m2} + C_{gs2} g_{m1}) + s^2 C_{gd1}(C_{gs2} + L_s g_{m1} g_{m2}) + sC_{gd1}(g_{m1} + g_{m2})} \tag{51}$$

Note that non dominant terms are eliminated for the sake of simplicity.
The equivalent impedance given by Miller effect is indicated in Fig. 12, whose values of individual components are:

$$C_{mil1} = \frac{C_{gd1}(g_{m1} + \alpha)}{\alpha} \tag{52}$$

$$C_{mil2} = \frac{C_{gd1}(g_{m1} + \alpha)}{g_{m1}} \tag{53}$$

$$L_{mil1} = \frac{L_s g_{m1}(C_{gs1}\alpha + C_{gs2} g_{m1})}{\alpha(g_{m1} + \alpha)} \tag{54}$$

$$R_{mil1} = \frac{g_{m1}(C_{gs2} + L_s g_{m1}\alpha)}{C_{gd1}(g_{m1} + \alpha)^2} \tag{55}$$

where $\alpha = 1/R_{Load}$.
Note that the resistive term R_{mil1} is related to the quality factor of the inductive term L_{mil1}, and it is relative small enough to be ignored.

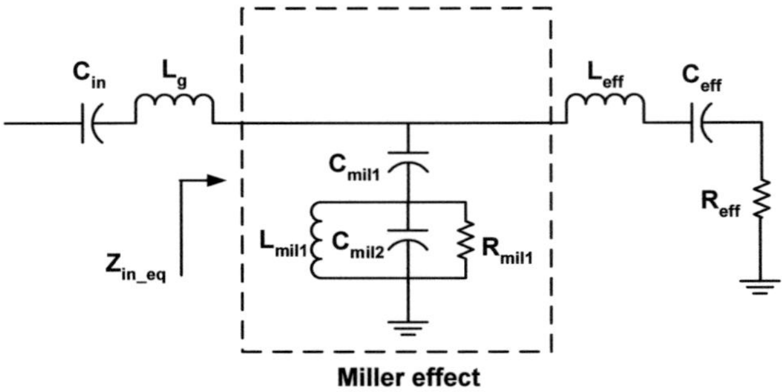

Fig. 12. Equivalent input circuit.

3.4 Modified input impedance by feedback

Now, the input impedance of the inductive degenerative topology including Miller effect must be re-evaluated.
The input impedance of the open circuit is well known as RLC series circuit, given as:

$$Z_{ino} = sL_s + \frac{1}{sC_{gs1}} + \frac{g_{m1}L_s}{C_{gs1}} \tag{56}$$

From the feedback system, the modified input impedance of the feedback system, as shown in Fig. 13, is given by:

$$Z_{inc} = \frac{Z_{ino}(Z_f + Z_{load})}{Z_{ino}(1 + G_m) + Z_L + Z_f} \tag{57}$$

Note that the close loop input impedance includes the Miller effect.
The feedback impedance Z_f is $(1/sC_{gd1})$, which is the gate-to-drain capacitor. By using the effective transconductance and load impedance as obtained above, the overall expression of the input admittance Y_{inc} of the close loop circuit after simplification is:

$$Y_{inc} = Y_{mil} + \frac{1}{\frac{1}{R_{eff}} + sC_{eff} + \frac{1}{sL_{eff}}} \tag{58}$$

where Y_{mil} is $1/Z_{mil}$, the admittance of the equivalent Miller circuit, and:

$$R_{eff} = \frac{g_{m1}L_s}{C_{gs1}}\left(1 + \frac{C_{gd1}R_{ds2} + 2C_{gs2}R_{ds2}}{g_{m1}L_s(1 + g_{m2}R_{ds2})}\right) \tag{59}$$

$$C_{eff} = C_{gs1} \tag{60}$$

$$L_{eff} = L_s + \beta \tag{61}$$

$$\beta = \frac{(R_{ds2} + Z_L)\left(C_{gd1}\left(g_{m1}(L_s + g_{m2}L_sR_{ds2}) + C_{gs2}(R_{ds2} + Z_L)\right) + C_{gs2}\left(2g_{m1}(L_s + g_{m2}L_sR_{ds2}) + C_{gs2}(R_{ds2} + Z_L)\right)\right)}{C_{gs1}(1 + g_{m2}R_{ds2})^2} \tag{62}$$

Thus, the actual RLC series circuit is changed by the feedback effect. The feedback effect effectively increases the inductive term L_{eff} and resistive term R_{eff} from the original open circuit input impedance Z_{ino}.

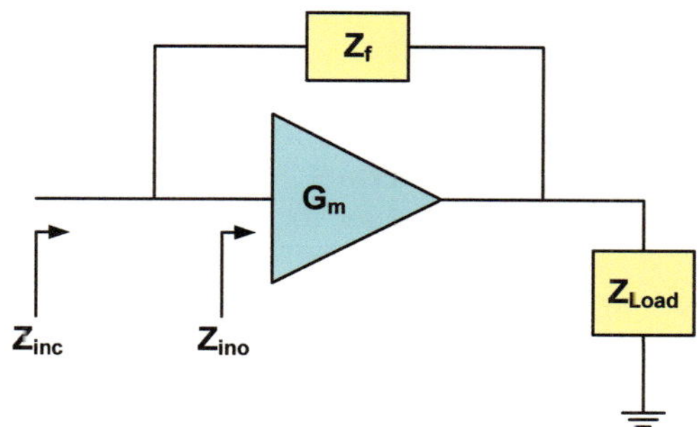

Fig. 13. Feedback system with effective transconductance.

For large R_{ds2}, the equivalent circuit can be further simplified as:

$$R_{eff} \approx \frac{g_{m1}L_s}{C_{gs1}}\left(1 + \frac{C_{gd1} + 2C_{gs2}}{g_{m2}}\right) \tag{63}$$

$$L_{eff} \approx L_s + \frac{C_{gd1}C_{gs2} + C_{gs2}^2 + C_{gd1}g_{m1}g_{m2}L_s + 2C_{gs2}g_{m1}g_{m2}L_s}{C_{gs1}g_{m2}^2} \tag{64}$$

Therefore, overall input impedance can be expressed as Fig. 12.
Note that the C_{mil1} can be ignored in high frequency and R_{mil1} also can be ignored due to its small value, so that the overall circuit can be considered as the combination of parallel LC and series LC circuits. The circuit also can be considered as a part of bandpass filter.

4. Mixer

Mixers are non-linear devices used in systems to translate from one frequency to another. All mixer types work on the principle that a large Local Oscillator - LO drive will cause switching/modulating the incoming RF into an Intermediate Frequency - IF, or in opposite direction.
There are two types of mixer, passive and active. Generally the passive types have better IM3 performance, but present higher conversion losses and hence higher noise Fig.s than active mixers.
Additionally, mixers can also be classified as single balanced mixers and double balanced mixers. Single balanced mixers are much less complex, but have inferior performance in terms of RF to IF and LO to IF rejection, compared to double balanced mixers.
The advantages of double balanced mixers are:

a. Both LO and input signals are balanced, providing both LO and input rejection at the output.
b. All ports of the mixer are inherently isolated from each other.
c. Higher linearity, compared to singly balanced.
d. Improved suppression of spurious products (all even order products of the LO and/or the input are suppressed).
e. Higher intercept points.
f. Less susceptible to supply voltage noise due to differential topology.

The disadvantages are:

a. Require a higher LO drive level.
b. Require differential input and LO signal.
c. Ports are highly sensitive to reactive terminations.

The Gilbert double-balanced mixer configuration is widely used in RFIC applications because of its compact layout and moderately high performance. This section will walk through the design of a CMOS Gilbert mixer focusing on the parameters that influence the linearity of the signal path, the noise, and therefore the spurious-free dynamic range of the mixer. Finally, some techniques to enhance the bandwidth of the Gilbert mixer will be also presented, so to be suitable for UWB applications.

4.1 Design guidelines

Depending on the application, the mixer may be designed with a low Single Side Band - SSB noise Fig., a particular gain or a high linearity. A good starting point is to use the differential LNA and add the switching transistors with the same W/L ratios.

As in the case of LNA design, the linearity of the mixer source can be increased by adding degeneration resistors (or inductors). As an example consider Z_S inserted in the sources of M_1 and M_2 in the circuit of Fig. 14.

There are several parameters to be achieved during the design process, such as device width, biasing, linearity of transconductance amplifier (input circuit), stability, input matching network, gain compression, Inter Modulation Distortion - IMD, noise Fig. and spurious free dynamic range.

Though the design method introduced here emphasizes the distortion-limited (large-signal) performance over noise-limited (small-signal) performance, there are many design choices possible. In Fig. 14, one may have to decide proper bias current and device width W_1, and W_2. Proper selection of W_1 should provide high g_m, saturation at low V_{DS} (for low power supply operation) and low noise. Large widths are preferred for noise, but the optimum width for both noise and power constraints can be estimated from the MOS device parameter [1]. Large widths also require large bias currents to obtain high g_m. Choosing $W_1 = W_2$ is typically the best approach.

The minimum current required to keep all devices in saturation must also be considered. Additionally, once the bias is determined, the linearity of signal path must be verified. The signal path from the transconductance amplifier through the source resistance and inductance is the dominant for the sake of linearization. As the resistance increases the linearity also increases, but the conversion gain also decreases to some degree. Source inductance is used mainly to guarantee stability by forcing a positive real component into the input impedance. This also helps to make the input impedance easier to match.

4.2 Device width and bias current

From Fig. 14, the voltage gain of the mixer with source degeneration is given by:

$$\frac{V_{out}}{V_{in}} \approx \frac{2}{\pi}\left(\frac{Z_L}{Z_s + \frac{1}{g_m}}\right) \tag{65}$$

This equation implies lower conversion gain with larger impedance at the source of M_1 and M_2, as expected. However, this equation does not provide a clue to determine the device width.

From the analysis of noise optimization, the optimal width can be found as [4]:

$$W_{opt} = \frac{1}{3\omega L C_{ox} R_{gen}} \tag{66}$$

where R_{gen} is the resistance of the source connected to the mixer input, typically 50 Ω, but sometimes determined by LNA output impedance.

For this width, I_{DS} must be large enough to saturate the MOSFET ($V_{DS} > V_{dsat}$). At the same time, large V_{DS} is undesirable, specially for low V_{DD} operation. Finally, large V_{DS} will increase hot electron effects at the drain, thereby increasing noise.

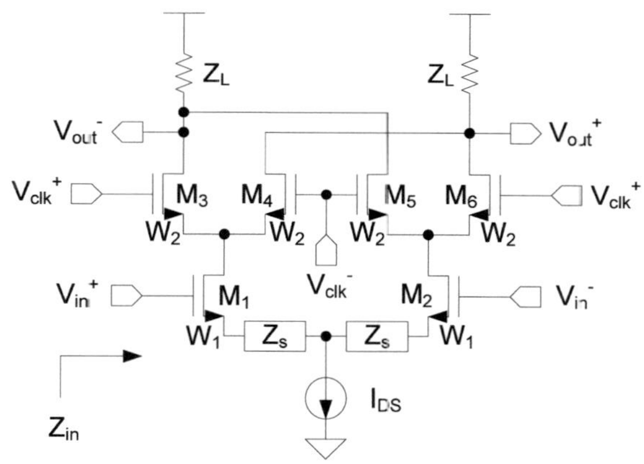

Fig. 14. Basic circuit of the Gilbert Cell Double Balanced (DB) Mixer.

4.3 Linearity of signal path

In order to investigate the linearity of the signal path, a transfer characteristic can be simulated by sweeping the input DC voltage. Consider the example given in Fig. 15. Note that the DC input voltage V_{Din} is $V_{in} - V_{ref}$.

It is expected that by increasing the resistance R_s, which increases negative feedback, the transfer characteristic would be linearized, by exchanging gain for linearity. In the simulation shown in Fig. 16, it can be seen that the gain (slope) becomes more linear over a wider input voltage range as R_s is increased.

A popular technique in low voltage RFIC design is to substitute resistors by inductors. This has the advantages that the ideal inductor does not add noise to the circuit, and it reduces the supply voltage requirement for the circuit. The effectiveness of this approach is somewhat frequency dependent. At low frequency, the gain degeneration and linearity improvement for reasonable sized inductors is limited, but it becomes more effective at higher frequencies.

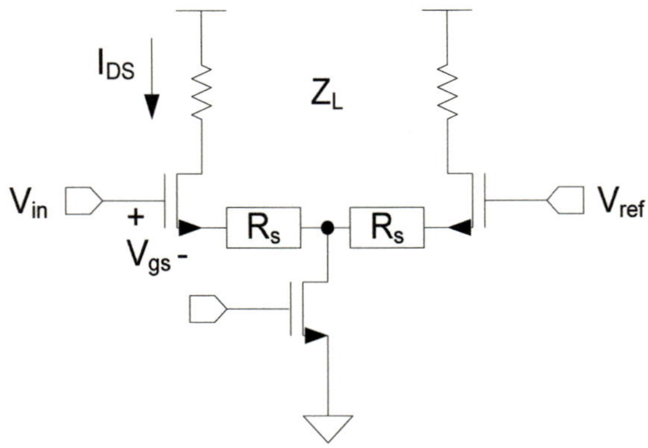

Fig. 15. Setup for transfer characteristic simulation.

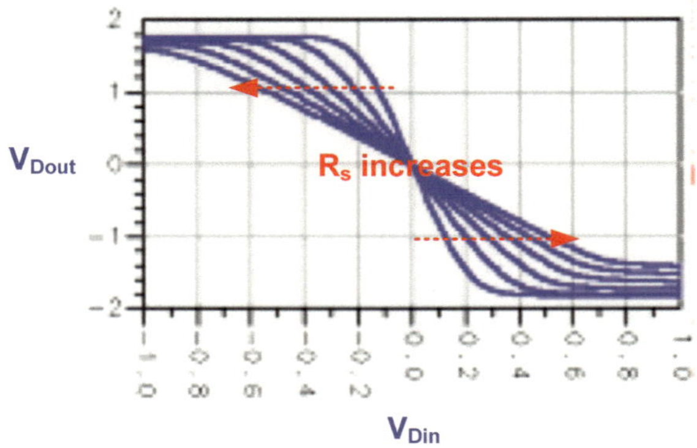

Fig. 16. DC input voltage sweeping for linearity simulation.

Also, inductors on Si substrates have low Q, on the order of 2 to 3. For a Q of 2.5, for example, a 5 nH inductor at 4GHz would have a series resistance of about 50Ω, thus, in fact both resistance and inductance are being added to the circuit. Therefore, it is valuable to investigate the effect of both inductor and resistor as Z_s.

4.4 Input impedance and stability

As explained earlier, the input impedance seen at gate of source degenerative topology with impedance Z_s is:

$$Z_{in}(j\omega) = \frac{1}{j\omega C_{gs}} + Z_s + \frac{\omega_T Z_s}{j\omega} \tag{67}$$

where $\omega_T = g_m / C_{gs}$.

Expression (67) was derived from a simple small-signal analysis; it neglected C_{gd} and assumed that the node between the source resistors is at virtual ground. As summarized in Table 2, if the source noise impedance Z_s is purely resistive, it is equivalent to an R and two series capacitors. If R is large, the equivalent input series capacitive reactance is large and has a large effect on Z_{in}. The real part is clearly positive.

Similarly, a series inductance L produces a non-frequency dependent positive real part and a series LC resonant network. Only the capacitor produces a negative resistance, a condition desirable for oscillators, not mixers, and with unusual frequency dependence. Therefore, negative input resistance can be avoided eliminating the possibility of using a capacitor.

Z_s	$Re[Z_{in}] + Im[Z_{in}]$
R	$R+\left(\frac{\omega_T R}{j\omega}+\frac{1}{j\omega C_{gs}}\right)$
L	$\omega_T L+\left(\frac{1}{j\omega C_{gs}}+j\omega L\right)$
C	$-\frac{\omega_T}{\omega^2 C}+\left(\frac{1}{j\omega C_{gs}}+\frac{1}{j\omega C}\right)$

Table 2. Summary of input impedance according to impedance at source.

Unfortunately, however, there is some parasitic capacitance between source and bulk of the transistors, as indicated in Fig. 17. Therefore, as R_s increases, the shunt C_{SB} effect on the source impedance increases, thus driving the input impedance negatively. If $\omega_T R_s C_{SB} > 1$, a negative real Z_{in} will show up. For this reason, it may be necessary to add some series inductance to compensate the negative resistance.
Expression (68) describes the resistive input impedance by considering the presence of C_{SB}.

$$\mathrm{Re}\{Z_{in}\}=\frac{R_s\left(1-\omega_T R_s C_{SB}\right)}{1-\omega^2 R_s^{\,2} C_{SB}^{\,2}} \tag{68}$$

An extrapolation of i_D - v_{DS} intercepts the v_{DS} axis at $v_{DS} = -V_A$, known as Early voltage. For a given process, V_A is proportional to L, selected by the designer. Typically, VA is in the range of 5 V/µm to 50 V/µm.

4.5 Output resistance

So far, only inside of Gilbert cell mixer has been discussed. In fact, signal bandwidth at both input and output is another critical problem for UWB mixer. Therefore, input and output bandwidth enhancements are also necessary.
For integrated circuits, there is no restriction of intermediate impedance between blocks. In fact, the shunt-peaking method is widely used for bandwidth enhancement and interconnection between blocks. However, it is sometimes necessary to provide a specific impedance value for both input and output (in many cases 50 Ω), thus the wideband impedance matching methods can be applied. The applicable methods for bandwidth enhancement are:

a. *Shunt-peaking: suitable* for conjugate matching with non-standard intermediate impedance.
b. *Wideband matching method*: suitable for both conjugate matching and standard impedance matching, but requires more passive components.
c. *Cascode topology*: applicable for both previous methods, in addition by reducing RC constant time.

Since cascode topology reduces voltage gain between gate and drain of transconductance amplifier, it reduces the effect of the gate-drain capacitance, the so called Miller effect. However, if cascode topology is applied to reduce Miller effect, one have to consider reduced overhead voltage by voltage drop through drain to source of the cascode device.

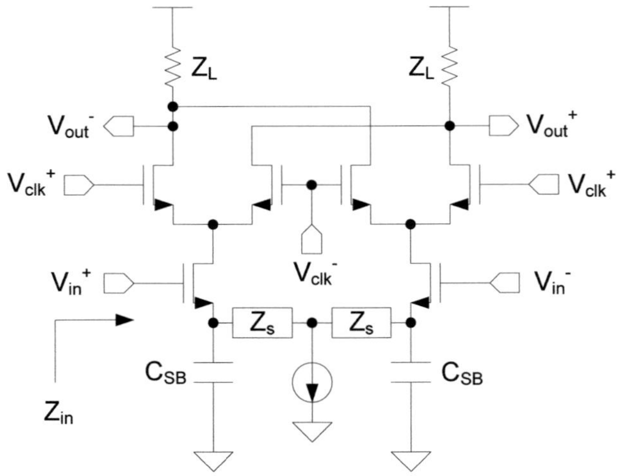

Fig. 17. Gilbert cell mixer with source to bulk capacitance.

5. Conclusions

This chapter provided the background foundations for the analysis and design of low noise amplifiers and mixers, along with their interconnections to other structures. Low noise amplifiers and mixers are among the most used structures in RF IC.

The performance of them may be compromised without proper interconnection. This chapter also presented the approaches to implement AC and DC coupling to interconnect structures, by taking into account performance and noise isolation.

6. References

IEEE 802.15.1 (2002) IEEE Standard for Local and Metropolitan Area Networks - IEEE Standard for Telecommunications and Information Exchange Between Systems.

Sedra, A. S. & Smith, K. C. (2009) *Microelectronic Circuit - 6th Ed.*, Oxford University Press, ISBN 0195323033.

Lee, T. H. (2004). *The Design of CMOS Radio-Frequency Integrated Circuits – 2nd Edition*, Cambridge University Press, ISBN 0521835399.

Coleman, C. (20040 *An Introduction to Radio Frequency Engineering*, Cambridge University Press, ISBN 0521834813.

Gilmore, R. & Besser, L. (2003) *Practical RF Circuit Design for Modern Wireless Systems – Vol. II*, Artech House Publishers, ISBN 1580535224.

Rogers, J. & Plett, C. (20030 *Radio Frequency Integrated Circuit Design*, Artech House Inc, ISBN 1607839792.

Ziel, A. (1986) *Noise in Solid State Devices and Circuits*, John Wiley and Sons, ISBN 0471832340.

3

RF CMOS Background

Tales Cleber Pimenta, Robson L. Moreno and Leonardo B. Zoccal
Universidade Federal de Itajuba
Brazil

1. Introduction

The Metal-Oxide-Semiconductor Field-Effect-Transistor (MOSFET) (or just MOS) is widely used and presents many advantages over the bipolar transistors (BJT) in many applications. It requires less silicon area and its fabrication process is relatively simpler. It is possible to implement most analog and digital circuits using almost exclusively MOS transistors. All these properties allow packing a large number of devices in a single integrated circuit.
Additionally, and most important, its operation requires less power, making it extremely suitable to RFID circuits.
This chapter aims to provide background on MOS transistors, from its physical operation to modeling, including RF modeling. The basic knowledge is essential to analyze and to design RFID circuits implemented using CMOS transistors. The chapter also presents noise analysis which is essential to low voltage signal, as it is the case of RFID circuits.

2. Physical CMOS operation

Fig. 1 shows the physical structure of the *n*-channel MOS transistor, or just *n*MOS transistor. The transistor is fabricated in a *p*-type silicon substrate. Two heavily doped *n*-type regions, indicated as n^+, are created in the substrate and will act as the source and drain (in terms of structure, source and drain can be interchanged). A thin layer of silicon oxide (SiO_2), of thickness t_{ox} (typically between 2 and 50 nm), is formed on the surface of the substrate, between the drain and the source regions. The silicon oxide is an excellent electrical isolator. Metal (or polysilicon, which is conductor) is deposited on top of the oxide layer to form the gate electrode. Metal contacts are also made in the source and drain regions, in addition to contact to the bulk, also known as the substrate or body. Therefore, the four contacts were formed: D-drain, S-source, G-gate and B-bulk.
The gate region has a length L and a width W, which are two important design parameters of the MOS transistor. Usually L is in the range of 0.1µm to 3µm while W is in the range of 0.2µm to 100µm.
There is also the *p*-channel MOS transistor, or just *p*MOS transistor, in which the dopings are reversed to the *n*MOS transistor.

2.1 Forming the channel

As can be observed from the Fig. 1, the substrate forms *pn* junctions with the drain and the source. In normal operation both junctions must be kept reverse-biased, or at least out of the

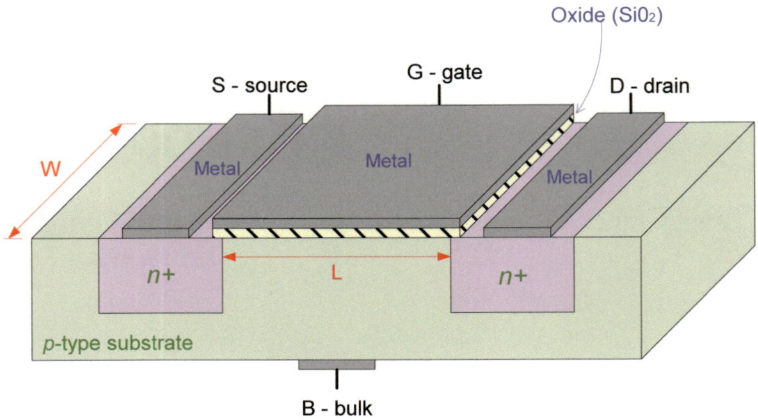

Fig. 1. Physical structure of an *n*MOS transistor.

forward condition all the time. Since the drain is biased at a positive voltage, it is only necessary to connect the bulk to the ground in order to keep both junctions cut off.

With no bias applied to the gate, there are two back-to-back diodes between drain and source, and consequently, there is no current. This is true since each *pn* junction forms a diode. In fact, the resistance between drain and source under this circumstance is in the range of $10^{12}\Omega$.

When a positive voltage is applied between gate and source - v_{GS}, holes (which are positively charged) are repelled from the surface of the substrate. As the voltage increases, the surface becomes completely depleted of charge. The voltage at which this occurs is known as threshold voltage – V_t.

If v_{GS} is further increased, electrons (which are negative charges) accumulate near the surface, under the gate, and an *n* region is created, thus forming a channel between drain and source, as indicated in Fig. 2. The channel was formed by inverting the substrate surface from *p* type to *n* type. Fig. 2 also shows the depletion region that forms around the channel and the two junctions.

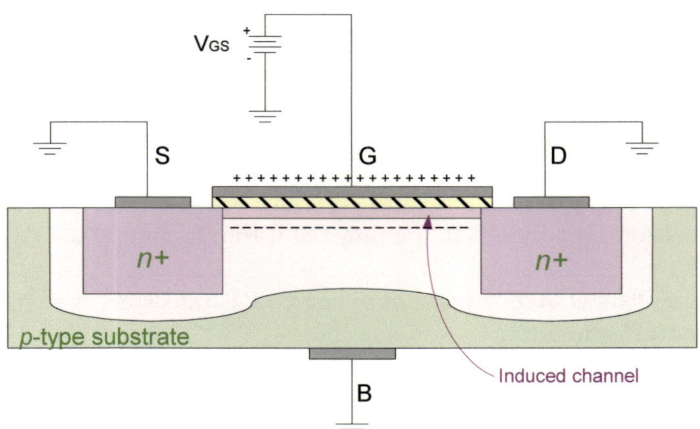

Fig. 2. *n*MOS with an induced channel.

The symbols for the nMOS transistor are given in Fig. 3, although other symbols may be found in the literature. The symbol in Fig. 3.a corresponds to the four terminal connection, and the symbol in Fig. 3.b corresponds to the three terminal connection, where source and substrate are shorted.

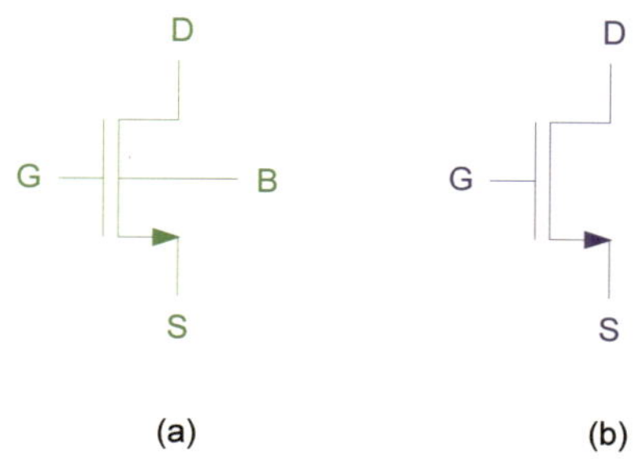

Fig. 3. Symbols for nMOS transistor; (a) four terminals and (b) three terminals.

2.2 Triode condition

Now, if a very small voltage v_{DS} is applied between drain and source, as indicated in Fig. 4, there will be a current flow through the channel. The current through the channel, named drain current - i_D is directly dependent on the voltage v_{GS} and the voltage v_{GS}. If v_{GS} increases, the channel becomes deeper and more current can flow. If v_{DS} is increased, based on Ohm`s Law, there will be more current, since the channel behaves as a resistance. If follows that the transistor is operating as a linear resistance whose value is controlled by v_{GS}. The resistance is very high for $v_{GS} \leq V_t$ and it decreases as v_{GS} increases.
This condition of operation is known as ohmic, linear or triode.

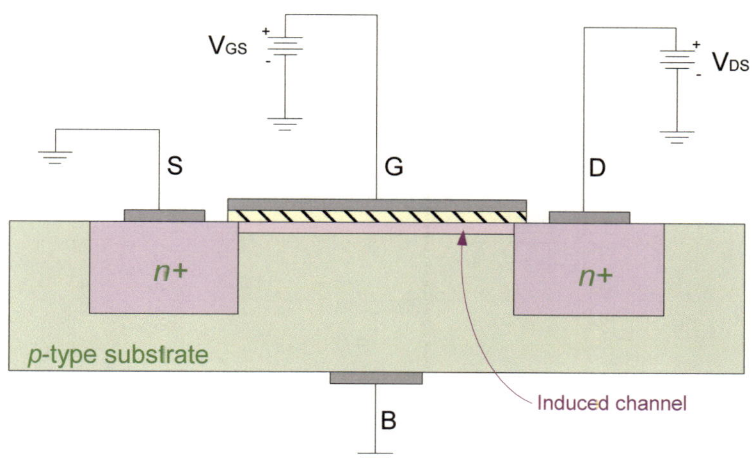

Fig. 4. Conduction under very small v_{DS}.

2.3 Saturation condition

As v_{DS} increases, the difference v_{DS} - v_{DS} becomes smaller at the edge between the gate and the drain diffusion, and therefore the channel becomes shallow. Therefore, the channel assumes a tapered shape, as indicated in Fig. 5. Since the channel becomes smaller at the drain end, its resistance increases, and therefore, the transistor does not operate ideally as a linearly controlled resistor.

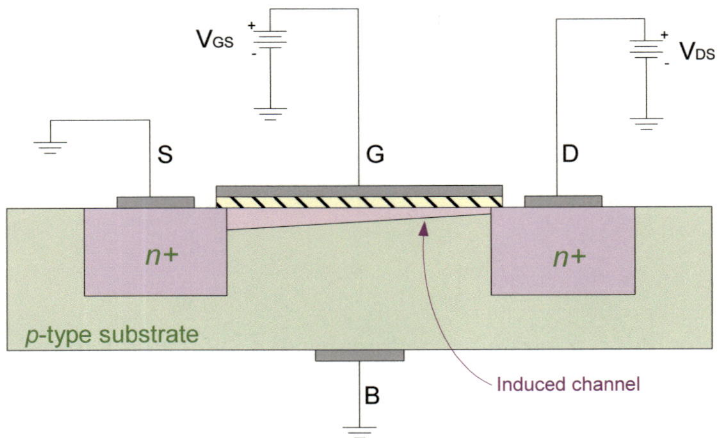

Fig. 5. Conduction under $0 < v_{DS} < v_{GS} - V_t$.

At the condition $v_{DS} = v_{GS} - V_t$, the channel ceases to exist at the drain side, as shown in Fig. 6. This situation is known as pinch off. At this point, further increases in v_{DS} moves the end of the channel further away from the drain, as presented in Fig. 7. This condition of operation is referred as saturation, therefore v_{DS} is referred as $v_{DSSAT} = v_{GS} - V_t$.

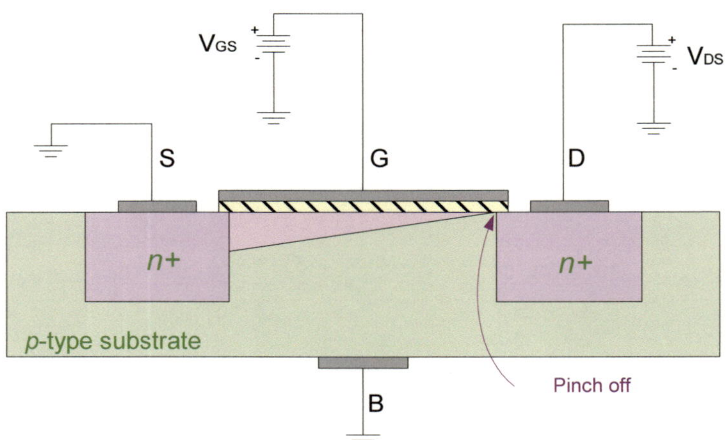

Fig. 6. Conduction under $v_{DS} = v_{GS} - V_t$..

Once the transistor enters the saturation region of operation, the drain current i_D becomes independent of the v_{DS}.

Fig. 8 summarizes the conditions of operation of an *n*MOS transistor. Close to $v_{DS} = 0$, current i_D is directly proportional to v_{DS}, with slope proportional to $v_{GS} - V_t$. As v_{DS} approaches $v_{DS} = v_{GS} - V_t$, the curve of bends because the channel resistance increases. After the $v_{DS} = v_{GS} - V_t$, the current becomes independent of v_{DS}.

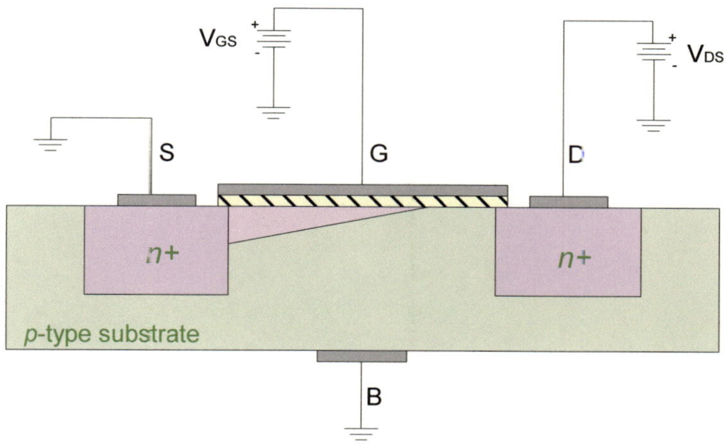

Fig. 7. Conduction under $v_{DS} > v_{GS} - V_t$.

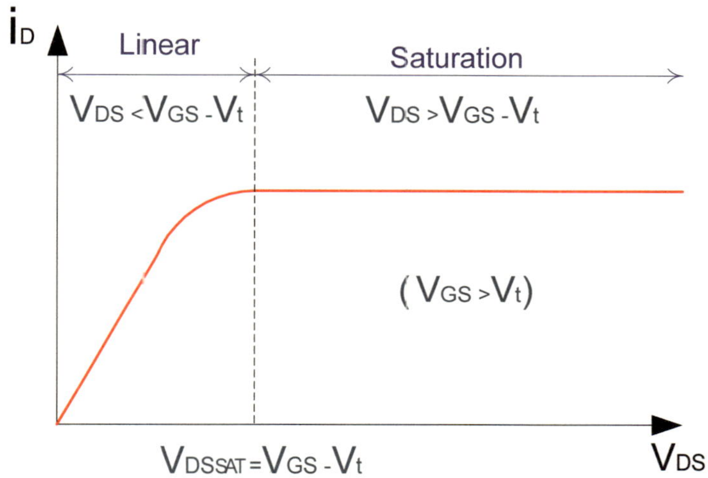

Fig. 8. Operation condition of an *n*MOS transistor.

2.4 Deriving the i_D - v_{DS} relationship

Consider the biasing depicted in Fig. 9. Since the channel potential varies from zero at the source to v_{DS} at the drain, the local voltage difference between gate and the channel varies from v_{GS} to v_{GS} – v_{DS}. Therefore, the channel density, or charge per unit length, is given as:

$$Q_d(x) = WC_{ox}[\mathcal{V}_{GS} - \mathcal{V}(x) - V_t] \tag{1}$$

where $v(x)$ is the potential at x and C_{ox} is the capacitance, per unity area, formed by the gate and the channel.

Since, by definition, current is proportional to charge times velocity, and considering the current is the same along the channel, then:

$$i_D = -WC_{ox}[\mathcal{V}_{GS} - \mathcal{V}(x) - V_t]v \quad (2)$$

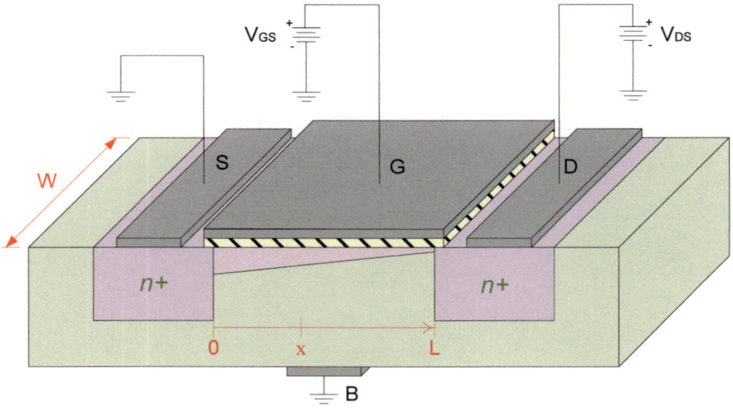

Fig. 9. Biasing of an *n*MOS.

The minus signal is due to the negative charge of electrons. The velocity of carriers at low fields is the product of mobility (μ) and the electric field (E). Noting that $E(x) = -dV / dx$ and representing the electrons mobility by μ_n, then expression (2) can be rewritten as:

$$i_D = WC_{ox}[\mathcal{V}_{GS} - \mathcal{V}(x) - V_t]\mu_n \frac{dV(x)}{dx} \quad (3)$$

Now integrating along the channel, one obtains:

$$\int_0^L i_D dx = \int_0^{V_{DS}} WC_{ox}[\mathcal{V}_{GS} - \mathcal{V}(x) - V_t]\mu_n dV(x) \quad (4)$$

Thus, the expression for the drain current in the triode region is:

$$i_D = \mu_n C_{ox} \frac{W}{L}[(\mathcal{V}_{GS} - V_t)\mathcal{V}_{DS} - \frac{\mathcal{V}_{DS}^2}{2}] \quad (5)$$

The value of the current for the saturation operation can be obtained by replacing $v_{DS} = v_{GS} - V_t$ into expression (5), as:

$$i_D = \frac{1}{2}\mu_n C_{ox} \frac{W}{L}(\mathcal{V}_{GS} - V_t)^2 \quad (6)$$

As described earlier, the current does not depend on v_{DS}. It can be observed from expressions (5) and (6) that the current is proportional to the ratio W / L, which is know as the aspect ratio. The designer can alter the aspect ratio to obtain the desired *i-v* characteristic.

Observe that expression (6) was obtained using the value of L, as given in Fig. 9. Nevertheless, when the transistor is saturated, the channel becomes shorter, as shown in Fig. 7. A reduction in the length of the channel, known as channel length modulation, means a variation in the resistance, and therefore a variation in the current i_D.

Expression (6) can be modified in order to include the variation in the channel length, represented as $L\text{-}\Delta L$, as:

$$i_D = \frac{1}{2}\mu_n C_{ox}\frac{W}{L-\Delta L}(v_{GS}-V_t)^2$$
$$i_D = \frac{1}{2}\mu_n C_{ox}\frac{W}{L}\frac{1}{1-(\Delta L/L)}(v_{GS}-V_t)^2 \quad (7)$$

which can be approximated to:

$$i_D \cong \frac{1}{2}\mu_n C_{ox}\frac{W}{L}\left(1+\frac{\Delta L}{L}\right)(v_{GS}-V_t)^2 \quad (8)$$

Since $\Delta L/L$ is proportional to v_{DS} (the larger v_{DS} the larger will be ΔL), then:

$$i_D \cong \frac{1}{2}\mu_n C_{ox}\frac{W}{L}(v_{GS}-V_t)^2(1+\lambda v_{DS}) \quad (9)$$

where λ is the parameter of proportionality.
The effect of channel length modulation can be seen in the i_D - v_{DS} characteristic of a MOS transistor shown in Fig. 10. The dependence of v_{DS} on i_D in the saturation region can be seen is represent by $(1+\lambda v_{DS})$ in expression (9) and can be observed in Fig. 10.

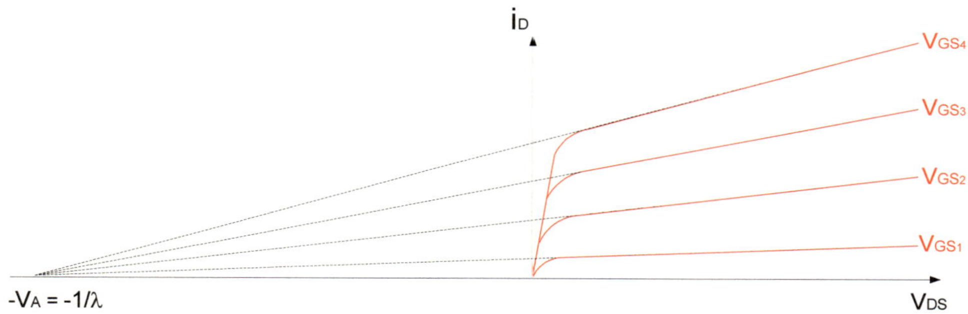

Fig. 10. Effect of channel modulation on saturation current.

An extrapolation of i_D - v_{DS} intercepts the v_{DS} axis at $v_{DS} = -V_A$, known as Early voltage. For a given process, V_A is proportional to L, selected by the designer. Typically, VA is in the range of 5 V/µm to 50 V/µm.

2.5 Output resistance

Fig. 10 and expression (9) show that an increase in v_{DS} causes an increase in i_D, meaning a resistive behavior. The value of the resistance is given as:

$$r_o = \left[\frac{\partial i_D}{\partial v_{DS}}\right]^{-1} = \left[\lambda\frac{1}{2}\mu_n C_{ox}\frac{W}{L}(v_{GS}-V_t)^2\right]^{-1} \quad (10)$$

which can be simplified to:

$$r_o = \frac{1}{\lambda i_D} = \frac{V_A}{i_D} \tag{11}$$

Therefore, a MOS transistor in the saturation region is not totally independent of v_{DS} and presents an output impedance given by (11)
Considering the transistor operating in the triode region, as given by expression (5), if the value of v_{DS} is sufficiently small, v_{DS}^2 can be neglected, and therefore:

$$i_D \cong \mu_n C_{ox} \frac{W}{L}[(v_{GS} - V_t)v_{DS}] \tag{12}$$

This relationship represents the behavior of the MOS transistor as a linear resistance whose value is controlled by v_{GS}, as given by:

$$r_{ds} = \frac{v_{DS}}{i_D} = [\mu_n C_{ox} \frac{W}{L}(v_{GS} - V_t)]^{-1} \tag{13}$$

2.6 Transconductance

The large signal behavior of a MOS transistor in the saturation region is given by expression (6). Nevertheless, for a given biasing, the designer may be interested in the small signal behavior of the transistor. For a given small variation in the v_{GS}, around the biasing, there will be a variation in the i_D current, given by the transconductance, as:

$$g_m = \left. \frac{\partial i_D}{\partial v_{GS}} \right|_{v_{GS}=V_{GS}} \tag{14}$$

which results in:

$$g_m = \mu_n C_{ox} \frac{W}{L}(v_{GS} - V_t) \tag{15}$$

Observe the transconductance depends on the ratio W/L and on the value of v_{GS}, and they can be controlled by the designer. By using expression (6), then expression can be written as:

$$g_m = \sqrt{2\mu_n C_{ox} \frac{W}{L} i_D} \tag{16}$$

In this case, the transconductance depends on the ratio W/L and the i_D current. That expression can be written also as:

$$g_m = 2\frac{i_D}{(v_{GS} - V_t)} \tag{17}$$

It clearly does not depend on ratio W/L but it depends on both v_{GS} and i_D.

2.7 Body effect

In many circuits, the substrate and the source are not at the same potential, as it is possible to stack transistors. In that case, the substrate it is at lower potential than the source, and

therefore the source-substrate junction becomes reversed biased. This reverse biasing widens the depletion layer, which in turn reduces the channel depth.
The effect of the bulk-source voltage V_{SB} can be easily represented by a change in the threshold voltage - V_t, as given by:

$$V_t = V_{t0} + \gamma[\sqrt{2\phi_f + V_{SB}} - \sqrt{2\phi_f}] \tag{18}$$

where V_{t0} is the threshold voltage for $V_{SB} = 0$, ϕ_f is a physical parameter (usually $2\phi_f = 0.6V$) and γ is a fabrication-process parameter given by:

$$\gamma = \frac{\sqrt{2qN_A\varepsilon_s}}{C_{ox}} \tag{19}$$

where q is the electron charge (1.6×10^{19} C), N_A is the doping concentration of the substrate and ε_s is the permissivity of silicon ($1.17\varepsilon_0 = 1.17 \times 8.854 \times 10^{-14} = 1.04 \times 10^{-12}$ F/cm).
Any signal between substrate and source promotes a drain current component. The substrate acts as a second gate, and in turn will present a corresponding trasnconductance, named body transconductance, given as:

$$g_{mb} = \left.\frac{\partial i_D}{\partial v_{DS}}\right|_{\substack{v_{GS}=V_{GS} \\ v_{BS}=V_{BS}}} \tag{20}$$

From expressions (6), (17) and (18), then it is possible to state that:

$$g_{mb} = \chi g_m \tag{21}$$

where χ is given by:

$$\chi = \frac{\partial V_t}{\partial V_{SB}} = \frac{\gamma}{2\sqrt{2\varphi_f + V_{SB}}} \tag{22}$$

And it is in the range of 0.1 to 0.

2.8 Small signal model

Considering the output impedance, the transconductance and the body effect, the small signal model of a nMOS transistor is given by Fig. 11, known as hybrid-п model.

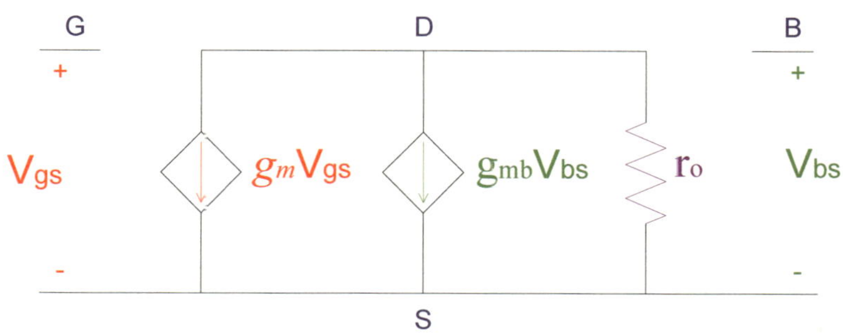

Fig. 11. Hybrid-п model.

If the source and the substrate are at the same potential, then the model can be simplified, as the term $g_{mb}v_{bs}$ goes to zero. The simplified hybrid-π model is shown in Fig. 12.

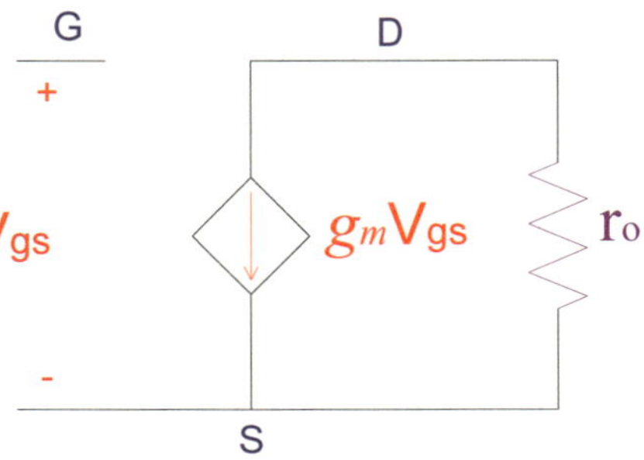

Fig. 12. Simplified hybrid-π model.

2.9 Summary

Table 1 summarizes the main *n*MOS equations.

Saturation	Condition	$v_{DS} \geq v_{GS} - V_t$.
	i-v characteristic	$i_D \cong \frac{1}{2}\mu_n C_{ox}\frac{W}{L}(v_{GS} - V_t)^2$
	Output resistance	$r_o = \frac{1}{\lambda i_D} = \frac{V_A}{i_D}$
	Transconductance	$g_m = \sqrt{2\mu_n C_{ox}\frac{W}{L}i_D}$ $g_m = \mu_n C_{ox}\frac{W}{L}(v_{GS} - V_t)$ $g_m = 2\frac{i_D}{(v_{GS} - V_t)}$
	Body transconductance	$g_{mb} = \chi g_m = \frac{\gamma}{2\sqrt{2\phi_f + V_{SB}}}g_m$

Triode	Condition	$v_{DS} < v_{GS} - V_t$.
	i-v characteristic	$i_D = \mu_n C_{ox}\frac{W}{L}[(v_{GS} - V_t)v_{DS} - \frac{v_{DS}^2}{2}]$
	Output resistance	$r_{linear} = \frac{v_{DS}}{i_D} = [\mu_n C_{ox}\frac{W}{L}(v_{GS} - V_t)]^{-1}$
	Threshold voltage	$V_t = V_{t0} + \gamma[\sqrt{2\phi_f + V_{SB}} - \sqrt{2\phi_f}]$

Table 1. Summary of *n*MOS equations.

2.10 *p*MOS transistor

In a *p*MOS transistor, a *p* channel is formed on an *n* substrate. Therefore, its operation is virtually the same as the *n*MOS transistor, except that all voltages and currents are opposite as in the nMOS transistor. Fig. 13 shows the symbols for the nMOS transistor, although other symbols may be found in the literature. The symbol in Fig. 13.a corresponds to the four terminal connection, and the symbol in Fig. 13.b corresponds to the three terminal connection, where source and substrate are shorted.

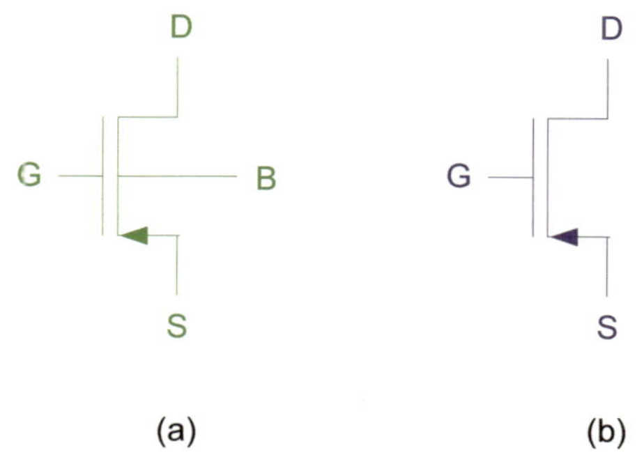

Fig. 13. Symbols for pMOS transistor; (a) four terminals and (b) three terminals.

3. RF CMOS model

Unfortunately, the structure and the operation of a MOS transistor present parasitic capacitances that limit its frequency of operation. The parasitic capacitances may result from the capacitor formed between the gate and the channel, between gate and source/drain, and between drain/source and substrate.

3.1 Gate capacitances

The gate, the dielectric and the channel form a capacitor. When the transistor is working in the triode region with a small voltage v_{DS}, the channel will be of uniform depth, as shown in Fig. 4. Therefore, the gate-channel capacitance can be considered equally divided between the source and the drain, and their values are:

$$C_{gs} = C_{gd} = \frac{1}{2} WLC_{ox} \text{ (triode region)} \tag{23}$$

When the transistor is working in the saturation region, the channel presents a tapered shape and it is pinched off at the drain end, as presented in Fig. 7. It can be seen that the gate to channel capacitance is almost entirely modeled at the source, since the drain does not present a channel. It can be shown that the capacitances are:

$$\begin{aligned} C_{gs} &= \frac{2}{3} WLC_{ox} \\ C_{gd} &\approx 0 \end{aligned} \quad \text{(saturation region)} \tag{24}$$

If the transistor is cut off, there is no capacitance between gate and channel, since there is no channel for cut off. The entire capacitance is then between the gate and the substrate, therefore:

$$\begin{aligned} C_{gs} &= C_{gd} = 0 \\ C_{gb} &= WLC_{ox} \end{aligned} \quad \text{(cut off)} \tag{25}$$

As can be observed from Fig. 1, the gate extends over the drain and the source areas. Therefore, there is an overlapping capacitance between the gate and the drain/source. Denoting the overlapping length by L_{ov}, then the overlap capacitance can be seen to be:

$$C_{gs_{ov}} = C_{gd_{ov}} = WL_{ov}C_{ox} \tag{26}$$

For modern processes, L_{ov} is usually in the range of 5% to 10% of L.

3.2 Junction capacitances

As shown by Fig. 2 there are two reversed biased junctions formed between the substrate and source/drain. Each junction consists of two semiconductors (drain/source and the substrate) and the depletion layer, thus forming a capacitor. The source-substrate capacitance can be found to be:

$$C_{sb} = \frac{C_{sb0}}{\sqrt{1+\frac{V_{SB}}{V_0}}} \tag{27}$$

where V_0 is the junction built-in voltage (0.6 V to 0.8 V), V_{SB} is the magnitude of the reversed bias voltage and C_{sbo} is the capacitance at zero reverse bias voltage.
By the same way, the drain-substrate capacitance is given by:

$$C_{db} = \frac{C_{db0}}{\sqrt{1+\frac{V_{DB}}{V_0}}} \tag{28}$$

3.3 The high frequency model

The small signal model of the MOS transistor given in Fig. 11 can be update to include the gate and the junction capacitances, as presented in Fig. 14. Although this model represents the transistor for high frequencies, it is very complex for manual analysis.

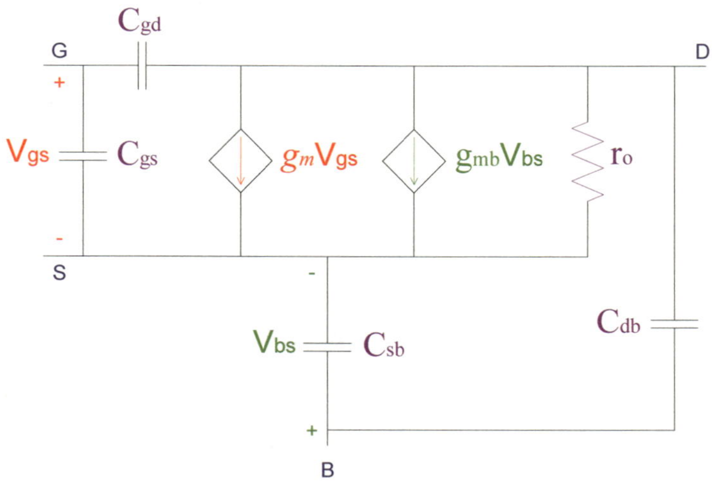

Fig. 14. Hybrid-π model including the parasitic capacitances.

If the source and the substrate are shorted, the model can be greatly simplified, as shown in Fig. 15.

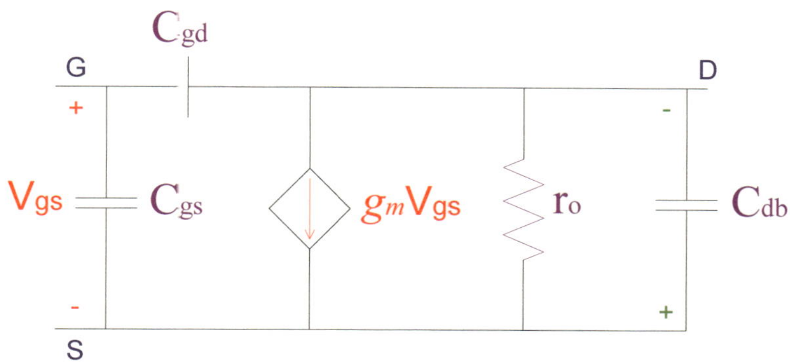

Fig. 15. Simplified high frequency model for source and substrate shorted.

4. Unity gain frequency

An important Fig. of merit for the MOS transistor is the unit gain frequency that is defined as the frequency in which the short circuit current gain becomes unit. This definition is based in the common source configuration, as shown in Fig. 16.

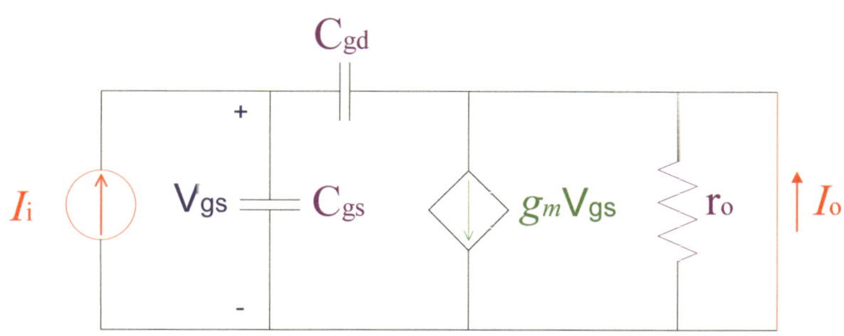

Fig. 16. Circuit model used to obtain the unit gain frequency.

The current I_o in the short circuit is given by:

$$I_o = g_m V_{gs} - sC_{gd}V_{gs} \cong g_m V_{gs} \tag{29}$$

The approximation is due to the fact that C_{gd} is very small and can be neglected. Also, from the circuit, V_{gs} can be expressed as:

$$V_{gs} = \frac{I_i}{s(C_{gs} + C_{gd})} \tag{30}$$

Therefore, from expressions (29) and (30):

$$\frac{I_o}{I_i} = \frac{g_m}{s(C_{gs} + C_{gd})} \tag{31}$$

Since the magnitude of I_o/I_i should be 1, as per definition, and considering physical frequencies ($s=j\omega$), then:

$$\omega_T = \frac{g_m}{(C_{gs} + C_{gd})} \tag{32}$$

Therefore, the unit gain frequency is:

$$f_T = \frac{g_m}{2\pi(C_{gs} + C_{gd})} \tag{33}$$

As can be observed, the unit gain frequency is directly proportional to g_m and inversely proportional to the internal capacitances. Therefore, in terms of frequency response the transistor should have large g_m and small capacitances.

4. RF CMOS noise model

The two most important types of noise in MOS devices are the *1/f* noise and the thermal noise.

4.1 Thermal noise

The main source of thermal noise in a MOS transistor is due to the resistive channel in the active region, and has a value of:

$$i_d^2 = 4kT\gamma g_m \tag{34}$$

where k is the Boltzmann's constant (about 1.38×10^{-23} *J/K*), T is the absolute temperature in kelvins and γ is a constant that is approximately *2/3* for long channel transistors and increase to the range *1-2* for short channel devices.
The other source of thermal noise is the gate. Fluctuation in the channel potential couples capacitively into the gate terminal, which in turn translates into a noise gate current. Noise gate current can also be produced by the resistive material of the gate. This total noise gate can be ignored at low frequencies but becomes significant at high frequencies as it is the case of RF circuits. It has been shown the gate noise may be expressed as:

$$i_g^2 = 4kT\delta g_g \tag{35}$$

where δ is approximately *4/3* for long channel transistors and increase to the range *2-4* for short channel devices, and g_g is given by:

$$g_g = \frac{\omega^2 C_{gs}^2}{5 g_m} \tag{36}$$

Mostly of the time, instead of using a current source at the gate, it is more convenient to consider an equivalent voltage source. The equivalent voltage source of expressions (31) and (32) is given by:

$$v_g^2 = 4kT\delta r_g \quad (37)$$

where r_g is given by:

$$r_g = \frac{1}{5g_m} \quad (38)$$

4.2 *1/f* noise

The *1/f* noise, also known as flicker noise or pink noise, arises mainly due to the surface imperfections that can trap and release charges. Since MOS devices are naturally surface devices, they produce much more *1/f* than bipolar devices (which are bulk devices). This noise is also generated by defects and impurities that randomly trap and release charges. The trapping times are statistically distributed in such a way that lead to a *1/f* noise spectrum.
The *1/f* noise can be modeled by a voltage source in series with the gate, of value:

$$v_f^2 = \frac{\beta}{WLC_{ox}f} \quad (39)$$

For *p*MOS devices, β is typically about $10^{-28}C^2/m^2$, but it can be up to 50 times larger for *n*MOS devices.
As can be observed from expression (53), the *1/f* noise is smaller for larger devices. This occurs because the large capacitance smoothes the fluctuation in the channel charge. Therefore, in order to achieve good *1/f* performance, larger devices should be used.
The *1/f* can also be modeled as a current source at the drain whose value is:

$$i_f^2 = \frac{\beta g_m^2}{WLC_{ox}^2 f} \cong \frac{\beta}{f}\omega_T^2 A\Delta f \quad (40)$$

where A is the area of the gate.

4.3 Noise model

The noise model of an *n*MOS transistor is presented in Fig. 17, where the transistor is considered noiseless. The decision of placing the noise sources as a voltage source at the gate, or as a current source at the drain is just a matter of convenience according to the circuit under analysis. As an example, the values of Fig. 17 could be:

$$\begin{aligned} v^2 &= v_g^2 = 4kT\delta r_g \\ i^2 &= i_d^2 + i_f^2 = 4kT\gamma g_m + \frac{\beta g_m^2}{WLC_{ox}^2 f} \end{aligned} \quad (41)$$

5. Conclusions

The proper understanding of physical operation to modeling of CMOS transistors is essential to the analysis and design of RFID circuits. Among its advantages, the CMOS transistors demands lower power consumption than other transistors.
Noise analysis of CMOS transistors is also fundamental to analysis and design of any circuit, including RFID.

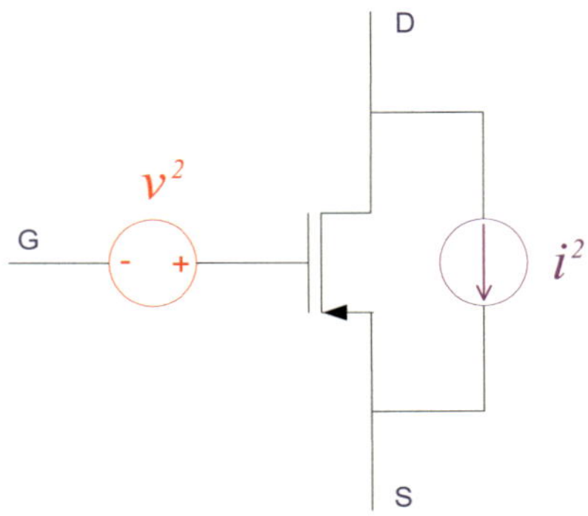

Fig. 17. Noise model of an *n*MOS transistor.

6. References

Allen, P. E. & Holberg, D. R. (2002) *CMOS Analog Circuit Design - 2nd Ed.*, Oxford University Press, ISBN 0195116445.

Johns, D. A. & Martin, K. (1997) *Analog Integrated Circuit Design*, John Wiley & Sons, ISBN 0471144487.

Sedra, A. S. & Smith, K. C. (2009) *Microelectronic Circuit - 6th Ed.*, Oxford University Press, ISBN 0195323033.

Lee, T. H. (2004). *The Design of CMOS Radio-Frequency Integrated Circuits – 2nd Edition*, Cambridge University Press, ISBN 0521835399.

Coleman, C. (20040 *An Introduction to Radio Frequency Engineering*, Cambridge University Press, ISBN 0521834813.

Gilmore, R. & Besser, L. (2003) *Practical RF Circuit Design for Modern Wireless Systems – Vol. II*, Artech House Publishers, ISBN 1580535224.

Rogers, J. & Plett, C. (20030 *Radio Frequency Integrated Circuit Design*, Artech House Inc, ISBN 1607839792.

Ziel, A. (1986) *Noise in Solid State Devices and Circuits*, John Wiley and Sons, ISBN 0471832340.

4

Structural Design of a CMOS Voltage Regulator for an Implanted Device

Paulo C. Crepaldi[1], Luis H. de C. Ferreira[1], Tales C. Pimenta[1], Robson L. Moreno[1], Leonardo B. Zoccal[1] and Edgar C. Rodriguez[2]
[1]*Federal University of Itajubá*
[2]*University of São Paulo*
Brazil

1. Introduction

There is a great interest in the development of equipment and devices that can accurately and efficiently monitor biological signals such as blood pressure, heart beat and body temperature, among others. It is highly desirable to have those devices operating in an environment free of wires, where the information can be accessed remotely and processed in real time by external equipments.

When the equipments are connected to communication network they form a telemedicine system by which the patients can be monitored remotely (biotelemetry), even over the internet, thus indicating the portability of these instruments (Miyazaki, 2003; Puers, 2005; Scanlon et al, 1996).

Microelectronics has become a powerful tool when used in this scenario. In recent years, integrated circuits are being fabricated with large densities and endowed with intelligence. The reliability of those systems has been increasing and the costs are lowering. The interaction between medicine and technology, as it is the case of microelectronics and biosensor materials, allows the development of diagnosing devices capable of monitoring pathogens and deceases. The design of sensors, signal conditioners and processing units aims to find solutions in which the whole system can be placed directly in the patient or, more desirable, implanted. It becomes a Lab-on-Chip and Point-of-Care device (Colomer-Farrarons, 2009). Since the implanted device becomes part of a biological data acquisition system it must meet few requirements such as reduced size, low power consumption and the possibility of being powered by an RF link, then it operates as a passive RFID tag (Landt, 2005).

The low power restriction is extremely important for the patient safety, by avoiding heating due to the increase of current density in the tissues surrounding the implant that could cause tissue damage. The power restrictions mean also limited power of RF transmitter that can, as well, to induce dangerous electromagnetic fields – EMF.

The focus in this chapter is to discuss the implementation of a Linear Voltage Regulator – LVR by considering the use of a low cost CMOS process, low-power, low silicon area and simple circuit topology.

The LVR is an ASIC structure whose electrical characteristics depend on the specific load conditions. Therefore, the idea is to discuss few structural solutions.

2. Implanted Device - Smart Biological Sensors

A typical CMOS front-end architecture of an in-vivo Biomedical Implanted Device – BID is shown in Figure 1. The system consists, basically, of the sensitive biological element, the transducer or detector element, the associate electronics and signal processors, and the RF link to establish a communication with the manager unit. The combination of the implanted device, the local wireless link and a communication network forms the Wireless Biosensor Network – WBSN (Guennoun, 2008).

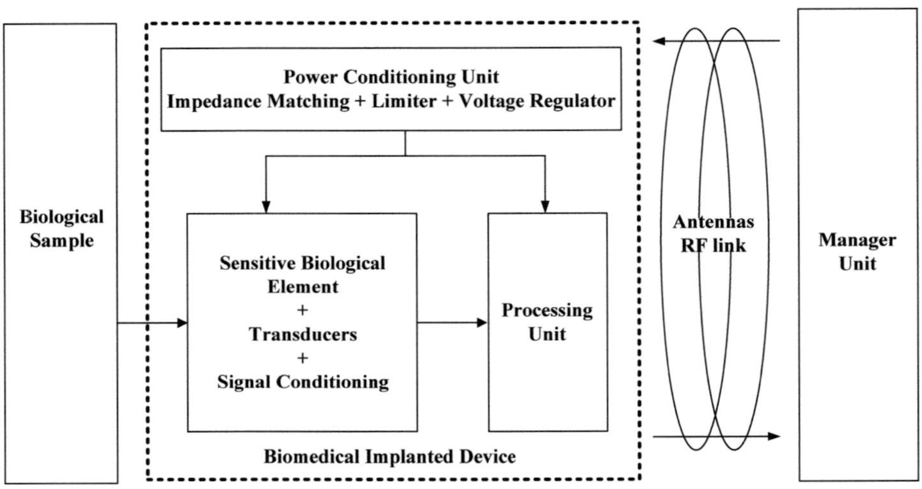

Fig. 1. Typical Implanted Biomedical Device acting as a RFID Tag.

Linear systems based on semiconductor devices demand a stable power supply voltage for proper operation. Fluctuations on the input line voltage, load current fluctuations and temperature variations may cause the circuit to deviate from its optimum operation bias point and even loose its linearity. Therefore, the power supply system must experience minimum impacts on the linearity due to those variations. Nevertheless, the impact of temperature variations in implantable devices is minimized since the body temperature is kept stable at approximately 37°C (Mackowiak, 1992).

The LVR is part of the power conditioning block that is responsible to supply a stable voltage to the sensors/transducers and its associated electronics.

Unlike the general voltage regulator application, an implantable device does not suffer a large range, but it is more limited. This condition minimizes the impact of load regulation specification.

The tag operation frequency is one of the most important considerations when designing a solution to suit the requirements. The operation frequency has enormous effect on price, performance, range and suitability for RFID projects. The general bands used to broadly classify the RFID tag families are low, high, and ultra high.

The low frequency range (typically between 125 kHz and 134 kHz) is most commonly used for access control, animal tracking and assets tracking. It offers low cost.

The high frequency range (typically 13.56MHz) is used for medium data rate transfer and reading range of up to 1.5 meters, usually for passive tagging. This frequency has also the advantage of not being susceptible to interference from the presence of water or metals. Since the user of an implantable monitoring system is exposed to a RF source near the skin, few safety considerations must be taken into account. The main biohazards and risks due to the RF exposure is mainly the heating from the electromagnetic field distribution on biological tissues (Osepchuk, J.M. & Petersen R. C., 2001). This frequency provides a good tradeoff between power level and human tissue penetration (Sauer, 2005; Vaillantcourt, 1997).
The ultra high band (typically between 850MHz and 950MHz) offers the largest reading ranges, of up to approximately 3 meters for passive tags and 100 meters for active tags. Relatively high reading speeds can be achieved at that band.

3. The topology of a voltage regulator

Classic topologies used in voltage regulators can be classified as linear or switched. Switched regulators present complex circuitry, mainly due to control unit, thus frequently requiring larger power consumption and larger silicon area. Furthermore they provide larger noise at the output due to the switched operation (Rincon-Mora & Allen, 1997).
Low dropout - LDO voltage regulators is one of the most popular power converters used in power management and is more suitable for implanted systems (Rincon-Mora, 1998, 2000). The basic topology of an LDO is presented in Figure 2.

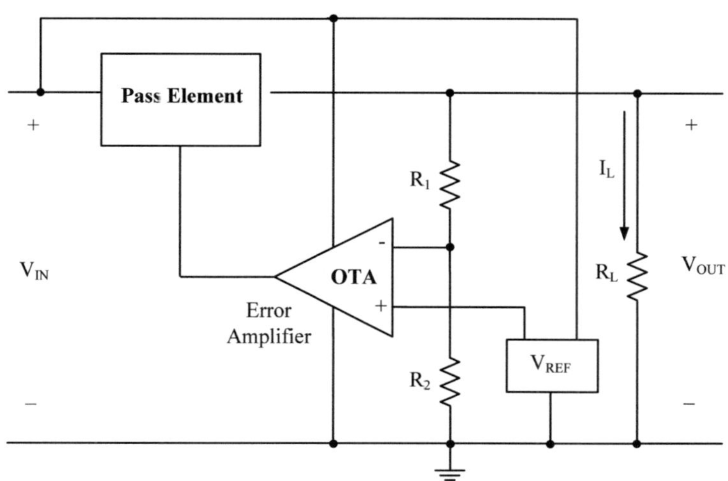

Fig. 2. Basic LDO topology.

The pass element can be implemented using bipolar or MOS transistors. Since a MOS transistor is controlled by its gate voltage, it offers the advantage of smaller power consumption and consequently higher efficiency for the voltage regulator. The MOS transistor can be either N or P type. The NMOS transistor requires a gate voltage higher than the source voltage, and therefore it may be necessary a charge pump to increase the voltage level. The proper choice for low voltage systems, such as implantable devices, it is the use of a PMOS LDO, as indicated in Figure 3 (Kugelstadt, 1999; Simpson, 1997). A

NMOS LDO without charge pump is reported in (Ahmadi & Jullien, 2009) using native transistors (zero threshold) and an internal capacitor to improve the stability, but two external capacitors are required.

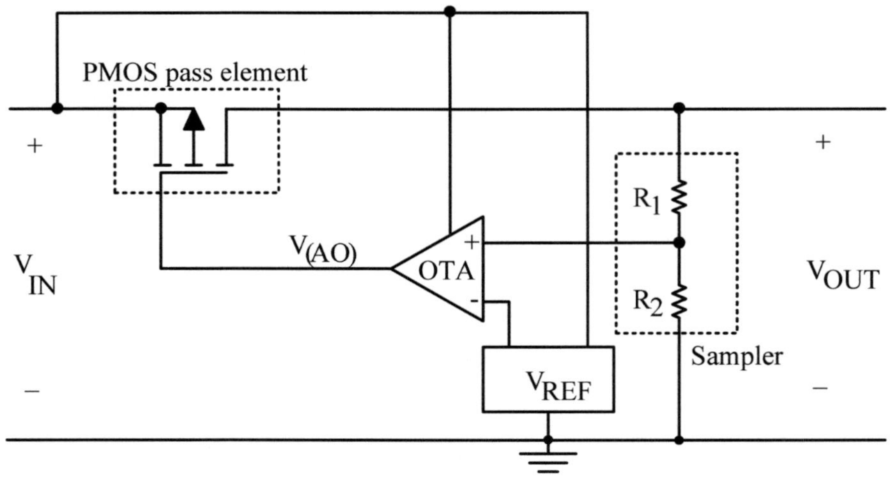

Fig. 3. PMOS based LDO.

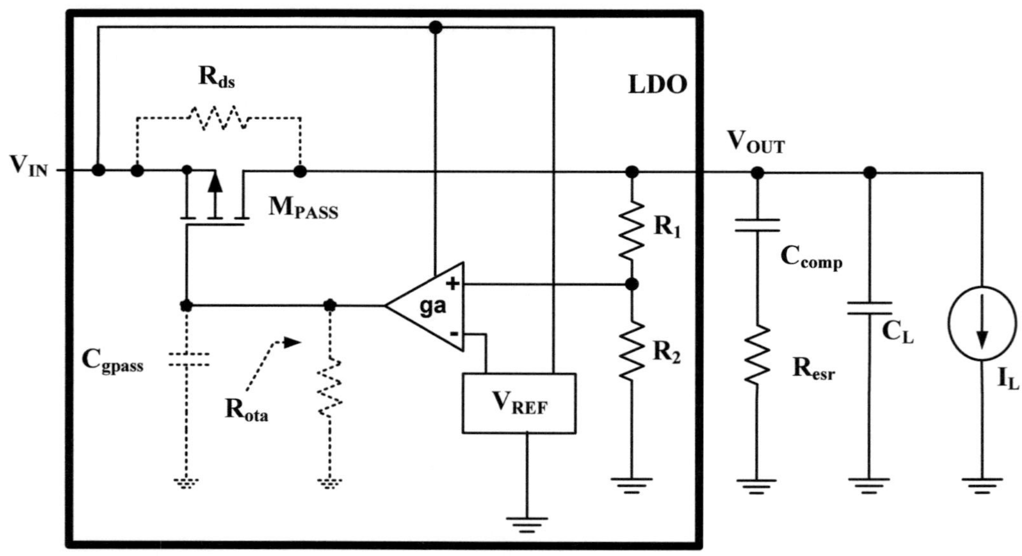

Fig. 4. Classic PMOS LDO with discrete frequency compensation scheme.

The closed loop system output voltage can be found to be:

$$V_{OUT} = \left(1 + \frac{R_1}{R_2}\right) V_{REF} \quad [V] \tag{1}$$

The use of an LDO circuit requires the stability analysis since it forms a closed loop system. The frequency response is degraded by the presence of two poles besides the dominant pole that can lead to an unstable condition. It is necessary to add a zero between these two poles to achieve a frequency compensation. The insertion of this zero is normally implemented by adding a discrete electrolytic capacitor (C_{comp}) at the output node that also contributes with an additional resistance R_{esr}, as represented in Figure 4. Additionally, R_{ota} is the output resistance of the transconductance amplifier, C_{gpass} is the gate capacitance of the PMOS pass transistor and R_{ds} is the channel resistance of the PMOS pass transistor.
The frequencies of these poles and zero are given by (Rogers, 1999):

$$f_{P0} = \frac{-1}{2\pi\left(R_{ds} + R_{esr}\right)C_{comp}} \approx \frac{-1}{2\pi R_{ds} C_{comp}} \quad [\text{Hz}] \tag{2}$$

$$f_{P1} = \frac{-1}{2\pi\left(R_{ds} // R_{esr}\right)C_{L}} \approx \frac{-1}{2\pi R_{esr} C_{L}} \quad [\text{Hz}] \tag{3}$$

$$f_{Z0} = \frac{-1}{2\pi R_{esr} C_{comp}} \quad [\text{Hz}] \tag{4}$$

$$f_{P2} = \frac{-1}{2\pi R_{ota} C_{gpass}} \quad [\text{Hz}] \tag{5}$$

Equation (1) shows that the dominant pole frequency depends on the drain-source resistance, which in turn depends on the drain current. As a consequence, the dominant pole can change its position according to the load. To overcome this situation, the zero must follow the pole. It is common to establish not just a single value for R_{esr} but a range of values as a function of load current.

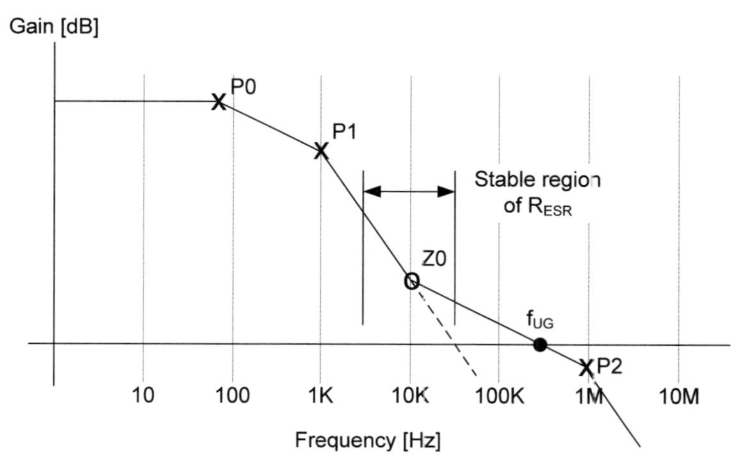

Fig. 5. Frequency response of a PMOS LDO regulator with external compensation capacitor PMOS based LDO.

Figure 5 presents the frequency response of a PMOS LDO. Unfortunately, the use of an external capacitor, such as an electrolytic capacitor, is prohibitive for an implantable device. Thus, the literature provides many contributions to solve the LDO stability problem. Few approaches maintain the external capacitor and modify the internal feedback loop by using buffers (Stanescu, 2003) and Miller compensation capacitor (Huang et al, 2006). Other approaches insert and internal zero, discarding the compensation capacitor, by using controlled sources and even Miller compensation (Huang et al, 2006).

Load Conditions: I_L = 500µA, C_L = 5pF	
V_{IN}	2.2V±10%
V_{OUT}	1V±5%
V_{BIAS}	2V
V_{REF}	200mV*
P_D	1mW**

* A lower value of 200mV was adopted to provide a wider range of output values, as stated by eq. (1)
** A safe value for the RF link power transfer is 10mW/cm² (Lazzi, 2005). The LVR power dissipation should be taken as just 10% of it, corresponding to 1mW, which represents twice as much as required by the load (0.5mW). Reported voltage regulators for implanted devices list a power dissipation range that can be as high as tents of mW (Zheng & Ma, 2010).

Table 1. LVR target values for an implanted blood pressure monitoring system.

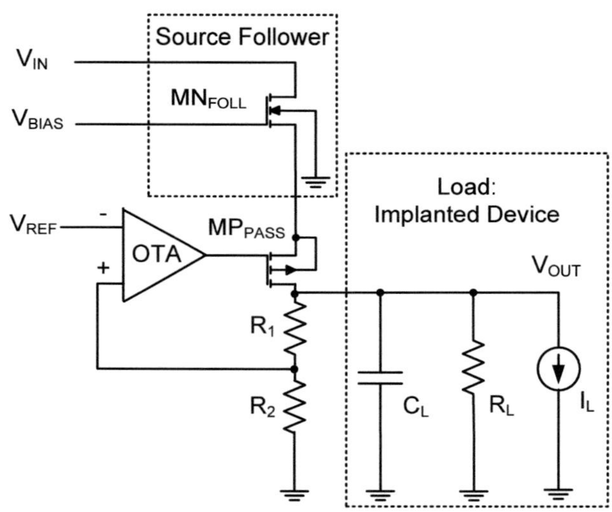

Fig. 6. LVR architecture.

The solution proposed here is the introduction of a source follower (MN_{FOL}) stage in between the input voltage and the LDO block, and the removal of the compensation capacitor C_{comp}, as shown in Figure 6. The source follower maintains the PMOS pass element in the triode region, which leads to an unconditionally stable system, as it will be described later.

The introduction of the extra source follower represents a disadvantage since it introduces extra power consumption and requires additional silicon area. The overall efficiency is also

affected, nevertheless the advantages overpasses de disadvantages, mainly for implanted devices.
Table 1 shows the target values for a project example. The load is an implanted physiological signal system that is used to monitor the blood pressure.

4. Frequency response analysis

The frequency analysis of the LVR can be evaluated by finding initially the open loop gain (Aβ) Figure 7. The originally closed loop is broken at a particular point, and the loop gain is given by:

$$A\beta = -\frac{v_r}{v_x} \quad [-] \tag{6}$$

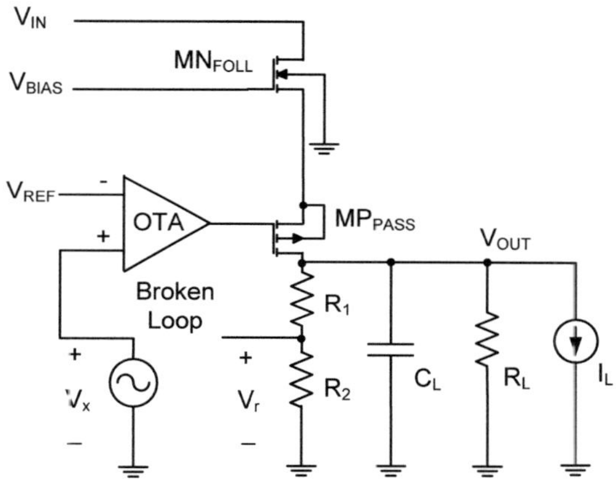

Fig. 7. Feedback broken to analyze the open loop gain.

In Figure 8 the OTA and the pass transistor (MP_{PASS}) are replaced by the small signal model.

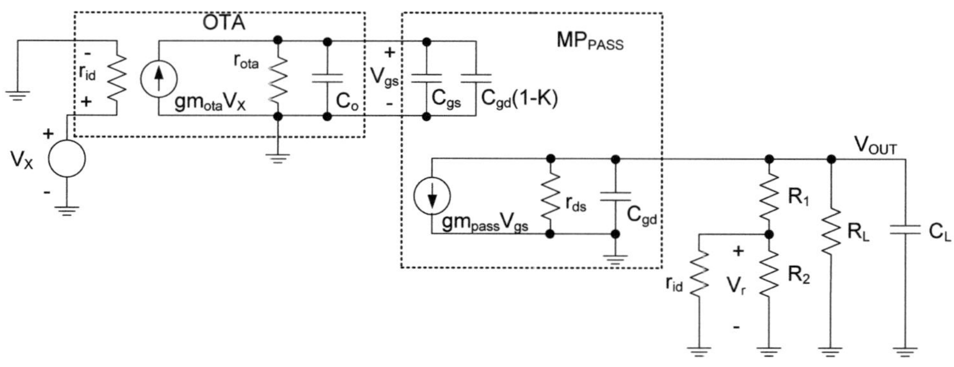

Fig. 8. Small signal equivalent circuit of the LVR

The total load resistance is minimized by the low value of r_{ds}, therefore the drain-gate voltage gain of MP_{PASS} is:

$$K = -\frac{v_{out}}{v_{gs}} = -gm_{pass} r_{ds} \quad [-] \tag{7}$$

The output voltage is:

$$v_{out} = -\frac{gm_{pass} r_{ds} \, gm_{ota} r_{ota}}{\left(1+\frac{S}{p_1}\right)\left(1+\frac{S}{p_2}\right)} v_x \quad [V] \tag{8}$$

Considering that r_{id} is much larger than R_2, then v_r is:

$$v_r = v_{out} \frac{R_2}{R_1 + R_2} \quad [V] \tag{9}$$

By combing (7) and (8), the loop gain is:

$$A\beta = \frac{gm_{pass} r_{ds} \, gm_{ota} r_{ota}}{\left(1+\frac{S}{p_1}\right)\left(1+\frac{S}{p_2}\right)} \frac{R_2}{R_1 + R_2} \quad [V] \tag{10}$$

It can be observed from Equation (9) that the feedback gain β is $R_2/(R_1+R_2)$. It is compatible with Equation (1) that states the relationship between V_{OUT} and V_{IN} is given by the factor 1/β.
The poles p_1 and p_2 are:

$$f_{P1} = \frac{-1}{2\pi\left(C_{gd} + C_L\right) r_{ds}} \quad [Hz]$$
$$f_{P2} = \frac{-1}{2\pi\left[C_o + C_{gs} + C_{gd}\left(1 + gm_{pass} r_{ds}\right)\right] r_{ota}} \quad [Hz] \tag{11}$$

Pole p_2 is the dominant one since r_{ota} is in the range of MΩ and can be at least 10^5 times larger than r_{ds}, which is the range of tens of Ohms. So the frequency stability of the regulator is a function of the OTA design, the geometric aspect ratio of MP_{PASS} and the load. As an ASIC application, the load current (I_L), resistance (R_L) and capacitance (C_L) can be stated as constants without impacting in the pole frequencies. The OTA output capacitance C_O can be neglected since the PMOS pass transistor has a larger geometric aspect and, consequently, larger C_{gs} and C_{gd}.
Equation (9) shows that at low frequencies (DC), the gain A is given by:

$$A = gm_{pass} r_{ds} gm_{ota} r_{ota} \quad [-] \tag{12}$$

Considering typically g_m in the range of 10^{-3} [V/A], tens of Ohm to r_{ds} and 10^6 Ohm for r_{ota}, than the gain is greater than 40 [dB]. The dominant pole will have a frequency in the range of tens of H_z and the unit frequency gain in the range of hundreds of KH_Z.

5. The sampler circuit

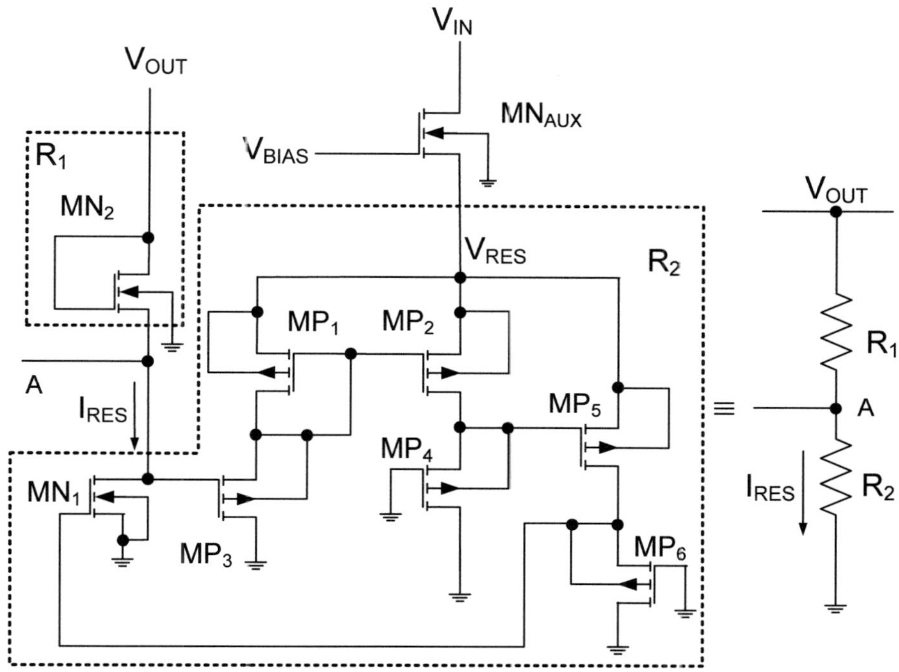

Fig. 9. Sampler Circuit for the LVR.

Figure 9 presents the sampler circuit. In order to implement the whole circuit in a single CMOS chip, R_1 is realized as a MOS diode (transistor MN_2) and R_2 is implemented through an interesting topology, a grounded MOS resistor (Dejhan, 2004). The use of the source follower transistor MN_{AUX} guarantees that the grounded MOS resistor is isolated from V_{IN}, thus avoiding a significant transference of ripple voltage to the output voltage. MN_{AUX} also imposes a smaller effective voltage to the MOS resistor, thus reducing the sampler current.
The power supply voltage of the sampling circuit (PMOS array) is reduced by approximately 1V, thus settling V_{RES} to 1.2V. This is important to reduce the ground current and to maximize the LVR efficiency and improving the overall power dissipation. The relationship R_1/R_2 is optimized by the adjustments of the aspect ratio of transistor MN_1 and MN_2.
The sampler circuit current I_{RES} is designed to be ≈1% of the maximum current load (≈ 5µA). The voltage at point A is virtually V_{REF}, due to the OTA virtual short circuit. Therefore, the R_1 equivalent resistance is given as:

$$R_2 = \frac{200\text{mV}}{5\mu\text{A}} = 40 \quad [\text{K}\Omega] \tag{13}$$

The aspect ratio of MN_1 was adjusted in order to set I_{RES} as close as to the target value of 5µA. So, R_1 (transistor MN_2) will be adjusted as a 160KΩ resistor.
The additional capacitances introduced by the grounded MOS resistor and MN_2 are smaller enough so that can be discarded in the previous frequency response analyses. All those transistors have small source and drain areas leading to capacitances in the range of fF. The eventual poles will be far away from the dominant one and the unit frequency gain.

6. The voltage references

On designing any system that requires a voltage reference, the temperature and power supply sensitivity must be taken into account.
Classical voltage references are based on the bandgap voltages, where two distinct voltages with opposite thermal coefficients (PTAT and CTAT) are summed to obtain an overall near zero coefficient. Besides, their bias circuits must be robust to guarantee a low sensitivity to the power line fluctuations. The bandgap voltage is about 1.12V for silicon at room temperature (Tzanateas, 1979).
Nevertheless, the evolution of fabrication process is pushing down the supply voltages. For instance, it is about 1.2V for a CMOS 0.13µm process. So there is a demand for new voltage references topologies to produce values bellow the classical bandgap value of 1.2V.
A literature revision shows the trends into this challenge (Koushaeian & Skafidas, 2010). However, these contributions show one or more of these aspect: complex circuits topologies with an elevated number of components, the need of special components that are not ready available from the CMOS common process, the need of trimming procedures, use of external components and use of MOS transistors that are not operating in classical modes. An alternative mode is the weak inversion in which the MOS transistor behavior approaches the bipolar ones.

6.1 Current mirror core

The core to produce the voltages references are the self biased current mirror illustrated in Figure 9. The use of a parasitic vertical PNP bipolar transistor Q_1 in a CMOS digital technology is justified since it presents known V_{BE} voltage and temperature behavior. The temperature does not represent the main impact factor since the whole system will be implanted.
Equations (12) and (13) are the starting point to establish the values of the currents I_E and I_D. The currents values are set to approximately 5µA (1% of maximum load current) in order to improve the LVR overall efficiency.

$$I_d = \frac{KP}{2(1+\delta)}\left(\frac{W}{L}\right)\left(V_{gs} - \left|V_{th0}\right|\right)^2 = \beta\left(V_{gs} - \left|V_{th0}\right|\right)^2 \quad [A] \tag{14}$$

$$I_e = I_{cs} \exp\left(\frac{V_{be}}{U_T} - 1\right) \quad [A] \tag{15}$$

where KP is the MOS transconductance given in [µA²/V], δ is a dimensionless fitting parameter for short channel devices, (W/L) is the geometric aspect ratio, V_{th0} is MOS the threshold voltage given in [V], I_{CS} is the bipolar saturation current given in [nA] and U_T the thermal voltage that is about 26.7 [mV] at 37ºC.

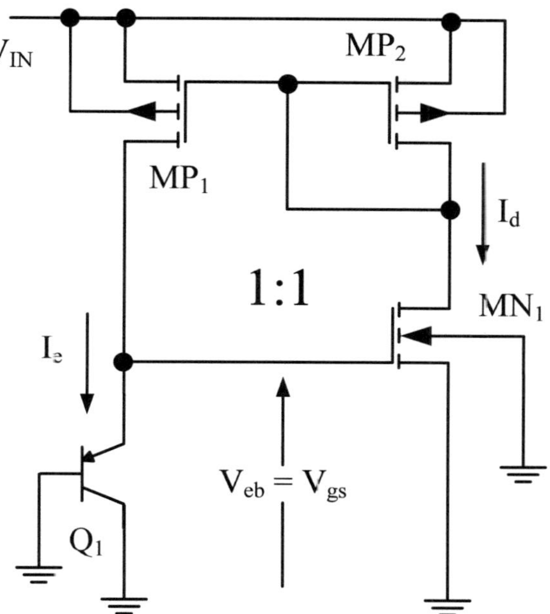

Fig. 10. Self biased current mirror.

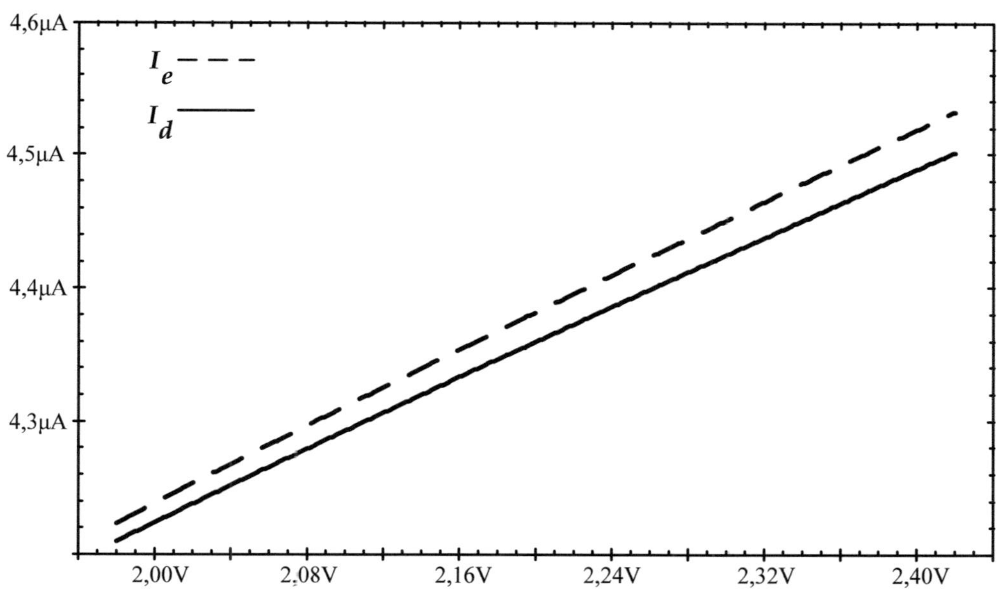

Fig. 11. Simulated results for the mirror currents @ T=37º.

There is no closed solution for both equations and it is necessary to develop an interactive simulation process to reach the optimum result for I_d, which is equal to I_e. The target value for these simulation is the geometric aspect ratio of the MOS transistors, since it is used a vertical PNP bipolar with a 100μm² emitter area. To minimize the short channel effects, the

channel length was fixed to 1µm for MN_1 and 2µm for MP_1 and MP_2 to improve the mirroring matching. The PMOS geometric aspects are also optimized by simulation.
Figure 11 shows the simulated currents for an input voltage variation of ±10% around to the ideal value of 2.2V. The temperature was fixed in 37ºC.
The relative error between the mirror currents, at the ideal operating point of V_{IN}=2.2V, can be calculated as:

$$E_{rr}(\%) = \frac{I_{eQ} - I_{dQ}}{I_{eQ}} 100 = \frac{4,38.10^{-6} - 4,36.10^{-6}}{4,38.10^{-6}} 100 \approx 0,45 \quad [\%] \tag{16}$$

It is important to evaluate the power supply dependence of those currents. The sensitivity is an adequate parameter to measure it and is given by (Gray & Meyer, 1993):

$$S_{V_{IN}}^{I_d} = \left.\frac{V_{IN}}{I_d}\right|_Q \left.\frac{\partial I_d}{\partial V_{IN}}\right|_Q \quad [-] \tag{17}$$

The derivative term can be found directly from the circuit topology to be:

$$\frac{\partial I_d}{\partial V_{IN}} = \frac{I_{dQ}\lambda_n}{1 - \frac{2U_T}{\left(V_{eb} - V_{th0(N)}\right)}} \tag{18}$$

where λ_n is the channel length modulation coefficient that is obtained by simulation and $V_{th0(N)}$ is the NMOS threshold voltage. Substituting (17) in (16) leads to:

$$S_{V_{IN}}^{I_d} = \frac{\lambda_n V_{INQ}}{\left(1 - \frac{2U_T}{V_{eb} - V_{th0(N)}}\right)} \quad [-] \tag{19}$$

An alternative way to evaluate the current sensitivity is by using Figure 11. The following equation offers a derivative approximation. It considers the variation of I_d due to variations on V_{IN}:

$$S_{V_{IN}}^{I_d} \approx \frac{V_{INQ}}{I_{dQ}} \frac{\Delta I_d}{\Delta V_{IN}} \quad [-] \tag{20}$$

Table 2 resumes the calculated and simulated results for the current sensitivity.
Consequently, for ±10%variation in V_{IN} around the quiescent value, the mirror currents will change approximately ±3%. Simulations results also point out that for the voltage references circuits discussed next, than V_{eb} voltage will play an important rule and suffers a 1.8 [mV] variation for the entire V_{IN} range, representing a deviation of ±0.13% from the 676 [mV] quiescent value. It indicates a power line rejection rate - PSRR better than 45 [dB] at low frequencies.

Body Temperature: 37°C	
Calculated	Simulated
V_{INQ}=2.2 [V]	V_{INQ}=2.2 [V]
I_{dQ}=5 [µA]	I_{dQ}=4.4 [µA]
-	λ_n=0.096 [V^{-1}]
V_{eb}=680 [mV]	V_{eb}=676 [mV]
-	$V_{th0(N)}$=523 [mV]*
$S_{V_{IN}}^{I_d}$ Eq. (19) = 0.316	$S_{V_{IN}}^{I_d}$ Eq. (20) = 0.331

* The threshold voltage value was indicated by a CMOS process.

Table 2. Id sensitivity: calculated and simulated values

6.2 The start up circuit

As a self biased circuit, the current mirror core needs a start up circuit to ensure the correct operating point. It is implemented by the circuit shown in Figure 12.

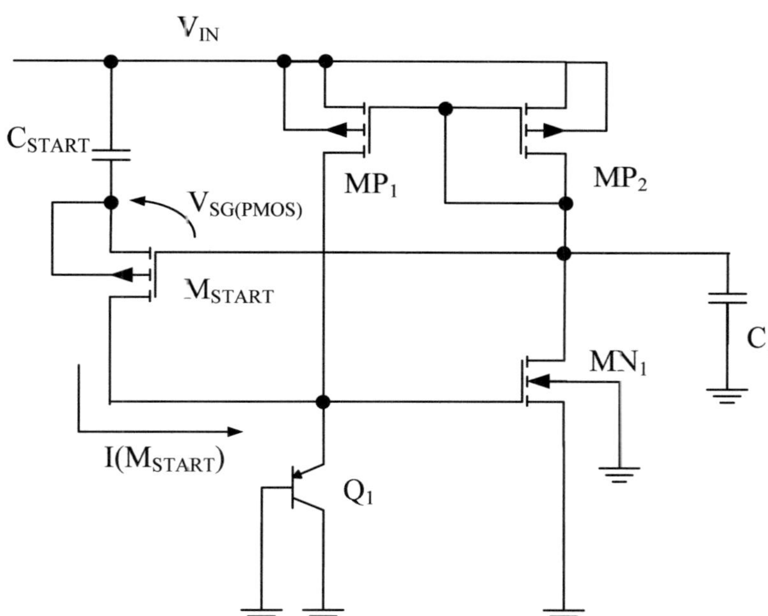

Fig. 12. The start up circuit added into the self biased current mirror.

C_{START} and C_1 are small capacitors (0.5pF) and M_{START} is a PMOS transistor, similar to those used in the current mirror. When the circuit is energized, assuming that the capacitors are discharged, the V_{sg} of M_{START} is greater than its threshold voltage. This will cause a transitory current to flow into Q_1 leading the system to desired operating point. At same time, C_{START} is charged toward V_{IN} reducing the V_{sg} of M_{START} and, consequently, turning it off. Figure 13 shows a simulating that validates the described action. The transitory current spends only 20 [ns] that is very low for a biomedical application.

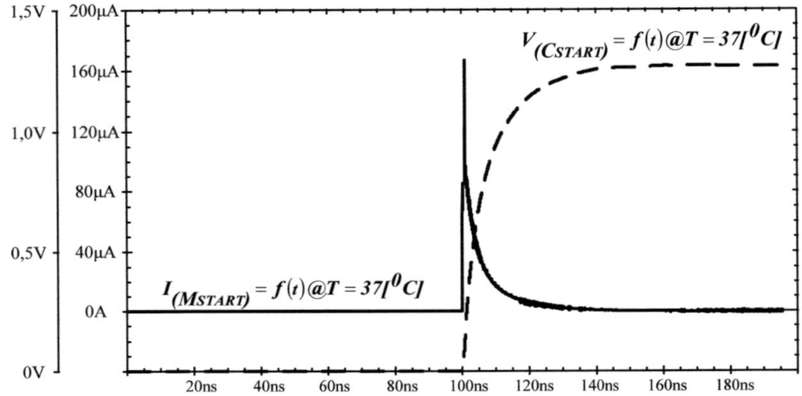

Fig. 13. The Start Up transient current.

6.3 V_{REF} voltage reference

The topology presented in Figure 14 is used to generate the V_{REF} voltage reference. The target value for this reference is 200 [mV] as discussed previously. The current I_d is mirrored to the composite transistor (Ferreira & Pimenta, 2006) formed by MN_{REF1} and MN_{REF2}. The gate bias comes from Q_1 collector and represents only a capacitive charge for the current mirror core since the gate currents are virtually zero. This capacitive effect contributes to improve the V_{eb} PSRR.

It is important to observe that the composite transistor exhibits different modes of operation for each transistor. MN_{REF2} has a nominal V_{gs2} voltage of 676 [mV] leading to strong inversion operation since $V_{th0(N)}$ is approximately 523 [mV]. However, voltage V_{gs1} of transistor MN_{REF1} is subtracted by 200 [mV] (the target output voltage). Thus, the effective value of V_{gs1} is 476 [mV], leading it to operate in weak inversion.

The adopted geometric aspect of MN_{REF2} is similar to current mirror transistor MN_1, W=2µm and L=1µm. It is necessary to evaluate the ideal geometric aspect of MN_{REF1} to guarantee the reference voltage of 200 [mV].

By equating the drain current of both NMOS transistors of the composite topology, then:

$$I_X\left(\frac{W}{L}\right)_{MN_{REF1}} \exp\left[\frac{V_{eb}-V_{REF}-V_{th(N)}}{nU_T}\right] = \beta_n\left(V_{eb}-V_{th0(N)}\right)^2\left(1+\lambda_n V_{REF}\right) \quad (21)$$

where I_X is the weak inversion characteristic current and n the weak inversion coefficient. In the strong inversion, the term $(1+\lambda_n V_{REF})$ can be approximate to unity. Note that the M_{NREF1} threshold voltage is presented as $V_{th(N)}$ since it suffers from body effect.

Solving the equality for V_{REF}:

$$V_{REF} = V_{eb} - V_{th(N)} - nU_T \ln\left[\frac{\beta_n\left(V_{eb}-V_{th0(N)}\right)^2}{I_X\left(\frac{W}{L}\right)_1}\right] \quad [V] \quad (22)$$

Equation (22) shows that V_{REF} can be adjusted by the geometric aspect of MN_{REF2} considering that all other parameters are assumed constant under the corporal temperature. Using interactive simulation, with V_{IN}=2.2V, the optimized geometric aspect ratio is 173. To improve the reference PSRR, the channel length of this transistor is doubled to 2 [μm].

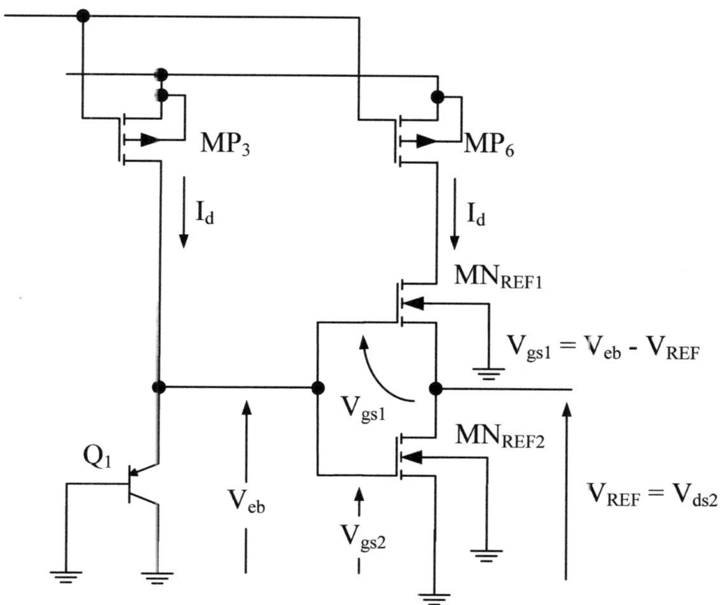

Fig. 14. The topology used to generate de voltage reference V_{REF}

6.4 V_{REF} sensitivity

Using a similar concept discussed in section 4.1, the V_{REF} sensitivity, related to the input voltage V_{IN}, can be expressed as:

$$S_{V_{IN}}^{V_{REF}} = \left.\frac{V_{IN}}{V_{REF}}\right|_Q \left.\frac{\partial V_{REF}}{\partial V_{IN}}\right|_Q \quad [-] \tag{23}$$

The derivate term can be evaluated directly from the circuit topology as:

$$\frac{\partial V_{REF}}{\partial V_{IN}} = \frac{U_T \lambda_n}{1 - \dfrac{2U_T}{\left(V_{eb} - V_{th0(N)}\right)}} \quad [-] \tag{24}$$

Combining equation (23) and (24) and using the known values, the V_{REF} sensitivity is 0.0415. Thus, for a ±10% variation in the input voltage V_{IN}, V_{REF} suffers just ±0.415%. Figure 15 shows the simulation result of those variations. As can be observed, the nominal values for V_{REF} and V_{IN} are, respectively, 200 [mV] and 2.2 [V]. Using this simulation to evaluate the sensitivity, results in:

$$S_{V_{IN}}^{V_{REF}} = \frac{2,2}{200.10^{-3}} \frac{\Delta V_{REF}}{\Delta V_{IN}} = 11 \frac{1,6.10^{-3}}{440.10^{-3}} \approx 0.04 \quad [-] \qquad (25)$$

Those results lead to a PSRR better than 40 [dB] at low frequencies.

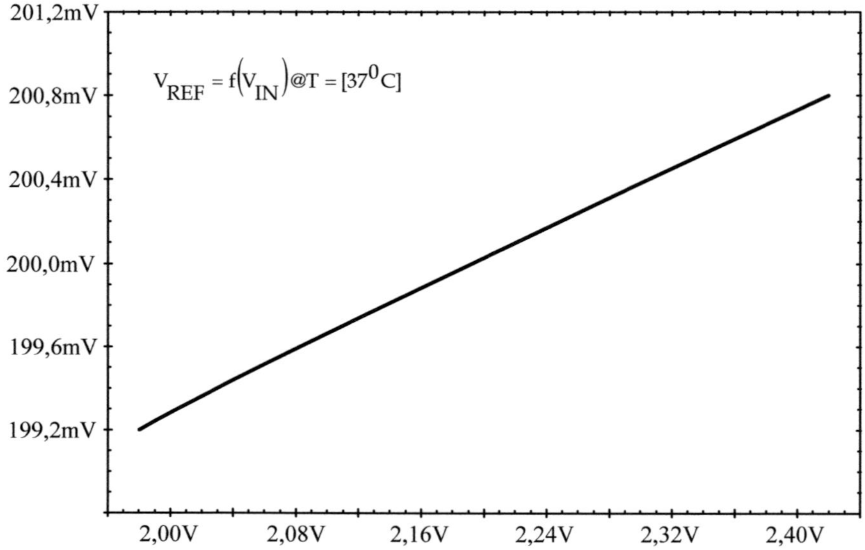

Fig. 15. Simulation of V_{REF} variations due to V_{IN}

6.5 V_{BIAS} voltage reference

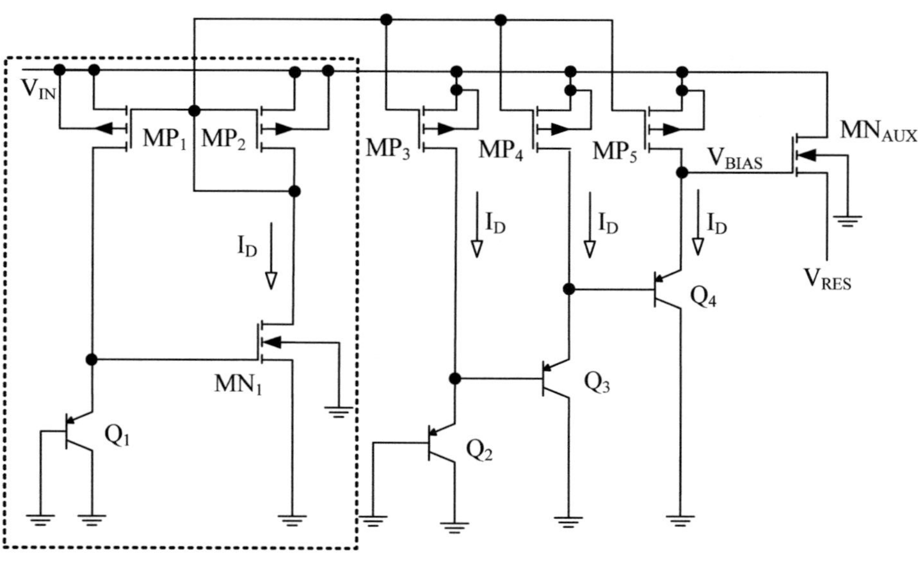

Fig. 16. Three stacked bipolar transistors used to generate V_{BIAS} voltage reference.

The circuit used to generate the V_{BIAS} is illustrated in Figure 16. The use of three stacked bipolar transistors generate a voltage of ≈ 2 [V], i. e. 3 times the quiescent value of V_{eb} (676 [mV]). The bias currents for Q_2, Q_3 and Q_4 are mirrored from the current mirror core with unity gain.

6.6 V_{BIAS} sensitivity

The V_{BIAS} sensitivity can be derived from the circuit topology. It is interesting to evaluate, at first, the V_{eb} for Q_1 transistor. The final result for V_{BIAS} will be three times larger. These formulations are:

$$S_{V_{IN}}^{V_{BIAS}} = \left.\frac{V_{IN}}{V_{BIAS}}\right|_Q \left.\frac{\partial V_{BIAS}}{\partial V_{IN}}\right|_Q \quad [-] \tag{26}$$

$$S_{V_{IN}}^{V_{BIAS}} = \frac{V_{IN}}{V_{BIAS}} \frac{3U_T\lambda_n}{1 - \dfrac{2U_T}{V_{eb} - V_{th0(N)}}} \quad [-] \tag{27}$$

For the know values, the V_{BIAS} sensitivity is calculated as ≈0.012, leading to a variation of ±1.2% for a variation of ±10% at the input voltage line V_{IN}.

6.7 PMOS pass transistor and NMOS follower geometric aspect ratios

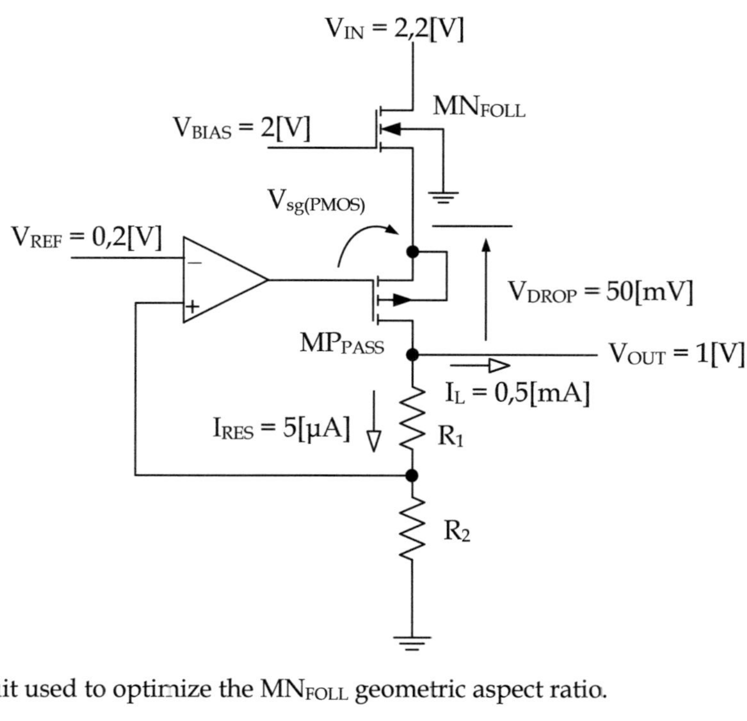

Fig. 17. Circuit used to optimize the MN_{FOLL} geometric aspect ratio.

Figure 17 shows the main current and voltages values used to estimate MP_{PASS} and MN_{FOLL} geometric aspects.

For MP_{PASS} transistor, two considerations are important. First, its geometric aspect must be larger enough to support the total nominal load current plus the sampler current. Second, its operation must be kept in the triode region to guarantee a low r_{ds} value. In the triode region, the resistance is given as:

$$R_{ds}(MP_{PASS}) = \frac{1}{\frac{KP}{2(1+\delta)}\left(\frac{W}{L}\right)\left(V_{sg} - V_{th0(P)} - V_{sd}\right)} \quad [\Omega] \tag{28}$$

The aspect ratio (W/L) design parameter should be raised to lower the drain-source resistance. Figure 17 also shows a suggested 50 [mV] (V_{DROP}) in order to keep the low dropout concept in the LDO circuitry. The V_{ROP} voltage corresponds to V_{sd} voltage in Equation (28).

By replacing MN_{FOLL} transistor by a 5.05 [μA] current source, interactive simulations lead to a MP_{PASS} geometric aspect of 2500/1. In order to evaluate the aspect ratio of MN_{FOLL} it must be noticed first that MN_{FOLL} suffers from body effect and its threshold voltage is corrected by using:

$$V_{th(N)} = V_{th0(N)} + \gamma\left(\sqrt{2\varphi_F + \left|V_{bs}\right|} - \sqrt{2\varphi_F}\right)$$
$$V_{th(N)} = 0{,}523 + 0{,}4\left(\sqrt{0{,}6 + 1{,}05} - \sqrt{0{,}6}\right) \approx 727 \quad [mV] \tag{29}$$

By using this result in the drain current equation, the MN_{FOLL} geometric aspect is given as:

$$I_d = \beta_n\left(V_{gs} - V_{th(N)}\right)^2$$
$$0{,}505.10^{-3} = 95{,}3.10^{-6}\left(\frac{W}{L}\right)\left[(2 - 1{,}05) - 0{,}727\right]^2 \Rightarrow \left(\frac{W}{L}\right) \approx 106 \tag{30}$$

6.8 PMOS pass transistor and NMOS follower capacitances

Those two transistors (MN_{FOLL} and MP_{PASS}) have a large geometric aspect ratio leading to relative large gate capacitances. The PMOS pass transistor gate capacitance is important since it is responsible to determine the OTA dominant pole.

The NMOS and PMOS S_iO_2 thickness (T_{OX}) can be used to obtain the gate capacitances per unit area as:

$$COX_{NMOS} = \frac{\varepsilon_{OX}}{T_{OX}} = \frac{3{,}45.10^{-13}}{7{,}5.10^{-9}}\frac{1}{10^6.10^4} \approx 4{,}6.10^{-15} \quad \left[\frac{F}{\mu m^2}\right]$$
$$COX_{PMOS} = \frac{\varepsilon_{OX}}{T_{OX}} = \frac{3{,}45.10^{-13}}{7{,}7.10^{-9}}\frac{1}{10^6.10^4} \approx 4{,}48.10^{-15} \quad \left[\frac{F}{\mu m^2}\right] \tag{31}$$

As the PMOS pass transistor operates in the triode region, the gate to source and gate to drain capacitances are:

$$C_{gs} = C_{gd} = \frac{1}{2}(WL)_{PMOS} C_{OX} \approx 5{,}75.10^{-12} \quad [F] \tag{32}$$

7. The Operational Transconductance Amplifier (OTA)

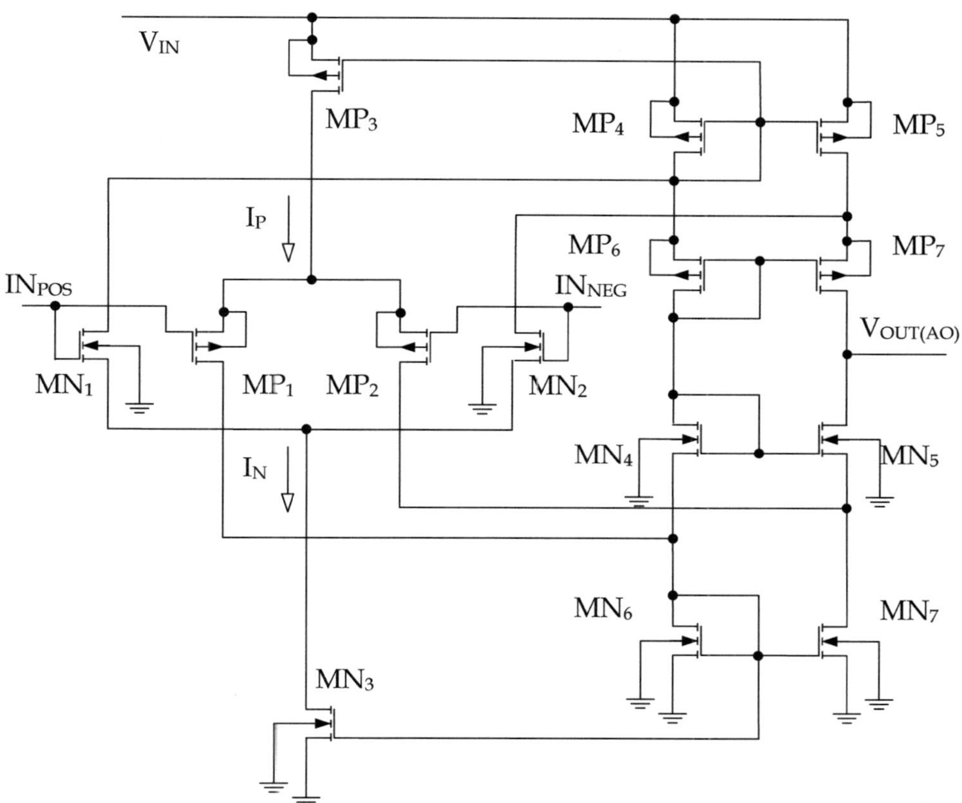

Fig. 18. Operational Transconductance Amplifier (OTA).

There are some features that must be taken into account in order to design the OTA:

1. To validate the closed loop properties, the OTA must have an open loop gain larger than 1000 (60 [dB]);
2. Since the OTA is powered by the input voltage V_{IN}, it must exhibit a good power supply rejection ratio. A target value of 40 [db] is used as a reference;
3. The OTA must have a low offset voltage. The offset voltage has a direct impact in equation (1) and can deviate from the nominal output voltage. A target value of 5 [mV] was adopted. It is very important to observe the matching on the OTA stage to minimize the systematic offset and the use of layout technique to minimize the random offset;

4. The total quiescent bias current must be kept as low as possible to improve the OTA overall efficiency. A target value of 3 [μA] was adopted, representing less than 1% of load current. As the OTA has three currents branches, it is assigned a current of 1 [μA] to each one;
5. The dominant pole discussed in the previous sections is a function of the OTA output resistance;
6. The OTA frequency response must lead to a stable system over the entire band. A margin phase of 70° degrees is a target value.
7. The OTA does not need fast responses due the physiological application. The slew rate and settling time targets are, respectively, 0.1 [V/μs] and 10 [μs].

One recommended topology is the folded cascade. It offers high output resistance and, as in this particular case, the dominant pole is fixed by the capacitive load. This is important to reduce the silicon area and extra power consumption by using additional compensation circuits.

It is used the self biased Operational Transconductance Amplifier - OTA topology (Mandal & Visvanathan, 1997). It provides additional reduction of silicon area and power consumption by using other biasing circuits. The OTA circuit is depicted in Figure 18.

As can be observed, the OTA has a rail-to-rail input stage. It is not absolutely necessary in this project, but it is interesting to have the possibility to generating output voltages near the rail lines V_{IN} and Ground. The OTA can be suitable for other applications that require different input voltage values.

The OTA open loop gain is:

$$AV_{OL} = gm_{ota} r_{ota} = \left(gm_P + gm_N\right) r_{ota} \quad [-] \tag{33}$$

where gm_N and gm_P are the NMOS and PMOS input differential pair transconductance, respectively. In the case of general purpose application, it should be used an additional circuitry to compensate their transconductances since they exhibits different values depending on the region of operation.

In this project, the gm variations, that can be as large as 100%, do not have a significant impact on the LVR stability. The dominant pole is far away enough from the other poles by several orders of magnitude.

7.1 OTA transistors geometric aspect

Figure 19 shows the lower half cascode from Figure 18 and the quiescent output voltage of 1.1 [V].

That voltage is considered split equally between the two NMOS transistor pairs. Observe that it is necessary to consider the total NMOS tail current I_N for $MN_{6,7}$. Using Equation 14:

$$\begin{gathered} 1.10^{-6} \approx 95,3.10^{-6} \left(\frac{W}{L}\right)_{MN6,7} (0,55 - 0,523)^2 \\ \left(\frac{W}{L}\right)_{MN6,7} \approx 14 \end{gathered} \tag{34}$$

Figure 20 shows the PMOS and NMOS differential voltage considerations.

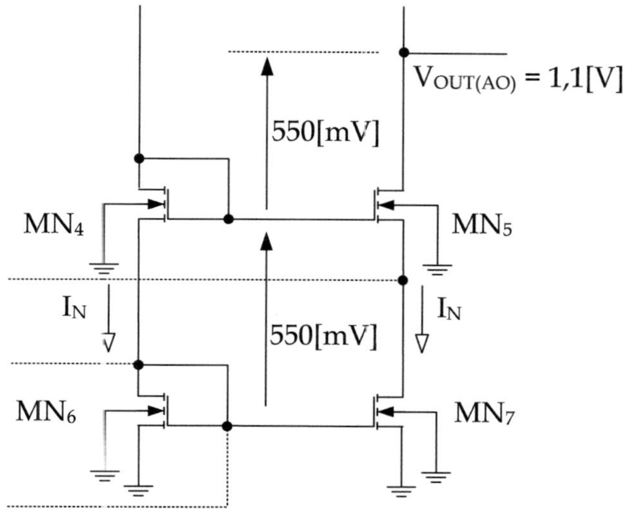

Fig. 19. Lower half used to evaluate the geometric aspect ratios.

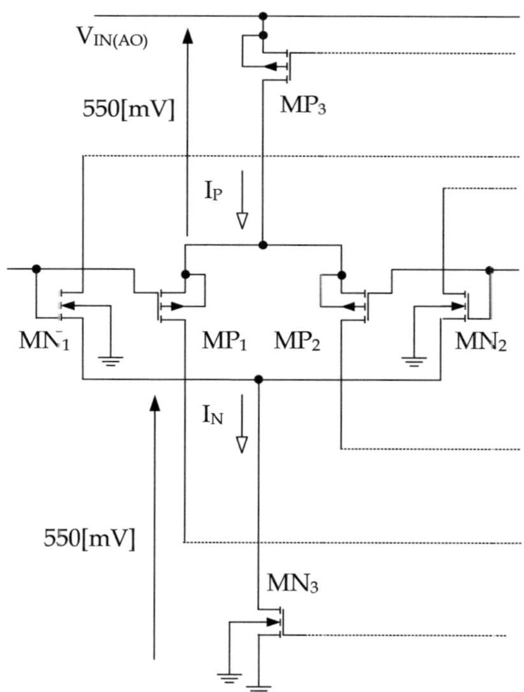

Fig. 20. NMOS and PMOS differential input pairs.

For transistors $MN_{1,2}$ it is necessary to consider the threshold voltage correction since they suffer from body effect and operate in weak inversion.

$$V_{th(N)} = 0{,}523 + 0{,}4\left[\sqrt{0{,}6+0{,}55} - \sqrt{0{,}6}\right] \approx 642 \quad [mV] \tag{35}$$

Therefore, by using the current formulation, then:

$$I_d = I_X\left(\frac{W}{L}\right)\exp\left(\frac{V_{gs} - V_{th(N)}}{nU_T}\right) \quad [A]$$
$$1.10^{-6} \approx 103{,}1.10^{-9}\left(\frac{W}{L}\right)\exp\left(\frac{0{,}55 - 0{,}642}{45.10^{-3}}\right) \Rightarrow \left(\frac{W}{L}\right) \approx 74 \tag{36}$$

NMOS transistors $MN_{4,5}$ operate similarly to $MN_{1,2}$. The only difference is they carry half of tail current. Thus, the geometric aspect of these transistors is divided by 2 (37). All PMOS transistors have their geometric aspect ratios adjusted by interactive simulations. Table 3 resumes OTA aspect ratios where the channel length and width are expressed in [µm].

Corporal Temperature: 37ºC	
$(W/L)_{MN1} = (W/L)_{MN2}$	74/1
$(W/L)_{MP1} = (W/L)_{MP2}$	158/1
$(W/L)_{MN3} = (W/L)_{MN6} = (W/L)_{MN7}$	14/1
$(W/L)_{MN4} = (W/L)_{MN5}$	32/1
$(W/L)_{MP3} = (W/L)_{MP4} = (W/L)_{MP5}$	158/1
$(W/L)_{MP6} = (W/L)_{MP7}$	272/1

Table 3. OTA geometric aspect ratios, in [µm]

7.2 OTA simulations results

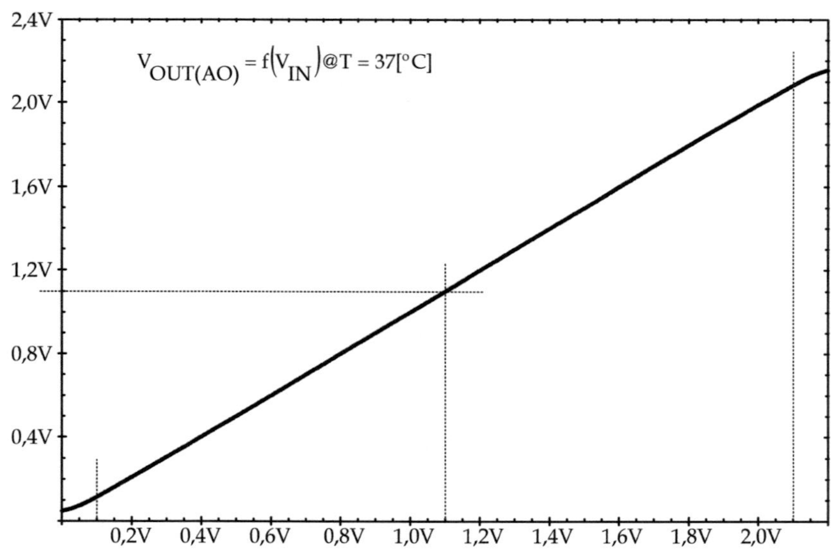

Fig. 21. OTA common mode range indicating a rail-to rail operation.

The following figures show the most relevant OTA parameters simulations. The common mode range - CMR is depicted in Figure 21. The OTA is buffer connected and the input signal is linear over the entire range, thus characterizing a rail-to-rail operation. The OTA analog ground is 1.1 [V] and the simulation shows that the systematic offset is minimum, thus representing a good matching between the OTA stages.
Figure 22 shows a configuration to analyze the OTA frequency response. The auxiliary capacitor and inductor (C_{AUX}, L_{aux}) guarantees a closed loop for DC signal and an open loop for AC signal. Thus the OTA will be properly biased since the DC path configures a buffer connection.

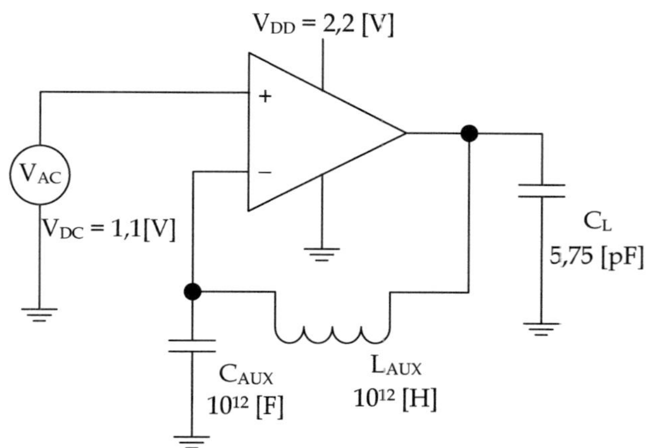

Fig. 21. Buffer configuration used to simulate the OTA frequency response.

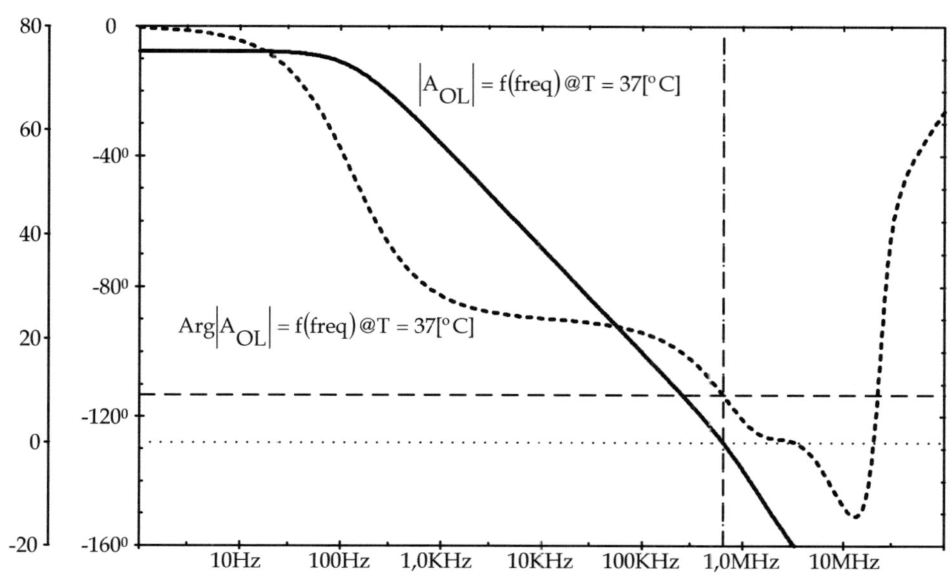

Fig. 22. OTA Frequency response.

The load capacitance C_L is represented by the PMOS pass transistor gate capacitance, evaluated according to Equation (32). The dominant pole (P_2) is located at ≈130 [H_Z] and the unit frequency gain (f_U) is located at ≈640 [KH_Z]. The phase margin (Φ_F) is ≈66°. Figure 22 show these results.
On a buffer configuration, the OTA is excited by a square wave to obtain the transient parameters. Figure 23 shows the resultant simulation considering a fluctuation of approximately ±10% around the 1.1 [V] analog ground. That simulation can be used to obtain the falling and raising slew rates (SR) and the settling time.

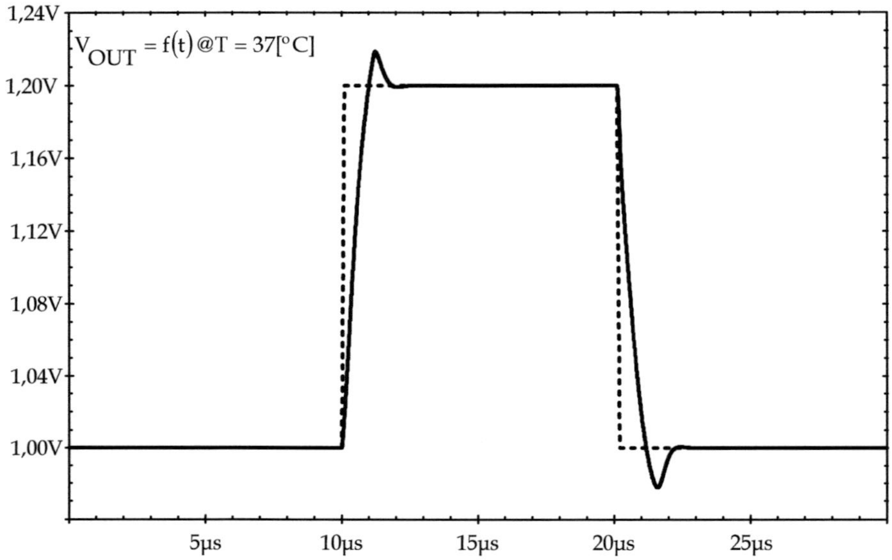

Fig. 23. OTA transient response.

Table 4 resumes the main parameters obtained from the interactive simulations.

Corporal Temperature: 37°C Input Voltage Supply: 2.2 [V]	
I_{DD} [µA]	3,5
P_D @ I_{DD} [µW]	7,7
CMR	V_{SS}+100 [mV] V_{DD} – 100 [mV]
OTA dominant pole [H_Z]	130
f_{UG} [MH_z]	0,64
Φ_M [^{0}]	66,6
T_{SET} @ 0,1% [µS] raise and fall	3
SR^+ e SR^- [V/µS]	0,2
PSRR @ 100H_Z [dB]	-81
PSRR @ 10MH_Z [dB]	-26,5

Table 4. OTA main parameters.

8. Layout considerations

Even by using the most advanced microfabrication techniques, it is not possible to guarantee that all the devices implemented in the same chip will have the same electrical characteristics. The aspect ratio of two similar devices can be controlled to a precision of approximately ±1% and, in the many cases, it can be better than ±0.1%. Therefore, the layout project of an integrated circuit, mainly analog application fields, must take mismatches into account (Shyu, 1984).

The layout, as a backend step, plays an important rule to fabricate matched devices in the integrated circuit. It is not possible to cancel the mismatch completely; nevertheless there are ways to minimize it.

The objective of component matching is to reduce the error introduced by the deviations in the fabrication process; therefore it is necessary to use layout techniques.

The mismatch can be classified as systematic and random (Ramos, 2007). The main sources of systematic mismatch are the process polarization (difference between the designed dimensions and actual dimensions), contact resistances, non-uniform current flow, interaction between diffusions, temperature gradient and stress gradient.

As an example, the silicon presents stress gradients, meaning that is a piezoresistive material and presents variation in its characteristic resistance due to mechanical stress. This gradient can be represented by isobaric lines along the die that show the different levels of intensity. It is minimum in the central region and maximum along the four corners. Figure 24 shows an example of those isobaric lines.

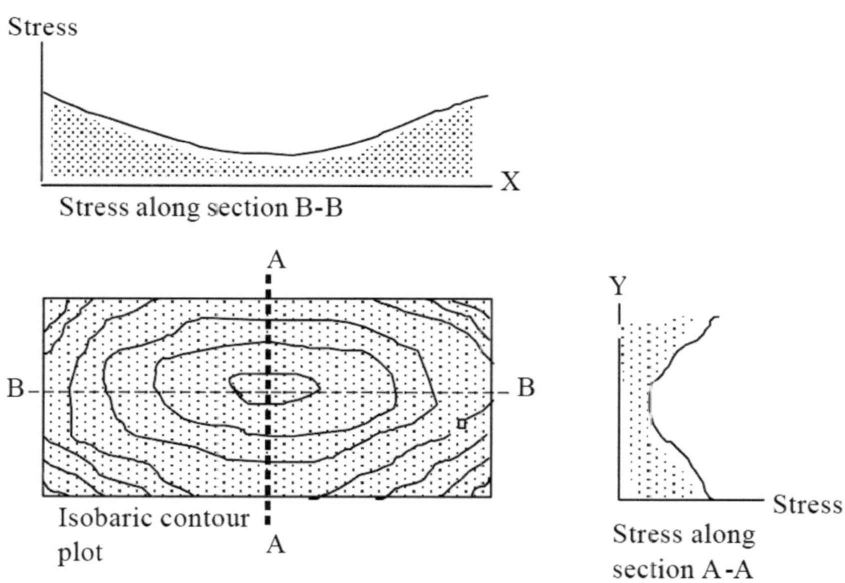

Fig. 24. Example of isobaric lines.

Consequently, it is recommended that components to be matched are placed near to each other to minimize the mechanical stress. The mechanical stress difference between two matched components is proportional to the stress gradient and their distance. For calculations purposes, the location of the component is determined as the average contribution of each section of the component as a whole. The resultant location is called centroid of the component. It is important that any symmetric axis crosses the centroid of the device or component. Some examples of centroid configurations are depicted in Figure 25.

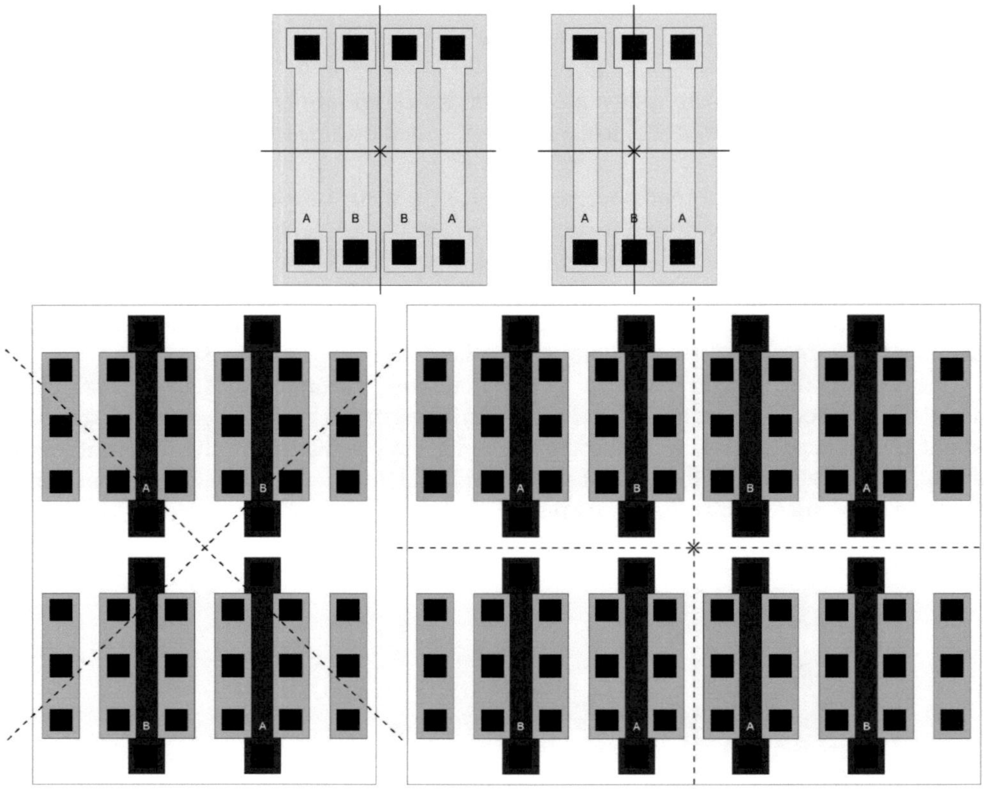

Fig. 25. Examples of centroid layout configurations.

The effects of mechanical stress in integrated resistors are quantified in terms of piezoresistivity, position of the centroids and stress gradients. These effects can be minimized by the proper choice of a low piezoresistivity material or by the resistor orientation in the wafer according to the minimum stress gradients. Other recommendation is the reduction of the distance between the centroids.

The temperature gradients can be analyzed in the same way as the stress gradients. Temperature gradients are obtained by isothermal lines and are separated from each other by a predefined temperature difference (ΔT). These temperature gradients are maximum around the perimeter of the component, and gradually, decrease towards the center.

The temperature effects are minimized in a similar way as the stress gradient: by using low linear temperature coefficient, by using minimal lines of the temperature gradient and by reducing the distance between the centroids.

Mainly for analog integrated circuits, the layout of two matched components is usually implemented by dividing each component into identical sections, placed symmetrically in a matrix array.

The common centroid layout, along with matched components placed in a matrix array in identical and symmetrical sections, is essential to reduce or even eliminate the systematic mismatching. Since the distance between the centroids is null, the mismatching caused by mechanical and temperature stress will be null. As an example, transistors on differential pair are placed in a cross coupled pattern.

In order to properly generate centroid layout, some rules must be observed (Hastings 2001):

1. Coincidence: Matched devices must have a common centroids or as close as possible;
2. Symmetry: The component matrix must be symmetrical in both X and Y axis. Ideally, the symmetry must be a consequence of the components placement and not the symmetry of each one individually;
3. Dispersion: The matrix must offer the greatest dispersion level, in other words, each component must be placed with high possible symmetry along the matrix array;
4. Compression: The matrix should be as compact as possible, ideally close to a square shape.

The random mismatch is different on each device and it is caused by microcospical irregularities in the materials or fabrication process. It can be reduced using the proper geometric aspect of the matched components. This geometric aspect is based in physical and statistical models that are characterized by the fabrication process Patrick & McAndrew, 2003).

Physically, the microscopic irregularities results from the material granularity (ex. polysilicon), photolithography errors, doping injections, thickness and permittivity of the gate oxide, etc. The effect of those errors may decreased as the components geometric aspect increase, since these parameters reach averaged values for large widths and areas.

Two parameters are considered in order to model the random mismatch: the process and the electrical parameters. Process parameters are physically independent and control the device electrical characteristics. In the case of a MOS transistor, the process and electrical parameters that have must be taken into account to matching purposes are listed in Table 5.

Process Parameters	Electrical Parameters
Flat band voltage	Drain current
Mobility	Gate-Source voltage
Substrate Doping Concentration	Transconductance
Chanel length variation	Output resistance
Chanel width variation	
Short channel effect	
Narrow channel effect	
Gate oxide thickness	
Source/Drain sheet resistance	

Table 5. Process and electrical parameters for component matching.

A CMOS process allows also the fabrication of bipolar transistors. Those transistors are also subject to matching rules. A lateral bipolar transistor does not have a good matching when compared with a vertical one. The poor matching of the lateral transistors are due to the surface effects and impossibility using large emitter areas.

Some rules for bipolar transistor matching are:

1. Identical geometric aspects for the emitter and collector since they affect the current flow in lateral transistors;

2. Minimum emitter area for matched transistors, otherwise there will be a degradation in the current gain (β);
3. Guard ring around the base to ensure that electrostatics charges will not influence the current flow in the neutral base;
4. Use of multiple collectors for lateral PNP transistors. A moderate match can be reached when the collectors are identical and out of the saturation condition;
5. The matched transistors should be close to each other in order to minimize the impact of the thermal gradient.
6. The matched transistors should be placed in gradients lines of minimum stress;
7. The transistor must be aligned with the wafer axis;
8. Place as many metal contacts as possible in the emitter (following the emitter geometry) to reduce the contact resistance and to distribute the current flow uniformly;
9. Use emitter degeneration. Lateral PNP transistors are often more benefited with emitter degeneration compared to the NPN vertical counterparts due to the Early voltage and the large emitter area. They are commonly used in current mirrors.

The matching over integrated components reflects the overall performance of the entire circuit or system. Depending on the matching accuracy, the circuits may present:

1. Minimum: In the range of ± 1% (representing 6 to 7 bits of resolution). Used for general use components in an analog circuit, such as current mirrors and biasing circuits;
2. Moderate: In the range of ± 0.1% (representing 9 to 10 bits of resolution). Used in bandgap references, operational amplifiers and input stage of voltage comparators. This range is the most appropriate for analog designs.
3. Severe: In the range of ± 0.01% (representing 13 t0 14 bits of resolution). Used in high precision analog to digital converters (ADCs) and digital to analog converters (DACs). Analog designs that use capacitors ratio reach this range easer then those that using resistors ratios.

Figure 26 shows an example of a PNP vertical bipolar transistor layout.

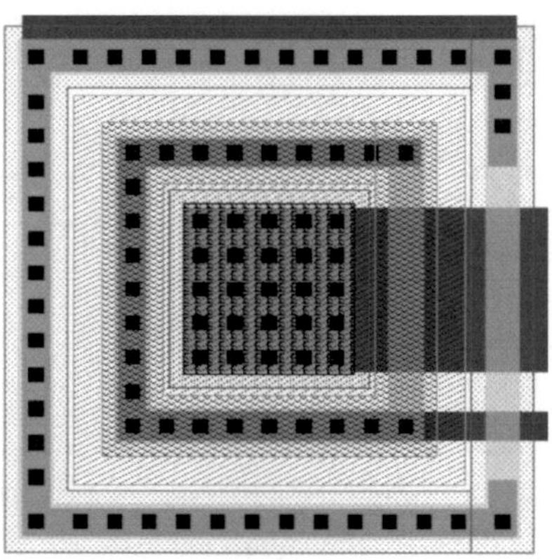

Fig. 26. PNP vertical bipolar transistor example.

9. LVR measurements

The example LVR was diffused in a 0.35µm standard CMOS process. It took an area of approximately 0.25 [mm²].

Figure 27 depicts the testing structure utilized to measure the main LVR parameters.
It is used a commercial operational amplifier (LM318) as a buffer to isolate the chip. The load current can be adjusted by potentiometer P_1 and the total load capacitance, considering the all parasitic, was measured as 30 [pF].
Before any LVR measure, the LM318 offset voltage was compensated through the procedure provided by the manufacturer. All the power supply lines are decoupled by 10 [μF] capacitors.

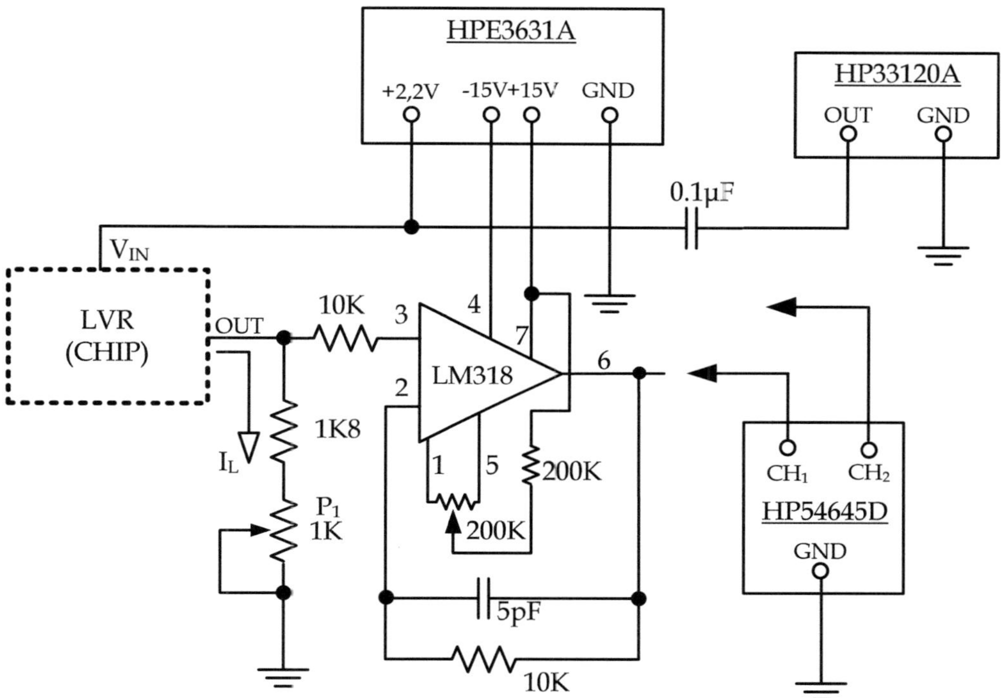

Fig. 27. The test structure to measure the LVR parameters.

Parameters	Simulated	Measured
T_{NOM}	37[°C]	37[°C]
V_{IN}	2.2[V]	2,218[V]
$I_{L(NOM)}$	0.5[mA]	0.5[mA]
$P_{D(NOM)}$	1.17[mW]	1.186[mW]
V_{OUT}	1[V] @ I_L = 0.5mA	1.038[V] @ I_L = 5[μA] 1.004[V]@ I_L = 0.5[mA]
I_Q	30[μA]	39[μA]
PSRR @ 10MHz	-42.6dB	-38dB
E_{FF} related to V_{IN}	42.8[%]	42.3[%]
T_{SET} @ 0,1%	14.87[μs]	18.6[μs]
OTA dominant pole	130[Hz]	126[Hz]

Table 6. Main LVR simulated and measured parameters.

Figure 28 shows the LVR response to a voltage step input and reveals a BIBO (bounded input - bounded output) system, in other words, the system is unconditionally stable and there is no need of any extra external component.
Table 6 is a comparison between the simulated and measured parameters.

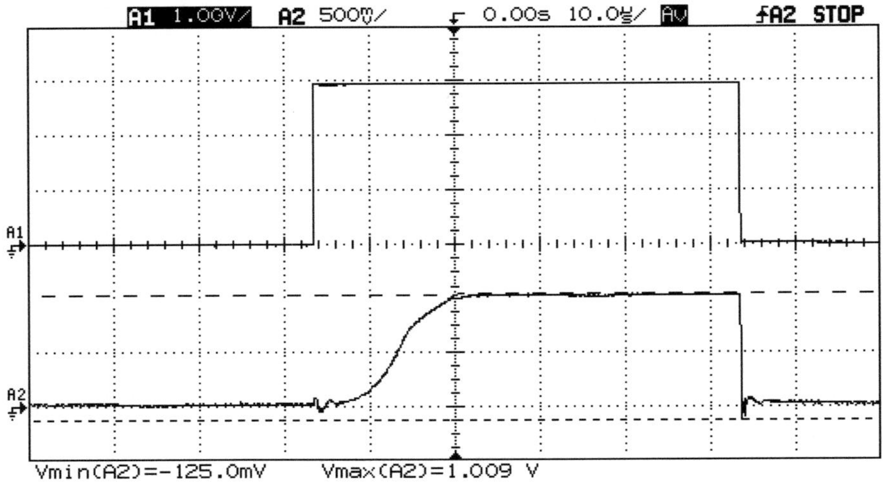

Fig. 28. LVR step response indicating a BIBO system.

The measured values show a good conformity with the simulated ones indicating proper design considerations.

10. Conclusions

We are witnessing the great revolution that has been imposed since the manufacture of the first bipolar transistor in the late 50s of the twentieth century. Electronics solutions are going to microelectronics and microelectronics is evolving to nanoelectronics. All these developments bring with them the yearning of the human being to access more efficient equipment. So, in virtually all branches of activities we will find what is called "High-Tec".
Medicine and its related sciences could not stay apart from this explosion of technology and intelligently sought the partnership with this powerful tool for circuit design.
Some solutions point to implantable systems (which would reduce the use of invasive techniques) that can be taken up on an outpatient basis and connected into a means of communication for a distance evaluation by a health professional.
The main objective of this chapter was the development of a voltage regulator for implantable applications. Some boundary conditions allow classic Figures of Merit, such as the temperature dependence, to be less severe, since the body temperature is kept constant. Another key issue was to search for solutions that avoid the presence of any external component. This is an essential boundary condition since the topology of classical LDO regulators depends on the presence of a capacitor (usually electrolytic and therefore too large for this application) connected in parallel with the load. Other regulators reported in the literature uses complex circuits or circuits that requires large silicon area.

The circuit described is a compromise of additional power dissipation in the source follower stage and unconditional stability. Even with the additional dissipation, the total power of the regulator (about 1.2 [mW]) is within a safe limit.

11. References

[1] Ahmadi, M.M. & Jullien, G.A., (2009). A Wireless Implantable Microsystem for Continuous Blood Glucose Monitoring. *IEEE Trans. Biomedical Circuits and Systems*, 3(3), pp.169-180.

[2] Colomer-Farrarons J., Miribel-Catala P., Rodríguez I., S.J., (2009). CMOS front-end architecture for In-Vivo biomedical implantable devices. In *Industrial Electronics, 2009. IECON '09. 35th Annual Conference of IEEE*. p. 4401 - 4408.

[3] Dejhan, K. et al., (2004). A CMOS Voltage-Controlled Grounded Resistor Using a Single Power Supply,. In *Communications and Information Technology, 2004. ISCIT, IEEE International Symposium on*. pp. 124-127.

[4] Ferreira, L.H.C. & Pimenta, T.C., (2006). A Weak Inversion Composite MOS Transistor for Ultra-Low-Voltage and Ultra-Low-Power Applications. In *Proceedings of 13th International Conference Mixed Design Intregated Circuit Systems*. pp. 10-12.

[5] Gray, P.R. et al., (1993). *Analysis and Design of Analog Integrated Circuits* Fourth Edi., John Wiley and Sons.

[6] Guennoun M., Zandi M., E.-K.K., (2008). On the use of Biometrics to Secure Wireless Biosensor Networks. In *Information and Communication Tecnologies: From Theory to Applications, 2008. ICTTA 2008*. pp. 1-5.

[7] Hastings, A., (2001). *The Art of Analog Layout*, Prentice Hall.

[8] Huang, W.-J., Liu, S.-H. & Lu, S.-I., (2006). A Capacitor-Free CMOS Low Dropout Regulator with Slew Rate Enhancement. In *VLSI Design, Automation and Test, 2006 International Symposium on*. pp. 1-4.

[9] Huang, W.-J., Liu, S.-H. & Lu, S.-I., (2006). CMOS Low Dropout Regulator with Single Miller Capacitor. *Electronics Letters*, 42(4), pp.216-217.

[10] Koushaeian, L. & Skafidas, S., (2010). A 65nm CMOS low-power, low-voltage bandgap reference with using self-biased composite cascode opamp. In *Low-Power Electronics and Design (ISLPED), 2010 ACM/IEEE International Symposium on*. pp. 95-98.

[11] Kugelstadt, T., (1999). Fundamental Theory of PMOS Low-Dropout Voltage Regulators. *Texas Instruments Incorporated. Application Note SLVA068*, pp.1-5.

[12] Landt, J., (2005). The History of RFID. *Potentials, IEEE*, 24(4), pp.8-11.

[13] Lazzi, G., (2005). Thermal Effects of Bioimplants. *Engineering in Medicine and Biology Magazine, IEEE*, 24(5), pp.75-81.

[14] Mackowiak, P.A., Wasserman, S.S. & Levine, M.M., (1992). A Critical Appraisal of 98.6°F, the Upper Limit of the Normal Body Temperature, and Other Legacies of Carl Reinhold August Wunderlich. *The Journal of American Medical Association*, 268(12), pp.1578-1580.

[15] Mandal, P. & Visvanathan, V., (1997). Self Biased High Performance A Folded Cascode CMOS Op-Amp. In *VLSI Design, 1997. Proceedings, Tenth International Conference on*. pp. 429-434.

[16] Miyazaki, M., (2003). The Future of e-Health – Wired or not Wired. *Business Briefing: Hospital Engineering & Facilities Management*, pp.1-5.

[17] Osepchuk, J.M. and P.R.C., (2001). Safety Standards for Exposure to RF Electromagnetic Fields. *IEEE Microwave Magazine*, 2(2), pp.57-69.

[18] Patrick, G.D. & McAndrew, C.C., (2003). Understanding MOSFET Mismatch for Analog Design. *IEEE Journal of Solid-State Circuits*, 38(3), pp.450-456.

[19] Puers, R., (2005). Implantable Sensor Systems. In *DISens Symposium Book*. pp. 1-14.

[20] Ramos, F.G.R., (2007). *Uma Referência de Tensão Programável Para Aplicações em Gerenciamento de Potência*. Master Tesis at Universidade Federal de Itajubá, 2007.

[21] Rincon-Mora, G.A., (2000). Active multiplier in Miller-compensated circuits. *IEEE J. Solid-State Circuits*, 35, pp.26-32.

[22] Rincon-Mora, G.A. & Allen, P., (1998). A low-voltage, low quiescent current, low drop-out regulator. *IEEE J. Solid-State Circuits*, 33, pp.36-44.

[23] Rincon-Mora, G. & Allen, P.E., (1997). Study and Design of Low Drop-Out Regulators. *School of Electrical and Computer Engineering – Georgia*.

[24] Rogers, E., (1999). Stability Analysis of Low-Dropout Linear Regulators with a PMOS Pass Element. *Texas Instruments Incorporated. Analog Applications Journal*, pp.10-12.

[25] Sauer C., Stanacevic M., Cauwenberghs G., Thakor, N., (2005). Power Harvesting and Telemetry in CMOS for Implanted Devices. *IEEE Trans. On Circuits and Systems I: Regular Papers*, 52(12), pp.2605-2613.

[26] Scanlon W G, Evans N E, C.G.C. and M.Z.M., (1996). Low-power radio telemetry: the potential for remote patient monitoring. *Journal of Telemedicine and Telecare*, 2(4), pp.185-191.

[27] Shyu, J.-B., Temes, G.C. & Acher, F.K., (1984). Random Error Effects in Matched MOS Capacitors and Current Sources. *IEEE Journal of Solid-State Circuit*, sc-19(6), pp.948-955.

[28] Simpson, C., (1997). A User's Guide to Compensating Low-Dropout Regulators. In *Wescon/97, Conference Proceedings*. pp. 270-275.

[29] Stanescu, C., (2003). Buffer Stage for Fast Response LDO. In *8th International Conference on Solid-State and Integrated Circuit Tecnology, ICSICT'06*. pp. 357-360.

[30] Tzanateas, G., Salama, C.A. & Tsividis, Y.P., (1979). A CMOS Bandgap Voltage Reference. *IEEE journal of Solid-State Circuits*, 14(3), pp.655-657.

[31] Vaillantcourt, P., Djemouai A., Harvey J. F., Sawan, M., (1997). EM radiation behaviour uponbiological tissues in a radio-frequency power transfer link for a cortical visual implant. In *Proc. IEEE Int. Conf. Engineering in Medicine and Biology*. pp. 2499-2502.

[32] Zheng, C. & Ma, D., (2010). Design of Monolithic Low Dropout Regulator for Wireless Powered Brain Cortical Implants Using a Line Ripple Rejection Technique. *IEEE Transactions On Circuits And Systems - II: Express Briefs*, 57(9), pp.686-690.

Part 2

Antennas/Tags

5

RFID Technology: Perspectives and Technical Considerations of Microstrip Antennas for Multi-band RFID Reader Operation

Ahmed Toaha Mobashsher[1], Mohammad Tariqul Islam[1] and Norbahiah Misran[2]
[1]*Institute of Space Science (ANGKASA), Universiti Kebangsaan Malaysia*
[2]*Dept. of Electrical, Electronic and Systems Engineering*
Universiti Kebangsaan Malaysia
Malaysia

1. Introduction

This chapter presents a comprehensive review of RFID technology concerning the antennas and propagation for multi-band operation. The technical considerations of antenna parameters are also discussed in details in order to provide a complete realization of the parameters in pragmatic approach to the antenna designing process, which primarily includes scattering parameters and radiation characteristics. The antenna literature is also critically overviewed to identify the possible solutions of the multi-band microstrip antennas to utilize in multi-band RFID reader operation. In the literature dual-band antennas are principally discussed since they are ideal to realize and describe multi-band antenna mechanism. However, it has been seen that these techniques can be combined to enhance multi-band antennas with wider bandwidths. Last but not least, the high gain dual-band antennas and limitations have been described and it is realized that the conventional feeding technique might limit the performance of multi-band antennas to only one frequency.

2. Radio frequency identification

The idea of early radio frequency identification (RFID) system was invented by Scottish physicist Sir Robert Alexander Watson-Watt in 1935. With the supervision of Watson-Watt, the British government developed the first active identify friend or foe (IFF) system. This prototype of RFID concept was modified in 1950s and 60s by using radio frequency (RF) energy for commercialization purpose. The first US patent in this field was published on January 23, 1973 for the invention of an active RFID tag with rewritable memory by M. W. Cardullo (Cardullo 1973). That same year, C. Walton received another RFID patent for a passive transponder used to unlock a door without a key. In the recent days, the low power ultra high frequency (UHF) RFID system research has gained a lot of importance after some of the biggest retailers in the world, e.g., Albertsons, Metro, Target, Tesco, Wal-Mart and the

US Department of Defense, have said they plan to use electronic product code (EPC) technology to track goods in their supply chain (Mitra 2008).
RFID is an emerging technology for the identification of objects and/or personnel. RFID is recognized as one of the technologies capable of realizing a complete ubiquitous computing network due to its strong benefits and advantages over traditional means of identification such as the optical bar code systems. Comparing with barcode, RFID has some advantages of rapid identifying, flexible method and high intelligent degree (Wang et al. 2007; Xiao et al. 2008). Furthermore, it can function under a variety of environmental conditions (Intermec Technologies Corporation 2006). It has recently found a tremendous demand due to emerging as well as already existing applications requiring more and more automatic identification techniques that facilitate management, increase security levels, enhance access control and tracking, and reduce labor force. A brief listing of RFID applications that find use on a daily basis is:

- Warehouse Management Systems
- Retail Inventory Management
- Toll Roads
- Automatic Payment Transactions
- High Value Asset Tracking and Management
- Public Transportation
- Automotive Industry
- Livestock Ranching
- Healthcare and Hospitals
- Pharmaceutical Management Systems
- Military
- Marine Terminal Operation
- Manufacturing
- Anti-counterfeit

2.1 RFID system

Basically RFID is a contact-free non-line-of-sight type identification technology using radio frequency consisting of a RFID transponder (tag), a RFID interrogator (reader) with an antenna and data processing unit (host computer). In case of the handheld RFID reader, the reader itself contains the feature of data processing unit. The typical block diagram of RFID system is shown in Fig. 1.

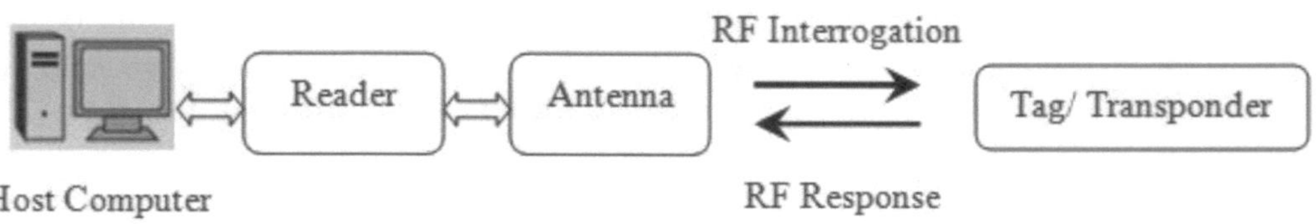

Fig. 1. Block diagram of RFID system

The interrogation signal coming from the reader antenna must have enough power to activate the transponder microchip by energizing the tag antenna, perform data processing and transmit back the data stored in the chip up to the required reading range (typically 0.3–

1m). The reader antenna receives the modulated backscattered signal from the tags in field of antenna and examines the data.

2.1.1 RFID tags

The tag is the basic building block of RFID. Each tag consists of an antenna and a small silicon chip that contains a radio receiver, a radio modulator for sending a response back to the reader, control logic, some amount of memory, and a power system. Tags contain a unique identification number called an Electronic Product Code (EPC), and potentially additional information of interest to manufacturers, healthcare organizations, military organizations, logistics providers, and retailers, or others that need to track the physical location of goods or equipment. All information on RFID tags, such as product attributes, physical dimensions, prices, or laundering requirements, can be scanned wirelessly by a reader at high speed and from a distance of several meters. According to the energizing power system, the tags can be classified into three types:

a. Passive tag - These tags (shown in Fig. 2 (a)) use the signal received from the reader to power the IC, and vary their reflection of this signal to transmit information back to the reader. Passive tags are the most common in cost-sensitive applications, because, having no battery and no transmitter, they are very inexpensive (Dobkin 2007). In this research we will consider only passive tags, the most commonly-encountered, and range-challenged, of the three types.

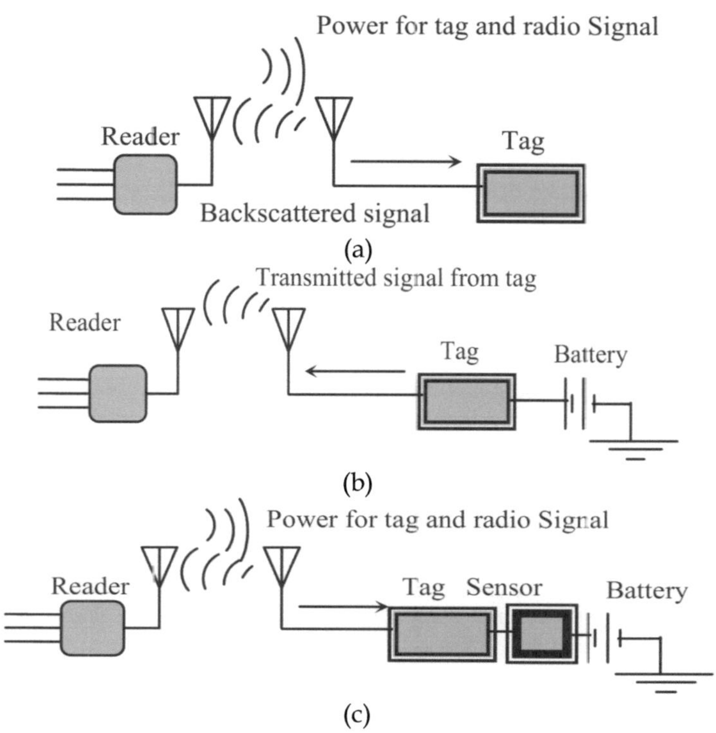

Fig. 2. Communication between (a) reader and passive tag, (b) reader and active tag, (c) reader and semi-passive tag (Khan et al. 2009)

b. Active tags - These tags are full-featured radios with their own transmitting capability independent of the reader. The primary advantages of active tags are their reading range and reliability. The typical communication between the reader and an active tag is shown in Fig. 2 (b). The tags also tend to be more reliable because they do not need a continuous radio signal to power their electronics. But due to the decay of battery life, the active tags have the disadvantage of shorter shelf life than passive tags, normally a few years after manufacturing (Garfinkel & Holtzman 2005).
c. Semi-passive tags - These tags, sometimes known as battery-assisted passive tags, (as shown in Fig. 2 (c)) have a battery, like active tags, but still use the reader's power to transmit a message back to the RFID reader using a technique known as backscatter. These tags thus have the read reliability of an active tag but the read range of a passive tag. They also have a longer shelf life than a tag that is fully active.

2.1.2 RFID reader

The RFID reader sends a pulse of radio energy to the tag and listens for the tag's response. The tag detects this energy and sends back a response that contains the tag's serial number and possibly other information as well. In simple RFID systems, the reader's pulse of energy functioned as an on-off switch; in more sophisticated systems, the reader's RF signal can contain commands to the tag, instructions to read or write memory that the tag contains, and even passwords (Garfinkel & Holtzman 2005).

RFID readers are usually on, continually transmitting radio energy and awaiting any tags that enter their field of operation. However, for some applications, this is unnecessary and could be undesirable in battery-powered devices that need to conserve energy. Thus, it is possible to configure an RFID reader so that it sends the radio pulse only in response to an external event. For example, most electronic toll collection systems have the reader constantly powered up so that every passing car will be recorded. On the other hand, RFID scanners used in veterinarian's offices are frequently equipped with triggers and power up the only when the trigger is pulled.

Like the tags themselves, RFID readers come in many sizes. The largest readers might consist of a desktop personal computer with a special card and multiple antennas connected to the card through shielded cable. Such a reader would typically have a network connection as well so that it could report tags that it reads to other computers. The smallest readers are the size of a postage stamp and are designed to be embedded in mobile telephones.

2.2 Near & far field concept & the selection of RFID operating bands

There are only two possible physics concepts used by RFID technology for the detection of RF tags as depicted in Fig. 3: near field concept (magnetic coupling) and far field concept. In the far field, electric and magnetic fields propagate outward as an electromagnetic wave and are perpendicular to each other and to the direction of propagation. The fields are uniquely related to each other via free-space impedance and decay as $1/r$. In the near field, the field components have different angular and radial dependence (e.g. $1/r^3$). The near field region includes two sub-regions: radiating and reactive. In radiating region, the angular field distribution is dependent on the distance. And in the reactive near field, energy is stored in the electric and magnetic fields very close to the source but not radiated from them. Instead, energy is exchanged between the signal source and the fields (Lecklider 2005).

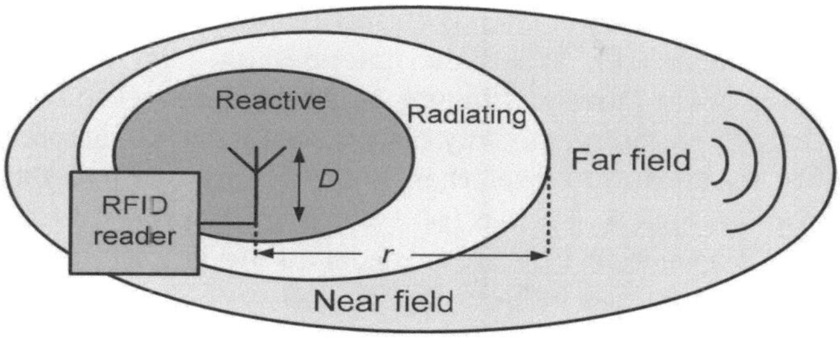

Fig. 3. Antenna near and far field region (Nikitin et al. 2007)

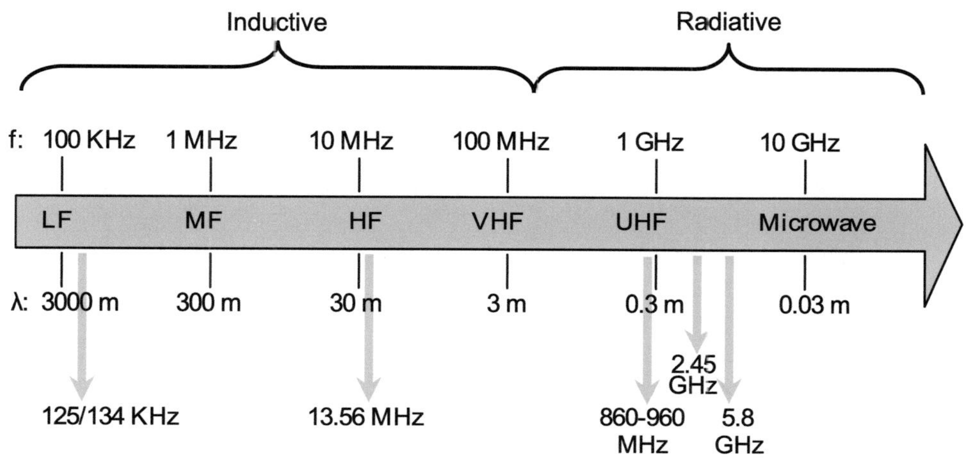

Fig. 4. Frequency-ranges used for RFID-systems

As shown in Fig. 4, several frequency bands have been assigned to RFID applications: 125/134 KHz, 13.56 MHz, 860-960 MHz, 2.450 (2.400–2.483) GHz and 5.800 (5.725–5.875) GHz. Several issues are involved in choosing a frequency of operation (Dobkin 2007).

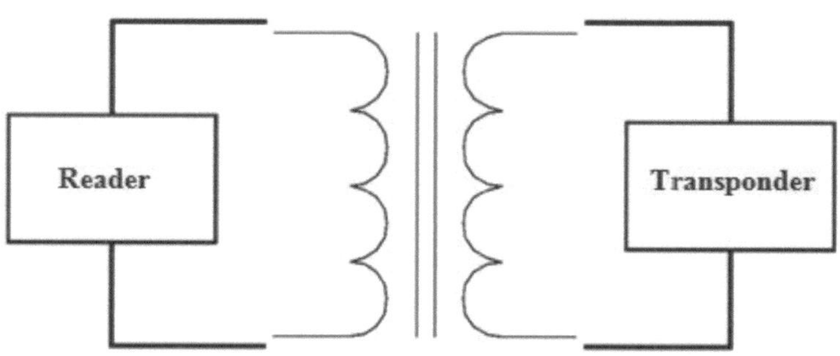

Fig. 5. Inductive coupling or near field detection of RFID reader

The most fundamental, as indicated in the diagram, is whether inductive or radiative coupling will be employed. The distinction is closely related to the side of the antennas to be used relative to the wavelength. When the antennas are very small compared to the wavelength, the effects of the currents flowing in the antenna cancel when viewed from a great distance, so there is no radiation. Only objects so close to the antenna that one part of the antenna appears significantly closer than another part can feel the presence of the current. As depicted in Fig. 5, in case of inductive coupling, the antennas act like transformers and the propagation time from reader to tag is fraction of cycle time. Thus, these systems, which are known as inductively-coupled systems, are limited to short ranges comparable to the size of the antenna. In practice, inductive RFID systems usually use antenna sizes from a few cm to a meter or so, and frequencies of 125/134 KHz (LF) or 13.56 MHz (HF). Thus the wavelength (respectively about 2000 or 20 meters) is much longer than the antenna.

Fig. 6. Radiative coupling or far field detection of RFID reader

Radiative systems use antennas comparable in size to the wavelength. The very common 900 MHz range has wavelengths around 33 cm. Reader antennas vary in size from around 10 to >30 cm, and tags are typically 10-18 cm long. These systems use radiative coupling, and are not limited by reader antenna size but by signal propagation issues. In these systems, the reader antenna launches an electromagnetic wave (exhibited in Fig. 6) and use backscattering from tag to reader. However, the propagation time from reader to tag is longer than a single RF cycle

A second key issue in selection of frequency bands is the allocation of frequencies by regulatory authorities. In essentially every country in the world, the government either directly regulates the use of the radio spectrum, or delegates that authority to related organizations.

RFID systems are typically operated in unlicensed bands. In the US, unlicensed operation is available in the Industrial, Scientific, and Medical (ISM) band at 902-928 MHz, among others. However, for Malaysia the UHF RFID band is 919-923MHz. The UHF RFID frequency allocation statuses are pictured in Fig. 7, where it is realized that, the 900-MHz ISM band is a very common frequency range for UHF RFID readers and tags in all over the world. That's why in this research, the frequency band of 902-928 MHz is aimed for the operation of UHF RFID band.

The practical consequence of UHF band being in proximity to other bands of different wireless applications is the possibility of interference: for example, a nearby cell phone transmitting tower may interfere with the operation of RFID readers, due to the finite ability of the reader receiver to reject the powerful cell signal. (Cellular base stations may sometimes use transmit powers of 10's to hundreds of watts.) Other users of the ISM band may also interfere with RFID readers, or encounter interference due to them: examples are cordless phones and older wireless local area networks.

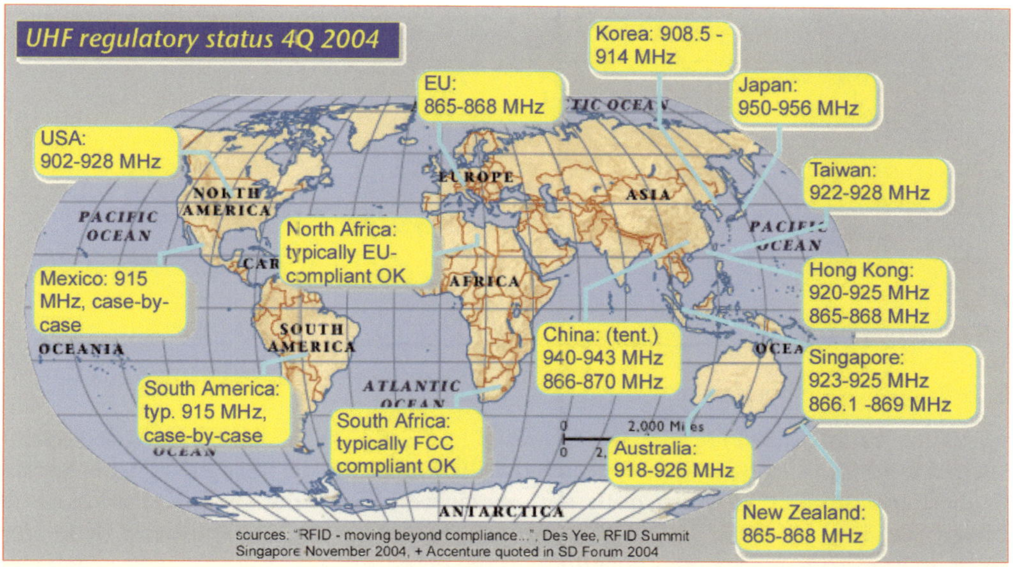

Fig. 7. UHF RFID frequency allocation statuses from 2004 (www.mapquest.com)

Finally, changes in operating frequency affect the propagation characteristics of the resulting radiated fields. Lower frequencies diffract more readily around obstacles, but couple less well to small antennas. Radiated fields are absorbed by many common materials in buildings and the environment, particularly those containing water. The degree of absorption due to water increases gradually with increasing frequency. Tags immersed in water-containing materials (i.e. injected into or swallowed by animals or people) must use very low frequencies to minimize absorption: this is a typical 125 KHz application. For locating large objects or people outdoors, a relatively low frequency may be desirable to avoid obstacle blockage; when a clear line of sight from the antenna to the tag can be assured, a higher frequency may be useful to reduce the size of the antennas.

3. Antenna characteristics

Antennas are the key components of any wireless communication system (Balanis 1996; Kraus 1988). According to The IEEE Standard Definitions of terms for Antennas, an antenna is defined as "a means for radiating or receiving radio waves" (IEEE Std 145-1993 1993). In other words, they are the devices that allow for the transfer of a signal (in a wired system) to waves that, in turn, propagate through space and can be received by another antenna. The receiving antenna is responsible for the reciprocal process, i.e., that of turning an electromagnetic wave into a signal or voltage at its terminals that can subsequently be processed by the receiver.

In the following sections, some of the antenna parameters are described that necessary to fully characterize an antenna and determine whether an antenna is optimized for a certain application.

3.1 Impedance bandwidth, reflection coefficient, VSWR & return loss

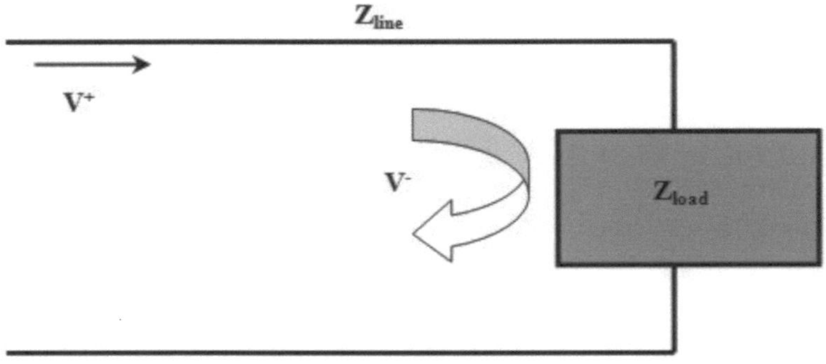

Fig. 8. Transmission line model

Impedance bandwidth indicates the bandwidth for which the antenna is sufficiently matched to its input transmission line such that 10% or less of the incident signal is lost due to reflections. Impedance bandwidth measurements include the characterization of the Voltage Standing Wave Ratio (VSWR) and return loss throughout the band of interest. VSWR and return loss are both dependent on the measurement of the reflection coefficient Γ. Γ is defined as ratio of the reflected wave V_o^- to the incident wave V_o^+ at a transmission line load as shown in Fig. 8. Transmission Line Model, and can be calculated by equation 2.1 (Balanis 1996; Stutzman 1998; Pozar 2001):

$$\Gamma = \frac{V_0^-}{V_0^+} = \frac{Z_{line} - Z_{load}}{Z_{line} + Z_{load}} \tag{1}$$

Z_{line} and Z_{load} are the transmission line impedance and the load (antenna) impedance, respectively. The voltage and current through the transmission line as a function of the distance from the load, z, are given as follows:

$$V(z) = V_0^+ e^{-j\beta z} + V_0^- e^{j\beta z} = V_0^+ (e^{-j\beta z} + \Gamma e^{j\beta z}) \tag{2}$$

$$I(z) = 1/Z_0 (V_0^+ e^{-j\beta z} - V_0^- e^{j\beta z}) = V_0^+ / Z_0 (e^{-j\beta z} - \Gamma e^{j\beta z}) \tag{3}$$

where $\beta = 2\pi/\lambda$.
The reflection coefficient Γ is equivalent to the S_{11} parameter of the scattering matrix. A perfect impedance match would be indicated by Γ = 0. The worst impedance match is given by Γ = -1 or 1, corresponding to a load impedance of a short or an open.
Power reflected at the terminals of the antenna is the main concern related to impedance matching. Time-average power flow is usually measured along a transmission line to determine the net average power delivered to the load. The average incident power is given by:

$$P^{i}_{ave} = \frac{|V_0^+|^2}{2Z_0} \tag{4}$$

The reflected power is proportional to the incident power by a multiplicative factor of $|\Gamma|^2$, as follows:

$$P^{r}_{ave} = -|\Gamma|^2 \frac{|V_0^+|^2}{2Z_0} \tag{5}$$

The net average power delivered to the load, then, is the sum of the average incident and average reflected power:

$$P_{ave} = \frac{|V_0^+|^2}{2Z_0}[1 - |\Gamma|^2] \tag{6}$$

Since power delivered to the load is proportional to $(1-|\Gamma|^2)$, an acceptable value of Γ that enables only 10% reflected power can be calculated. This result is Γ= 0.3162.

When a load is not perfectly matched to the transmission line, reflections at the load cause a negative traveling wave to propagate down the transmission line. Ultimately, this creates unwanted standing waves in the transmission line. VSWR measures the ratio of the amplitudes of the maximum standing wave to the minimum standing wave, and can be calculated by the equation below:

$$VSWR = \frac{V_{max}}{V_{min}} = \frac{1+|\Gamma|}{1-|\Gamma|} \tag{7}$$

The typically desired value of VSWR to indicate a good impedance match is 2.0 or less. This VSWR limit is derived from the value of Γ calculated above.

Return loss is another measure of impedance match quality, also dependent on the value of Γ, or S_{11}. Antenna return loss is calculated by the following equation:

$$\text{Return Loss} = -10\log|S_{11}|^2 \text{, or } -20\log(|\Gamma|) \tag{8}$$

A good impedance match is indicated by a return loss greater than 10 dB. A summary of desired antenna impedance parameters include Γ<0.3162, VSWR<2, and Return Loss > 10 dB.

3.2 Radiation pattern

One of the most common descriptors of an antenna is its radiation pattern. Radiation pattern can easily indicate an application for which an antenna will be used. For example, fixed indoor RFID reader applications, such as a ware-house, would necessitate a nearly omni-directional antenna which could be hung in the ceiling, since the position of the detectable object might not be known. Therefore, radiation power should be spread out uniformly around the user for optimal reception. However, for high range RFID detection applications, a highly directive antenna would be desired such that the majority of radiated power is directed to a specific, known location. According to the IEEE Standard Definitions of Terms for Antennas, an antenna radiation pattern (or antenna pattern) is defined as: "a mathematical function or a graphical representation of the radiation properties of the antenna as a function of space coordinates. In most cases, the radiation pattern is determined in the far-field region and is represented as a function of the directional coordinates. Radiation properties include power flux density, radiation intensity, field strength, directivity phase or polarization (IEEE Std 145-1993 1993).

In most cases, it is determined in the far-field region where the spatial (angular) distribution of the radiated power does not depend on the distance. Usually, the pattern describes the normalized field (power) values with respect to the maximum values. The radiation property of most concern is the two-or three-dimensional (2D or 3D) spatial distribution of radiated energy as a function of the observer's position along a path or surface of constant radius. In practice, the three-dimensional pattern is some-times required and can be constructed in a series of two-dimensional patterns. For most practical applications, a few plots of the pattern as a function of ϕ for some particular values of frequency, plus a few plots as a function of frequency for some particular values of θ will provide most of the useful information needed, where ϕ and θ are the two axes in a spherical coordinate.

There are two common portions used to describe the characteristic of a radiation pattern of an antenna:

a. **Co-polar pattern:** diagram representing the radiation pattern of a test antenna when the reference antenna is similarly polarized, scaled in dBi or dB relative to the measured antenna gain
b. **Cross-polar pattern:** diagram representing the radiation pattern of a test antenna when the reference antenna is orthogonally polarized, scaled in dBi, or dB relative to the measured antenna gain

3.3 Antenna polarization

Polarization is a property of a single-frequency electromagnetic wave; it describes the shape and orientation of the locus of the extremity of the field vectors as a function of time. In antenna engineering, the polarization properties of plane waves or waves that can be considered to be planar over the local region of observation are of interest. For plane waves, it is sufficient to specify the polarization properties of the electric field vector since the magnetic field vector is simply related to the electric field vector. The plane containing the electric and magnetic fields is called the plane of polarization and is orthogonal to the direction of propagation (Volakis 2007).

The polarization of an electromagnetic wave may be linear, circular, or elliptical (Kumar & Ray 2003). The instantaneous field of a plane wave, traveling in the negative z -direction, is given by

$$E(z,t) = E_x(z,t)\hat{x} + E_y(z,t)\hat{y} \tag{9}$$

The instantaneous components are related to their complex counter-parts by

$$E_x(z,t) = E_x \cos(\omega t + \beta z + \phi_x) \tag{10}$$

and

$$E_y(z,t) = E_y \cos(\omega t + \beta z + \phi_y) \tag{11}$$

where E_x and E_y are the maximum magnitudes and ϕ_x and ϕ_y are the phase angles of the x and y components, respectively, ω is the angular frequency, and b is the propagation constant. For the wave to be linearly polarized, the phase difference between the two components must be

$$\Delta\phi = \phi_y - \phi_x = n\pi \text{ , where n=0, 1, 2, ...} \tag{12}$$

The wave is circularly polarized when the magnitudes of the two components are equal (i.e., $E_x = E_y$) and the phase difference $\Delta\phi$ is an odd multiple of $\pi/2$; in other words,

$$\Delta\phi = \phi_y - \phi_x = \begin{cases} +(2n+1/2)\pi \textit{ for RHCP} \\ \textit{or} \\ -(2n+1/2)\pi \textit{ for LHCP} \end{cases} \tag{13}$$

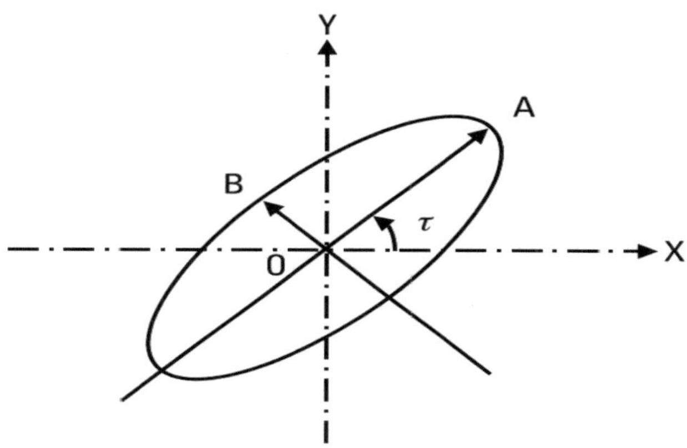

Fig. 9. Elliptically polarized wave

If $E_x \neq E_y$ or $\Delta\phi$ does not satisfy (11) and (12), then the resulting polarization is of elliptical shape as shown in Fig. 9. The performance of a circularly polarized antenna is characterized by its AR. The AR is defined as the ratio of the major axis to the minor axis; in other words,

$$AR = \frac{\text{major axis}}{\text{minor axis}} = \frac{OA}{OB} \tag{14}$$

where

$$OA = \left[\frac{1}{2}\left\{E_x^2 + E_y^2 + \left[E_x^4 + E_y^4 + 2E_x^2E_y^2\cos(2\Delta\phi)\right]^{1/2}\right\}\right]^{1/2} \tag{15}$$

and

$$OB = \left[\frac{1}{2}\left\{E_x^2 + E_y^2 - \left[E_x^4 + E_y^4 + 2E_x^2E_y^2\cos(2\Delta\phi)\right]^{1/2}\right\}\right]^{1/2} \tag{16}$$

The tilt angle τ of the ellipse is given by

$$\tau = \frac{\pi}{2} - \frac{1}{2}\tan^{-1}\left[\frac{2E_xE_y}{E_x^2 - E_y^2}\cos(\Delta\phi)\right] \tag{17}$$

For CP, OA = OB (i.e., AR = 1), whereas for linear polarization, AR → ∞. The deviation of AR from unity puts a limit on the operating frequency range of the circularly polarized antennas. Generally, AR = 3–6 dB (numerical value 1.414 to 2) is acceptable for most of the practical applications.

3.4 Directivity & gain

Directivity of an antenna, D is defined as the ratio of the radiation intensity U in a given direction from the antenna to the radiation intensity averaged over all directions, i.e. an isotropic source. It is introduced to describe the directional properties of antenna radiation pattern. For an isotropic source, the radiation intensity U_0 is equal to the total radiated power P_{rad} divided by 4π. So the directivity can be calculated by:

$$D = \frac{U}{U_0} = \frac{4\pi U}{P_{rad}} \tag{18}$$

If not specified, antenna directivity implies its maximum value, i.e. D_0.

$$D_0 = \frac{U|_{\max}}{U_0} = \frac{U_{\max}}{U_0} = \frac{4\pi U_{\max}}{P_{rad}} \tag{19}$$

Antenna gain G is closely related to the directivity, but it takes into account the radiation efficiency e_{rad} of the antenna as well as its directional properties, as given by:

$$G = e_{rad} D \tag{20}$$

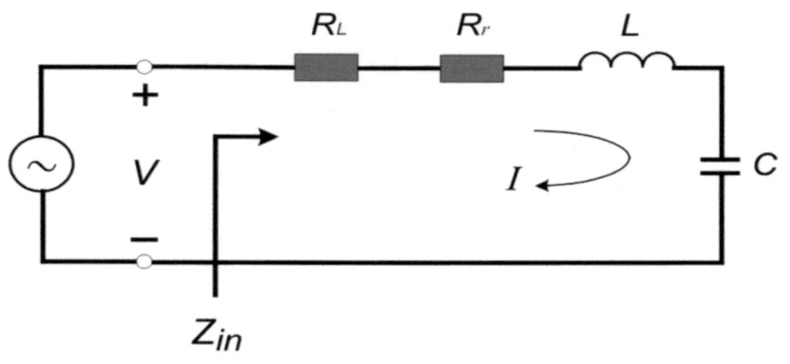

Fig. 10. Equivalent circuit of antenna

Fig. 10 shows the equivalent circuit of the antenna, where R_r, R_L, L and C represent the radiation resistance, loss resistance, inductor and capacitor, respectively. The radiation efficiency e_{rad} is defined as the ratio of the power delivered to the radiation resistance R_r to the power delivered to R_r and R_L. So the radiation efficiency e_{rad} can be written as:

$$e_{rad} = \frac{\frac{1}{2}|I|^2 R_r}{\frac{1}{2}|I|^2 R_r + \frac{1}{2}|I|^2 R_L} = \frac{R_r}{R_r + R_L} \tag{21}$$

According to the IEEE Standard Definitions of Terms for Antennas (IEEE Std 145-1993 1993), the antenna absolute gain is "the ratio of the intensity, in a given direction, to the radiation intensity that would be obtained if the power accepted by the antenna were radiated isotropically" (IEEE Std 145-1993 1993). The maximum gain G_0 is related the maximum directivity D_0 mathematically as follows:

$$G_0 = e_{rad} D_0 \text{ (Dimensionless)} \tag{22}$$

Also, if the direction of the gain measurement is not indicated, the direction of maximum gain is assumed. The gain measurement is referred to the power at the input terminals rather than the radiated power, so it tends to be a more thorough measurement, which reflects the losses in the antenna structure.
Gain measurement is typically misunderstood in terms of determining the quality of an antenna. A common misconception is that the higher the gain, the better the antenna. This is only true if the application requires a highly directive antenna. Since gain is linearly proportional to directivity, the gain measurement is a direct indication of how directive the antenna is (provided the antenna has adequate radiation efficiency).

4. Multi-band antenna techniques: review

When the antenna operates only at more than one spot frequency, then it is known as a multi-frequency antenna. When it operates over a finite BW at all of the frequencies, it is known as multi-band antenna. When two or more resonance frequencies of a MSA are close to each other, one gets broadband characteristics. When these are significantly separated, dual-band or multi-band operations are obtained. In literature, numerous multi-band antennas are available. However, in order to understand the technique of multi-band operation, it is worthy to understand the mechanism of dual-band antennas which could be extended to more than two bands employing the same or combination of other techniques.
For dual-band operations, various single and multilayer microstrip antennas configurations are possible. In the single-layer microstrip antenna, dual-band operation can be achieved by utilizing the multi-resonance characteristics of a single patch, by reactively loading the patch with quarter-wavelength stubs, by using shorting posts, by cutting slots, and by adding lumped elements, among other techniques. Multi-resonators in both planar and stacked configurations yield dual-band operations. Both electromagnetic as well as aperture coupling mechanisms are used in multilayer configurations (Kumar & Ray 2003).

4.1 Higher order or orthogonal mode microstrip antennas

As is well-known, a simple rectangular patch can be regarded as a cavity with magnetic walls on the radiating edges. The first three modes with the same polarization can be indicated by TM_{100}, TM_{200}, and TM_{300}, where TM denotes the magnetic field transverse with respect to the interface normal. TM_{100} is the mode typically used in practical applications; TM_{200} and TM_{300} are associated with a frequency approximately twice and triple of that of the TM_{100} mode. This provides, in principle, the possibility to operate at multiple frequencies. In practice, the TM_{200} and the TM_{300} modes cannot be used. Indeed, owing to

the behavior of the radiating currents, the TM_{200} pattern has a broadside null, and the TM_{300} pattern has grating lobes.
The simplest way to operate at dual frequencies is to use the first resonance of the two orthogonal dimensions of the rectangular patch, i.e., the TM_{100} and the TM_{101} modes. In this case, the frequency ratio is approximately equal to the ratio between the two orthogonal sides of the patch. The obvious limitation of this approach is that the two different frequencies excite two orthogonal polarizations. This simple method is very useful in low-cost short-range applications, where polarization requirements are not pressing (Maci & Gentili 1997).

4.1.1 Single feed dual-band microstrip antenna

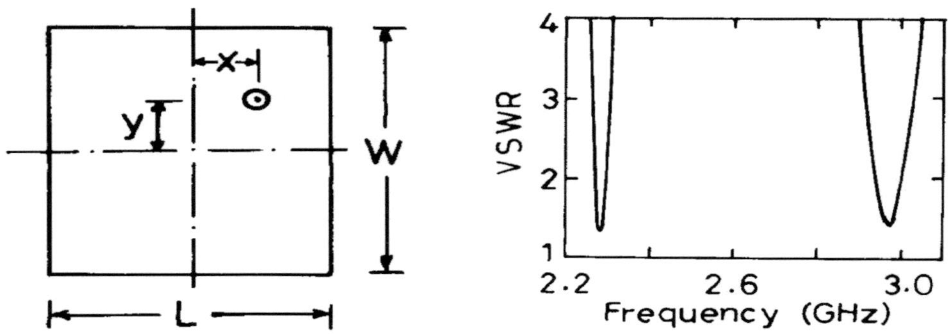

(a) (b)

Fig. 11. (a) Rectangular microstrip antenna with a single feed for orthogonal dual-band operation and its and (b) VSWR plots (Chen & Wong 1996)

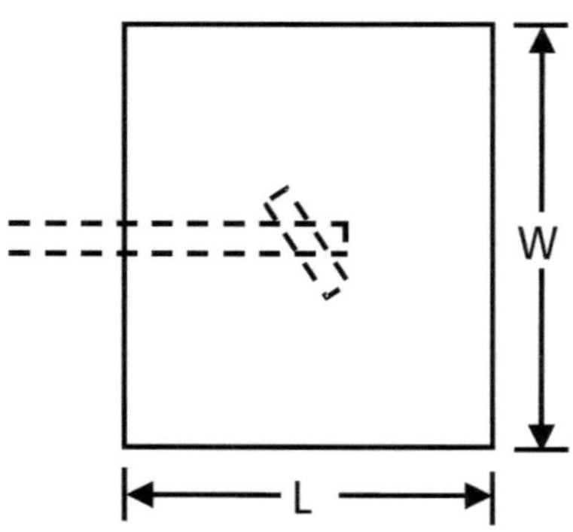

Fig. 12. Aperture coupled RMSA with an inclined slot

An interesting feature of these antennas is their capability of simultaneous matching of the input impedance at the two frequencies with a single feed structure (denoted by "single-point" in Fig. 11). This may be obtained with a probe-fed configuration, which is displaced from the two principal axes of the patch. As demonstrated in literature (Chen & Wong 1996), the performance of this approach in terms of matching level and bandwidth is almost equal to that of the same patch fed separately on the two orthogonal principal axes. This provides

the possibility of using the well-known design formula for standard feeds. It is also worth noting that the simultaneous matching level for structures that provide the same polarizations at the two frequencies is, in general, worse with respect to the case relevant to orthogonal polarization.
Instead of using a single coaxial feed, similar results are obtained by using an aperture coupled rectangular microstrip antenna, in which an inclined slot is cut in the ground plane with respect to the microstrip feed line as shown in Fig. 12 to give proper matching at both the frequencies (Antar et al. 1995). The required slot length and inclination angle can be approximately obtained by projecting the slot onto the two orthogonal directions. The two projections can be thought of as the length of two equivalent slots that excite the patch at the two separate polarizations. The inclination of the slots may also be adjusted, in order to compensate for error introduced by the matching stub, which is designed to be a quarter of a wavelength for only one frequency.

4.1.2 Dual feed microstrip antennas

The use of a circulator or diplexer that should be used in single fed dual-band microstrip antenna to isolate reception from transmission may be avoided by feeding the RMSA at two orthogonal points as shown in Fig. 13(a) (Srinivasan et al. 2000a). Since these feed points are at null locations of the respective orthogonal modes, the loading of one feed point does not affect the input impedance at the other feed point. The isolation between the two modes using orthogonal feeds is nearly 30 dB and 40 dB at the lower and higher resonance frequencies, respectively.

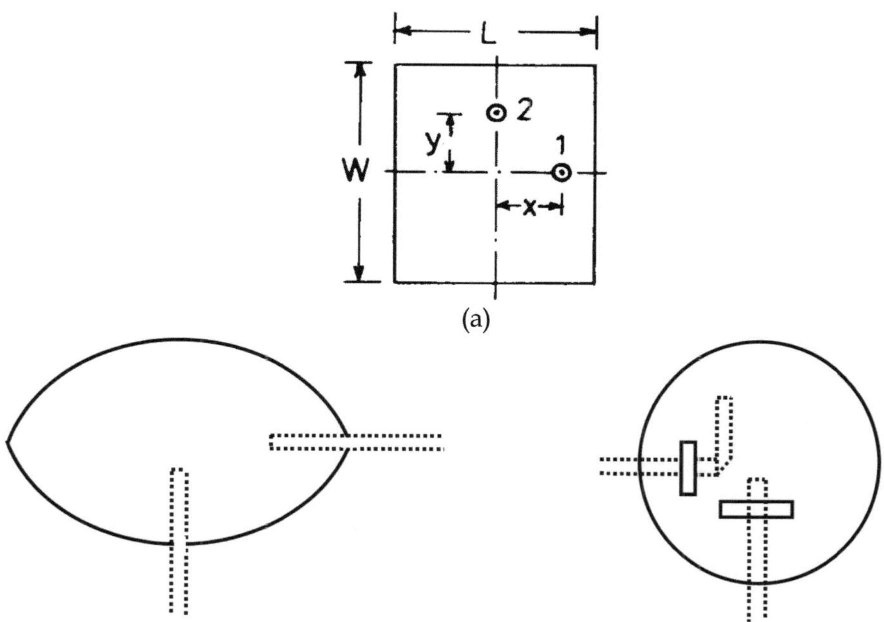

Fig. 13. (a) Rectangular microstrip antenna with two orthogonal feeds for dual-band operation, (b) Elliptical microstrip antenna with two orthogonal feeds, (c) Circular microstrip antenna with two orthogonal slots (Kumar & Ray 2003)

Similar results are obtained for an ellipse with two orthogonal feed points. This configuration is fed with two orthogonal electromagnetically coupled microstrip lines (Deepukumar et al. 1996). As before, the frequency ratio of dual-band operation is approximately equal to the ratio of the orthogonal dimensions in the two planes. The isolation between the two ports is 27 dB.
Another variation using a circular patch is shown in Fig. 13 (c). It is excited by two orthogonal microstrip lines through the two orthogonal slots cut in the ground plane. By changing the slot dimensions, the two orthogonal resonance frequencies can be changed (Murakami et al. 1993).

4.2 Multi-patch antenna design approach

It is also a common practice to utilize two or more patches to accomplish multi-band. This section describes two main multi-patch techniques for dual-band or multi-band antennas.

4.2.1 Multi-patch stacked antennas

The dual-frequency behavior of these antennas is obtained by means of multiple radiating elements, each of them supporting strong currents and radiation at the resonance. This category includes multi-layer stacked patches (Fig. 14) that can use circular (Long & Walton 1979; Dahele & Lee 1982; Bennegueouche et al. 1993; Iwasaki & Suzuki 1995), annular (Dahele et al. 1987; Tagle & Christodoulous 1997), rectangular (Wang, et al. 1990; Yazidi et al. 1993), and triangular (Bhatnagar et al. 1986) patches. These antennas operate with the same polarization at the two frequencies, as well as with a dual polarization.

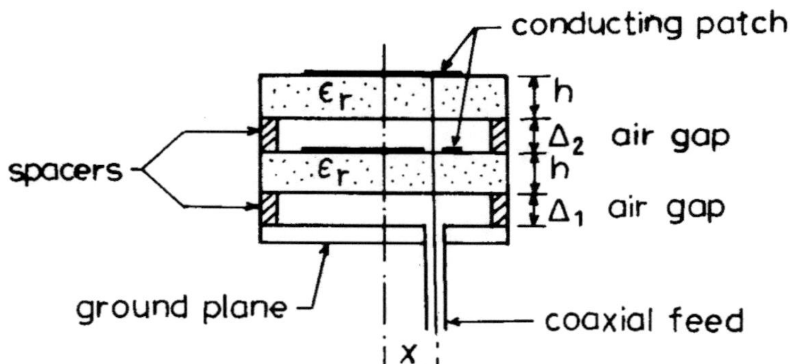

Fig. 14. A dual-frequency stacked circular-disc antenna (Long & Walton 1979)

The same multilayer structures can also be used to broaden the bandwidth of a single-frequency antenna, when the two frequencies are forced to be closely spaced. In this latter case, the lower patch can be fed by a conventional arrangement and the upper patch by proximity coupling with the lower patch (Wang et al. 1990). In order to avoid disappearance of the upper resonance, the sizes of the two patches should be close, so that only a frequency ratio close to unity may be obtained. A direct probe feed for the upper patch may also be used (Long & Walton 1979; Dahele et al. 1987). In this case, the probe passes through a clearance hole in the lower patch, and is electrically connected to the upper patch. This kind of configuration insures one more degree of freedom (the hole radius) in designing the optimum matching at the two frequencies, and allows a wider range of the frequency ratio with respect to the structure in which the upper patch is electromagnetically coupled. In comparison with the resonant frequencies of the two isolated patches, the frequency of the upper (smaller) patch increases, and the frequency of the lower (larger) patch decreases. In any case, due to the strong coupling between the two elements, simple design formulas cannot be found, so that a full-wave analysis is, in general, required in the first phase of the design.

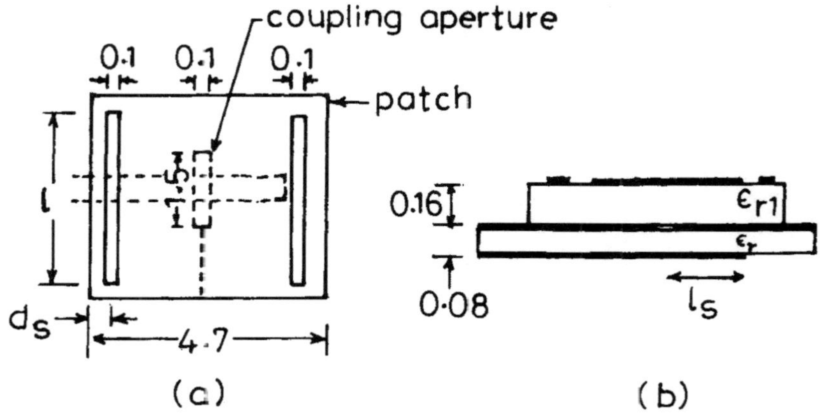

Fig. 15. An aperture-coupled rectangular microstrip antenna with two slots: (a) top and (b) side views (Yazidi et al. 1993)

4.2.2 Multi-patch co-planar antennas

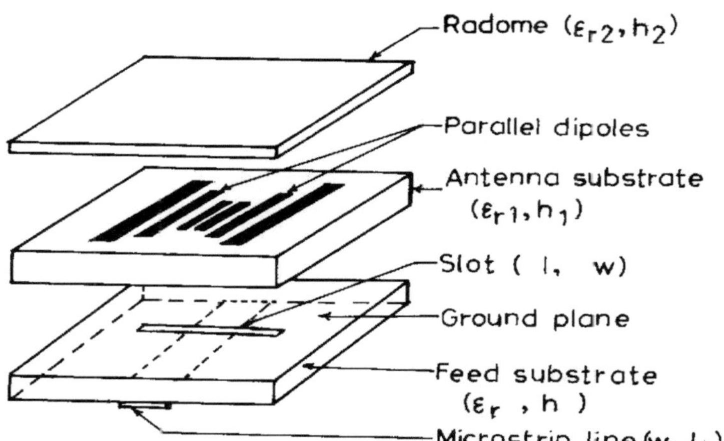

Fig. 16. Aperture-coupled coplanar parallel dipoles for multi-frequency operation (Croq & Pozar 1992)

Coplanar parallel dipoles fed by aperture coupling could be used to obtain multi-frequency operation. The dipoles of different lengths are fed by a microstrip line through a rectangular slot cut in the ground plane. In general, this antenna consists of 2N dipoles of N different lengths, which are symmetrically excited through the aperture at N frequencies (Croq &

Pozar 1992). Either the longest identical pair of dipoles could be placed in the center of the slot and smallest identical pair close to the edges of the slot, or the smallest dipoles could be placed in the center and the longest at the edge. For the latter case, six symmetrical dipoles are shown in Fig. 16. Since there are three pairs of dipoles, there will be three resonance frequencies. The radiation pattern is in the broadside direction at all the three frequencies, and the antenna is attractive for its simplicity.

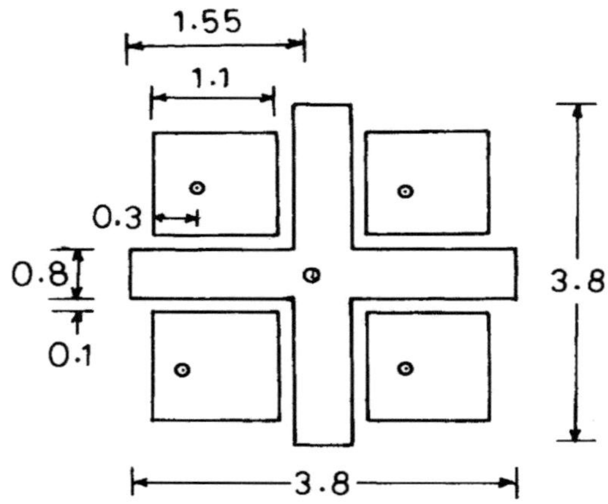

Fig. 17. Dual frequency sub-array microstrip antenna (Salvador et al. 1995)

Many radar and communication systems often require a large separation between the two frequencies, so the multi-resonator configuration requires patches of very different resonant lengths. A simple example of this concept is shown in Fig. 17 (Salvador et al. 1995). It consists of a cross-shaped patch at the S-band and a sub-array of four patches at the X-band. The resonance frequency of the cross-patch is only slightly perturbed by the addition of four square patches, since the radiating edges of the cross patch are away from the four square patches. However, the resonance frequency of the square patches is affected by the presence of the cross patch, which causes a reactive loading to the square patch. Therefore, the upper resonance frequency corresponding to the four square patches is slightly lower than that of the isolated square patches. The decrease of this upper frequency is noticeable when the spacing is less than the substrate thickness because of increased gap coupling. In designing the antenna, one should carefully choose the distance between the square patches, which should be less than 0.71 λ_0 to avoid scan blindness at the upper frequency.

4.3 Loaded multi-band antennas

The patch can be loaded for multi-band operation of microstrip antennas. The loading could be primarily stubs, notches, pins, lumped elements like resistors or capacitors and slots. Nevertheless, combination of these loading is also possible. In the following section the load antenna techniques are described in brief.

4.3.1 Stub loaded microstrip antennas

The reactive-loading approach was first used in Richards et al. (1985), where an adjustable coaxial stub was employed. This structure may provide both tuning and design of the frequency ratio in a simple manner; on the other hand, it is encumbering and not well-suited for high frequencies. In Davidson et al. (1985), a more practical configuration is presented, in which the stub is constituted by a microstrip. The tuning of the two frequencies was obtained by changing the length of the short-circuited coaxial line. Instead of short circuited coaxial line, a λ/2 short-circuited microstrip line is used as shown in Fig. 18(b), which can be etched on the same substrate along with the patch. Antennas with a single stub have slightly higher cross-polar level because it is asymmetrical configuration. To make the configuration symmetrical, not harming dual frequency operation, a λ /4 open-circuited stub is placed along both the radiating edges of the rectangular patch.

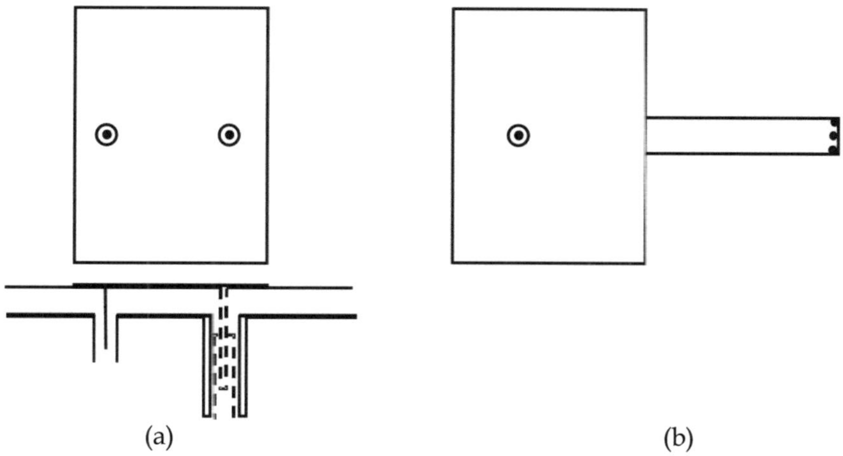

Fig. 18. Various dual-band stub-loaded RMSA configurations: (a) short-circuited coaxial line, (b) short-circuited λ/2 stub (Kumar & Ray 2003)

4.3.2 Notch-loaded dual-band microstrip antennas

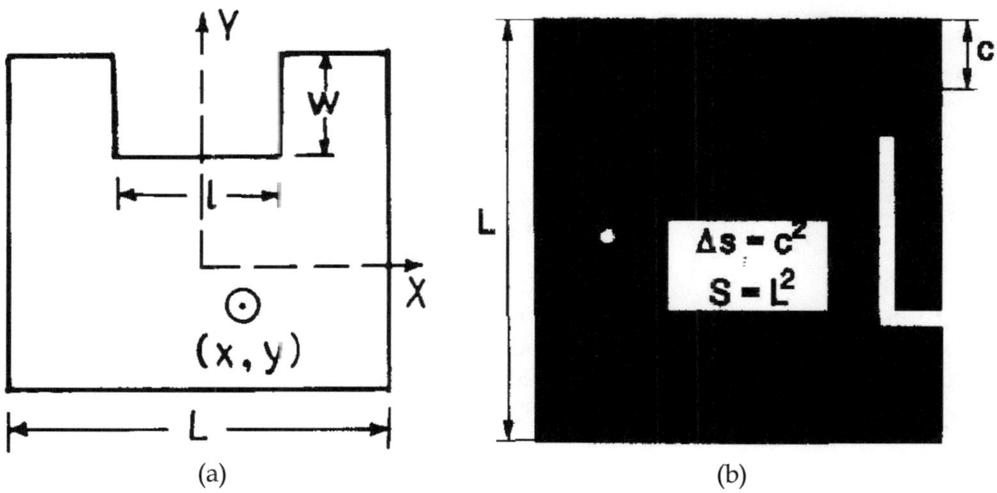

Fig. 19. (a) Notch loaded square microstrip antenna and (b) Dual-band circularly polarized microstrip patch antenna (Nakano & Vichien 1989; Hernandez & Robertson 1995)

Loading the radiating edge with an inset (Nakano & Vichien 1989; Palit et al. 1998) or a spur-line (Hernandez & Robertson 1995; Hernandez & Robertson 1993; Vaello & Hernandez 1998) ("notch loading") is an alterative way to introduce a dual-frequency behavior that creates the same effect as the microstrip-loading effect, with the advantage of reduced size. However, both with stubs and notches, the frequency ratio cannot be designed to be higher than 1.2 without introducing strong cross-polarization levels or pattern distortion at the additional frequency.

4.3.3 Pins and lumped elements loaded dual-band antenna

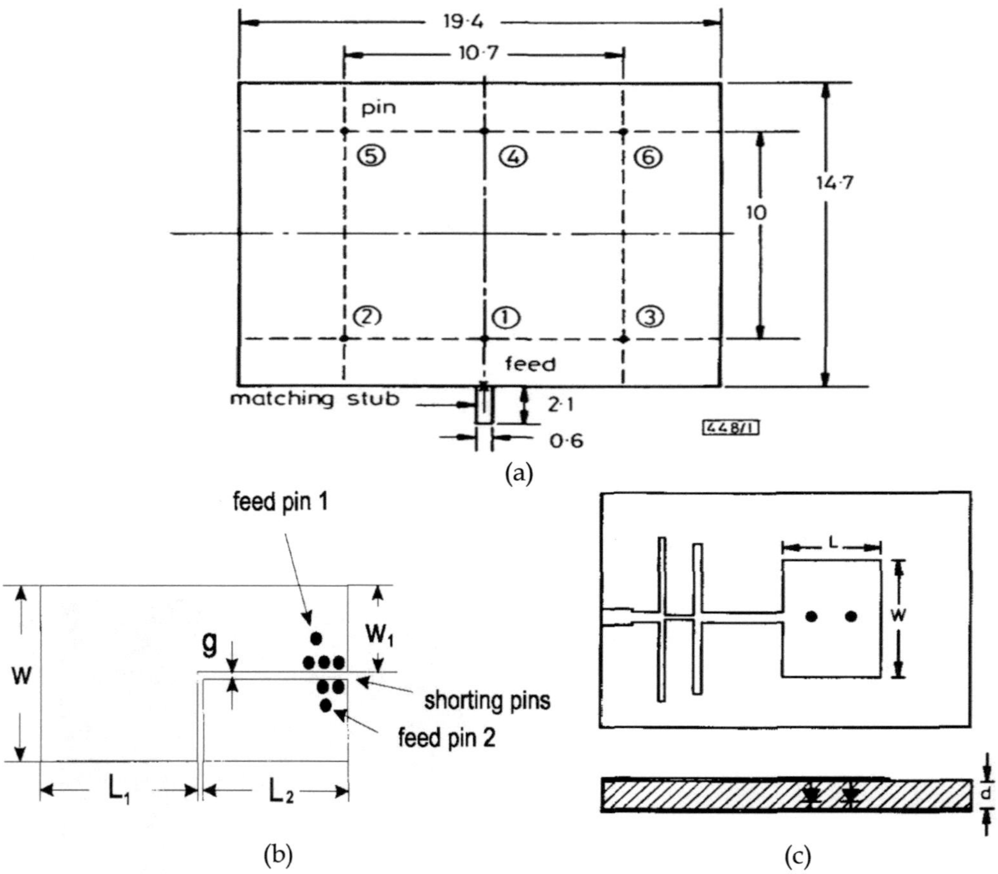

Fig. 20. (a) A rectangular microstrip antenna with shorting posts for dual-band operation, (b) top view of dual-band antenna mounted on the conducting telephone case, (c) Edge-Jed rectangular microstrip patch with double stub matching network and symmetrically loaded with two varactor diodes (Zhong & Lo 1983; Gao et al. 2002; Waterhouse & Shuley 1992)

A rectangular microstrip antenna operating in the TM_{10} and TM_{30} modes has a broadside radiation pattern with the same polarization at the two frequencies. The ratio of their resonance frequencies is approximately three. A shorting post placed at the null position of the TM_{30} mode will not change its corresponding resonance frequency but will have a strong effect on the TM_{10} mode frequency (Zhong & Lo 1983). An RMSA with six shorting posts is shown in Fig. 20 (a). Since all these posts are located at the nulls of the TM_{30} mode, f_2 remains constant at around 1,865 MHz, while f_1 varies from 613 MHz to 891 MHz. The ratio f_2/f_1 varies from 3.0 to 2.1, which could be lowered by using more shorting posts. However similar principle might be seen in other antennas in literature (Gao et al. 2002; Pan & Wong 1998; Liu et al. 1997; Srinivasan et al. 1998). The structure of the antenna might be circular (Tang et al. 1997; Pan & Wong 1997) or triangular. Very high values of the frequency ratio (4-5) can be obtained by means of lumped loaded elements like resistors (Srinivasan et al. 2000b), varactors (Waterhouse & Shuley 1992), connected from the patch to the ground plane.

4.3.4 Slot antenna technique

Another kind of reactive loading can be introduced by etching slots on the patch. The slot loading allows for a strong modification of the resonant mode of a rectangular patch, particularly when the slots are oriented to cut the current lines of the unperturbed mode. In particular, as shown in (Wang & Lo 1984), the simultaneous use of slots and short-circuit vias allows a frequency ratio of from 1.3 to 3, depending on the number of vias. Other kinds of slot-loaded patches have been independently introduced in (Maci et al. 1993) and (Yazidi et al. 1993), and consist of a rectangular patch with two narrow slots etched close to and parallel to the radiating edge. The same configuration has been investigated in ([58] Maci, S., 1995), and extended to dual polarization in (Piazzesi et al. 1995).

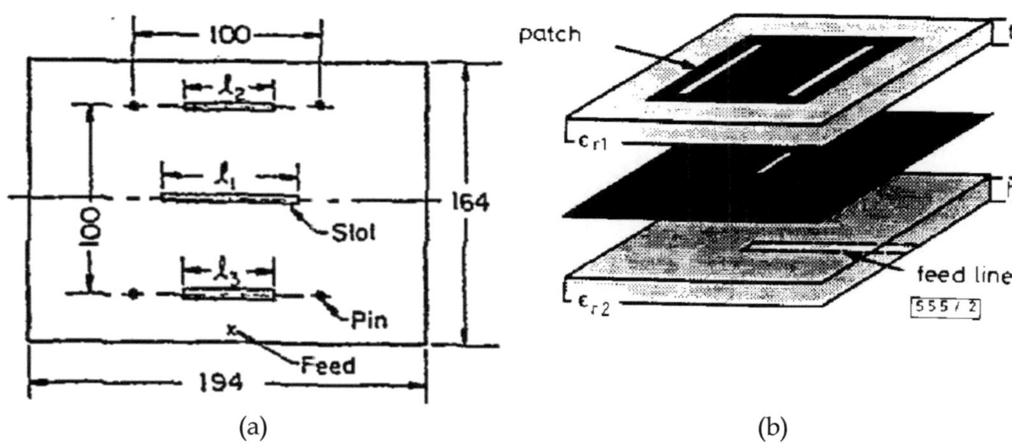

Fig. 21. (a) The microstrip antenna with shorting pins and slot, (b) Aperture coupled microstrip antenna for dual frequency operation (Wang & Lo 1984; Yazidi et al. 1993)

5. Dual-band high gain antennas & limitations

Many dual band antennas are developed and reported in the literature, especially for 2.4/5.8GHz. In order to achieve good radiation characteristics especially high gain, there are a lot of approaches taken. Printed dipoles (Kim et al. 2005; Lin et al. 2003), printed monopole (Wu et al. 2003; Jianhui et al. 2008; Wu et al. 2005), planar (Raj et al. 2005), slot (Wong et al. 2007) antennas are popularly used to provide dual frequency operation. But these antennas have complicated patch structures. Dielectric resonators (Chen et al. 2009; Ding & Leung 2009) and chip antennas (Moon & Park 2003) also provide dual band coverage which are very hard to fabricate. Rectenna (Suh & Chang 2002; Heikkinen & Kivikoski 2003), stacked patch (Yang et al. 2005), aperture coupled antenna (Yang et al. 2008), even though offer these two bands, occupy large space and difficult to integrate with handheld applications. However, the gain of these reported antennas are very low; lower than 7dBi, even printed simple element arrays (Lin et al. 2003; Wu et al. 2003) could not boost the gain higher.

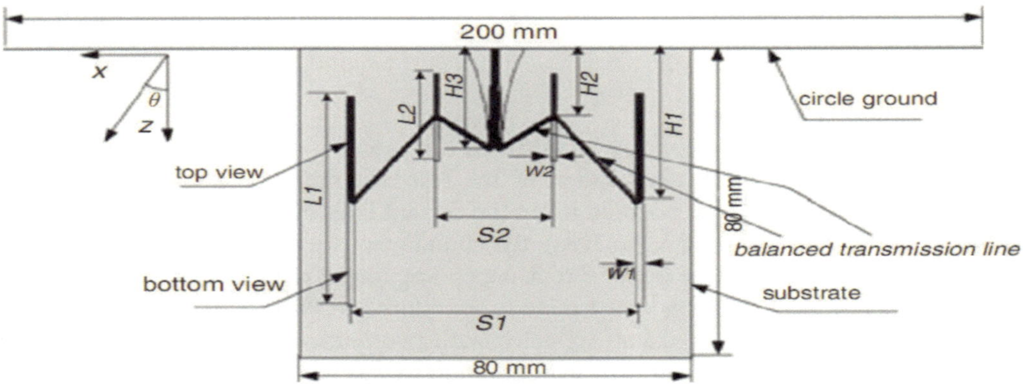

Fig. 22. Geometry of the dual-band bidirectional high gain antenna (Zhang et al. 2009)

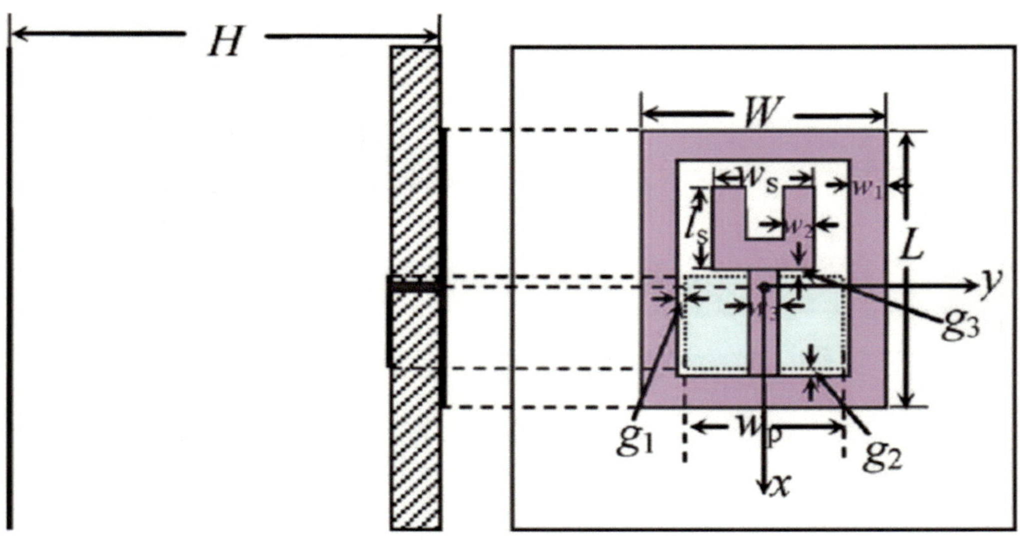

Fig. 23. Configuration of novel high-gain dual-band antenna (He et al. 2009)

Lately, to have high gain in these two bands a four-element printed dipole array antenna with balanced twin transmission line is reported (Zhang et al. 2009). Here the use of vertical patch increases the antenna volume. But still the gains for the 2.4 and 5.8 GHz bands are between 4.8–6 and 6–8.8 dBi, respectively.

Another novel high-gain dual band antenna is reported (He et al. 2009) shortly. The radiator is composed of three parts: the fork-like monopole, the rectangular ring, and the rectangular patch. A metal reflector with the same dimensions as the substrate is used behind the designed antenna, so the directivity/gain of the presented antenna is enhanced for both bands by suppressing the backside radiation. This antenna came up with a good peak gain; but still the antenna is inadequate to achieve desired bandwidth for lower frequencies and gain variation is severe over both the bands with an unlike radiation pattern in frequency bands. However taking the metal reflector into account requires bigger space for the antenna profile.

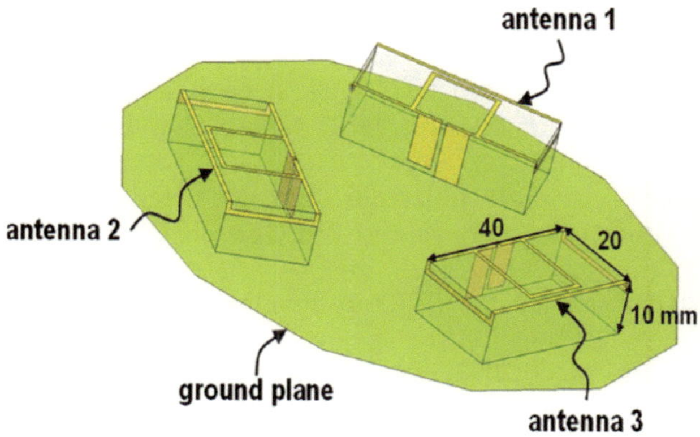

(a)

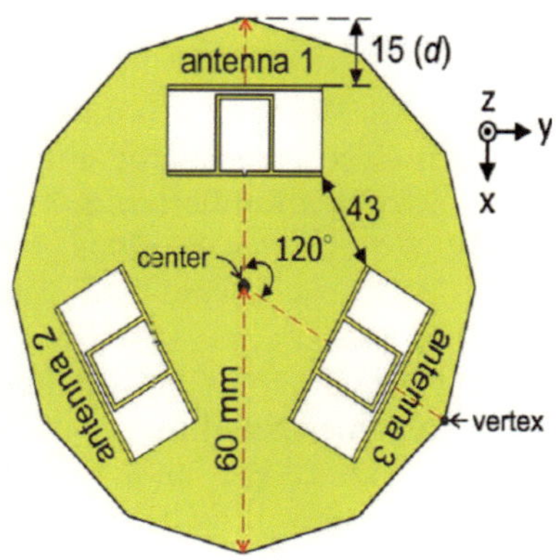

(b)

Fig. 24. (a) Configuration of the proposed, high-gain, dual-loop antennas for MIMO access-point applications (b) Top view of the proposed, three-antenna MIMO system (Su 2010)

Another novel, high-gain, dual-loop antenna design applied to a three-antenna system for MIMO access point (AP) applications is presented in Fig. 24 (Su 2010). The metal shape of the dual-loop antenna is configured to be affixed to the surfaces of a foam base occupying a moderate size, which allows the antenna to be surface-mountable on the ground plane and to be concealed in the casing of the AP at the height of 10 mm. The proposed design comprises two loop antennas of uniform width, namely a large 2.4-GHz outer loop and a small 5-GHz inner loop, both attached onto the rectangular foam base and operating at 1.0-wavelength resonant mode. Both loops also share common antenna feeding and grounding portions. However, the antenna shows peak gain of 7dBi over the operating bands.

From the literature review, it is realized that previously reported antennas limit good performances only to one frequency band or sometimes lack in consideration of compactness. This can be attributed to the conventional feeding techniques that the antennas are being fed. So it is necessitated to introduce a new feeding technique to have the best performances in the operating bands.

6. Conclusion

A review of RFID technology from the point of antenna specifications in presented in this chapter. The antenna theory is also described for proper convenience of antenna characterization. The parameters are mainly related to scattering parameters including return loss and VSWR as well as radiation characteristics like radiation patterns, antenna gain, polarization and so on. Moreover, a wide literature review has been done in order to identify the techniques to design multi-band microstrip antennas. Mostly dual-frequency operation is discussed since they mean the basics of multi-band operation. However, it has been seen that these techniques can be combined to enhance multi-band antennas with wider bandwidths. Finally, the high gain antennas and limitations have been described and it is realized that the conventional feeding technique might limit the performance of multi-band antennas to only one frequency.

7. Reference

Chen, H.-M., Wang, Y.-K., Lin, Y.-F., Lin, S.-C. & Pan, S.-C. 2009. A compact dual-band dielectric resonator antenna using a parasitic slot. IEEE Antennas and Wireless Propagation Letters 8: 173–176.

Ding, Y. & Leung, K.W. 2009. Dual-band circularly polarized dual-slot antenna with a dielectric cover. IEEE Transactions on Antennas and Propagation 57(12): 3757–3764.

Dobkin, D. M. 2007. The RF in RFID: Passive UHF RFID in Practice. Massachusetts: Newnes Newton.

Gao, S.-C., Li, L.-W., Yeo, T.-S. & Leong, M.-S. 2002. Small dual-frequency microstrip antennas. IEEE Transactions on Antennas and Propagation 51(1): 28-36.

Garfinkel, S., & Holtzman, H. 2005. Understanding RFID Technology. Upper Saddle River: Addison-Wesley.

He, X., Hong, S., Xiong, H., Zhang, Q. & Tentzeris, E.M.M. 2009. Design of a novel high-gain dual-band antenna for WLAN applications. IEEE Antennas and Wireless Propagation Letters 8: 798-801.

Heikkinen, J. & Kivikoski, M. 2003. A novel dual-frequency circularly polarized rectenna. IEEE Antennas and Wireless Propagation Letters 2: 330–333.

Intermec Technologies Corporation. 2006. RFID overview: introduction to radio frequency identification. Whitepaper, RFID Journal.

Jianhui, G., Shunshi, Z., Linglong, X. & Zhu, S. 2008. Dual-band monopole antenna for 2.45/5.8 GHz RFID applications. China-Japan Joint Microwave Conference, pp. 133–135.

Khan, M. A., Sharma, M. & Prabhu, B. 2009. A Survey of RFID tags. International Journal of Recent Trends in Engineering, 1(4): 68-71.

Kim, M.J. Cho, C.S. & Kim, J. 2005. A dual band printed dipole antenna with spiral structure for WLAN application. IEEE Microwave and Wireless Components Letters 15(12): 910–912.

Kim, M.J. Cho, C.S. & Kim, J. 2005. A dual band printed dipole antenna with spiral structure for WLAN application. IEEE Microwave and Wireless Components Letters 15(12): 910–912.

Kumar, G. & Ray, K.P. 2003. Broadband microstrip antennas. Boston: Artech House.

Lecklider, T. 2005. The world of the near field. Evaluation Engineering, http://www.evaluationengineering.com/index.php/solutions/emcesd/emerge

ncy-department-helps-nj-hospitals-achieve-top-ranking.html [04 September 2010]

Lin, C.-C. Su, C.-M., Hsiao, F.-R. & Wong, K.-L. 2003. Printed folded dipole array antenna with directional radiation for 2.4–5 GHz WLAN operation. Electronics Letters 39(24): 1698–1699.

Mitra A. 2008. A survey of recent patents on radio frequency identification systems and applications. Recent Patents Electrical Engineering 1: 40-6.

Moon, J.-I. & Park, S.-O. 2003. Small chip antenna for 2.4/5.8-GHz dual ISM-band applications. IEEE Antennas and Wireless Propagation Letters 2: 313- 315.

Nikitin, PV, Rao, KVS & Lazar, S. 2007. An overview of near field UHF RFID. IEEE International Conference on RFID, pp. 167-174.

Pozar, D. M. 2001. Microwave and RF design of wireless system. New York: John Wiley & Sons, Inc.

Raj, R.K., Joseph, M., Paul, B. & Mohanan, P. 2005. Compact planar multi band antenna for GPS, DCS, 2.4/5.8 GHz WLAN applications. Electronics Letters 41(6): 290–291.

Srinivasan, V., Kapur, R. & Kumar, G. 1998. MNM for compact dual frequency rectangular microstrip antenna. Proceedings of APSYM-98, pp.88–91.

Srinivasan, V., Malhotra, S. & Kumar, G. 2000. Multiport network model for chip-resistor-loaded rectangular microstrip antenna. Microwave and Optical Technology Letters. 24(1): 11–13.

Srinivasan, V., Ray, K.P. & Kumar, G. 2000. Orthogonal polarized microstrip antennas. Proceedings of NSAML-2000, pp. 43–46.

Su, S.-W. 2010. High-gain dual-loop antennas for MIMO access points in the 2.4/5.2/5.8 GHz bands. IEEE Transactions on Antennas and Propagation 58(7): 2412 - 2419.

Suh, Y.-H. & Chang, K. 2002. A high-efficiency dual-frequency rectenna for 2.45-and 5.8-GHz wireless power transmission. IEEE Transactions on Microwave Theory and Techniques 50(7): 1784–1789.

Volakis, J. L. 2007. Antenna engineering handbook. New York: McGraw-Hill.

Wang, H.-j., Wang, G. & Shu, Y. 2007 Design of RFID reader using multi-antenna with difference spatial location. Proceedings of WiCom International Conference, pp. 2070–3.

Wong, M., Sebak, A.R. & Denidni, T.A. 2007. Analysis of a dual-band dual slot omni directional stripline antenna. IEEE Antennas and Wireless Propagation Letters 6: 199–202.

Wu, J.-W., Hsiao, H.-M., Lu, J.-H. & Wang, Y.-D. 2005. Dual-broadband T-shaped monopole antenna for wireless communication. IEEE Antennas and Propagation Society International Symposium, pp. 470-473.

Wu, T.-Y., Fang, S.-T. & Wong, K.-L. 2003. Printed monopole array antenna for WLAN operation in the 2.4/5.2/5.8 GHz bands. Microwave and Optical Technology Letters 7(5): 370–372.

www.mapquest.com.

Xiao, Z.-h., Guan, Z.-q. & Zheng, Z.-h. 2008. The research and development of the highway's electronic toll collection system. Knowledge Discovery and Data Mining 3: 359–62.

Yang, G., Ali, M. & Dougal, R. 2005. A multi-functional stacked patch antenna for wireless power beaming and data telemetry. IEEE Antennas and Propagation Society International Symposium, pp. 359–362.

Yang, H., Yan, S., Chen, L. & Shi, H. 2008. Investigation and design of a modified aperture-couple fractal antenna for RFID applications. ISECS International Colloquium on Computing, Communication, Control, and Management, pp. 505 - 509.

Zhang, J., Zhang, X.-M., Liu, J.-S., Wu, Q.-F. Ying, T. & Jin, H. 2009. Dual-band bidirectional high gain antenna for WLAN 2.4/5.8 GHz applications. Electronics Letters 45(1): 6-7.

6

Low-Cost Solution for RFID Tags in Terms of Design and Manufacture

Chi-Fang Huang
Institute of Communication Engineering, Tatung University
Taiwan

1. Introduction

Even invented and applied initially during the World War II, RFID (Radio Frequency IDentification) technologies [1] have attracted much attention recently. Precisely speaking, RFID technologies have been applied very widely in some proprietary or closed systems, for example, animal control [2], portal control (access badges), etc. in last decades. The main advantages of RFID application are, storing item data in an electronic way even for further update, data access by electromagnetic wave in a wireless manner, and allowing quick multiple accesses to RFID tags. Based on the diverse applications, different spectrum bands are allocated, for example, LF (125 - 134.2 kHz and 140 - 148.5 kHz) for animal control, HF (13.56MHz) for electronic ticket, and UHF (868 MHz-928 MHz) for logistics, etc. Most of the frequencies are located in the ISM (Industrial, Scientific and Medical) bands [1].

However, RFID was emphasized again mainly because of the need of supply chain [3]. By proposing a standard for the format of electronic data used for goods items, of which EPC (Electronic Product Code) [4] is an example, the products can be registered at once when they are shipped out from the factories in one country, and be released when they are checked out at the counter of a supermarket in the other country in the world. These products might have been transferred through Customs of many countries and carried by different traffic means. When being through these check points, the related data stored in the tags are updated. This is called "product tracking" and is to be carried out in an "Internet of Thing (IOT)" [5].

This Chapter is to have a review on the technology theme – how to provide low-cost RFID Tags, when RFID technology is to be applied into the logistics area where the RFID tags are supposed to be not re-usable and to be as "zero-cost" as possible. Generally speaking, there are three major parts composing a RFID Tag's total cost, namely, antenna, chip and assembly for them. The cost of antenna, in addition to the design phase, is mainly dependent on the manufacturing process. Therefore, manufacturing process should be focused if antenna's cost, then the tag, is concerned. This is the theme of this Chapter.

Not like the other antenna applications, for example, wireless LAN or mobile phone, in which antennas need to be compliant to the end products' appearance by following the market trend. In the tag antenna industry, on the contrary, it does not need to design or modify the tag antenna often. The tag antenna just needs to electrically match the chip used in the beginning of design. It is not necessary for tag providers to prepare a wide product spectrum in the market. Again, not like the mobile phone industry, RFID tag's players just need few types of antenna to run their business. Therefore, they only need to pay their attention on the manufacture cost of tag antenna, because of the huge amount of worldwide supply.

For RFID tag chip, there is a key factor related to its cost-down, namely, reliability assurance. Since this kind of chip is very low-cost, possibly under sub-cent scale in the future, and is of huge amount in production, any means for total QC (Quality Control – checking any flaws in terms of chip's functions) in the manufacture procedure will raise their cost extremely. However, if not doing so, the risk of causing the chip silent or dead is very high, and under both of situation, the chip will not echo the reader's signal at all. Chip is always under high risk of being damaged from foundry to being packaged with antenna mentioned later. For example, electrostatics is one of killers, *i.e.* ESD (ElectroStatic Discharging) [6], in the whole procedure. Packaging the antenna and chip together is another potential bottleneck of the process of lowering the cost of a RFID tag, because that, both of production speed and reliable package is two important musts yet it seems a dilemma. Usually, this give hints that expensive and sophisticate machines are necessary, and that cost of each tag is raised again.

In this Chapter, focusing on the low-cost subject of RFID tags, the manufacture aspect of tag antennas is discussed. It has been believed that, applying the traditional printing technologies [7][8] to produce the antennas will lower the cost of the antenna part. One of the major efforts of this present work is to produce the tag antennas by traditional printing methods including offset printing, screen printing and a hybrid one based on gravure printing and vacuum deposition technology, to demonstrate the possibilities of making low-cost tags in high-volume. Fig. 1 is a demonstration of high-speed production of RFID tags by offset printing technology. There are several tens of printed tag antennas on each paper sheet.

Fig. 1. Demonstration of high-speed production of RFID tags by offset printing technology

Tags working both for UHF band [9][10] and HF band [10] are explained from the design phase to the performance evaluation in this Chapter. The designed passive tags of UHF and HF bands are to be responsible for the EM wave of 915MHz and 13.56MHz, respectively, from the reader.

Conclusively, this Chapter contributes to thoroughly outline the related issues and technologies for producing low-cost RFID tags. From the method details in design to the manufacturing technologies involved are mentioned and discussed. Specially focusing on the various printing technologies, the author explains the associated advantages and disadvantages when applying them from the point of industrial view. Moreover, the characteristics of used material are fully investigated and explained as for the design and production of this kind of low-cost RFID tags. To an engineer, the present content does provide a technical guide for the purpose claimed by the Chapter title.

2. Design of antenna for RFID tag

Referring to Figure 2, RFID tag antenna is a kind of planar antennas [11], in which the antenna metal layer is laminated on a dielectric substrate. Usually, even they look diverse in shape in RFID Tag industry; the type of dipole antenna [12] is used for the tags operating at frequency for UHF band and for higher bands. In designing such a kind of tags, the material parameters, for example, the conductivity σ of the antenna metal and the dielectric constant ε_r, are necessary to be given in the simulation phase. Usually, they are frequency-dependent, and practically, they should be given by real measurement in stead of consulting with literatures when the materials plus the used frequency are assigned. Measurement techniques for these two parameters are to be discussed later.

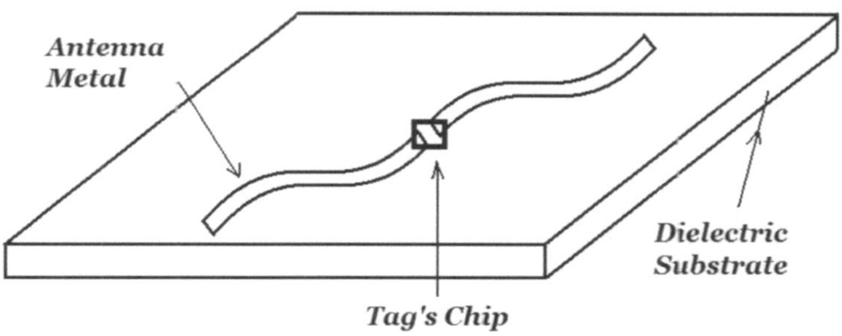

Fig. 2. The physical structure of a RFID tag.

The operation in a tag is that, the antenna receives the incoming EM energy and transfers into the chip; and chip sends back the data-modulated EM wave to the RFID reader. For passive tags, the chip specially makes use the incoming wave as the DC bias energy for itself in addition to interpreting the commands inside the wave from reader. As depicted in Figure 2, to ensure the efficiency of energy transfer in between chip and antenna, they should be in a "match" condition. In ordinary antenna industry, the antenna is designed with a standard input impedance, for instance, 50Ω or 75Ω, to have impedance match with transceivers or the other RF devices. However, in the RFID Tag industry, for the purpose of cost-down, usually the match network inside the chip is not offered. Consequently, it needs a complex conjugated matching [12] to ensure highest power transfer in between the chip and antenna, namely, to maximize the tag performance. Those two "X" marks show the input impedance positions of the chip and antenna on the Smith Chart in

Figure 3. Most of the cases, chip's is the lower "X", and antenna's is the other one. That means, usually the chip is capacitive; and the antenna for being designed should be inductive at the operating frequency. The present tag antennas are developed based on this fundamental theory.

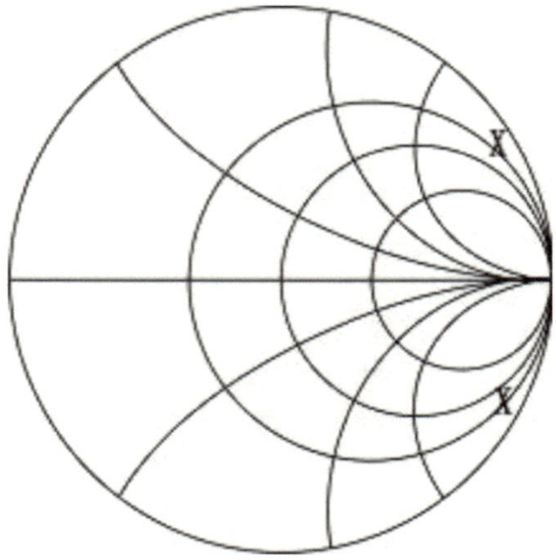

Fig. 3. Situation of complex conjugated impedance matching on the Smith Chart [12]

As an Electromagnetic design tool, CST [13] is employed to help design antenna prototype in this work. As mentioned above, dipole antenna is a good reference for designing RFID tag antennas, however, varied constraints may be usually applied for the commercial tags, for example, wider bandwidth, limited antenna size or different used materials, etc. Consequently, an antenna engineer actually has not many directions to design out a tag antenna, if he or she is not so experienced, even an expensive EM simulation package, say, CST, is available. Try-and-error approach is practical, but only for well-educated and experienced engineers, because he or she knows the antenna insight well. Under such a situation being lacking in much design experience, a systematic design methodology is probably useful.

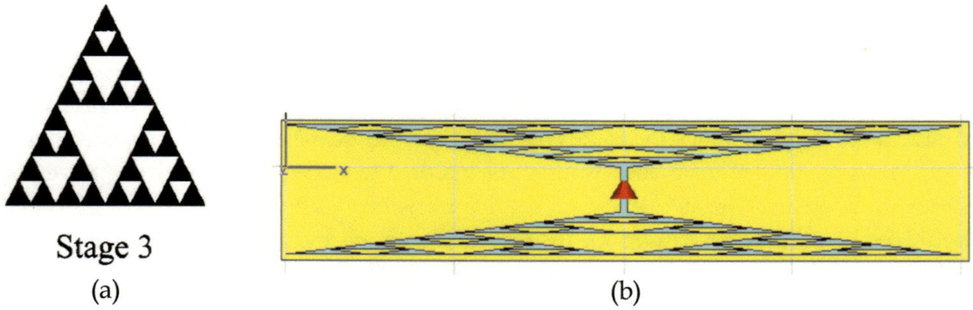

Fig. 4. (a) Sierpinski gasket fractal, (b)Simulation model of a tag antenna in the EM package CST

Antenna design based on fractals [7][14], see Fig. 4(a), has attracted attention recently in antenna industry or academics since it is quite easy to follow. Fig. 4(b) shows a simulation model of a tag antenna based on Sierpinski gasket fractals. In addition to generating fractals through different stages, the rectangular dimension of this tag is also under adjustment to search for the target input impedance of the antenna. A single RFID tag of UHF band designed by fractal methodology and made by offset printing technology is shown in Fig. 5. This tag antenna has also been printed by screen printing approach on PET (Polyethylene terephthalate). Usually, screen printing is able to offer thicker film and better performance, yet suffering with slower production speed.

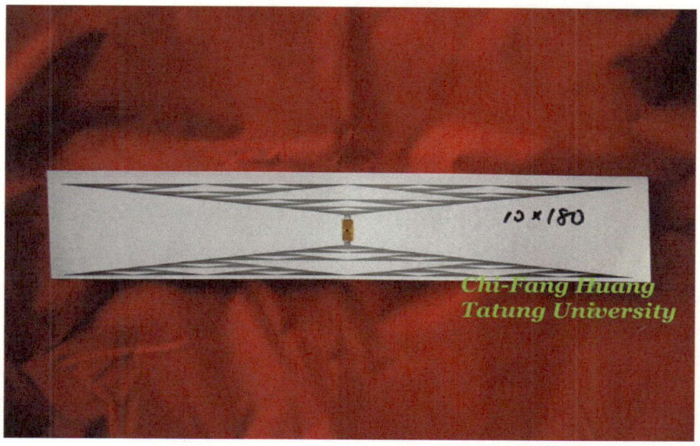

Fig. 5. A single RFID tag of UHF band made by offset printing technology

3. Review and application of the printing technologies for RFID tags

In the report [8], there have been many kinds of traditional printing technologies mentioned and discussed. For example, offset printing (lithography), flexography, gravure process, screen printing, etc. Each one has its unique advantages and associated drawbacks in terms of the combined factors of engineering and cost. For example, offset printing is fast, yet only provides thin printed layer not mentioning its expensive equipment investment. Fig. 6(a) shows an offset printing machine in a shop. Screen printing is usually considered to be capable of providing thicker layer, yet speed is not so competitive in production. In theory, the tag antenna should be full of metallic material to have highest receiving and radiating efficiency. However, constrained by the printing process, usually the ink used is with low conductivity (discussed below) because that the other non-conducting materials are added into ink. Fig. 6(b) shows its printing process [8].

Another issue is that, the printed layer provided by offset printing usually is of the order 1~2 μm which is not enough to be a good radiating metal for antenna considering the sufficient skin depth [12]. Fortunately, one can use the multi-stage of plate cylinders, see Fig. 6(b), and multiple printing procedure to increase the necessary thickness before the ink is not attachable. That means, there are three cylinders (three stages) at least in charge of three color inks in sequence in a normal printing machine, then the thickness increase can be achieved by putting the same conducting inks on the cylinders in different stages. If the thickness is still not satisfied after a printing running on the machine shown in the Fig. 6(a), feeding the printed sheets into machine from beginning again for multiple printing can be considered. Fig. 1 shows the resultant sheets by such an engineering approach.

(a)

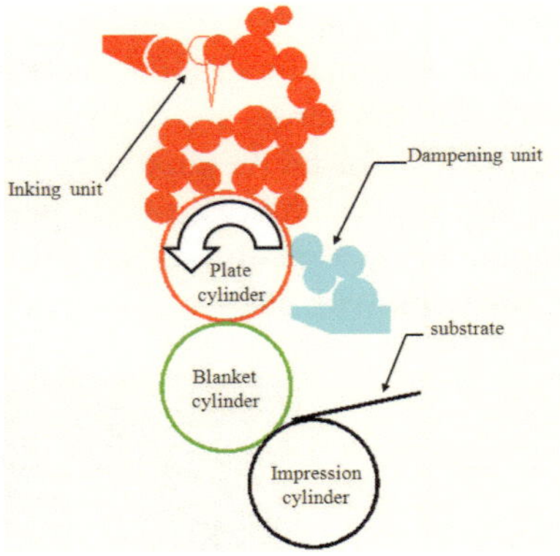

(b)

Fig. 6. (a) A high-speed offset printing machine; (b) the offset process [8]

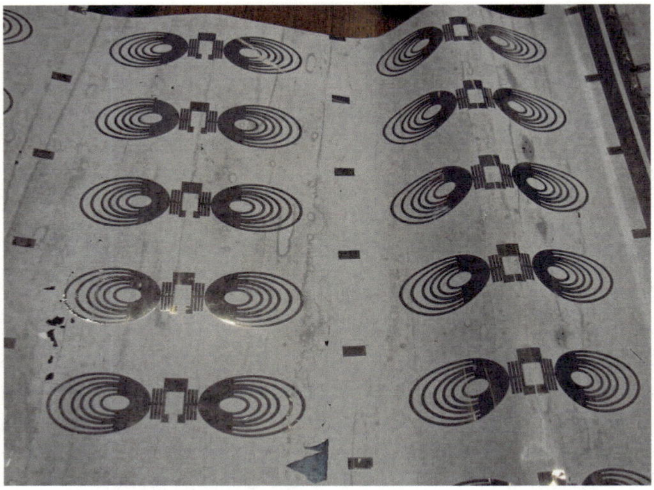

Fig. 7. A hybrid method with gravure printing and vacuum deposition technology

Traditionally, gravure printing is thought as a factory process for mass production of printing subjects on diverse substrates, for example, papers, plastics and metal films, etc. Furthermore, it is usually adopted to produce the goods bag; consequently, it seems a good idea that one can print the RFID tag on the bag with the same printing process to form a "smart bag". This is another thought of using traditional printing technology to promote RFID technology into the logistics, not mentioning the advantage of cost-down. A hybrid method with gravure printing and vacuum deposition technology has been proposed [10], in which the former is mainly to produce the printing mask and the latter functions to deposit metal film on the substrate. Such a method is implemented in a factory scale for mass production either producing tags only, see Fig. 7, or producing "smart bag" mentioned above.

Fig. 8 is a HF tag operating at 13.56MHz and is used to be embedded inside an ID card of students in Taiwan. It is made by such a hybrid process. Usually, the planar coil is used as the antenna structure for this band.

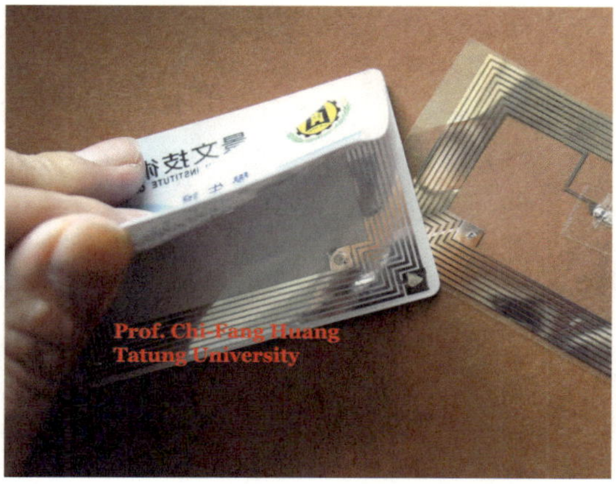

Fig. 8. A HF tag

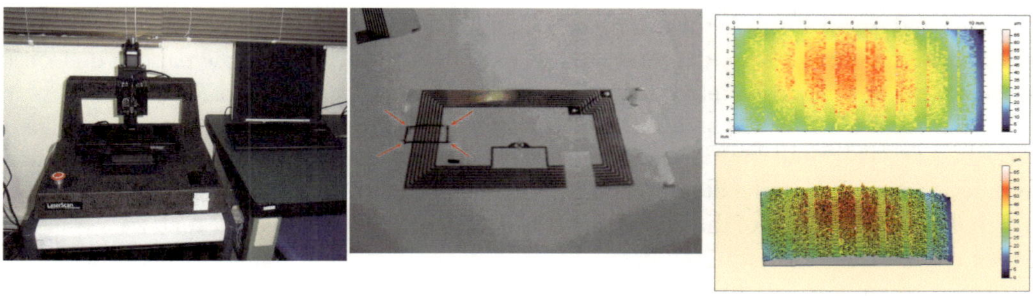

(a) (b) (c)

Fig. 9. (a) A confocal laser scanning microscope (b) Antenna film under measurement (c) measured thickness distribution

Unfortunately, this hybrid method is not able to offer thicker metal film as well, actually, what deposited is thinner, usually is about lower than half μm , even the layer is complete metal material. In industry, the thickness due to this process or by the other printing techniques all should be well monitored in terms of quality control. A confocal laser scanning microscope [15] has been suggested to measure the thickness of the RFID tag antenna made by this hybrid method as shown in Fig. 9(a). Fig. 9(b) is the antenna film under measurement and Fig. (c) shows the measured thickness distribution. It is indeed observed from Fig. 9(c) that requiring the uniformity of metal film is a main issue in this kind of production.

On the other hand, confocal laser scanning microscope is a kind of expensive equipment, on the contrary, economic ones for quick testing in manufacture lines are crucially necessary. A method of using the concept of eddy current [12] is also proposed [16]. Referring to the Fig. 10, a coil probe is designed to test the film sample which will affect the coil inductance because of the generation of eddy current on the circular conducting film. Such a deviation of inductance will be converted into a voltage reading by an electronic circuit to show the related thickness of printed film. This equipment and technique are very convenient for engineers to monitor the production line as for the film thickness from time to time.

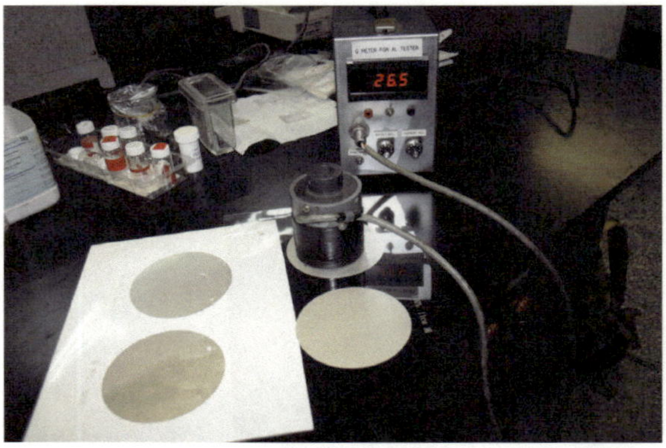

Fig. 10. An economic method to measure the conducting film's thickness

Material factors are very important in antenna design and should be studied thoroughly. Since there are two kinds of material being involved in the tag, and since this tag antenna is to be printed on a substrate, for example, the paper when using offset printing technology, before beginning the design, the conductivity σ of the conductive ink, the paper's dielectric constant ε_r and its associated loss tangent $\tan\delta$ should be given. The lithographic conductive ink used in this series of study of offset printing is CLO-101A purchased from Precisia LLC [17], and its corresponding conductivity σ was measured based on the techniques described in the literatures [18][19]. The measured conductivity is $3.85\times10^6\ S/m$, which is only 6.6% or so of the copper's $5.8\times10^7\ S/m$. As what expected, such a kind of ink is not as good as ordinary conductors to be antenna radiating material. This should be seriously taken into account when the tag performance is emphasized and they are produced by printing technologies.

(a) (b)

Fig. 11. (a) A resonating metal cavity following the theory in [19], (b) conducting ink on the wall

On the other hand, when applying the hybrid method of gravure printing and vacuum deposition technology, the different considerations are encountered. Firstly, PET (Polyethylene terephthalate) is always used as the antenna substrate for this method. Using the method mentioned in [20][21], Fig. 12 shows a closed metallic cavity, inside which the dielectric material under test is enclosed, for measuring layered PET's dielectric constant and loss tangent. The results are $\varepsilon_r = 3.733$ and $\delta = 0.0158$, respectively. On the other hand, the measured dielectric constant ε_r of paper used for offset and screen printing is 2.83, and $\tan\delta$ is 0.046 around the frequency 915MHz. This shows that the paper is with more loss than PET and should be carefully considered. That means PET is better than coated paper as the substrate of the tag antenna. Anyway, PET has an environmental pollution issue, if the printed tags are to be used for logistics. Also, even the vacuum deposition technology is usually not able to provide enough thickness of conducting film as the radiator of tag antenna, $1\,\mu m$ or so in our realization shown in Fig. 7 and Fig. 8, but it has equal conductivity as what the aluminum has. It has been found that, the performance made by it is quite better than that of offset printing on papers.

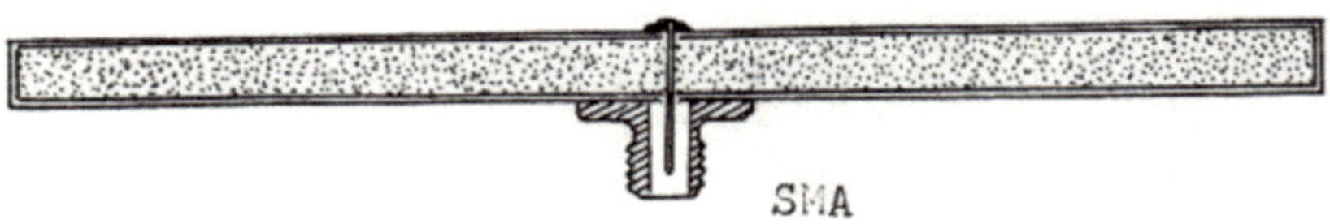

Fig. 12. Cavity method for measuring the PET's dielectric constant and loss tangent

Fig. 13. A tag using the company brand being antenna's arm

As for further application, usually text or company logo may be designed into the antenna shape. Following the idea published in [22], a tag antenna using the brand name of TATUNG COMPANY [23] is shown in Fig. 13, which is made by offset printing. Such a kind of design benefits the advantage without applying patent for the tag. However, because of the physical nature of antenna, for instance, its current distribution, normal computer fonts are not necessary to fit to the working shape of antenna.

Another example is shown in Fig. 14, where the logo of Taiwan Lamination Industries, Inc. [24], who is a gravure printing company, is to form one arm of the dipole antenna. This tag is made by the hybrid method of gravure printing and vacuum deposition technology, and produced by Taiwan Lamination Industries, Inc. TI's RFID chip [25] is used for this UHF tag shown in Fig. 14, which has input impedance $380 - j62.12\Omega$. Hence, the target impedance for the antenna is $380 + j62.12\Omega$ for a complex conjugated matching condition in theory. The simulation model established in the CST package for this tag antenna is shown in Fig. 15.

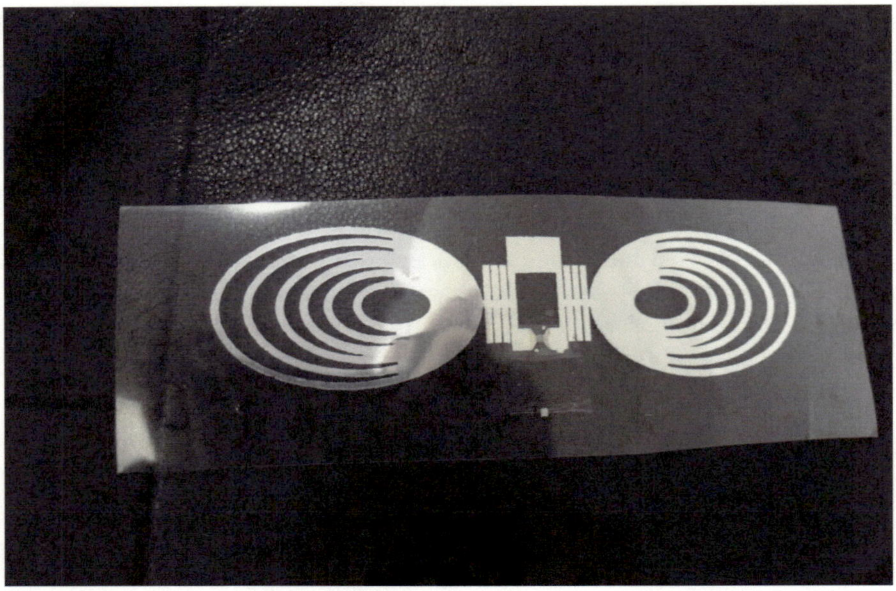

Fig. 14. A tag antenna using a company logo

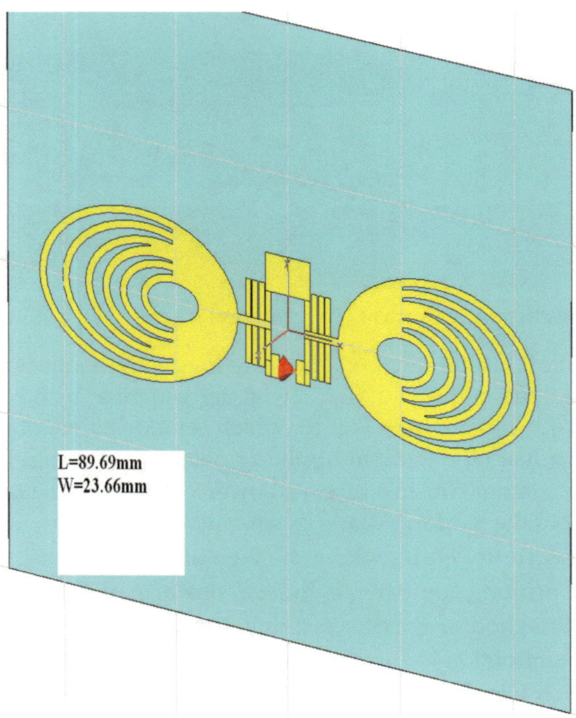

Fig. 15. Simulation Model of a UHF tag antenna

4. Performance analysis

As an example, back to the tag shown in Fig. 14 which is made by the hybrid method of gravure printing and vacuum deposition technology and has a size $85.8mm \times 23mm$, it can have a reading distance about 5 m when the measurement is carried in an antenna anechoic chamber in Tatung University. The tag shown in Fig. 5 has a dimension $10mm \times 180mm$. The reading distance of tags made by offset printing is always less than 2m. Less conductivity σ of conductive ink, thinner printed ink's layer and higher loss in substrate (coated paper) indeed make the tags produced by offset printing technology less efficiency. Anyway, both of these two different approaches have unique advantage of being able to produce tags in high-speed and in high volume, yet being low-cost. Anyway, sometimes the reading distance is not the absolute criterion to judge the tag performance. If the application focuses on the aspect of cost than the reading distance, the tags produced by the offset or screen printing on paper are more preferred.

5. Value-added application for RFID tags

As mentioned above, gravure printing is usually employed in making plastic bags, see Fig. 16. The concept of "smart bag" may be presented if the production both of bag and RFID tag can be combined together. Fig. 17 shows a new concept of embedding a RFID tag into the layer of a bag to form a "smart bag". In such a value-added application, however, some limitations should be considered. For example, thin metal foil and lossy paper (say, lossy Kraft paper) are not proper as the cover layers of the bag, because of their influence on the UHF wave transmission.

Fig. 16. Process of bag production in a gravure printing factory

Fig. 17. "Smart bag" – embedding a RFID tag into a plastic bag

6. Conclusion

This Chapter has outlined and demonstrated a complete procedure by which the offset printing technology or the hybrid method of gravure printing and vacuum deposition technology is applied to produce high volume and low-cost RFID tags. Based on the concept of complex conjugated matching, the design for tag antenna by the help of the EM simulation package is explained firstly. To precisely design the antenna by computer simulation, the techniques of measuring material parameters are also applied to obtain those parameters of conductive ink, paper and PET substrates. By the up-to-date offset printing and gravure printing and vacuum deposition machines, the tag antennas had been printed out by a high-speed manner to demonstrate its possibility to be a low-cost product.

7. Acknowledgements

This series of RFID tag project was initially granted by Tatung Company [23], Taipei, TAIWAN, who plays the main role offering long-term support for the academic-industrial projects being carried on in Tatung University, and then Taiwan Lamination Industries, Inc. [24], who is a gravure printing company and is involved now in the development of PET-based printed tags and "smart bags" mentioned above. The interactive experience between the authors and managers of this company has generated much new knowledge of the hybrid method of gravure printing and vacuum deposition technology. Both of these two companies are appreciated. Sun Sui Print Co., Ltd [26], Taipei, TAIWAN, is appreciated for their kind support to provide the offset machines in printing the RFID tags designed in the present work. Moreover, we want to specially thank Mr. Wen-Ho Wu, the factory manager of this company. Without his professional guide in the offset printing procedure, this present work would not be done completely.

8. References

[1] Klaus Finkenzeller, RFID Handbook: Fundamentals and Applications in Contactless Smart Cards and Identification, Wiley & Sons Ltd. New York, 2nd edition, 2003

[2] Q. Zong and W. Bao, "The dairy cattle data acquisition system based on PDA," World Automation Congress (WAC), 2010

[3] R. Bansal, "Coming soon to a Wal-Mart near you," IEEE Antennas Propag. Mag., vol. 45, pp. 05–106, 2003

[4] S. Sarma, D. Brock, and D. Engels, "Radio frequency identification and the electronic product code," IEEE Micro, pp. 50-54, 2001

[5] Z. Song, A. A. Cárdenas and R. Masuoka, "Semantic middleware for the Internet of Things," Internet of Things (IOT), pp. 1-8, 2010

[6] C. Duvvury, "ESD: design for IC chip quality and reliability," Proceedings on IEEE 2000 First International Symposium on Quality Electronic Design, ISQED 2000, pp. 251 - 259.

[7] Chi-Fang Huang, Jing-Qing Zhan and Tsung-Yu Hao, " RFID Tag Antennas Designed by Fractal Features and Manufactured by Printing Technology," The 1st International Workshop on RFID Technology - Concepts, Applications, Challenges Workshop, Funchal, Portugal, June, 2007

[8] Anne Blayo, and Bernard Pineaux, "Printing Processes and their Potential for RFID Printing," Joint sOc-EUSAI conference, 2005

[9] K. V. S. Rao, P. V. Nikitin, and S. F. Lam, "Antenna Design for UHF RFID Tags: A Review and a Practical Application," IEEE Trans. Antennas and Propagation, Vol. 53, No. 12, pp. 3870-3876, 2005

[10] Sung-Fei Yang, Design of RFID Tag Antenna Based on Gravure Printing and Vacuum Deposition Technology, Master Thesis, Tatung University, July, 2007

[11] J. R. James, P. S. Hall and C. Wood, Microstrip Antenna, IEE Electromagnetic Waves Series 12, 1981

[12] David K. Cheng, Field and Wave Electromagnetics, Addison-Wesley, 1992, 2nd Ed..

[13] http://www.cst.com

[14] Douglas H. Werner and Suman Ganguly, "An Overview of Fractal Antenna Engineering Research," IEEE Antennas and Propagation Magazine, Vol. 45, No. 1, pp. 38-57, 2003

[15] A. Diaspro, S. Annunziata, M. Raimondo, P. Ramoino and M. Robello, "A Single-Pinhole Confocal Laser Scanning Microscope for 3-D Imaging of Biostructures," IEEE Engineering in Medicine and Biology Magazine, Vol. 18, Issue: 4, pp. 106 – 110, 1999

[16] Yueh-Ching Lin, Design of Logo-Based Tag Antennas of RFID, Master Thesis, Tatung University, July, 2008

[17] http://www.precisia.net

[18] Tom Y. Otoshi, and Manuel M. Franco, "The Electrical Conductivities of Steel and Other Candidate Materials for Shrouds in a Beam-Waveguide Antenna System," IEEE Transactions on Instrumentation and Measurement, Vol. 45, No. 1, pp. 77-83, 1996

[19] R. Clauss, and P. D. Potter, "Improved RF Calibration Techniques - A Practical Technique for Accurate Determination of Microwave Surface Resistivity," JPL Technical Report 32-1526, Vol. XII, pp. 59-67.
[20] W. F. Richards, Y. T. Lo and J. Brewer, "A simple experimental method for separating loss parameters of a microstrip antenna," IEEE Trans. Antennas Propagat., vol. AP-29, pp. 150-151, 1981
[21] Chi-Fang Huang, "A Cascaded 2-D Array of Microstrip Antenna," Tatung Journal, Vol. XIV, pp. 69-83, 1984
[22] M. Keskilammi and M. Kivikoski,"Using Text as a Meander Line for RFID Transponder Antennas," IEEE Antennas and Wireless Propag. Letters, Vol. 3, pp. 372-374, 2004
[23] http://www.tatung.com
[24] www.twn-lami.com.tw
[25] http://www.ti.com
[26] http://www.sunsui.com.tw/

7

Conductive Adhesives as the Ultralow Cost RFID Tag Antenna Material

Cheng Yang[1, 2] and Mingyu Li[3]
[1]*Department of Mechanical Engineering*
The Hong Kong University of Science and Technology
[2]*Tsinghua University the Graduate School at Shenzhen*
[3]*School of Materials Science and Engineering*
HIT Shenzhen Graduate School
China

1. Introduction

Radio Frequency Identification (RFID) has rapidly expanded its market in recent years; until 2019, the market volume of RFID will probably reach 3.9 billion USD globally (for those passive tags). It will replace barcodes and find a lot more applications where barcodes cannot do today.[1] RFID takes the advantages such as the high-speed scanning, miniaturized size, high reliability, high memory volume, safe, and excellent read accessibility, as compared to barcodes. However, the high materials and fabrication costs are the major bottleneck for wider applications. Currently, the cost of chip is still the major part of the overall cost of a tag, which contributes about 30% to 70% of the total cost of a tag. The rest part is the sum of the materials cost including the antenna, substrate, and that for integrating them together. Since the cost of the chip keeps dropping due to the technical development, the need for reducing the other parts now is more urgent than ever. Therefore, it becomes a challenging part nowadays for reducing the cost of antenna, which takes the highest mass weight of all electrical components.

Currently, there are several alternative fabrication methods of the RFID tag antennas. For example, there are etched/punched antennas, wound antennas, which are based on metallic foils and printed antennas, which are based on the electrically conductive adhesives (ECAs). Even though each method has its pros and cons, printed antennas are currently regarded as the most promising one, primarily due to both productivity and cost concerns. Moreover, printability renders the antenna fabrication process integrated into the whole tag manufacturing system,[2] especially suitable for mass production of the RFID tags. It will also be indispensible for manufacturing the chipless tags, which eliminates the silicon chip from the tag, not only for saving cost, but there would be other benefits such as thinner in shape and more environmentally benign. However, printed RFID tags often have a shorter life-time than the etched tags (life span of more than ten years), which causes limitations in such as passports requirements; there are also concerns about the read range, which is related to the relatively low electrical conductivity. Thus there are still a lot of rooms for improvement for the printed materials.

There are a few alternative printing techniques which are applicable for printing the antenna materials (such as gravure, screen, roll to roll, flexography, and stencil etc.), which are briefly shown in Fig. 1. Here this chapter primarily elucidates the works which are based on the (flat-bed) screen printing method, which is very representative at the stage of lab prototyping. Screen printing is a low cost printing technique which has a very long history; as firstly appeared about 2000 years ago in the Qin dynasty in China. Screen printing technique uses a woven mesh to support an ink-blocking stencil. The attached stencil forms open areas of mesh that can transfer ink or other printable materials which can be pressed through the mesh as a sharp-edged image onto a substrate. A roller or squeegee is moved across the screen stencil, forcing or pumping ink to pass through the open areas in the woven mesh. There is a wide range of screen materials which include steel, polyester, glass fiber, silk fiber, and nylon etc. They form a smooth, porous, finely woven fabric which is stretched over a wood or aluminium frame. Areas of the screen are blocked off with a non-permeable material as a stencil. The open spaces of the screen allow the ink appear on the substrate beneath the screen. Generally, screen-printing method can render the printed resolution to be about 100 microns and above, which is determined by many factors such as the selection of the material of the screen mask and the automatic control of the processing conditions. The screen mask can be conveniently prepared by the ordinary photolithography method, thus it is a very promising and competitive printing method for producing the ultralow cost RFID antennas and even tags for prototyping. Moreover, screen printing can work on a large range of substrate materials such as textiles, ceramics, woods, papers, glasses, metals, and plastics. Fig. 2A shows a worker in a label printing company in Dongguan, China, whom is working on a flat-bed screen printer. The RFID tags printed in this way is shown in Fig. 2B. There were a few layers of inks including the hot-melt adhesive layer, the ECA layer, and the ink layers which were printed onto a piece of PET film consecutively. Then the printed pattern was heat-transferred onto a piece of fabric sample, which underwent dozens of washing cycles (e.g. 40 cycles) for evaluating the reliability of the sandwiched RFID tags. [3]

As the major component for the printed RFID antenna, ECAs are composed of two major parts: one is the conductive filler, such as silver, copper, and nickel; the other is the nonconductive polymer resin, which can be epoxy, polyester, polyurethane, ceramic, and other dispersants which can fit for the printing condition and some other factors. Nevertheless, high electrical conductivity of the printed antenna material is indispensible, so that the read range performance can match most of the applications of the tag.[4] Among all available printed materials including metals, carbon, and intrinsically conductive polymers, silver is considered as the most promising one, due to its high electrical conductivity (6.2×10^5 S/cm, which is the highest among all metals), relatively low material cost, and excellent reliability in long-term uses without the concern of electrochemical etches. Silver fillers are usually ground into micron-sized flakes when they are mixed with the resin dispersant; thus the overall electrical conductance of the ECA is not only determined by the intrinsic conductivity of silver, but also by the percolation effectiveness among them.[5] To improve the percolation of the silver fillers in the ECAs for practical uses, silver flakes with the diameter ranging from 30 micron to 3 micron are usually selected, which can be conveniently fabricated by mechanical machining methods such as ball-milling etc.[6] The anisotropic morphology renders the silver fillers more easily build up associated network inside the resin dispersant so that the percolation threshold (the minimum filler content requirement for achieving ohmic conductance) of the filler can be decreased.[7] Further decreasing the size of the fillers inevitably increases the viscosity of the filler-dispersant mixture, which may cause problems during printing.

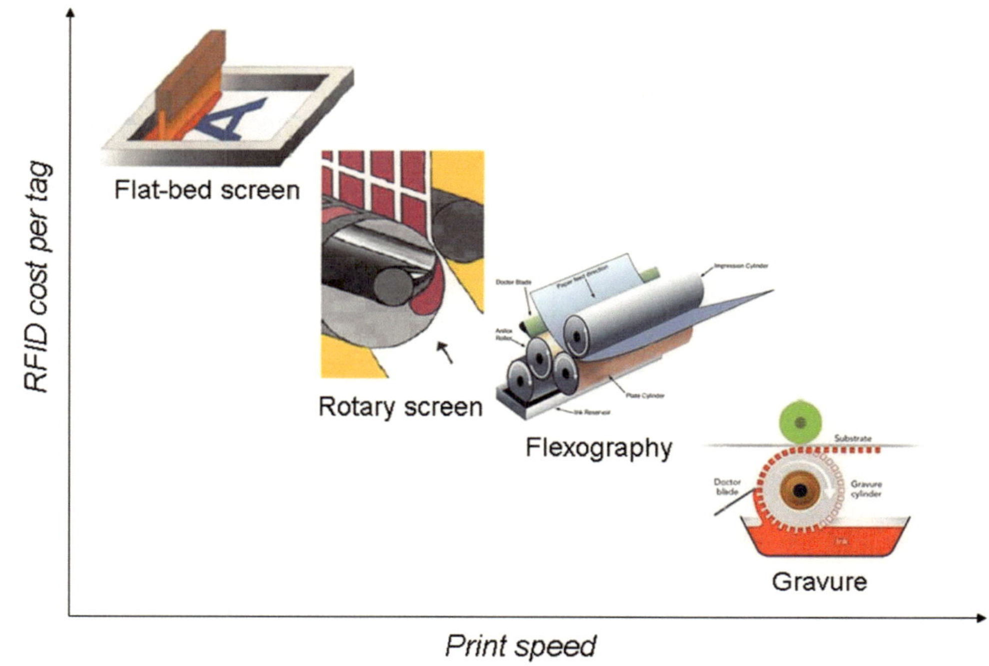

Fig. 1. A schematic comparing the printing speed and the RFID cost per tag.

Fig. 2. Photographic images showing the RFID tag incorporated high reliability hot-press labels for garments. A) A worker is screen-printing the ultralow cost ECA based antenna in his work line in a company in Dongguan, China; B) A group of labels ready for heat-transfer printing; C) Samples cotton fabric pieces with the labels heat-transferred onto them. (Upper: before washes; bottom: after washing for forty cycles.) The RFID read range performance remained the same after the heat-transfer process and the subsequent washing cycles.

There have been intensive studies about the ECAs in the last two decades,[8] majorly considered as the substitute for the Sn/Pb eutectic solders as an interconnect material in the traditional electronic packaging industry. This is not only because they have fewer troubles about environmental problem (no lead is involved), but they have lower processing temperature and more convenient processing procedures (the curing temperature of ECAs is normally lower than the melting point of the eutectic solders, i.e. 183 ^{o}C). However, a simple mixture of the conventional resin dispersant, such as bisphenol-A type of epoxy resin and silver fillers such as microflakes (at 75% by weight) can often result in the electrical resistivity of the ECA in the range of 10^{-4} Ω ·cm. As compared to the Sn/Pb eutectic solders, the electrical conductivity of the ECAs needs to be improved to cater for general application of electrical devices.

As a noble metal, silver suffers less from the electrochemical etching problem than many others such as copper and nickel etc. However, ECAs filled by silver flakes still exhibit a high contact resistance due to a variety of factors, including the contamination from the impurities and additives of the resin dispersant (such as the free radicals from the initiator, the organic ligands from the curing agents etc.). Moreover, silver oxide exhibits a very high electrical resistivity (i.e. about 10^{16} times higher than pure silver).[9] Unlike the eutectic solders, which have a much lower melting point, the melting point of silver is 962 ^{o}C, which makes it very difficult to be annealed or sintered by conventional processing conditions. Early studies majored in those methods which can improve the physical contact among the silver fillers;[10] for example, by selecting the highly contracted resins,[10] or through applying an additional hot-laminating step after curing the ECAs.[11, 12] These strategies were shown to be able to reduce the bulk electrical resistance of the printed ECA irreversibly.[13]

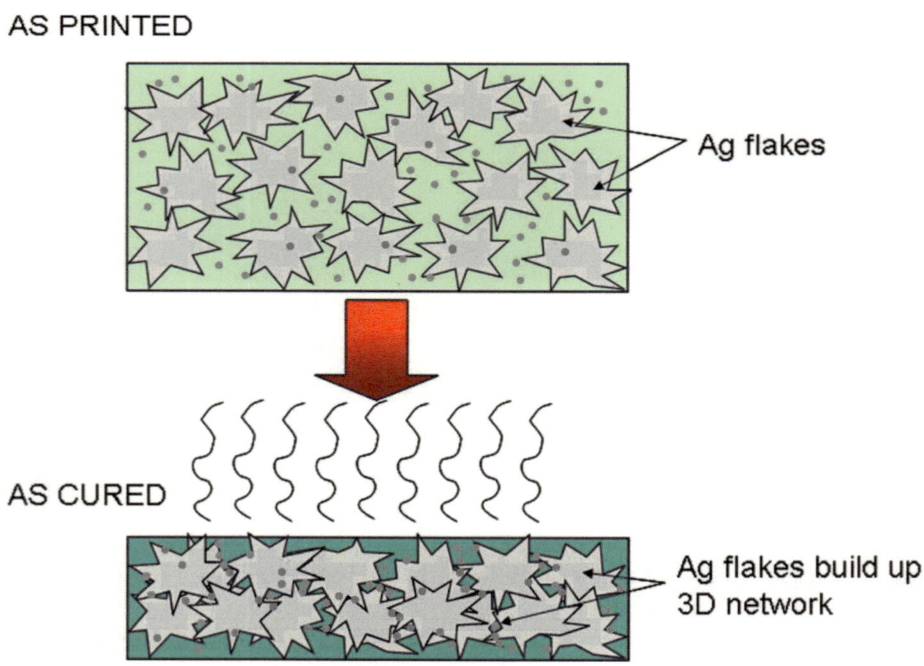

Fig. 3. A schematic showing the influence of the curing step of the ECAs, which is critical to the percolation of the fillers.

2. Recent progress of silver filler modifications

In recent years, Wong et al. conducted the researches on the self-assembled monolayer (SAM) protecting layers to the silver fillers. By seasoning a small quantity of the organic molecules (usually those which can form ligands with the metallic fillers) into the ECA formulations, the electrical resistivity of the ECAs can be drastically reduced.[14] The mechanism is rather complicated, which is supposed to be related to the red-ox process of the silver surface. Some of the SAM molecules exhibit a certain level of reducing property.[15, 16] There is a large range of the feasible compounds, including malonic acid etc., which can be used for this purpose.[15, 17, 18] On the other hand, Jiang et al. studied the effect of adding a certain ration of nano-sized silver particles to supplement the silver micro-flake fillers. By adding 40 wt% and 60 wt% of the nanosilver and microsilver, at 80 wt% filler content level, the electrical resistivity of a modified formulation can achieve ~5 x 10^{-6} Ω cm.[17] It was anticipated that the silver nanoparticles can benefit from the melting point depression effect due to the small size. Thus the silver fillers fuse with each other and build up a percolated network through ohmic contact.[19] Consequently, the electrical conductivity of the composite material approaches the lower limit of the conductive-nonconductive mixture (as shown in Fig. 3).

Yang et al. recently worked on a novel method to achieve better percolation of the silver fillers. An iodination step is applied to the silver microflakes prior to the mixing step of the ECAs, so that the electrical conductivity of the silver based ECAs can be significantly improved.[20] Silver has a strong interaction with iodine and the reaction results in the formation of silver iodide and some other compounds. Silver iodide is a semiconductive material which has indirect band-gap; the size of the silver cations is much smaller than that of iodide.[21] Silver cations can conveniently move around through the interstitial sites so as to exhibit a certain superionic conductivity.[21, 22] On the other hand, the solution-based silver microflake treatment process can eliminate the oxide layers from the silver surface.[23] After the iodination process, those iodinated regions occupy active sites such as the terraces and steps of the silver surface more selectively, and experience a subsequent ripening process,[24-27] leaving the remaining part a clean silver surface due to an electrochemical process,[26, 27] although the dynamic process still needs further investigation.

The reaction between the solid (Ag) and solute (I_2) is partially determined by the diffusion function, thus the resulting iodinated surface layer exhibits a level of nonstoichiometry. This part appears in the form of nano-islands, which are distributed on the silver flake surface. For example, TEM-EDS and SEM-EDS (Fig. 4) results both suggested that the nano-islands are distributed very sparsely on top of the silver flakes, which suggests that under optimum conditions for the best conductivity (i.e., when filled with A3) and there are excess amount of silver inside the nano-islands. The excess silver can actively involve in the charge transfer process and facilitates the reconstruction of the silver surfaces.[21]

As shown in Fig. 5A, four groups of samples were analyzed by TOF-SIMS: (1) bare silver wafer, (2) sparsely covered by the nano-islands (**1**: Ag : I = 100 : 0.2) (resembling to the surface of A3), (3) moderately covered by the nano-islands(2: Ag : I = 100 : 0.4) (resembling to the surface of A9), (4) fully iodinated surface (3: Ag : I = 100 : 20), respectively. The sum of the relative peak intensities of $^{107}Ag_2OH^+$ and $^{107}Ag_2O^+$ over that of the silver base peak ($^{107}Ag^+$) is used as the index to demonstrate the overall oxidation level of the surface. After experiencing the curing and purging processes, the surface oxidation level of the unmodified bare silver sputtered wafer samples increased 15.5%, which suggests the oxidation of the silver surface in the curing process in the absence of Ag/AgI nanoclusters. While the surface oxidation level of the modified silver decreased 60.4% in condition **1**, and 54.3% condition **2**, respectively. This is direct evidence that the Ag/AgI nanoclusters on the silver surface prevented the silver metal surface from oxidation in curing process. But on the sample which was fully iodinated (condition **3**), the curing process incurred an increase of the total oxidation level. Since the ratio of ($^{107}Ag_2OH^+$ + $^{107}Ag_2O^+$)/$^{107}Ag^+$ is an index of the overall oxidation level of the sample surface, it appears that the less the nanoclusters covering the surface, the more fragment signals from the exposed silver metal surface were collected. For those samples with low and medium coverage levels of nanoclusters (condition 1 and 2), after curing, the overall oxidation levels were lowered by 60% and 54%, respectively. Considering the surfaces of these two samples were partially covered by the nanoclusters, after the curing process, the oxidation of the silver surface (except for the nanoclusters) was greatly inhibited. It suggests that during the curing process, the nanoclusters influence oxygen adsorption on the silver surface and recover the part of the oxidized surface. This phenomenon may be attributed to excess amount of silver in the nanoclusters, which exhibit stronger reducing property than the bulk silver substrate.[28, 29]

Fig. 5B demonstrates the situation of the silver surface when it is saturated by iodine treatment (Ag : I = 100 : 20). We tentatively partitioned the depth into two regions to facilitate the study of this spectrum: The left side illustrates the region of the nano-islands and the right side the region of the silver metal. In Fig. 5B, this ratio ($^{107}AgIO^-$/$^{107}Ag^-$) decreases with the sputtered depth, showing that the deeper the sputtering the stronger the collected substrate signals (i.e. $^{107}Ag^-$). After curing, this ratio ($^{107}AgIO^-$/$^{107}Ag^-$) increased at the sample surface which is about several nanometers in depth. For example, at the depth of ~7 nm, it is 1.1 times higher than the ratio of the sample before cure, showing that the nano-islands are further oxidized after cure. This is quite different from the TOF-SIMS analysis on a control sample of pure silver iodide crystalline powder (Aldrich, [7783-96-2], 99.999%), which shows negligible $^{107}AgIO^-$ peak intensity (ratio $AgIO^-/Ag^-$ = 6.3 ± 0.88%). As an unstable and naturally rare substance, the observation of a large quantity of silver hyperiodite ($^{107}AgIO^-$) anions in the TOF-SIMS spectra indicates that in the nanocluster regions a large amount of oxygen incorporates into the Ag/AgI nano-islands.[24, 26, 28-30] Comparison of the spectra before and after the mimic curing process demonstrates that the nano-islands are reactive to ambient oxygen and the curing process can accelerate the oxidation process. The inter-conversion between AgI and $AgIO_x$ (x = 1, 3) species has been demonstrated to be a complicated charge transfer and oxidation process which is related to many factors.[31, 32] The redistribution of the silver surface species could alter the path of oxidation of the silver surface, which may play a key role in reducing the contact resistance of the silver microflake network in the ECAs. Both the concentration and amount of iodine are crucial factors in determining the coverage and morphology of the nano-islands on the silver surface. The experimental results suggested that the coverage of these nonstoichiometric nano-islands plays a key role in modulating the surface property of silver. The ECA samples filled with A3 showed the highest electrical conductivity among all listed conditions e.g., A1, A4, A5, A6, and A9, etc., as shown in Fig. 6 (this figure only shows the resistivity data of the ECA samples lower than 10^{-3} Ω·cm). The A3 filled ECA has a volume resistivity of 5.92 x 10^{-5} Ω·cm with a silver filler content of 40 wt% (6.5 v/v%). The volume resistivity increased to 4.81 x 10^{-4} Ω · cm when the silver filler content decreased to 27.5 wt% (3.8 v/v%). Further reduction of the filler content resulted in higher and unstable resistivity. For example, the resistivity of the ECA filled with 70 wt% of A1 is only 1.51 x 10^{-4} Ω·cm (not shown in this figure), and filled with 60 wt% of A5 is only 2.99 x 10^{-4} Ω·cm. While the resistivity of the ECA filled with 70 wt% of A3 is 6.90 x 10^{-6} Ω · cm, and filled with 60 wt% of A3 is 1.13 x 10^{-5} Ω·cm. When further decreasing the content of A3 in the ECAs to be lower than 27.5 wt%, i.e., 27 wt%, 26 wt%, 25 wt% etc., from the SEM analysis of the cross sections, sedimentations of the fillers were observed, which is due to the mismatch of the density between silver micro-flake and the epoxy resin. These sedimentations denote that when the silver filler content is lower than 27 wt%, the silver fillers can not form an associated network, which is crucial for electrical percolations. Even though this sedimentation effect may have problems in omnidirectional percolation; experimental evaluations suggest that the ultralow filler content ECAs all exhibit excellent 2D electrical conductivity in the form of printed thin film resistors.

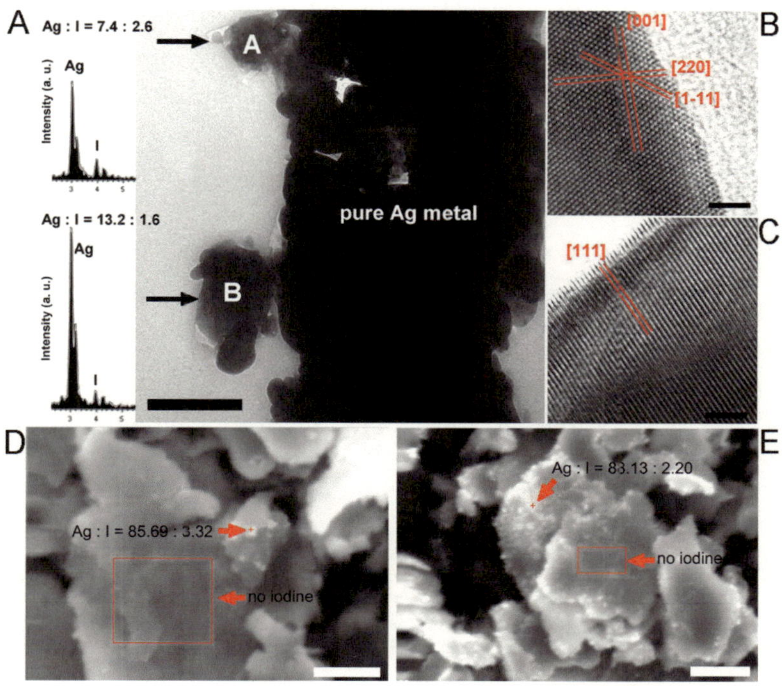

Fig. 4. A)-C): TEM-EDS analysis of the ECA cross sections. A) TEM-EDS of the nano-islands on a sectioned ECA sample (filled with A9). (Scale bar = 200 nm) EDS spectra are accompanied on the left. B) HRTEM image of bare silver micro-flake surface. (scale bar = 2 nm) C) HRTEM image of A9 filled ECA surface, except for the nanocluster parts. (scale bare = 2 nm) The crystal lattice of silver metal is marked in both the images of (b) and (c). (All samples are embedded in a resin filled with 75 wt% of the filler.) D)-E): SEM-EDS analysis of the iodinated silver flakes. D) Sample A3, the elemental ratios bentween silver and iodine are listed in this image; (scale bar = 2.5 μm) E) Sample A9, the elemental ratios bentween silver and iodine are listed in this image. (scale bar = 2.5 μm) (Copyright © 2010 WILEY-VCH)

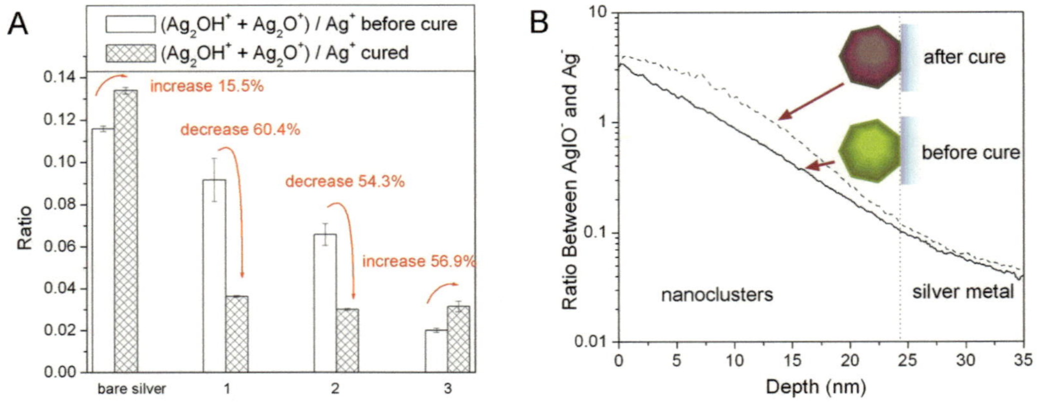

Fig. 5. TOF-SIMS analyses of the silver sputtered silicon wafer samples. A) Silver sputtered silicon wafers after treatment of different concentration of iodine solutions. (The cured samples refer to those experienced a curing and post-washing process prior to this analysis.) Conditions: 1. Ag : I = 100 : 0.2; 2. Ag : I = 100 : 0.4; 3. Ag : I = 100 : 20. B) Depth profile of the surface-modified sputtered wafer sample (treatment condition: Ag : I = 100 : 20; y-axis in logarithmic scale). (Copyright © 2010 WILEY-VCH)

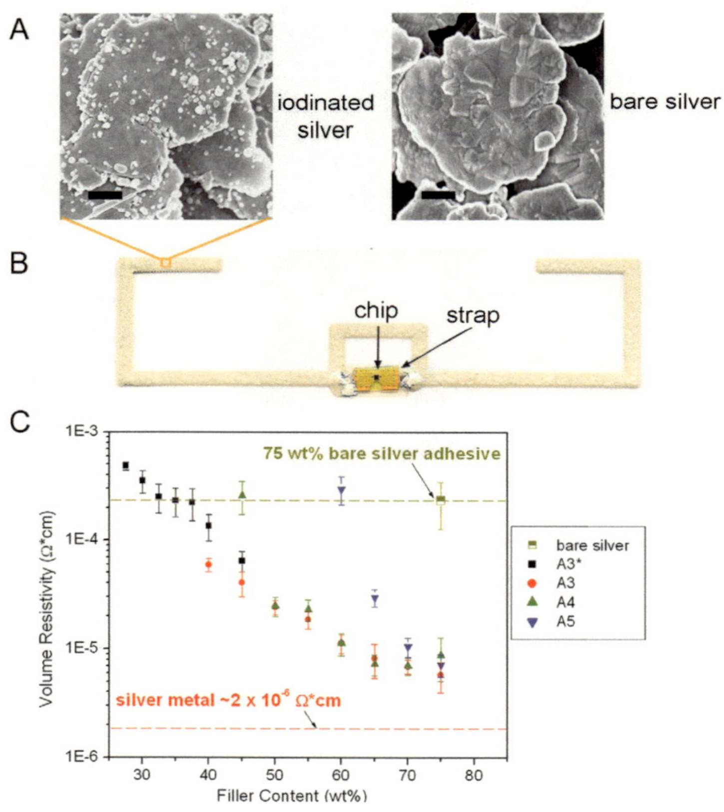

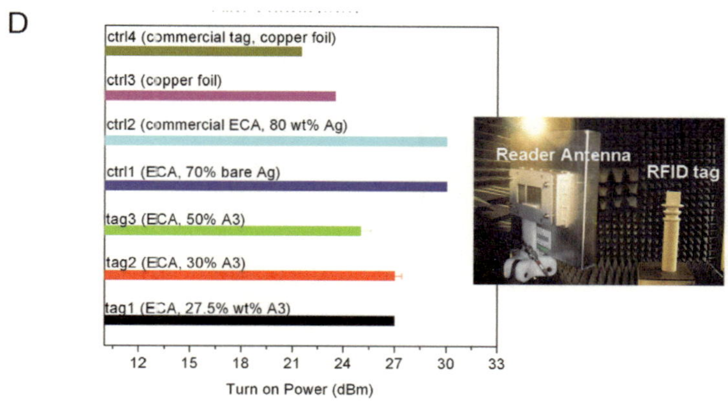

Fig. 6. A) SEM images of the iodinated silver micro-flake sample (left) and original bare silver micro-flake sample (right) (scale bar = 500 nm); B) A photographic image of a piece of UHF RFID tag using the A3-filled ECA as the antenna material. (A3, A4, and A5 are those silver fillers underwent different iodination conditions) C) The volume resistivity data of the modified ECAs with selected iodination conditions and a control bare silver-filled ECA (containing 75 wt% of silver filler). The partition lines drawn here is for comparing the resistivity of the modified ECAs versus the control silver adhesives (olive line) and silver metal (red line). (*This series of data of A3 filled ECAs are based on Novolac type epoxy resin to adjust the viscosity at low filler content.) A4: Ag : I = 0.5 : 100; A5: Ag : I = 1 : 100; D) Read range testing result of the RFID tags. Turn on power measurement on the A3 filled ECA RFID tag samples (tag 1, 2, 3) and the control samples (ctrl 1, 2, 3, 4). Except for tag2 and tag3, conformance was found when measuring all other tag samples. An EPCglobal Ultrahigh Frequency Class 1 Generation 2 RFID strap (Alien Technology Inc.) was adhered to the center of each tag antenna and the measurement was in a UHF RFID system (CSL CS461) in an anechoic chamber with a fixed reader-to-tag distance of 1 meter. Inset: A photographic image showing the turn-on power test.

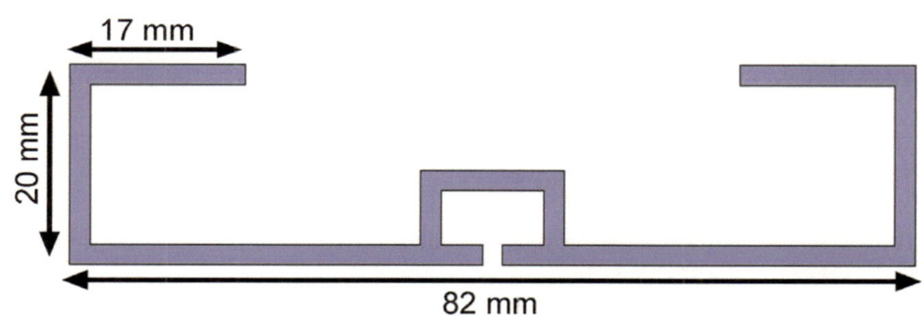

Fig. 7. Designed geometry of RFID tag antenna based on the ECA sample with 30 wt% silver.

3. Recent progress in the environmentally benign dispersants

Besides the electrical conductivity, there are also some other considerations which determine the overall performance of the ECAs. For example, the materials involved in the printed RFID must be safe to human health and environmentally benign. This issue is especially critical to the area of food packaging and logistics at item-level. Li et al. evaluated the performance of some biocompatible curing agents to substitute the toxic curing agents for epoxy resins.[33, 34] However, the use of conventional bisphenol-A type of epoxy resin still faces lots of environmental and health problems. Yang et al. recently investigated the blocked aliphatic polyurethane (PU) as the dispersant material for preparing the ECAs for the RFID application. Two types of the methyl ethyl ketone oxime (MEKO) -blocked polyurethane prepolymers were used as they have been evaluated by the European Food Safety Authority as a safe material for food can coatings recently.[35] Moreover, as the -NCO group of the resin is blocked, the ECAs exhibit a long shelf-life than the unblocked resins. On the other hand, their moderate viscosity (1000 cPa·s to 3000 cPa·s, dependent on the amount of the silver filler content) is suitable for general screen-printing processes. Bayermaterialscience Desmodur BL 4265 SN is a kind of MEKO blocked isophorone diisocyanate (IPDI) trimer (#1) while Desmodur BL 3175 SN is a kind of MEKO blocked hexamethylene diisocyanate (HDI) trimer (#2). Both of them can couple with polyols to form low viscosity, transparent, long shelf-life paste for preparing light-stable colorless flexible, colorfast and weather-stable stoving coatings. These blocked polyurethane prepolymers are stable at room temperature when being mixed with the polyol hardener and catalyst. After mixing the dispersant with silver microflakes, the resulting ECA paste were screen-printed to PET films and cured. The structure of the blocked polyurethane prepolymers is demonstrated in scheme 1.

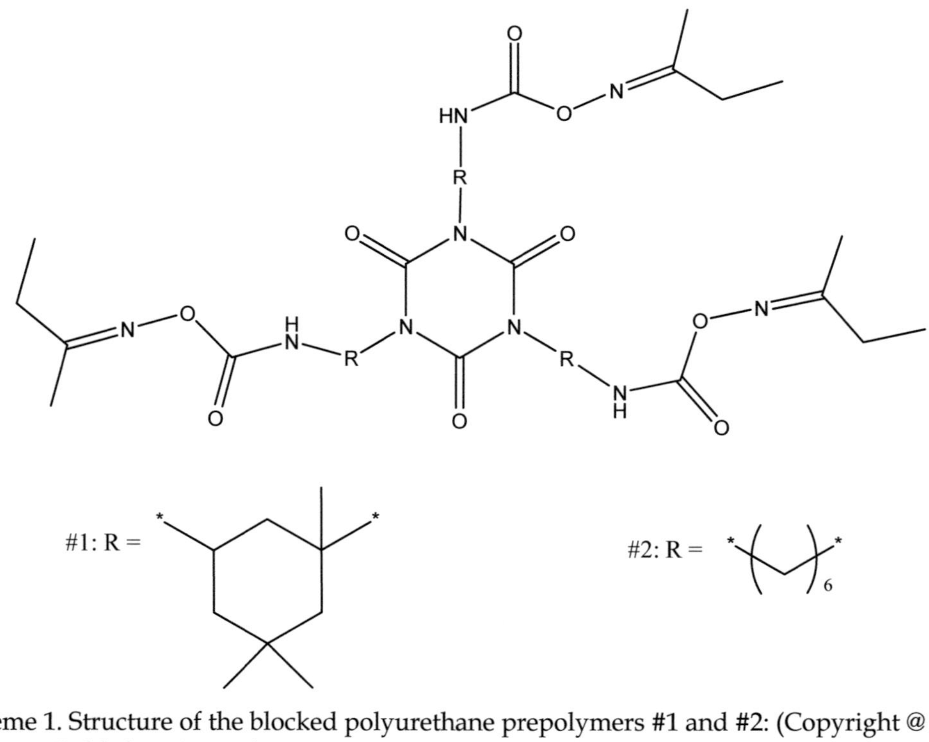

Scheme 1. Structure of the blocked polyurethane prepolymers #1 and #2: (Copyright @ 2011 Springer Publishing House)

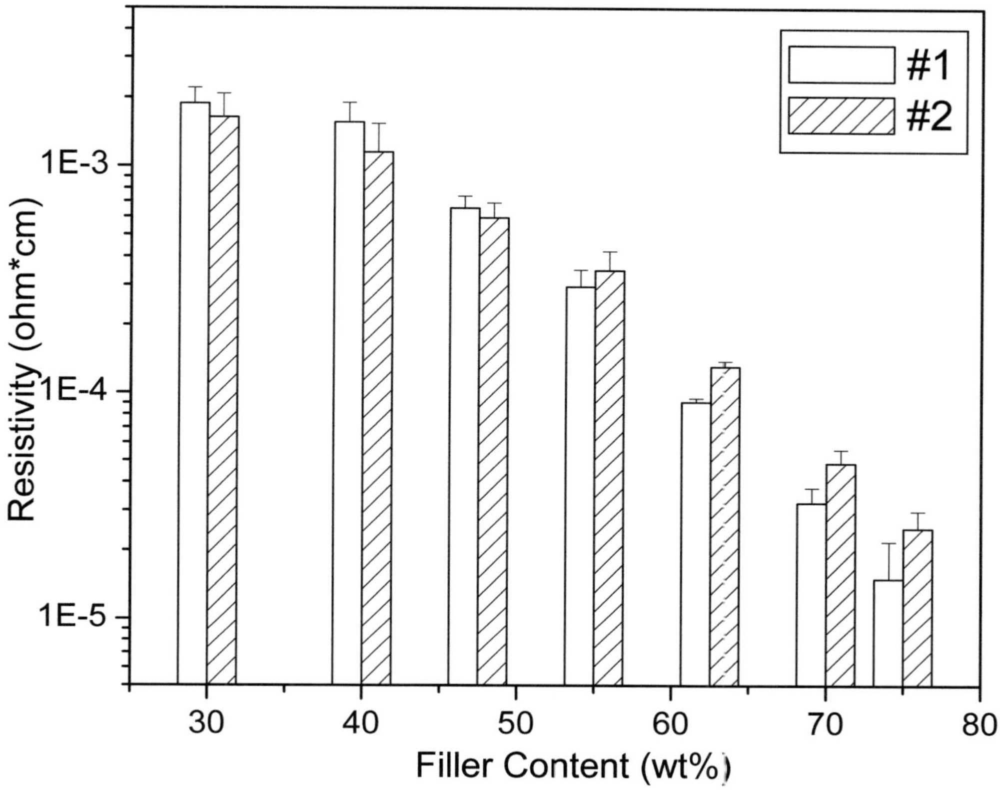

Fig. 8. Resistivity of the printed resistor of the ECAs (#1 and #2) with different silver filler contents. (Copyright @ 2011 Springer Publishing House)

Fig. 8 illustrates the electrical resistivity of the ECA antenna samples containing different silver contents (from 30 wt% to 75 wt%), which displays a series of optional conditions for the aimed cost-effectiveness. Here we can observe that with different content of silver fillers, the electrical resistivity of the ECAs varies in a range from about 2 x 10^{-5} Ω·cm to 2 x 10^{-3} Ω·cm. [3] When the silver content is lowered down to 40% and 30%, we observed that the resistivity reaches a plateau. This chart suggests that the conductivity of the PU based ECAs is comparable to those ECAs based on bisphenol-A epoxy, thus they are useful to general applications.

Fig. 9A and 9B show the changes of resistivity versus the aging time in a TERCHY MHU-150L humidity chamber (85°C/85%RH) for up to 720 hours of the two series of ECA samples. From these figures, we can observe that after the aging test, most of the electrical resistances are even lower than those before aging. For the ECA samples with relatively low filler content (30% and 40% of the filler content), the resistance even dropped about 20%. For the other samples, the variation of the resistance value is smaller than 10%, which suggest that these PU based ECAs having superior reliability up to 720 hours in their electrical conductivity.

From the SEM analysis (Fig. 11) of the cross sections of the ECA samples, we can observe that disregarding the variation of the silver content (e.g. from 30 wt% to 75 wt%), the silver microflakes can be homogeneously distributed in the PU dispersant. In order to evaluate the performance of the PU-dispersed ECA in high frequency applications, we conducted the read range examination of the ECA printed RFID tag antennas.

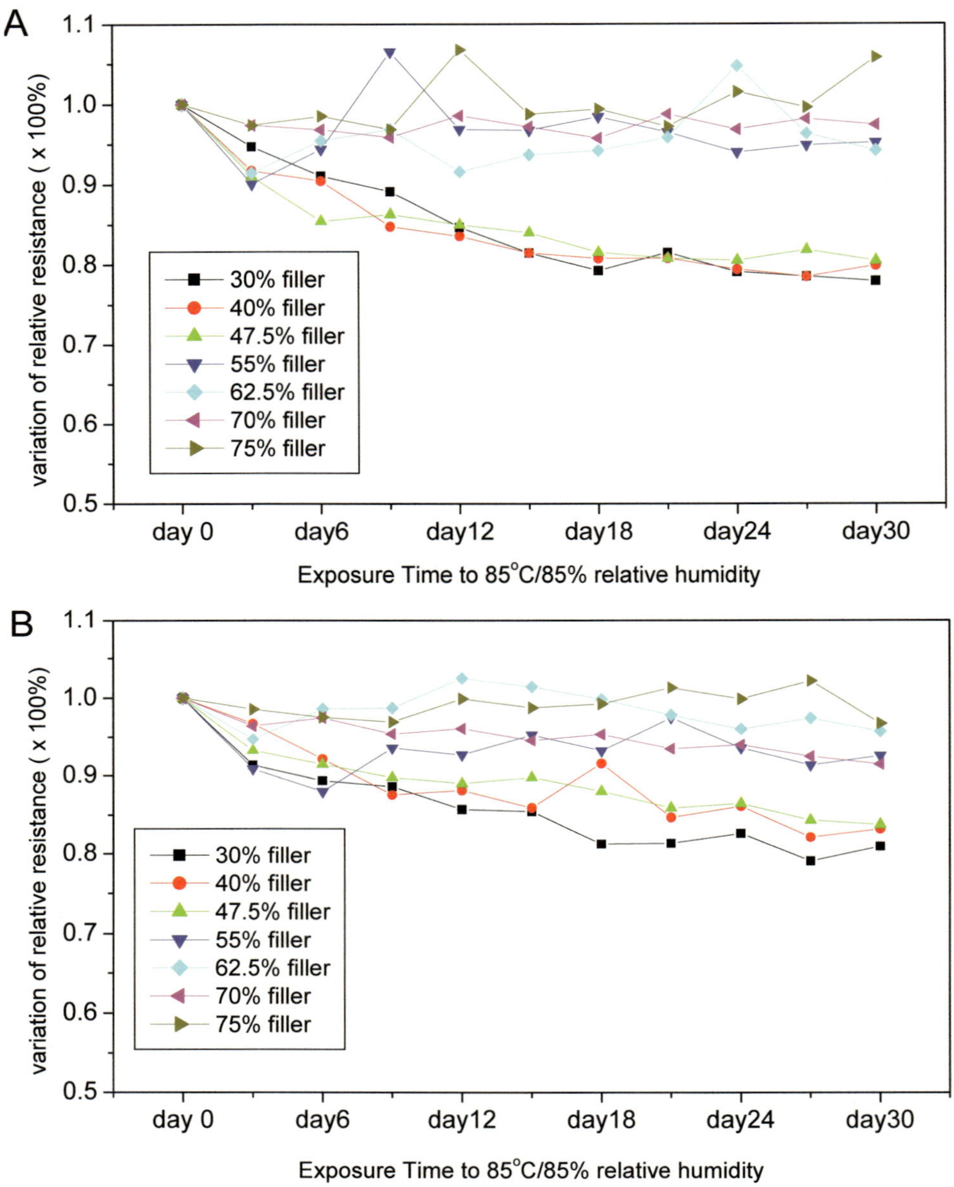

Fig. 9. Reliability analysis (85oC/85% relative humidity) of the ECAs. A) Variation of the relative resistance of the printed resistors of the ECA #1; B) variation of the relative resistance of the printed resistors of the ECA #2. . (Copyright @ 2011 Springer Publishing House)

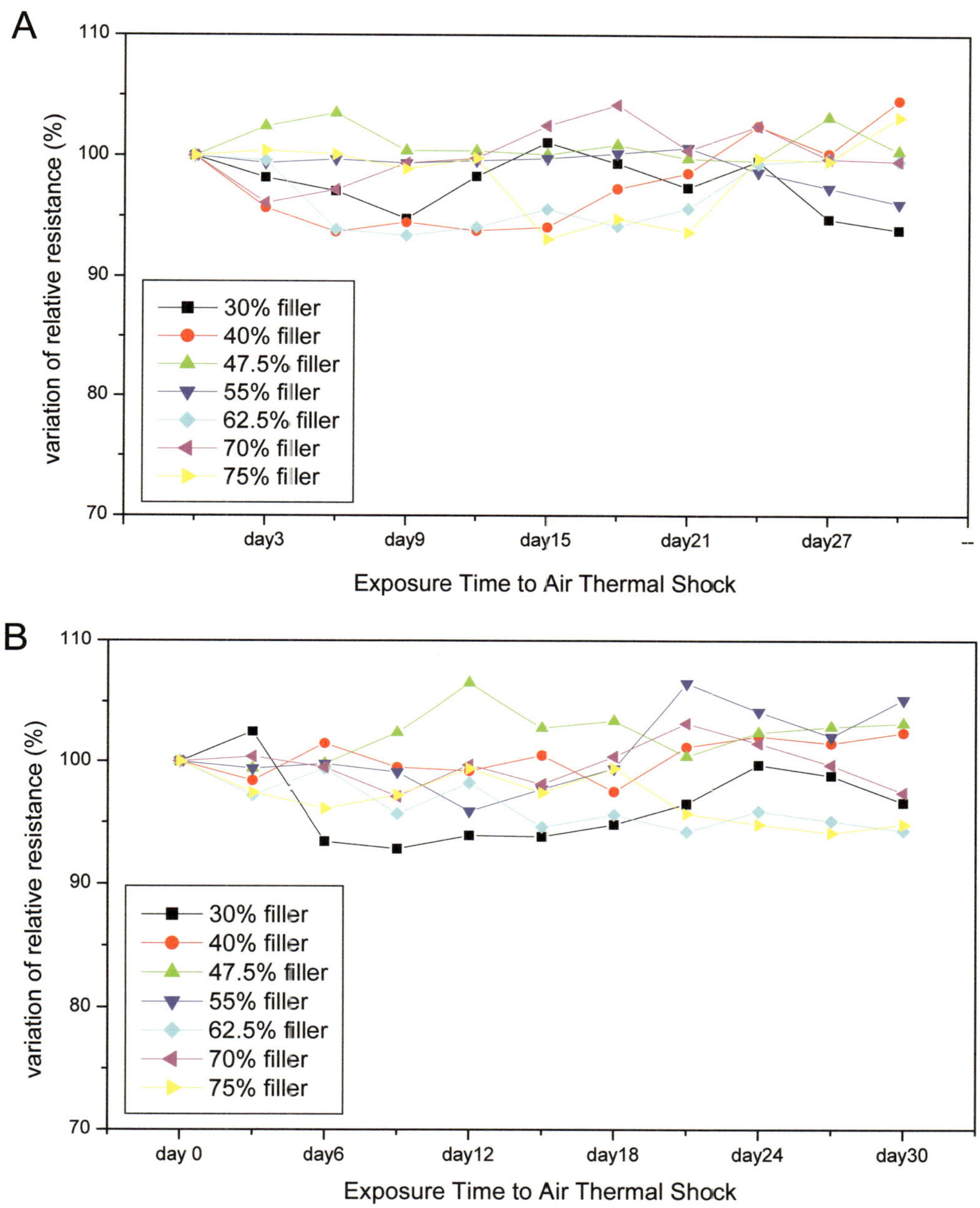

Fig. 10. Thermal cycling analysis of the ECAs. A) Variation of the relative resistance of the printed resistors of the ECA #1; B) Variation of the relative resistance of the printed resistors of the ECA #2. (Copyright @ 2011 Springer Publishing House)

Usually, the RFID antenna is designed based on the dipole antenna which is about half wavelength in length, as demonstrated in Fig. 8. Since the passive tag where the power is obtained solely from the received electromagnetic wave, the tag antenna must match the tag circuit to maximize the transfer of power into and out of it. We selected an Alien's Gen 2 RFID chip which has an impedance value of 30 - 110j Ω, so we designed the tag antenna with the impedance value of 30 + 110j Ω to conjugate match with the chip. In the simulation, we considered both the resistivity of the materials, surface roughness, and configuration of the antenna. Based on the simulation result, we designed a series of RFID tag antenna based on #1 and #2 series. The antenna is a 82 mm-long dipole with a short line connecting two parts, as shown in Fig. 9.[36] For example, the simulated impedance of the ECA antenna filled with 30 wt% of silver filler is 33 + 108j at 915 MHz which well matches the Alien's RFID strap (30 - 110j). The calculated return loss values is -24 dB, which means over 99% power is transmitted to RFID chip. We found that the -10 dB power transmission bandwidth of the antenna is 60 MHz which covers the operation frequency of North American, China, and Hong Kong standards.[37] Herein we use the minimum turn-on power of the reader as the index of the RFID tag antenna performance. The reader is located one meter in distance towards the RFID tag (a piece of EPCglobal Class 1 Gen 2 RFID Chip is adhered to the center of the antenna). From the experimental result, we can observe that the minimum turn-on power of the reader is consistent with the electrical resistivity of the ECA samples, i.e. with the increment of the resistivity of the antenna, the reader needs a higher minimum turn-on power to detect the tag (Fig. 12). Therefore, using the same antenna design, we can adjust the content of silver filler in the ECA to cater to different requirement of read range. As for the real application of RFID technique, the power out-put of the reader is often fixed to a certain value. Controlling the resistivity of the ECA can probably be a convenient way to cater to the different requirement of read range requirement. Apparently that by using the low silver filler content paste the cost of RFID tags can be dramatically reduced. Meanwhile, the environmentally benign polyurethane based ECAs take the advantage in food supply chain and medical applications etc.

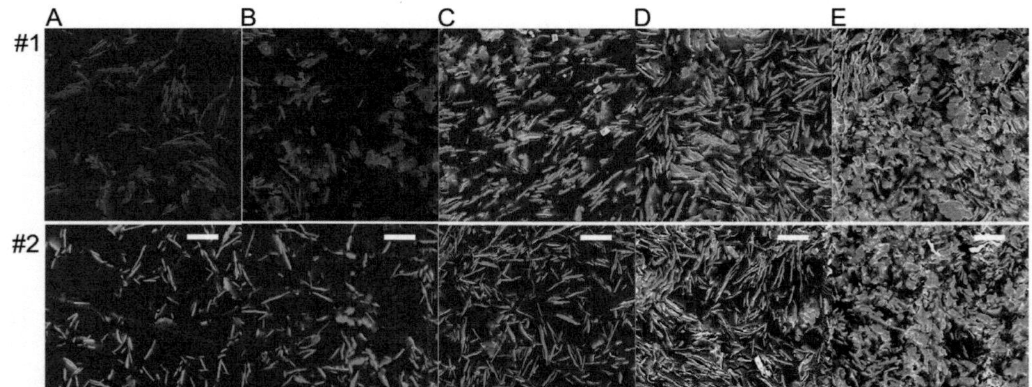

Fig. 11. SEM images of the cross sections of some of the ECA bulk samples. A) 30% of filler; B) 40% of filler; C) 55% of filler; D) 70% of filler; and E) 75% of filler. (Scale bar = 10 μm) (Copyright @ 2010 Springer Publishing House)

The ECA samples with different silver content were prepared, printed into pre-designed geometries and their performances such as electrical resistivity, adhesion strength to PET film, and high frequency performances were studied. From the experimental results, the

ECA with the silver content as low as 47.5% still maintain an acceptable conductivity (6.56 x 10^{-4} and 5.96 x 10^{-4} Ω·cm), which is efficient for high frequency applications. This suggests that by adjusting the silver content, the electrical and mechanical properties of the ECAs can be modulated. On the other hand, we observed that the silver content at 70% showed similar conductivity to those with higher silver content, which suggests that the silver content at this level reaches the summit of the conductivity. In a 720-hour 85 °C/85%RH aging test, we observed that in a large range of silver contents from 30% to 75%, the electrical resistivity of this PU based ECA was very stable. They also passed the 720-hour thermal cycling test for electrical conductivity. After all, blocked-PU based resin has been demonstrated efficient for fabricating the low-cost and flexible ECAs, which has also been demonstrated feasible in the ultra high frequency RFID tag antennas.

4. Water-based ECAs

PU displays various characters such as adjustable mechanical properties, shape-memory property, and excellent stability.[38-40] Moreover, many PU-based resins are biocompatible and can be obtained from renewable resources such as from vegetable oils.[41-43] The water-based PU resins exhibit even more advantages since there is no organic small molecule involved or released during the printing process. Recently, Yang et al. investigated the feasibility of applying the water-based PU resin as the dispersant material for the ECAs. Here cycloaliphatic PU is prepared in the emulsion based reaction. As shown in Scheme 2, the water-borne PU dispersant is prepared mainly in four steps: 1. polyether polyol (here is polytetrahydrofuran 2000), dihydroxylmethylpropionic acid (DHPA), and isophorone diisocyanate (IPDI) are mixed together for preparing the prepolymer; 2. chain extender (butylene diol) is added until the chain propagation is terminated; 3. triethylamine (TEA) is added to neutralize the system; 4. water is added dropwise so that the PU is transferred into aqueous solution. Finally, the organic solvent and the unreacted chemicals are removed by vacuum. The resulting PU emulsion is translucent bluish with long shelf-life and stable rheological property. The structure of the PU resin prepared in this way was confirmed by FT-IR spectrum. As shown in Fig. 13, the FT-IR spectrum of the dried film of the as-prepared water-borne PU is investigated. The peaks at 2933 cm^{-1} and 2854 cm^{-1} confirm the existence of the $-CH_2-$ group, the 1698 cm^{-1} the carbonyl group, and 1239 cm^{-1} and 1108 cm^{-1} confirm the C-O vibrations. The as-prepared PU has excellent thermal stability, which was confirmed by using thermalgravimetric analysis (TGA). The temperature of the sample was ramped from room temperature to 600 °C with the speed of 20 °C/min in the air (Fig. 14). The sample lost less than 10% weight before it reached 250 °C. Further raising the temperature resulted in the total decomposition, until the temperature reached 430 °C. This result suggests that the PU dispersant is suitable for the general solder reflow process as well when it is applied in the traditional packaging process.

The WBECAs were prepared by mixing the PU resin and a certain portion of the modified silver microflakes together by using a THINKY ARE250 mixer.[20] By adjusting the ratio between the two components we are able to achieve an optimum between the mechanical strength and electrical conductivity. $NaBH_4$ has been considered as a very powerful reducing agent for protecting many metals from oxidations. For example, addition of small amount of $NaBH_4$ has been demonstrated effective for improving the percolation among the copper and nickel powders via an in-situ reducing process for ink-jet printing conductive lines.[44] Here we tentatively added in 0.5% (by weight) and 1% (by weight) of $NaBH_4$ (*vs.*

Ag) into the WBECAs, as an agent for preventing the oxidation issue during the processing steps. The cross section images of the samples were studied on both transmission electron microscopy (TEM) and scanning electron microscopy (SEM). As shown in Fig. 15, the electrical resistivity of the printed resistor which is based on different silver content and $NaBH_4$ treatment condition are listed in Table 1. From Fig. 15, we can observe that the addition of $NaBH_4$ can effectively reduce the electrical resistivity of the printed resistors which were prepared by using the WBECAs. The improvement of the resistivity is about one order of magnitude.

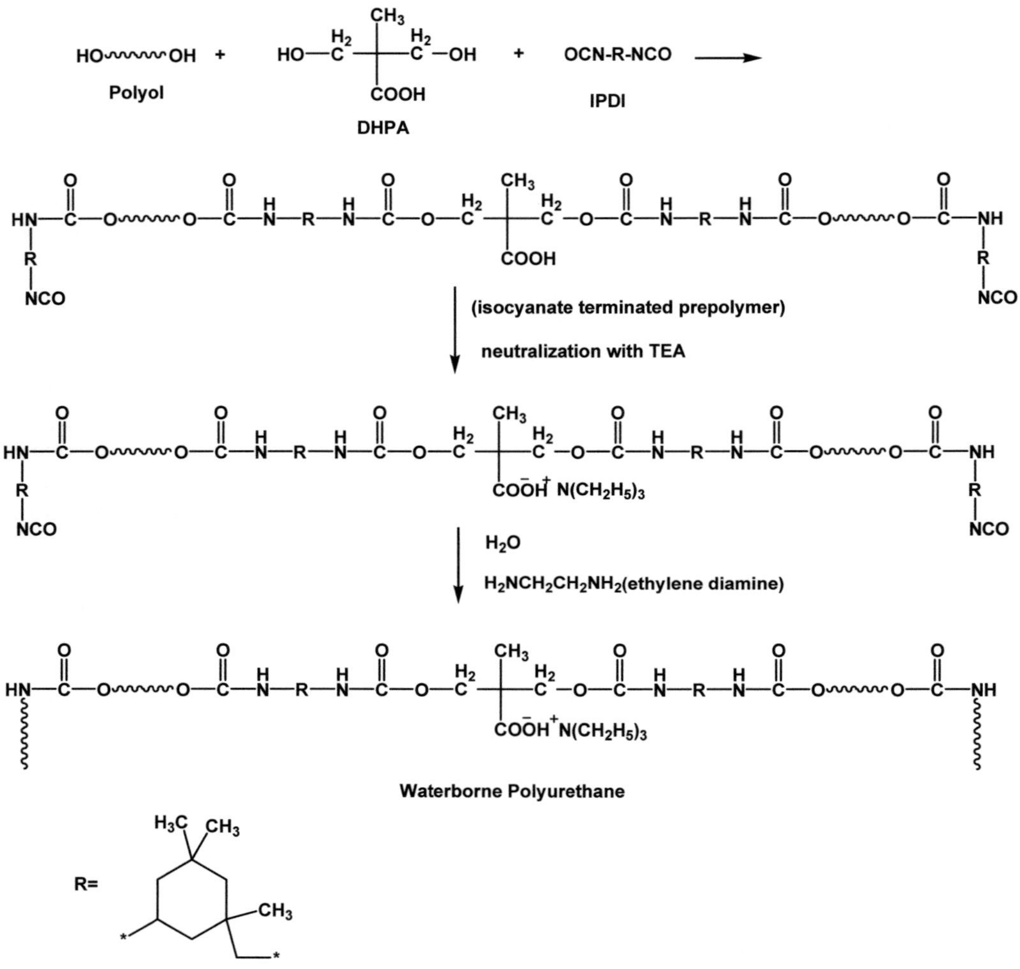

Scheme 2. Preparation route of the water-borne PU dispersant.

The measurement of the variation of electrical resistivity of the printed ECA samples were conducted in a TERCHY MHU-150L humidity chamber (85°C/85% relative humidity) for 60 days for the temperature-humidity testing (THT) (Fig. 16). As shown in Fig. 16, we can observe a trend of decrease of the electrical resistivity over the period of time. The reasons of the decrement of the electrical resistivity of all the samples are related to the following points: 1) the water-borne PU dispersant is intrinsically an emulsion which contains both the hydrophilic part and the hydrophobic part; water molecules trapped in the interstitial sites are eliminated during the aging process or thermal curing process which renders shrinkage of the total size; 2) since the glass transition temperature (T_g) of the water-borne PU dispersant is much lower than room temperature (~-20 °C), the creeping of the hydrophobic polymer chain enhances the phase separation of the hydrophobic/hydrophilic regions, which results in a stronger interaction among the polymer chains by hydrophobic interaction and hydrogen bond as well. These two factors take effect both in the thermal curing process (if there is any) and the aging process as well. Thus we observed kind of variation of the electrical resistivity. After all, we did not observe any increase of the electrical resistivity of all samples after the aging test, which suggests sufficient reliability for real applications. Since many rubbery substrates are very sensitive to the high temperature (due to their extremely low T_g), they can be used as the stretchable circuit boards and fabricated at room temperature by using the WBECAs as the circuits and interconnects.

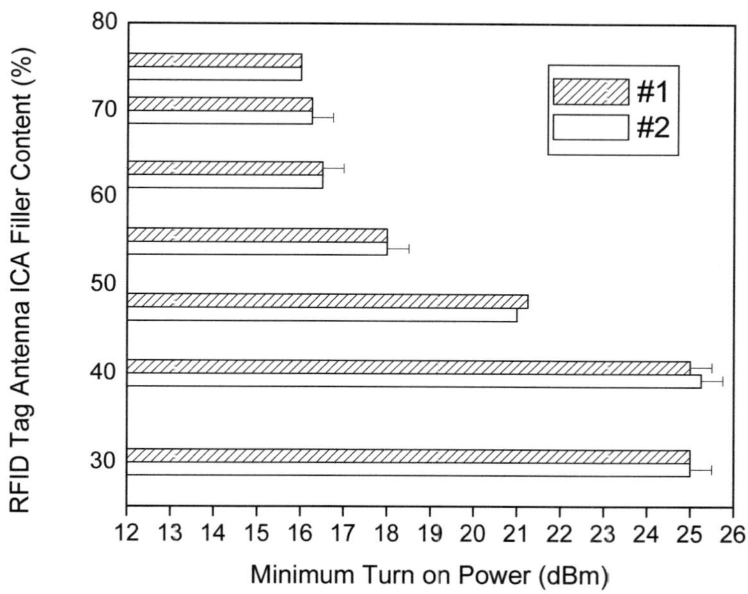

Fig. 12. Minimum turn-on power of the reader in detecting the RFID tags with the antenna printed using the ECAs. (Copyright @ 2011 Springer Publishing House)

The relation between the silver content and the tensile property of the WBECA thin film samples were investigated on an Advanced Rheometric Expansion System (ARES) (TA instruments, USA). The specimens were prepared on a piece of smooth low density polyethylene (LDPE) substrate, so that they could form an even and flat thin film. When they were naturally dried, they were peeled off carefully from the substrate and then cut into small strips with the dimension near 40 x 3 x 0.1 mm^3 (each was accurately confirmed by a caliper), and mounted onto ARES by a thin film tensile test fixture. The measurement was conducted at 25 °C with a 2000 g·cm transducer. The extension speed was 0.2 mm/s in a strain-controlled mode. As shown in Table 1, we can observe that the Young's modulus of all the three samples does not change significantly along with the different silver content level. This suggests that the addition of $NaBH_4$ does not have significant influence to the mechanical strength of the WBECA samples.

Compared to the other traditional dispersants for the ECAs, such as epoxy, polyester, and polyacrylates etc., water-borne PU as the resin dispersant displays a few advantages: 1. the resin is dispersed in water, thus the printing process does not involves toxic volatile materials and the residues can be conveniently removed by water; 2. the PU materials can be prepared from a large variety of sources such as from plants, thus PU has better environmental benign character and adjustable mechanical strength; 3. the urethane bond is relatively strong, thus the materials have a high reliability for general electronic packaging applications; 4. the curing step for the ECAs can take place at even room temperature (of course a higher temperature may help accelerate the process) thus it saves energy; 5. the WBECAs have adjustable rheological property thus they are suitable for many types of printing process such as screen printing, gravure printing, and roll-to-roll printing etc.

In summary, by sensitizing a small amount of $NaBH_4$, the electrical conductivity of the WBECAs can be effectively improved of about one order of magnitude; the percolation threshold of the silver filler is reduced as well. The lowest electrical resistivity ever measured in this material was in the order of 10^{-5} Ω · cm. The mechanical strength of the thin films of the free-standing WBECAs improves along with the PU dispersant amount. These WBECAs can be applied in the general printing process for general applications as ordinary ECAs can do, while they display many unique properties, such as amenity for processing, environmentally benign, excellent shelf-life and reliability in long-term storage and applications, water-proof, and the mechanical property can be adjusted by choosing different prepolymers.

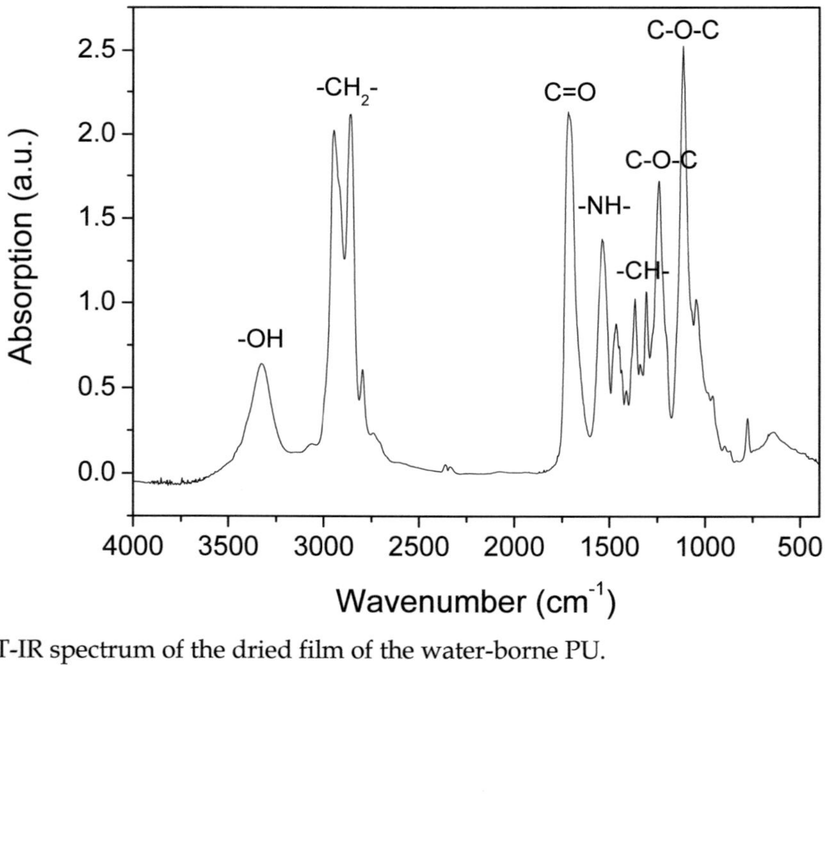

Fig. 13. FT-IR spectrum of the dried film of the water-borne PU.

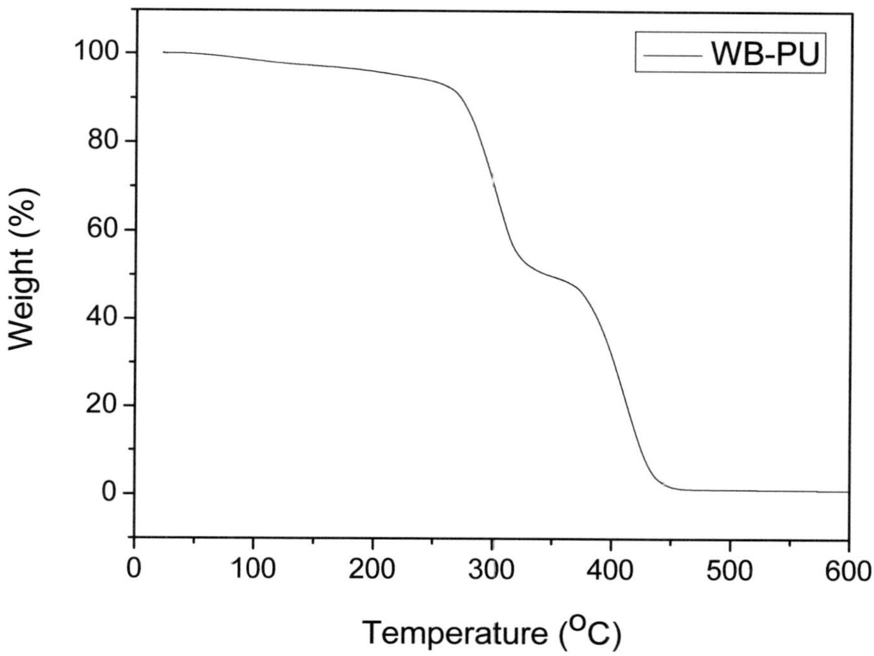

Fig. 14. TGA analysis of the PU dried film. The sample was ramped from 25 °C to 600 °C in the air.

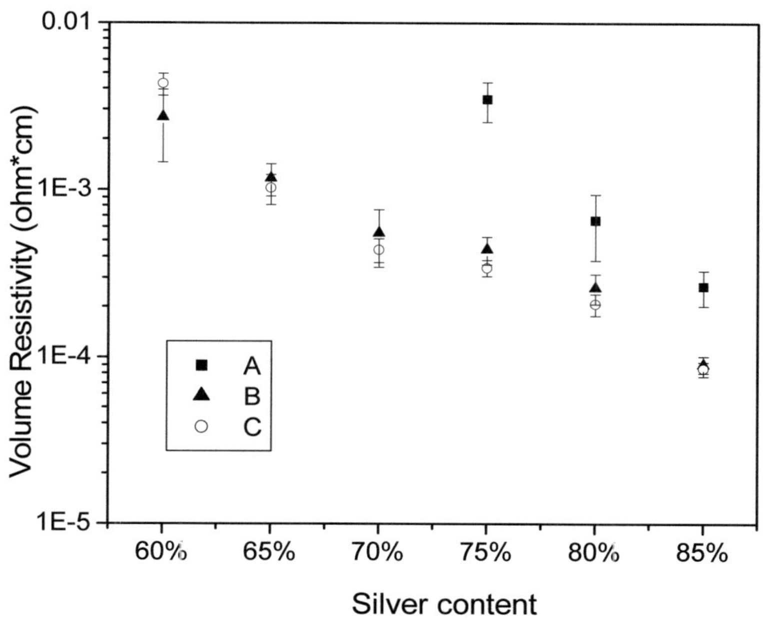

Fig. 15. Volume resistivity of the WBECAs (80 wt% of silver) versus different addition amount of $NaBH_4$. (A) no $NaBH_4$ addition; (B) 0.5% of $NaBH_4$; (C) 1% of $NaBH_4$.

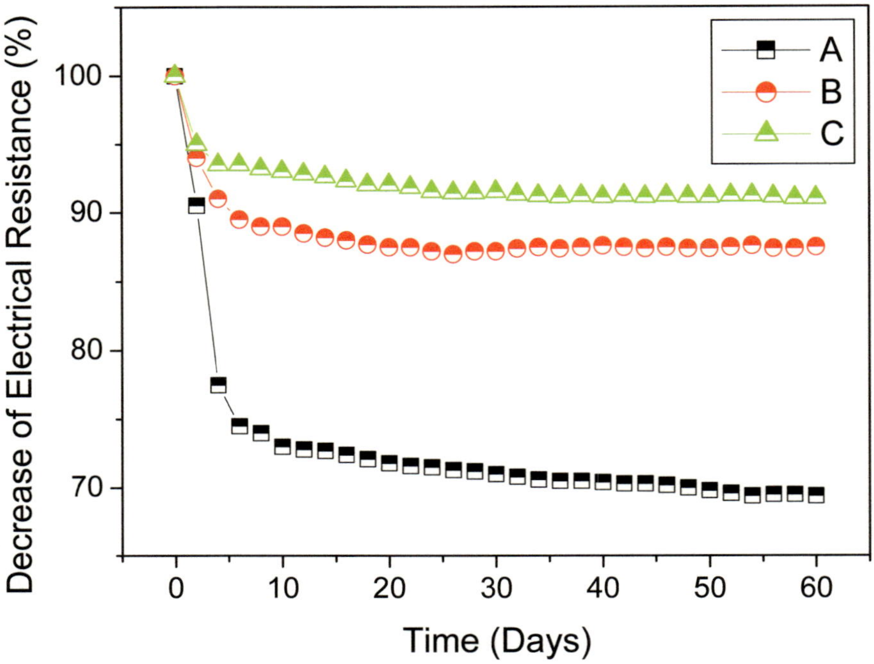

Fig. 16. Thermal-humidity reliability of the WBECAs versus aging time. (A) no $NaBH_4$ addition; (B) 0.5% of $NaBH_4$; (C) 1% of $NaBH_4$.

Young's modulus (MPa)	60% silver	70% silver	80% silver	85% silver
no treatment	0.291	0.322	0.311	0.309
0.5% $NaBH_4$	0.289	0.338	0.364	0.358
1% $NaBH_4$	0.297	0.319	0.347	0.339

Table 1. A table showing the Young's modulus of the WBECA thin film samples including the untreated, 0.5% of $NaBH_4$ treated, and 1% of $NaBH_4$ treated ones.

5. Conclusions

In summary, the authors introduced the recent progress of the silver microflake-filled ECAs as a candidate for the RFID tag antenna applications. ECAs exhibit many advantages such as printability and low-temperature processability as compared to the conventional antenna preparation methods, which render them significant in both the conventional Complementary Metal Oxide Semiconductor (CMOS) based and the organic all-printed ones. However, their electrical, mechanical, and environmental performances are still undergoing intensive investigations. In this chapter, the authors gave several simple introductions about how to improve the electrical conductivity of the ECAs and introduced some PU based resin dispersants for ECAs. By adjusting the balance between the electrical conductivity and the materials cost, ECAs could find a larger market in both far field and near field applications. Any significant advancement of the materials would enhance the widespread uses of the tags, which is benefit from both the lower cost and higher performances. The examples given in this article have their merit and limitations; we expect that they may give elicitations for developing techniques for manufacturing low-cost, flexible ubiquitous information terminals.

6. Acknowledgement

The authors acknowledge the financial support from the Tsinghua University the Graduate School at Shenzhen.

7. References

[1] Das, R. & Harrop, P. (2009). Printed, Organic & Flexible Electronics Forecasts, Players & Opportunities 2009-2029, IDTechEx, ISBN/SKU #:IDT6769, March, 2009

[2] Finkenzeller, K. (2002). RFID-Handbuch Hanser-Verlag, ISBN 978-344-6212-78-7, München, Germany

[3] Yang, C.; Xu, B. & Yuen, M. M. F. (2008). Using Novel Materials to Enhance the Efficiency of Conductive Polymer, The 58th IEEE Electronic Components and Technology Conference, Vol. 5, pp. 213, ISBN 978-1-4244-2230-2, Orlando, Florida, USA, May 27-30, 2008

[4] Syed, A.; Demarest, K. & Deavours, D. D. (2007). Effects of Antenna Material on the Performance of UHF RFID Tags, IEEE International Conference on RFID, pp. 57-62, ISBN 1-4244-1013-4, Grapevine, TX, USA, March 26-28, 2007

[5] Stauffer, D. & Aharony, A. (1992). Introduction to Percolation Theory, Taylor & Francis ISBN 0-850-663156, London, UK

[6] Chiang, H. W.; Chung, C. L.; Chen, L. C.; Li, Y.; Wong, C. P. & Fu, S. L. (2005). Processing and shape effects on silver paste electrically conductive adhesives (ECAs), Journal of Adhesion Science and Technology, Vol. 19, No. 7, (June 2005), pp. 565-578, ISSN 0169-4243

[7] Wu, H. P.; Wu, X. J.; Ge, M. Y.; Zhang, G. Q.; Wang, Y. W. & Jiang, J. Z. (2007). Effect analysis of filler sizes on percolation threshold of isotropical conductive adhesives, Composites Science and Technology, Vol. 67, No. 6, (May 2007), pp. 1116-1120, ISSN 0266-3538

[8] Li, Y.; Moon, K. S. & Wong, C. P. (2005). Electronics without lead, Science, Vol. 308, No. 5727, (June 2005), pp. 1419-1420, ISSN 0036-8075

[9] Chatterjee, K.; Banerjee, S. & Chakravorty, D. (2004). Metal-to-insulator transition in silver nanolayers grown on silver oxide nanoparticles, Europhysics Letters, Vol. 66, No. 4, (May 2004), pp. 592-599, ISSN 0295-5075

[10] Lu, D. Q. & Wong, C. P. (2000). Effects of shrinkage on conductivity of isotropic conductive adhesives, Int. J. Adhesion & Adhesives, Vol. 20, No. 3, (May 2000), pp. 189-193, ISSN 0143-7496

[11] Kim, K. D. & Chung, D. D. L. (2005). Electrically conductive adhesive and soldered joints under compression, Journal of Adhesion Science and Technology, Vol. 19, No. 11, (November 2005), pp. 1003-1023, ISSN 0169-4243

[12] Lu, D. D.; Tong, Q. & Wong, C. P. (1999). Conductivity Mechanisms of Isotropic Conductive Adhesives (ICA's), IEEE Transactions on Electronics Packaging Manufacturing, Vol. 22, (July 1999), pp. 223-227, ISSN 1521-334X

[13] Su, B. & Qu, J. (2004) A micro-mechanics model for electrical conduction in isotropically conductive adhesives during curing, Proceedings - Electronic Components & Technology Conference, Vol. 2, (June 2004), pp. 1766-1771, ISBN: 0-7803-8365-6

[14] Li, Y.; Yim, M. J.; Moon, K. S. & Wong, C. P. (2009). Electrically conductive adhesives, Smart Materials, (November 2009), pp. 11/12, CRC Press, ISBN 978-1-4200-4372-3, Boca Raton, Fl, USA

[15] Yim, M.; Li, Y.; Moon, K. & Wong, C. P. (2007). Oxidation prevention and electrical property enhancement of copper-filled isotropically conductive adhesives, J. Elect. Mater., Vol. 36, No. 10, (August 2007), pp. 1341-1347, ISSN 1573-4803

[16] Li, Y.; Whitman, A.; Moon, K. S. & Wong, C. P. (2005). High performance electrically conductive adhesives (ECAs) modified with novel aldehydes, Proceedings - Electronic Components & Technology Conference, Vol. 2 , (May, 2005) pp. 1648-1652, ISBN 0-7803-8906-9

[17] Jiang, H. J.; Moon, K. S.; Li, Y. & Wong, C. P. (2006). Surface functionalized silver nanoparticles for ultrahigh conductive polymer composites, Chem. Mater., Vol. 18, No. 13, (May 2006), pp. 2969-2973, ISSN 0897-4756

[18] Li, Y.; Moon, K.; Whitman, A. & Wong, C. P. (2006). Enhancement of electrical properties of electrically conductive adhesives (ECAs) by using novel aldehydes, IEEE Transactions on Components and Packaging Technologies, Vol. 29, No. 4, (October 2006), pp. 758-763, ISBN 0-7803-8906-9

[19] Lu, D. D.; Li, Y. & Wong, C. P. (2008). Recent advances in nano-conductive adhesives, J. Adhes. Sci. & Tech., Vol. 22, No. 8-9, (August 2008), pp. 801-834, ISSN 0169-4243

[20] Yang, C.; Xie, Y. T.; Yuen, M. M. F.; Xu, B.; Gao, B.; Xiong, X. M. & Wong, C. P. (2010). Silver Surface Iodination for Enhancing the Conductivity of Conductive Composites, Adv. Func. Mater., Vol. 20, No. 16, (August 2010), pp. 2580-2587, ISSN 1616-301X

[21] Matsunaga, K.; Tanaka, I. & Adachi, H. (1998). Electronic mechanism of Ag-cluster formation in AgBr and AgI, Journal of the Physical Society of Japan, Vol. 67, No. 6, (December 1998), pp. 2027-2036, ISSN 0031-9015

[22] Hull, S. (2004). Superionics: crystal structures and conduction processes, Rep. Prog. Phys., Vol. 67, No. 7, (July 2004), pp. 1233-1316, ISSN 0034-4885

[23] Bardi, U. & Rovida, G. (1983). Leed, AES and Thermal Desorption Study of Iodine Chemisorption on the Silver (100), (111) and (110) Faces, Surface Science, Vol. 128, No. 2-3, (January 1983), pp. 145-168, ISSN 0039-6028

[24] Zhang, X.; Stewart, S.; Shoesmith, D. W. & Wren, J. C. (2007). Interaction of aqueous iodine species with Ag2O/Ag surfaces, J. Electrochem. Soc., Vol. 154, No. 4, (February 2007), pp. F70-F76, ISSN 0013-4651

[25] Thiel, P. A.; Shen, M.; Liu, D. J. & Evans, J. W. (2009). Coarsening of Two-Dimensional Nanoclusters on Metal Surfaces, J. Phys. Chem. C, Vol. 113, No. 13, (March 2009), pp. 5047, ISSN 1932-7447

[26] Hasse, U., Fletcher, S. & Scholz, F. (2006). Nucleation-growth kinetics of the oxidation of silver nanocrystals to silver halide crystals, J. Solid State Electrochem., Vol. 10, No. 10, (May 2006), pp. 833-840, ISSN 1432-8488

[27] Andryushechkin, B. V.; Zhidomirov, G. M.; Eltsov, K. N.; Hladchanka, Y. V. & Korlyukov, A. A. (2009). Local structure of the Ag(100) surface reacting with molecular iodine: Experimental and theoretical study, Phys. Rev. B, Vol. 80, No. 12, (September 2009), pp. 125409 1-10, ISSN 1098-0121

[28] Bonacic-Koutecky, V. & Mitric, R. (2005). Theoretical exploration of ultrafast dynamics in atomic clusters: Analysis and control, Chem. Rev., Vol. 105, No. 1, (January 2005), pp. 11-66, ISSN 0009-2665

[29] Hagen, J.; Socaciu, L. D.; Le Roux, J.; Popolan, D.; Bernhardt, T. M.; Woste, L.; Mitric, R.; Noack, H. & Bonacic-Koutecky, V. (2004). Cooperative effects in the activation of molecular oxygen by anionic silver clusters, J. Am. Chem. Soc., Vol. 126, No. 11, (February 2004), pp. 3442-3443, ISSN 0002-7863

[30] Sibbald, M. S.; Chumanov, G. & Cotton, T. M. (1997). Reductive properties of iodide-modified silver nanoparticles, J. Electroanal. Chem., Vol. 438, No. 1-2, (November 1997), pp. 179-185, ISSN 0022-0728

[31] Fourcade, F.; Tzedakis, T. & Bergel, A. (2003). Electrochemical process for metal recovery from iodized silver derivatives in liquid/solid mixture: Experimental and theoretical approaches, Chem. Eng. Sci., Vol. 58, No. 15, (August 2003), pp. 3507-3522, ISSN 0009-2509

[32] Patil, K. C.; Rao, C. N. R.; Lacksone, J. & Dryden, C. E. (1967). Silver nitrate-iodine reaction-iodine nitrate as reaction intermediate, J. Inorg. & Nucl. Chem., Vol. 29, No. 2, (February 1967), pp. 407-412, ISSN 0022-1902

[33] Li, Y.; Xiao, F.; Moon, K. & Wong, C. P. (2006). Amino acid as a novel curing agent for epoxy resins in electronic materials, PMSE Preprints, Vol. 94, pp. 873-874, ISSN 1550-6703

[34] Li, Y.; Xiao, F. & Wong, C. P. (2007). Novel, environmentally friendly crosslinking system of an epoxy using an amino acid: tryptophan-cured diglycidyl ether of bisphenol a epoxy. J. Polym. Sci. A: Polym. Chem., Vol. 45, No. 2, (January 2007), pp. 181-190, ISSN 0887-624X

[35] (2005). EFSA, Parma, Italy

[36] Chen, S. L. & Lin, K. H., Performance of a Folded Dipole with a Closed Loop for RFID Applications, in "Progress In Electromagnetics Research Symposium" (2007) pp. 329-331, ISBN 978-1-934142-01-1, Prague, Czech Republic, August 27-30, 2007

[37] Barthel, H. (2008). Regulatory status for using RFID in the UHF spectrum, EPCglobal Inc., ISBN 978-1-4244-2041-4, October 2008

[38] Subramani, S.; Park, Y. J.; Lee, Y. S. & Kim, J. H. (2003). New development of polyurethane dispersion derived from blocked aromatic diisocyanate, Prog. Org. Coat., Vol. 48, No. 1, (November 2003), pp. 71-79, ISSN 0300-9440

[39] Yang, C.; Tang, Y. H.; Lam, W. M.; Lu, W. W.; Gao, P.; Zhao, C. B. & Yuen, M. M. F. (2010). Moisture-cured elastomeric transparent UV and X-ray shielding organic-inorganic hybrids, J. Mater. Sci., Vol. 45, No. 13, (July 2010), pp. 3588-3594, ISSN 0022-2461

[40] Wicks, D. A. & Wicks, Z. W. (2001). Multistep chemistry in thin films; the challenges of blocked isocyanates, Prog. Org. Coat., Vol. 43, No. 1-3, (November 2001), pp. 131-140, ISSN 0300-9440

[41] Sharma, V. & Kundu, P. P. (2008). Condensation polymers from natural oils, Prog. Polym. Sci., Vol. 33, No. 12, (December 2008), pp. 1199-1215, ISSN 0079-6700

[42] Guner, F. S.; Yagci, Y. & Erciyes, A. T. (2006). Polymers from triglyceride oils, Prog. Polym. Sci., Vol. 31, No. 7, (July 2006), pp. 633-670, ISSN 0079-6700

[43] Petrovic, Z. S. (2008). Polyurethanes from vegetable oilsm Polym. Rev., Vol. 48, No. 1, (January 2008), pp. 109-155, ISSN 1558-3724

[44] Li, D.; Sutton, D.; Burgess, A.; Graham, D. & Calvert, P. D. (2009). Conductive copper and nickel lines via reactive inkjet printing, J. Mater. Chem. Vol. 19, (April 2009), pp. 3719-3724, ISSN 0959-9428

8

Key Factors Affecting the Performance of RFID Tag Antennas

Yung-Cheng Hsieh[1], Hui-Wen Cheng[2] and Yu-Ju Wu[3]
[1]*Department of Graphic Communication Arts*
Dean, Research and Development
National Taiwan University of Arts
[2]*Department of Graphic Communication Arts*
Research Assistant
National Taiwan University of Arts
[3]*School of Technology*
Assistant Profeddor
Eastern Illinois University
[1,2]*Taiwan*
[3]*USA*

1. Introduction

Bar codes and Radio Frequency Identification (RFID) both belong to a group of technologies called Automatic Identification and Data Capture. People have all become very aware of bar codes as they have permeated our existence in the last 25 years. In fact, it is tough to buy something in a store that does not use bar codes these days. But bar codes have four disadvantages: you have to be able to see them, the bar code cannot be written on or defaced, you cannot change the data once they are printed, and they take up space on the object they are printed on. To eliminate those disadvantages, RFID is the solution. RFID is a means of capturing data about an object without using a human to read the data. Along with Smart cards, and Magnetic Stripe technology and a host of others, this is a method of automating our need for data. Recently, the technique of RFID grabs people's attention because it captures data about an object without using a human to read the data.
Individual RFID tags must be cost-efficiency for these applications (usually less than one to two cents). The cost of antennas is a crucial factor in the mass production of antennas. To reach this goal, emphasis has been placed on the development of printed electronics technologies to enable the manufacturing of RFID tags in an economically competitive way (Hodgson, n.d.; Björninen, et al., 2009). Various printing processes has been or is currently being used for producing a number of electronic components such as printed circuits, displays, RFID antennas, batteries, etc. Printing techniques such as flexographic, offset and gravure are suited for mass production, while screen printing and ink-jet printing have been identified as processes that could be employed for printing the antennas in order to bring down the cost of RFID tags (Sangoi, 2004; Subramanian, 2005). Screen printing enables very thin printing and also very thick films. It has been used for a long time to print circuits and remains interesting for electronic printing. In the future, different printing methods are

likely to co-exist in the printed electronics market. The choice of printed electronics technologies will base on the normal parameters such as run length, feature size and variable data requirements (Blayo & Pineaux, 2005; Parashkov, et al, 2005).
Three requirements of printed electronics are resolution, accuracy of position, and amount of material deposited (i.e., thickness and content of active particles). Although the achievable resolution with screen printing (usually under 50 lines per centimeter) is not sufficient for high-performance electronics, it is still applicable to print gates for TFTs, dielectrics, and semiconductors. In printed electronics, silver particles are often used to form the conductive layer. Thin conducting layers are preferred to maintain low manufacturing costs while maintain good radiation efficiency (Parashkov, et al, 2005; Björninen, et al., 2009). Therefore, the amount of silver and the thickness of the conductive layer need to be well defined. Previous works have shown that decreasing conductor thickness increases losses and thereby decreases efficiency and results to weaker backscatter from the tag. Gao and Yuen's paper (2009) exam the effects of printing thickness on the performance of UHF RFID tags and found out that the 10 µm thick RFID antenna exhibits relatively good radiation efficiency. Koptioug et al.'s paper on "On the Behavior of Printed RFID Tag Antennas, Using Conductive Paint" indicated that with conductive layers of thickness beneath 10 µm, a commercially available silver-based paint with finite conductivity showed low radiation efficiency at high frequency. The thinner printed silver paste RFID tag antenna is a potential solution for low cost RFID tags. However, the print quality needs special attention when RFID tags are printed using very thin conducting layers.

1.1 Needs of the study

RFID technology has been around for many years, but it is only in the past few years that we have seen a surge in its acceptance and a massive growth in its use. However, RFID has not been able to replace the current bar code system yet because of the high production cost of RFID tags, especially the cost of printing RFID tag antennas. Printing the antennas is the most critical part of producing an RFID tag. The high production cost problem of printing RFID tag antennas can be eliminated if the conventional screen printing process can be applied to perform the printing tasks effectively. According to literatures, screen printing technology can be used for RFID tag printing, providing significant time and cost savings compared to traditional etching technology. Therefore, there is a great need to investigate the possibility of applying screen printing method to print RFID tag antennas to perform the task of automatic identification and data capture.

1.2 Purposes of the study

This study was a true experimental research in nature and aimed to investigate the process consistency and accuracy of printing RFID tag antennas via the screening printing method with a conductive ink, silver-based (Ag) ink, on PET, PVC, and Wet Strength paper. The target values of RFID frequency in this study were set at 13.56 MHz (HF). The purposes of the study were triple fold:

1. to establish the specifications of antenna ink film thickness and ink density,
2. to compare the solid ink density, ink film thickness, and impedance differences in process consistency and capability of printing RFID antennas on the three different substrates, and
3. to determine the optimal substrates for RFID tags using screen printing technology with conductive inks, in terms of process capability.

The reason of selecting PET and PVC as substrates is that they have high transparency and rigidity. Currently, PET and PVC have been frequently used as substrate materials of RFID tags. The reason of choosing Wet Strength paper is that it is commonly used in the package industry, and its low cost is also suitable for mass production of RFID tags.

1.3 Limitations and assumptions of the study

The following limitations must be considered when interpreting the results of this study:

4. The RFID antenna used in this study was not randomly selected; instead it was specially designed for the study.
5. The company taking part to help the screen printing production for the study had their own experienced printing crews; the authors did not actually perform the printing process in every detail. This study assumes that there were no operator effects on solid ink density and ink film thickness, although only one experienced operator ran the press during the experiment.
6. The make, ages, and physical conditions of the press machine used to run the experiment were not studied. Their effects on the results were therefore not discussed.
7. The type of Ag inks, three substrates, and chips were held as constants. This research did not investigate the consistency of the materials; and therefore, their effects on the results of this study were not explored.
8. Since the pressroom temperature and relative humidity were well controlled, their effects on the experimental results were not studied. It is assumed that there were no temperature and humidity effects on the results of the study.

2. Methodology

This study was a true experimental research in nature and aimed to investigate the process consistency and capability of printing RFID tag antennas via the screening printing process

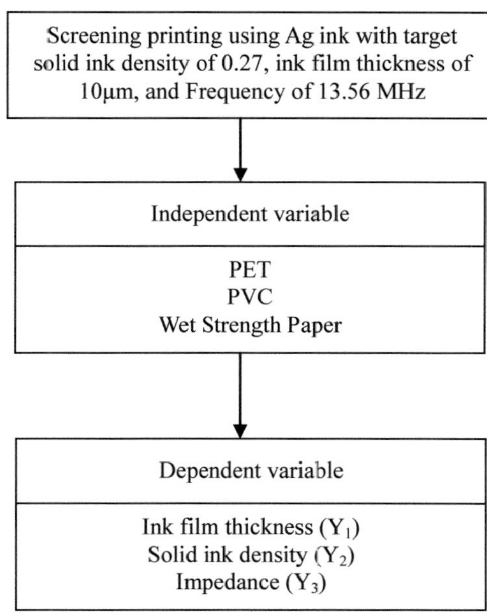

Fig. 1. Research framework

with a conductive ink, silver-based (Ag) ink, on PET, PVC, and Wet Strength paper. The research framework is displayed in Figure 1. The three factors were PET, PVC, and Wet Strength paper. The dependent variables were the solid ink density (SID), ink film thickness (IFT), and impedance (IMPED) of the printed RFID tag antennas.

2.1 The test form

A single color test form for the tag antenna was designed for this study (as shown in Figure 1). The test form is 45mm x 76mm in size and was designed for the frequency of 13.56MHz.

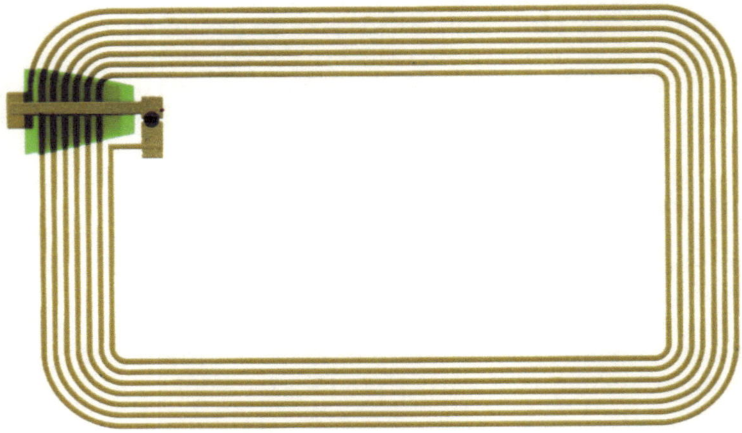

Fig. 2. Antenna design (13.56MHz, 45mm x 76mm) for the study

2.2 Experimental materials

This section describes the experimental procedures for the study. It consists of the screen printing plate materials (see Table 1), substrates (see Table 2) and press setting (see Table 3) for the experiment.

Materials	Description
Fabric Material	PET
Mesh Counts	300 meshes / inch
Mesh Angle	45 degree
Screen Tension	25 N/cm
Thickness of Sensitized Emulsion	25μm

Table 1. Screen plate-making material used for the experiment

Substrates	Manufacturer	Specification
PET (Polyethylene Terephthalate)	NAN YA Plastic Corporation	Thickness: 200μm
PVC (Polyvinyl Chloride)	NAN YA Plastic Corporation	Thickness: 300μm
Wet Strength Paper	HO Zone Paper Inc.	gsm: 80
Silver-based (Ag) Ink,	Flint Conductive Ink for Screen Printing	

Table 2. Substrates and ink used in the study

Item	Description
Press (semi-automatic)	Liang-Chen Mechanical Company
Screen Printer	Mini-Angel Company in Taipei
Press Operator	Mr. Lou
Relative Humidity	46~50%
Temperature	25°C
Blade hardness	70 degree
Squeegee Angle	75 degree
Squeegee Speed	30 m/min

Table 3. Screen printing press setting for the study

After receiving the test form, the participating screen printer was asked to print the test form based on their in-house standard operating procedures and conditions. During the press runs, the research team was present all the time to monitor the whole operation process to make sure that the press run was well-controlled.

2.3 Experimental procedure

Two print tests were run with the first operation serving as a pilot test to familiarize the press operator with printing the test form, while the second operation served as the actual printing experiment where printed RFID tag antennas were sampled. After the first press run, the press was shut down and cleaned, the run counter was set to zero, and the desired materials and conditions were made ready for the next run.

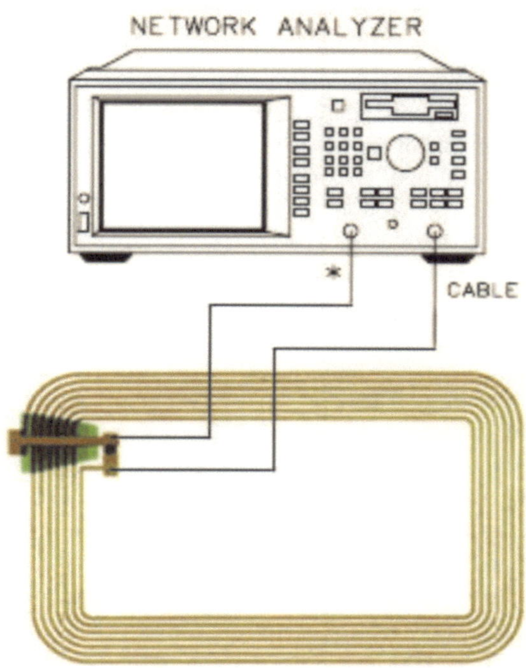

Fig. 3. The diagram of antenna impedance measurement

One hundred printed tags were collected for each press run after the press was determined to be at equilibrium and the desired solid ink density of .27 and ink film thickness of 10 microns (μm) (according to the practical experience of the participating screen printer of the study) were achieved. Consequently, a total of 300 printed tags were gathered for the three runs; and then, 50 printed tags were systematically sampled for each of the three substrates for a total sample size of 150 (3*50). Finally, an X-Rite® 530 reflective spectrodensitometer using Murray-Davies equation (n=1) was applied to measure solid ink density (SID) of the printed tags for this study. It is important to note that each specific measured area on the sampled tag was read five times to reduce the measuring error. Thus, the final data entered onto computer for the analysis was a mean of five readings from the X-Rite® 530. The ink film thickness of the printed antennas was measured by a high-accuracy digimatic indicator. The impedance of the printed tag antennas was read using a HP 8714ET RF Network Analyzer (T/R) (300 kHz to 3 GHz) (see Figure 3. below). The target frequency to be achieved was 13.56 MHz. Finally SPSS 14 and Minitab 14 statistical software packages were used for data analyses.

3. Results and findings

This section describes the overall results and findings obtained through data analyses. The first sub-section exhibits the descriptive statistics for all the measurements. The second sub-section shows the analyses of variance to test the hypotheses whether there was a significant difference in solid ink density, ink film thickness, and impedance of the antennas among the three substrates of the study. The last sub-section analyzes the process consistency and capability for printing RFID antennas on PET, PVC, and Wet Strength paper, respectively.

3.1 Descriptive statistics

Solid ink density (SID) refers to the light-stopping power of color on substrates, measured through the complementary-colored filter. In conventional printing workflows, the setup of solid ink density is a vital factor to achieve an optimum print. Once the right amount of solid ink density is determined, the RIP software automatically optimize the steps for the target linearization, that is, enables a printer to deliver ink on a particular media optimally so that an image's tones can be correctly reproduced. Different linearization settings and profile combinations will affect the final prints. Solid ink density measurement provides an effective means of monitoring and controlling ink film thickness (Tritton, 1997, pp.95-96).

Ink film thickness (IFT) is the most significant of the process variables and the one most easily adjusted during printing: it can be seen affect many print attributes such as tone transfer and print density (Tritton, 1997, pp.141-142).

Impedance is a measure of opposition to a sinusoidal alternating electric current. The concept of electrical impedance generalizes Ohm's law to AC circuit analysis. Unlike electrical resistance, the impedance of an electric circuit can be a complex number, but the same unit, the ohm, is used for both quantities. (Wikipedia, Wikipedia. Retrieved February 26, 2007, from http://en.wikipedia.org/wiki/Electrical_impedance#Definition_of_electrical_impedance)

Table 4 shows the SID, IFT, and impedance basic statistics (mean, standard deviation, minimum, maximum, and 95% Confidence Interval of the mean) of the PET, PVC, and Wet

Strength paper. The overall average SID value of the PET was .266 with a standard deviation of .006, .280 for PVC with a standard deviation of .005, and .266 for Wet Strength paper with a standard deviation of .005. The average IFT value of PET was 8.860μm with a standard deviation of .783, 11.300 for PVC with a standard deviation of .741, and 8.670 for Wet Strength paper with a standard deviation of .688. As for the antenna impedance, the average number was 27.690 ohm with a standard deviation of 1.687 for PET, the average was 26.135 with a standard deviation of 1.142 for PVC, and the average was 27.428 with a standard deviation of 1.183 for Wet Strength paper. It is important to note that the 95% confidence intervals (95% C.I.) of the means of SID, IFT, and impedance for the three substrates are listed in the very right-hand side column of Table 4. However, Table 4 could be used for the specifications for screen printers to print RFID tag antennas using Ag ink.

Observed Attribute	N	Mean	Std. Dev.	Min.	Max.	95% C.I. of Mean
PET_SID	50	0.266	0.006	0.255	0.280	(0.264, 0.267)
PVC_SID	50	0.280	0.005	0.270	0.290	(0.279, 0.282)
wet_SID	50	0.266	0.005	0.260	0.275	(0.264, 0.267)
PET_IFT	50	8.860	0.783	7.250	10.500	(8.638, 9.082)
PVC_IFT	50	11.300	0.741	10.000	13.000	(11.090, 11.510)
wet_IFT	50	8.670	0.688	7.500	10.250	(8.475, 8.866)
PET_IMPED	50	27.690	1.687	24.858	31.034	(27.211, 28.170)
PVC_ IMPED	50	26.135	1.142	25.719	30.960	(25.810, 26.460)
wet_ IMPED	50	27.428	1.813	25.051	31.034	(26.913, 27.944)

Table 4. Descriptive statistics of solid ink density, ink film thickness, and antenna impedance on the different substrates

3.2 Hypothesis testing

In this section, One-way ANOVA and Box-plot statistical procedures were employed to determine whether the differences in solid ink density (SID), ink film thickness (IFT), and impedance readings of the RFID tag antennas printed using screen printing with Ag ink on the PET, PVC, and wet strength paper were significant. The hypothesis being tested was whether the reading difference among the substrates was equal to zero. The significant level (α) was set at .05 for all tests. The results for the SID, IFT and impedance are exhibited in Table 5, Table 6, and Table 7, respectively.

Hypothesis testing on the SID difference for the three substrates

The hypothesis for testing the SID reading difference on the three different tag antennas is:

$$\text{Ho}: \mu_{\text{PET\_SID}} = \mu_{\text{PVC\_SID}} = \mu_{\text{wet\_SID}}$$

$$\text{Ha}: \mu_{\text{PET\_SID}} \neq \mu_{\text{PVC\_SID}},\ \text{or}\,\mu_{\text{PET\_SID}} \neq \mu_{\text{wet\_SID}},\ \text{or}\,\mu_{\text{PVC\_SID}} \neq \mu_{\text{wet\_SID}}$$

As shown in Table 5, the significant value of p is .000 < .05 (α) and therefore the ANOVA suggests that Ho be rejected, i.e., at least one pair of the mean SID values is significantly different at .05 level. Examining the bottom part of Table 5 (95% Confidence Interval for Mean) in detail, one can conclude that there existed significantly different SID readings between the pair of PET and PVC tags and the pair of PVC and Wet Strength paper tags. In addition, the differences in SID readings were not significant at .05 level between PET and Wet Strength paper tags.

Source	DF	SS	MS	F	P
Factor	2	0.007	0.004	120.090	0.000
Error	147	0.004	0.000		
Total	149	0.011			

S = 0.005420 R-Sq = 62.03% R-Sq(adj) = 61.52%

```
                                   Individual 95% CIs For Mean Based on Pooled StDev
Level     N     Mean    StDev   --+---------+---------+---------+-------
PET_SID  50  0.26560  0.00586   (--*--)
PVC_SID  50  0.28010  0.00539                              (--*--)
wet_SID  50  0.26550  0.00497   (--*--)
                                --+---------+---------+---------+-------
                                0.2650    0.2700    0.2750    0.2800
```

Pooled StDev = 0.00542

Table 5. Hypothesis testing on the SID difference among the three substrates

Likewise, the two straight lines originated from PVC_SID box in Figure 4 (the box plot of SID readings for the three substrates) indicate the two pairs substrates with significantly different SID reading were (PET, PVC) and (PVC, Wet). Among the three substrates, PVC has the highest SID mean values than the other two substrates have.

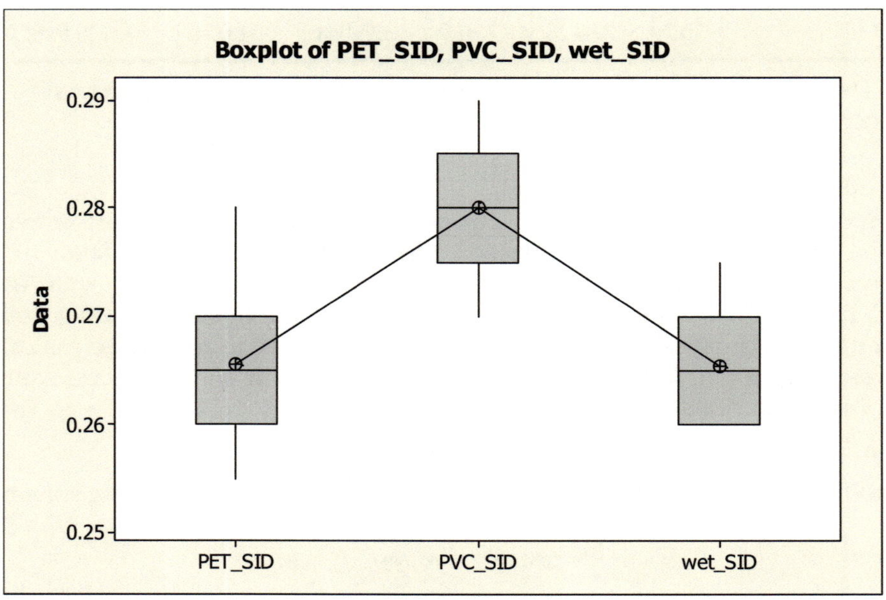

Fig. 4. Box plot of SID readings for the three substrates

Hypothesis testing on the IFT difference for the three substrates

The hypothesis for testing the IFT reading difference on the three different tag antennas is:

$$Ho: \mu_{PET\_IFT} = \mu_{PVC\_IFT} = \mu_{wet\_IFT}$$

$$Ha: \mu_{PET\_IFT} \neq \mu_{PVC\_IFT}, \text{ or } \mu_{PET\_IFT} \neq \mu_{wet\_IFT}, \text{ or } \mu_{PVC\_IFT} \neq \mu_{wet\_IFT}$$

Source	DF	SS	MS	F	P
Factor	2	215.110	107.555	197.450	0.000
Error	147	80.075	0.545		
Total	149	295.185			
S = 0.7381 R-Sq = 72.87% R-Sq(adj) = 72.50%					

```
                                 Individual 95% CIs For Mean Based on Pooled StDev
Level     N    Mean  StDev  ----+---------+---------+---------+-----
PET_IFT  50   8.860  0.783    (--*-)
PVC_IFT  50  11.300  0.741                                (-*--)
wet_IFT  50   8.670  0.688  (-*--)
                            ----+---------+---------+---------+-----
                              8.80      9.60     10.40     11.20
```

Pooled StDev = 0.738

Table 6. Hypothesis testing on the IFT difference among the three substrates

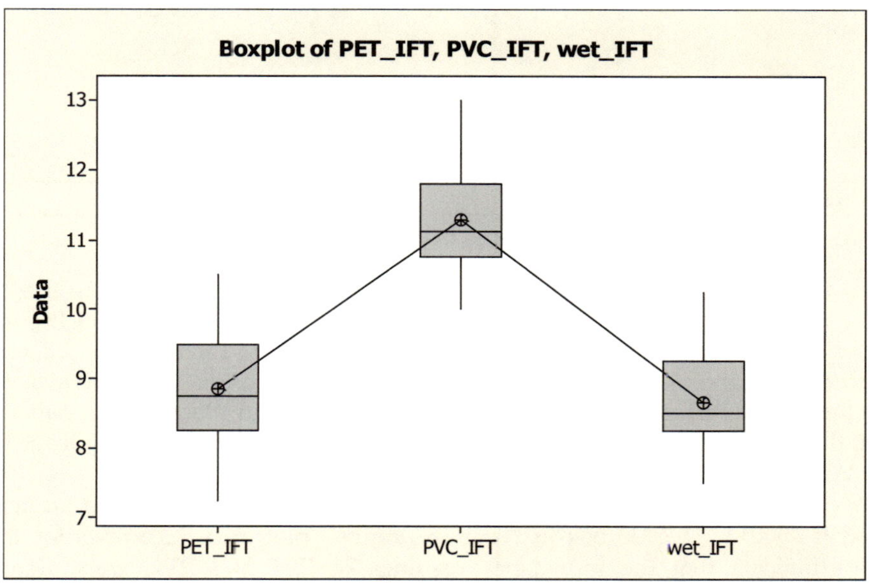

Fig. 5. Box plot of IFT readings for the three substrates

As shown in Table 6, the significant value of p is .000 < .05 (α) and therefore the ANOVA suggests that Ho be rejected. That means that at least one pair of the average IFT values is significantly different at .05 level. If we examine the bottom part of Table 6 (95% C. I. for

Mean), we can conclude that there were significantly different IFT readings between the pair of PET and PVC tags and the pair of PVC and Wet Strength paper tags. Moreover, the differences in IFT readings were not significant at .05 level between PET and Wet Strength paper tags.

The same conclusions could be drawn if we examine Figure 5 in detail: the two straight lines originated from PVC_IFT box in Figure 5 (the box plot of IFT readings for the three substrates) indicate that the IFT readings of PET and PVC were significantly different at .05level, and those of PVC and Wet Strength paper were also significantly different. Among the three substrates, PVC has the highest IFT mean values than the other two substrates have.

Hypothesis testing on the impedance (IMPED) difference for the three substrates

The hypothesis for testing the IFT reading difference on the three different tag antennas is:

$$\text{Ho}: \mu_{\text{PET\_IMPED}} = \mu_{\text{PVC\_IMPED}} = \mu_{\text{wet\_IMPED}}$$

$$\text{Ha}: \mu_{\text{PET\_IMPED}} \neq \mu_{\text{PVC\_IMPED}}, \text{ or } \mu_{\text{PET\_IMPED}} \neq \mu_{\text{wet\_IMPED}}, \text{ or } \mu_{\text{PVC\_IMPED}} \neq \mu_{\text{wet\_IMPED}}$$

Source	DF	SS	MS	F	P
Factor	2	69.33	34.66	13.98	0.000
Error	147	364.57	2.48		
Total	149	433.90			
S = 1.575 R-Sq = 15.98% R-Sq(adj) = 14.84%					

```
                                   Individual 95% CIs For Mean Based on Pooled St
Dev
Level       N   Mean   StDev   ---+---------+---------+---------+------
PET_IMPED  50  27.690  1.687                        (------*-----)
PVC_IMPED  50  26.135  1.142   (-----*------)
wet_IMPED  50  27.428  1.813                     (-----*-----)
                               ---+---------+---------+---------+------
                               25.90     26.60     27.30     28.00
```

Pooled StDev = 1.575

Table 7. Hypothesis testing on the IFT difference among the three substrates

As shown in Table 7, the significant value of p is .000 < .05 (α) and therefore the ANOVA suggests that Ho be rejected. That means that at least one pair of the mean IFT values is significantly different at .05 level. Examining the bottom part of Table 7 (95% C. I. for Mean) more closely, we can conclude that there were significantly different impedance readings between the pair of PET and PVC tags and the pair of PVC and Wet Strength paper tags. In addition, the differences in impedance readings were not significant at .05 level between PET and Wet Strength paper tags.

The same conclusions could be drawn if we examine Figure 6: the two straight lines originated from PVC_IMPED box in Figure 6 (the box plot of IFT readings for the three substrates) indicate that the impedance readings of PET and PVC were significantly different at .05level, and those of PVC and Wet Strength paper were also significantly different at .05 level. Among the three substrates, PVC has the lowest impedance mean values than the other two have. It is important to note that the box plot of PVC_IMPED in Figure 6 shows that the impedance variation (the height of the box in the middle of the PVC_IMPED) of PVC was extremely small compared with that of the other two substrates.

3.3 Capability study

The section is to discuss the process consistency and capability of the observed attributes for the three types of substrates. The tools used to analyze the consistency for each variable are Individual Control Chart (I Chart), Moving Range Charts (MR Chart), and Capability Analysis.

Interpretation of the relative PCR (Cp or Pp)

In capability analysis, overall capability depicts how the process is actually performing relative to the specification limits. Potential capability depicts how the process could perform relative to the specification limits, if shifts and drifts could be eliminated. The difference between the two represents the opportunity for improvement. Without both overall and potential estimates, it is hard to identify the size of the opportunity. Process capability is a measure of how capable a process is of meeting specifications. A Cp index (PCR) of 1 means that a process is exactly capable of meeting specifications, while less than 1 means that it is outside specification limits. Ideally, one would like to see a Cp much larger than 1, because the larger the index, the more capable the process. Some practitioners consider *1.33* to be a minimum acceptable value for this statistic, and few believe that a value less than 1 is acceptable (Ryan & Joiner, 1994).

Determination of the lower specification limits (LSL) and upper specification limits (USL)

Due to the lack of historical parameters of LSL and USL for the observed attributes (SID, IFT, and impedance) for RFID tag antennas using screen printing with Ag ink on the three substrates, a method of determining the proper LSL and USL is necessary. In this study, the LSL and USL for each attribute are determined based on the following procedures (Hsieh, 2003; Montgomery, 1997, pp. 180-229):

1. Construct the trial I and MR control chart of each attribute for the four plates.
2. Examine every control chart; if it is in control, then use the lower control limit (LCL) and upper control limit (UCL) as the LSL and USL. If it is in out-of-control condition (for most cases), reconstruct the control chart after eliminating all out-of-control points in the initial charts to obtain the revised values for mean, LCL, and UCL.
3. For each attribute, the difference between revised LCL and UCL of each plate obtained in the previous step is computed and named $6\sigma_{revised}$, i.e., $UCL_{revised} - LCL_{revised} = 6\sigma_{revised}$. Then $3\sigma_{revised}$ of each plate is computed for the purpose of obtaining the "average $3\sigma_{revised}$" of the four plates, $3\hat{S}_{revised}$ namely, i.e.,
$$3\hat{S}_{revised} = (3\sigma_{revised/PET} + 3\sigma_{revised/PVC} + 3\sigma_{revised/wet}) / 3.$$
4. For each attribute, the final LSL and USL are obtained by subtracting from and adding to the $3\hat{S}_{revised}$, the revised mean of each plate, i.e.,
$$LSL_{final} = Mean_{revised} - 3\hat{S}_{revised}$$
$$USL_{final} = Mean_{revised} + 3\hat{S}_{revised}$$
5. The LSL_{final} and USL_{final} were then used to assess the relative Process Capability Ration (PCR) for the *revised* individual measurement control chart (I-Chart) of each attribute for the three substrates.

The revised control limits ($UCL_{revised}$ and $LCL_{revised}$) for the three attributes (SID, IFT, IMPED) of the three substrates are displayed in Table 8. Table 9 shows the $3\hat{S}_{revised}$ of the attributes computed from Table 8 by taking the average $\sigma_{revised}$ of the three substrates. The LSL_{final} and USL_{final} of the attributes for the three substrates are then computed and exhibited in Table 10.

	PET		PVC		Wet Strength Paper	
	$LCL_{revised}$	$UCL_{revised}$	$LCL_{revised}$	$UCL_{revised}$	$LCL_{revised}$	$UCL_{revised}$
SID	0.249	0.282	0.266	0.294	0.253	0.278
IFT	6.960	10.760	9.455	13.145	6.811	10.529
IMPED	23.080	32.300	25.624	26.080	22.100	32.760

Table 8. The revised control limits of the attributes for the substrates

	$3\hat{S}_{revised}$
SID	$(3\sigma_{revised\_PET\_SID} + 3\sigma_{revised\_PVC\_SID} + 3\sigma_{revised\_wet\_SID}) / 3$ = (0.017 +0.014 +0.013) / 3 = 0.015
IFT	$(3\sigma_{revised\_PET\_IFT} + 3\sigma_{revised\_PVC\_IFT} + 3\sigma_{revised\_wet\_IFT}) / 3$ = (1.900 + 1.845 +1.859) / 3 = 1.868
IMPED	$(3\sigma_{revised\_PET\_IMPED} + 3\sigma_{revised\_PVC\_IMPED} + 3\sigma_{revised\_wet\_IMPED}) / 3$ = (4.610 + 0.228 + 5.330) / 3 = 3.389

Table 9. The $3\hat{S}_{revised}$ of the attributes computed from Table 8

	PET		PVC		Wet Strength Paper	
	LSL_{final}	USL_{final}	LSL_{final}	USL_{final}	LSL_{final}	USL_{final}
SID	0.251	0.281	0.265	0.295	0.251	0.281
IFT	6.992	10.728	9.432	13.168	6.802	10.538
IMPED	24.301	31.079	22.463	29.241	24.041	30.819

Table 10. The LSL_{final} and USL_{final} of the attributes for the substrates

Capability analysis for solid ink density (SID)

The capability analyses of solid ink density for the substrates are exhibited in Figure 7, Figure 8, and Figure 9. As shown in those figures, PVC has the highest relative PCR value (Cp = 1.04), followed by the Wet Strength paper (Cp = 1.02), and PET (Cp = .95). Therefore, this study concludes that the PVC and Wet Strength paper are barely acceptable substrates for printing consistent ink density because their relative PCR are only slightly higher than 1.00. Figure 7 also implies that PET is not an acceptable substrate for printing consistent SID for RFID tags due to the low Cp value (Cp = .95).

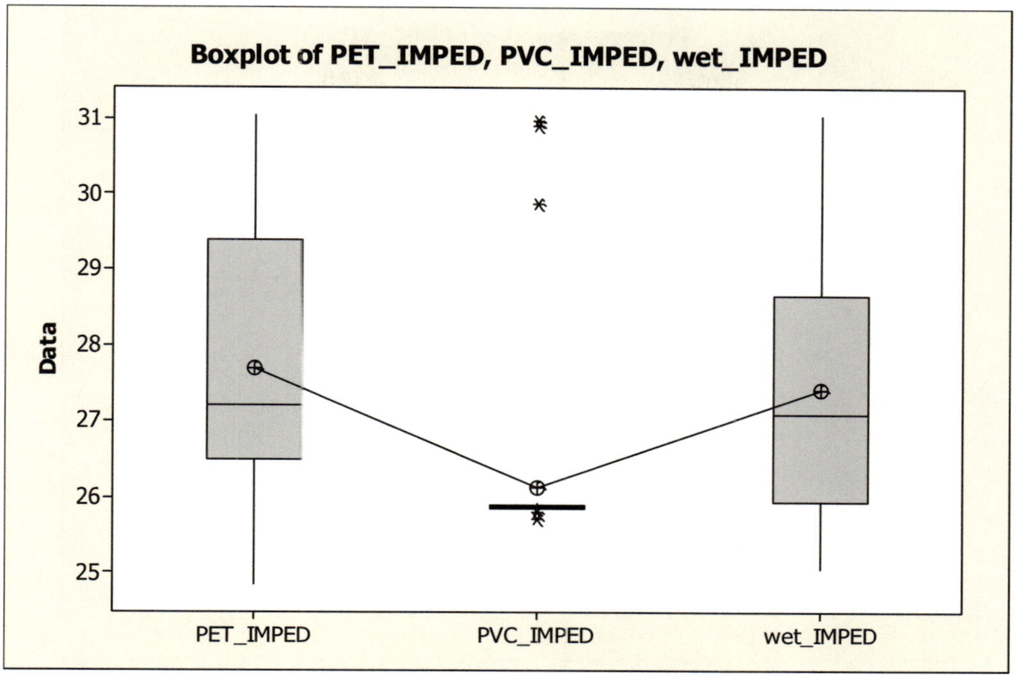

Fig. 6. Box plot of impedance readings for the three substrates

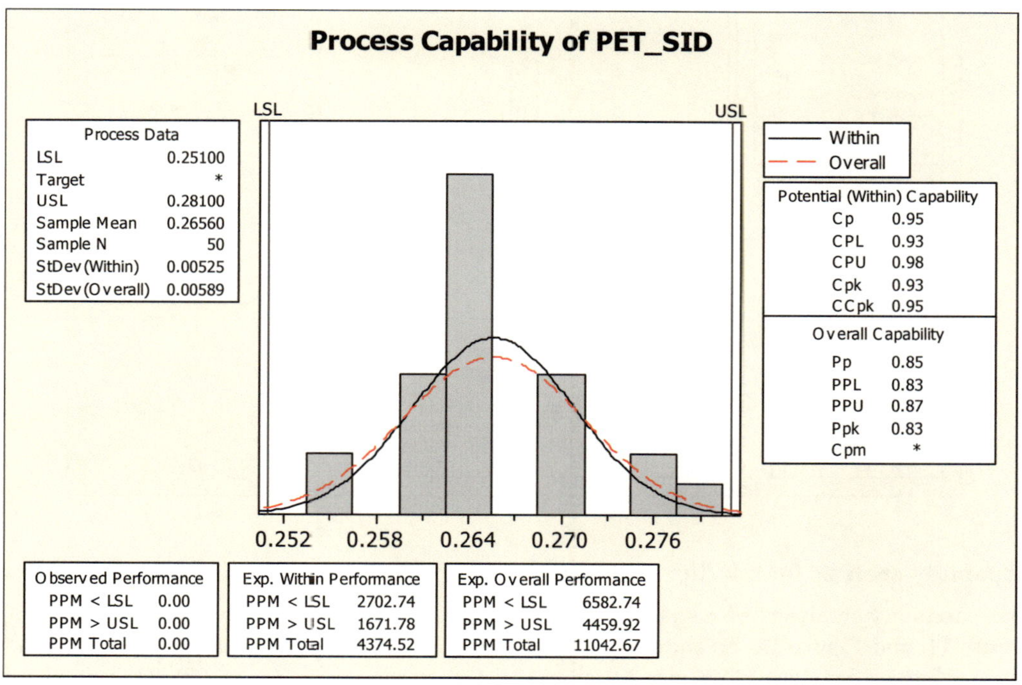

Fig. 7.

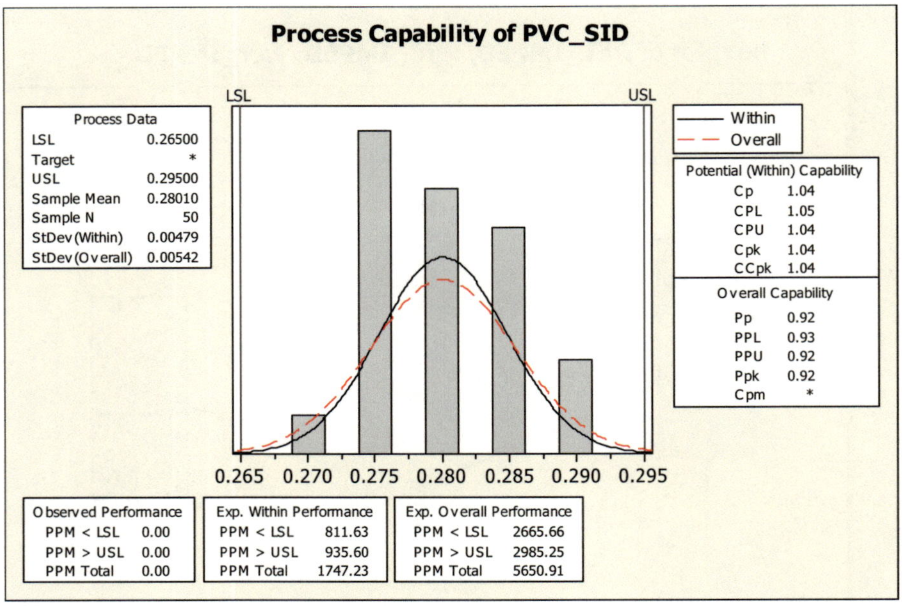

Fig. 8.

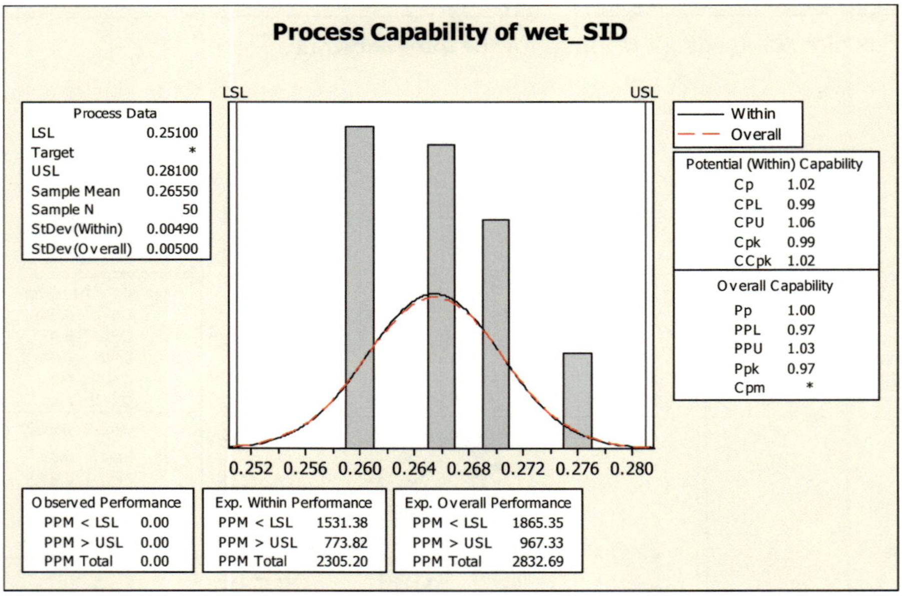

Fig. 9.

Capability analysis for ink film thickness (IFT)

The capability analyses of ink film thickness for the substrates are shown in Figure 10, Figure 11, and Figure 12. As shown in those figures, PET has the largest relative PCR (Cp = 1.10), followed by the PVC (Cp = 1.01), and wet strength paper (Cp = .98). Therefore, this study concludes that the PET was the most acceptable substrate for printing consistent ink film thickness among the three substrates in terms of relative PCR. Due to the small Cp value (.98) of Wet Strength paper, the study concludes that Wet Strength paper might not be an acceptable substrate for printing consistent ink film thickness. However, the Cp value of PET (1.10) is smaller than 1.33; that means that PET is only acceptable, but not necessary satisfied, as the substrate for printing consistent ink film thickness.

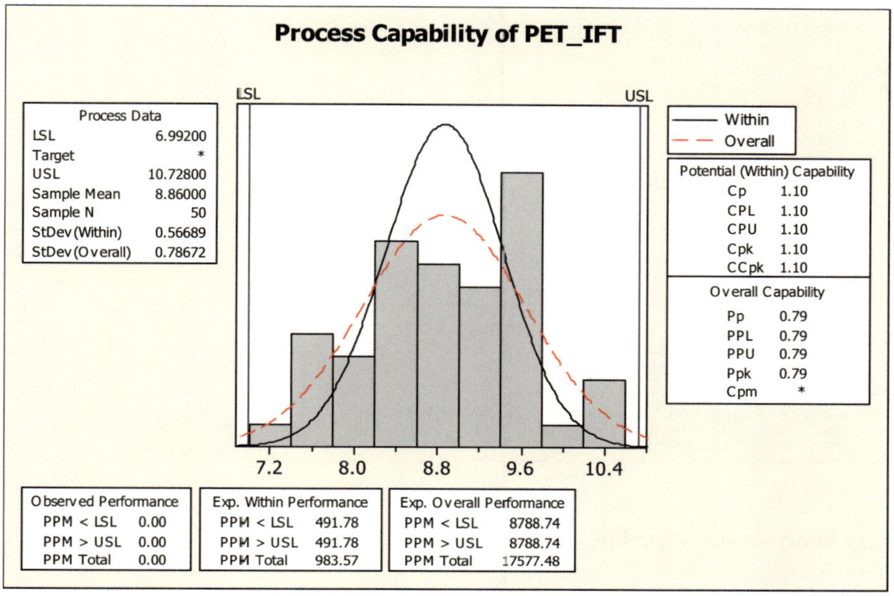

Fig. 10.

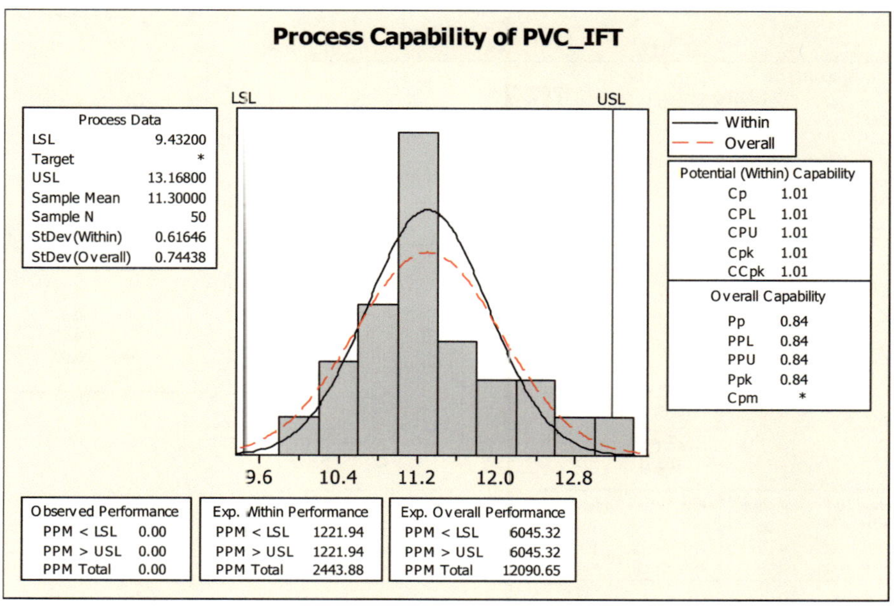

Fig. 11.

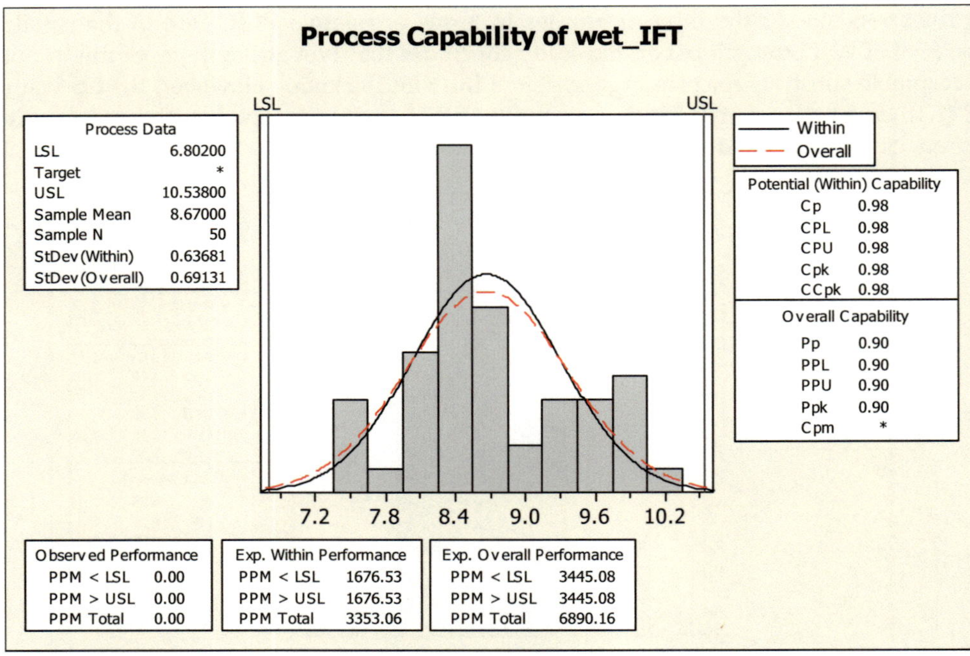

Fig. 12.

Capability analysis for impedance (IMPED)

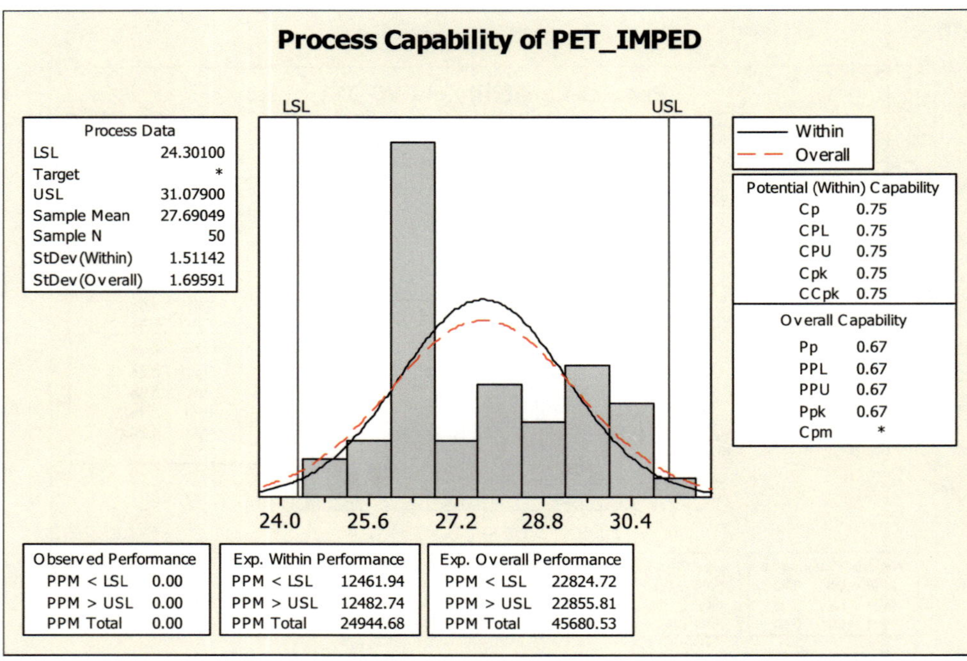

Fig. 13.

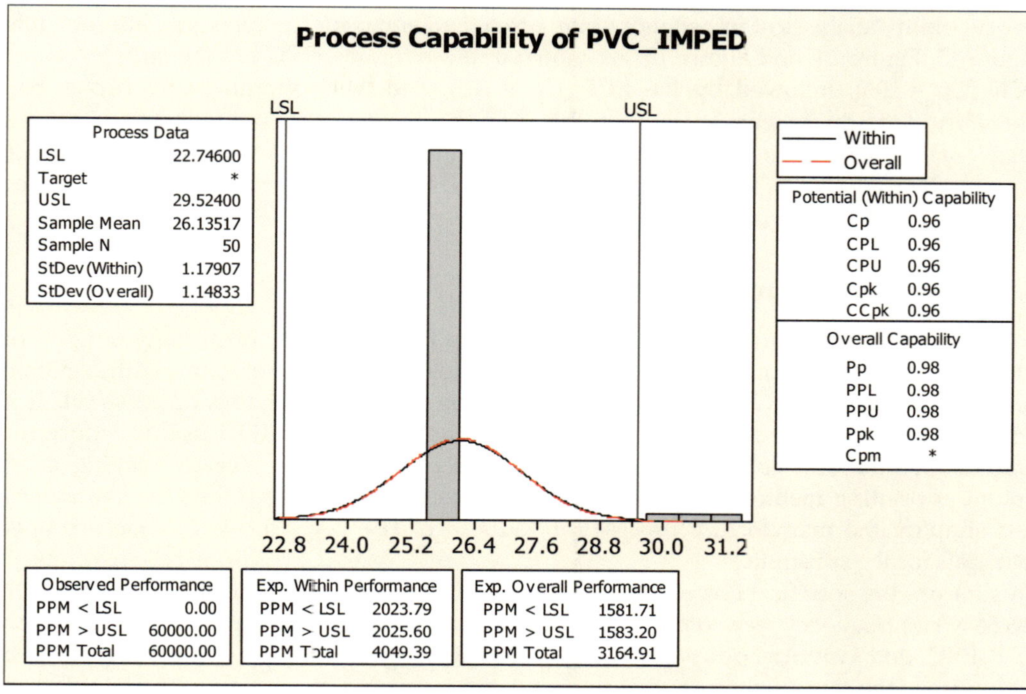
Process Capability of PVC_IMPED
LSL
USL
Process Data
LSL 22.74600
Target *
USL 29.52400
Sample Mean 26.13517
Sample N 50
StDev(Within) 1.17907
StDev(Overall) 1.14833
Within
Overall
Potential (Within) Capability
Cp 0.96
CPL 0.96
CPU 0.96
Cpk 0.96
CCpk 0.96
Overall Capability
Pp 0.98
PPL 0.98
PPU 0.98
Ppk 0.98
Cpm *
22.8 24.0 25.2 26.4 27.6 28.8 30.0 31.2
Observed Performance
PPM < LSL 0.00
PPM > USL 60000.00
PPM Total 60000.00
Exp. Within Performance
PPM < LSL 2023.79
PPM > USL 2025.60
PPM Total 4049.39
Exp. Overall Performance
PPM < LSL 1581.71
PPM > USL 1583.20
PPM Total 3164.91

Fig. 14.

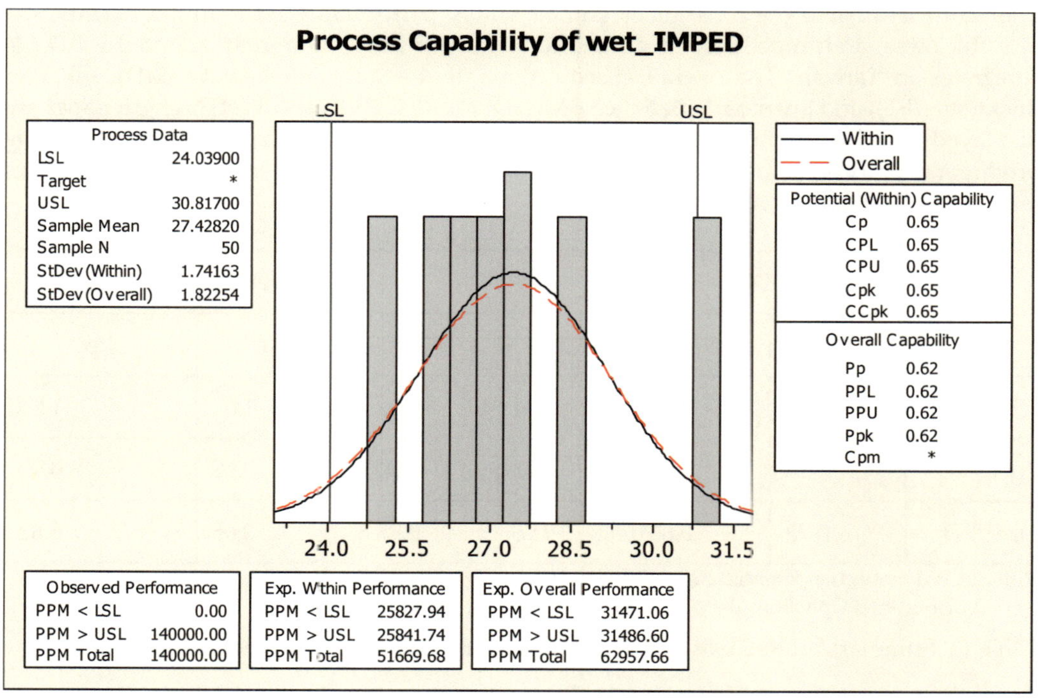
Process Capability of wet_IMPED
LSL
USL
Process Data
LSL 24.03900
Target *
USL 30.81700
Sample Mean 27.42820
Sample N 50
StDev(Within) 1.74163
StDev(Overall) 1.82254
Within
Overall
Potential (Within) Capability
Cp 0.65
CPL 0.65
CPU 0.65
Cpk 0.65
CCpk 0.65
Overall Capability
Pp 0.62
PPL 0.62
PPU 0.62
Ppk 0.62
Cpm *
24.0 25.5 27.0 28.5 30.0 31.5
Observed Performance
PPM < LSL 0.00
PPM > USL 140000.00
PPM Total 140000.00
Exp. Within Performance
PPM < LSL 25827.94
PPM > USL 25841.74
PPM Total 51669.68
Exp. Overall Performance
PPM < LSL 31471.06
PPM > USL 31486.60
PPM Total 62957.66

Fig. 15.

The capability analyses of impedance of the printed antennas for the substrates are shown in Figure 13, Figure 14, and Figure 15. As shown in those figures, PVC has the largest relative PCR (Cp = .96), followed by the PET (Cp = .75), and wet strength paper (Cp = .65). Therefore, this study concludes that the PET, PVC, and wet strength paper are all unacceptable substrates for printing RFID antennas to achieve consistent impedance because their relative PCR are all smaller than 1.0. The worst substrate to deliver consistent impedance of the antenna is Wet Strength paper (Cp = .65).

4. Conclusions and recommendations

Substrate is definitely a dominant variable affecting the printability of printing antenna of RFID tags. The results of study can assist the printing industry in determining the optimal substrates for printing RFID tags using screen printing technology with conductive ink. It is wise to apply screen printing technology to print the antennas of RFID tags to reduce the production time and cost. Based on the experience of the study, screen printing is an optimum printing method, in terms of process consistency and capability, to save money and simplify the manufacture process for RFID tags. The key issue is to choose proper materials for the substrate.

To sum up, the specifications of antenna ink film thickness and ink density to achieve HF (13.56 MHz) frequency were reported respectively when the RFID antennas were printed on PET, PVC, and Wet Strength paper using screen printing. In addition, the consistency and capability of the three processes were analyzed and compared. It is hoped that the study can assist screen printers in determining the optimum substrates for printing RFID tag antennas. This study evaluated the consistency and capability performance on solid ink density, ink film thickness, and impedance of printed antennas for three commonly adopted RFID tag substrates in Taiwan. The overall specifications of the solid ink density (SID), ink film thickness (IFT), and antenna impedance (IMPED) for PET, PVC, and Wet Strength paper are displayed in Table 4. These real-world specifications could be incorporated into RFID tag production process as an evaluation mechanism to ensure the screen printers in Taiwan meet required quality levels.

	PET		**PVC**		**Wet Strength Paper**	
	Cp	Pp	Cp	Pp	Cp	Pp
SID	0.95	0.85	1.04	0.92	1.02	1.00
IFT	*1.10*	0.79	1.01	0.84	0.98	0.90
IMPED	0.75	0.67	*0.96*	0.98	0.65	0.62

Note: 1 Cp denotes the potential capability; Pp denotes the overall capability
2.Underlined Cp values denote the largest Cp in the group.

Table 11. Summarized Relative PCR (Cp Value) of the attributes for the three substrates

Based the presentation of Figure 7 to 15, Table 11 summarizes the capability performance of the substrates, in terms of relative Cp indexes, in solid density (SID), ink film thickness

(IFT), and antenna impedance (IMPED). As shown in Table 11, the PVC and Wet Strength paper appear to be barely capable of yielding consistent SID because their SID Cp values are just over 1.00 (1.04 for PVC, 1.02 for Wet Strength paper). As to PET, its Cp value is .95 (<1.00) and that means PET is not an acceptable substrate for printing RFID tag antennas in terms of process consistency. According to Table 11, this study concludes that the Wet Strength paper is not a suitable substrate to print consistent IFT and impedance attributes for RFID tags because it had smallest IFT and IMPED Cp value among the three substrates, and its Cp values for IFT and impedance attributes are both smaller than 1.00. This study therefore concludes that the Wet Strength paper is the least capable substrate of producing consistent results in ink film thickness and antenna impedance. Table 11 also indicates that the PET is the most capable substrate for printing consistent ink film thickness among the three substrates in terms of relative PCR.

However, as shown in Table 11, none of the three substrates was capable of producing consistent antenna impedance. Therefore the study recommends that a further research be necessary to investigate the possible special variations in the screen printing process to print PET, PVC, and Wet Strength paper using Ag ink, based on the low Cp values of the antenna impedance for the three substrates.

5. Acknowledgements

We wish to express our sincere gratitude to all participants in this study who invested time and materials in the effort to help us finish this research. Sincere appreciation is also expressed to the Nation Science Council (NSC) of Taiwan who mainly provided financial support to this research. The completion of this research could not have been accomplished without the help from my research assistants, Hui-Wen Cheng, Hui-Ping Liou, Chung-Yin Yang, Mandy Wu, and Yi-Li Wang. I would like to take this opportunity to express my thanks to them.

6. Reference

Björninen, T., Merilampi, S., Ukkonen, L., Sydänheimo, L., & Ruuskanen, P. 2009 "The Effect of FabricationMethod on Passive UHF RFID Tag Performance," International Journal of Antennas and Propagation, Vol. 2009, 1-8.

Blayo, A. & Pineaux B. 2005 "Printing Processes and their Potential for RFID Printing," Proceedings of Joint sOc-EUSAI conference, 27-30.

Definition of electrical impedance.Wikipedia. Retrieved February 26, 2007, from http://en.wikipedia.org/wiki/Electrical_impedance#Definition_of_electrical_impedance

Eldred, Nelson R. & Scarlett, Terry. 1995 "What The Printers Should Know About Ink," Pittsburgh, PA: GATFPress.

Gao, B. & Yuen, M. F. 2009 "Optimization of Silver Paste Printed Passive UHF RFID Tags," Proceedings of International Conference on Electronic Packaging Technology & High Density Packaging, 512-515.

Hodgson, A. The role of paper in the future of printed electronics. Retrieved May 26, 2011, from http://www.copadd07.ethz.ch/papers/3.pdf Hoff, Samuel B. 1997 "Screen Printing: A Contemporary Approach," New York, NY: Delmar Publishers.

Hsieh, Yung-Cheng. 2003 "A Capability Study of Dot Reproduction for CTP Plates," Visual Communication Journal, 2003. pp. 27-40.

Koptioug, A., Jonsson, P., Sidén, J., Olsson, T., & Gulliksson, M. On the Behavior of Printed RFID Tag Antennas, Using Conductive Paint, Retrieved May 26, 2011, from http://www.sensiblesolutions.se/articles/antenn_03_-_on_the_behavior.pdf

Montgomery, Douglas C. 1997 "Introduction to Statistical Quality Control (3rd ed.)," New York: John Wiley & Sons, Inc.

Parashkov, R., Becker, E., Riedl, T. Johannes, H. H., & Kowalsky, W. 2005 "Large Area Electronics Using Printing Methods," Proceedings of the IEEE, Vol. 93, No. 7, 1321-1329.

Ryan, B. F. and Joiner, B. L. "Minitab Handbook," Belmont, CA: Duxbury Press.

Sangoi, R., Smith, C. G., Seymour, M. D., Venkataraman, J. N., Clark, D. M., Kleper, M. L., & Kahn, B. E.

2004 "Printing Radio Frequency Identification (RFID) Tag Antennas Using Inks Containing Silver Dispersions," Journal of Dispersion Science and Technology, Vol. 25, No. 4, 513–521.

Subramanian, V., Fréchet, J. M. J., Chang, P. C., Huang, D. C., Lee, J. B., Molesa, S. E., Murphy, A. R., Redinger, D. R., & Volkman, S. K.

2005 "Progress Toward Development of All-Printed RFID Tags: Materials, Processes, and Devices," Proceedings of the IEEE, Vol. 93, No. 7, 1330-1338.

9

Troubleshooting RFID Tags Problems with Metallic Objects Using Metamaterials

Mª Elena de Cos and Fernando Las-Heras
Universidad de Oviedo
España

1. Introduction

Radiofrequency Identification (RFID) is a technology that is being rapidly developed and that uses radiofrequency (RF) signals for the automatic identification of objects or persons. Although the first article regarding modulated electromagnetic backscattering (basic principle of passive RFID) was published in 1948 (Stockman, 1948) it has been a long way to progress for reaching today levels (Rao, 1999; Finkenzeller, 2004; Pozar, 2004). Nowadays RFID finds many applications in logistics, supply chain management, access control, electronic toll systems, targets identification, vehicle security, animals tracking and patients' identification in hospitals.

An RFID system is composed of a reader, a reader antenna (usually circularly polarized patch antenna), RFID 'tags' or transponders and a middleware or subsystem of data processing. A passive RFID tag consists of an antenna and an application specific integrated circuit (ASIC) chip. IC chips have complex input impedances, and their impedances vary with frequency. A key point for tag antenna design is that it must be conjugately matched with the desired IC chip for the maximum power transfer (Gevi, 2004; Rao et al, 2005).

The different types of RFID systems are distinguished by two major characteristics: the power source of the tag and the frequency of operation. With regards to the power source of the tag, they can either be active (powered by battery), passive (powered by the reader field) or semi-passive (battery assisted backscatter). According to the frequency of operation the RFID systems are generally distinguished into four frequency ranges; i.e., low frequency (LF) (125-134.2 kHz), high frequency (HF) (13.56 MHz), ultra high frequency (UHF) (433, 860-960 MHz) and microwave frequency (2.45, 5.8 GHz). In addition, the standards of the UHF RFID are different for each country: 866-869 MHz in Europe, 902-928 MHz in America and 950-956 MHz in Asia. The communication frequencies used depends to a large extent on the application. Regulations are imposed by most countries (grouped into 3 Regions: US, Europe and Asia) to control emissions and prevent interference with other Industrial, Scientific and Medical equipment (ISM).

The higher the frequency band the faster the speed of tag reading and also the larger the information storage capacity. This is the reason why UHF RFID has gained popularity in many applications and it can be expected that the same will happen in the near future with microwave RFID.

In a typical application tags are attached to objects (or persons). Each tag has a certain amount of internal memory (EEPROM) in the chip in which it stores information about the

object (or person), such as its EPC (electronic product code) or unique identification (ID) serial number and some other data depending on the application, i.e. manufacture date and product composition, (or personal information for access control or health care matters).
A passive back-scattered RFID system operates as follows: a modulated signal with periods of unmodulated carrier is transmitted by a reader and is received by the tag antenna. Then the RF voltage developed on antenna terminals during unmodulated period is converted to dc. The chip is powered up with this dc voltage and sends back the information by varying its front end complex RF input impedance. The modulation of the back-scattered signal is carried out by toggling the impedance between two different states, i.e., conjugate match and some other impedance (Rao et al, 2005)
The tag antenna, together with the chip sensitivity, plays a key role in the RFID system performance, such as the reading range (VanBladel, 2002) and compatibility with the tagged object. In sum, the requirements for RFID tag antennas are the following (Foster & Burberry, 1999):

- Good impedance matching for receiving maximum signals from the reader to power up the chip;
- Insensitive to the attached object to keep performance consistent;
- Required radiation patterns (omnidirectional, directional or hemispherical);
- Small enough and low profile to be attached to or embedded into the specified object (Rao et al, 2005);
- Robust in mechanical structure (since they could be bent in some applications);
- Low cost in both materials and fabrication.

Antennas do not operate independently of nearby objects. On the contrary, these objects can ruin the radiation properties of the antenna to different extent. In RFID systems, the material of the objects the tags are attached to should have minimum effect on tag antenna behaviour, so that the reading performances of tags, such as readable range and reading stability, do not change. However, the performance of a tag antenna varies when it is mounted on different objects (Dobkin & Weigand, 2005; Clarke et al, 2006). On the one hand if the object surface is made of a dielectric material, then the readable range is decreased due to frequency shift of the resonance frequency. On the other hand, metallic objects which are usually tagged in RFID applications seriously degrade the terminal impedance matching, bandwidth, radiation efficiency and readable range of the tag antenna. This is such a critical problem that global deployment of passive UHF RFID systems is being hindered by the performance degradation of tag antennas placed nearby metallic objects. As it has already said, in the vicinity of conductors, the antenna radiation parameters are modified; for example radiation efficiency is decreased. In addition, a metallic surface typically decreases the input impedance of the antenna (which makes that lower or not enough power can be supplied to the IC chip, so the reading range is reduced or even the tag is not read at all) and varies its resonance frequency. The electromagnetic wave is greatly reflected by the conductor surface yielding a significant reduction of the RFID tag operating distance or its total malfunctioning (Dobkin& Weigand, 2005; Clarke et al, 2006; Rao et al, 2005). These negative effects are increased at higher frequencies and so, RFID operation in the SHF band with tags attached to metallic objects presents an even more critical problem to be solved.
To overcome these problems and to obtain RFID tags usable with metallic objects, researchers have proposed different approaches:

- To design novel antennas rather than dipole based antennas (with the inconvenient of large thickness or with shorting planes). As for example patch antennas (Ukkonen et al,

2006) that already have a metallic ground plane but they show some shortcomings as narrow bandwidth and not negligible thickness. Another possibility that has been already explored are tag antennas using a planar inverted-F structure (Hirkonen et al, 2004; Kwon & Lee, 2005) that can operate well on metallic objects, since they already have large ground planes, but they have several important drawbacks such as high cost and difficulty in manufacturing, because they require multiple shorting pins and a large ground plane, as well as thick dielectric substrates.

- To use dipoles separated $\lambda/4$ from the metallic object (for example using foam, which leads to thick antenna designs and more complex manufacturing process)
- The adoption of ferroelectric material to insulate the tag from metal (which is rather expensive).
- To use Perfect Magnetic Conductors (PMCs) since they have a +1 reflection coefficient with magnitude of 1 (in the ideal lossless case) and a phase of 0°. So, they show in-phase reflection, which seems to be a proper solution to the destructive interference problem when the antenna is placed very close to the metallic plate. Thus, the PMC can be used as a barrier between the antenna and the metallic plate in order to electromagnetically insulate the antenna from the disturbing metallic plate effects. For this reason, this approach is going to be analyzed in this chapter. In addition, other advantages such as enhanced efficiency can be obtained as a reward for the use of PMCs. PMCs do not exist in nature and so they have to be synthesised. For this reason they are known as Artificial Magnetic Conductors (AMCs) and behave as PMCs over a certain frequency band.

2. Design of AMC structures for different RFID frequency bands

An Artificial Magnetic Conductor (AMC) is dual to a Perfect Electric Conductor (PEC) from an electromagnetic point of view. For design and analysis purposes, AMC condition is indicated by a reflection coefficient with magnitude of 1 (in the ideal lossless case) and a phase of 0°. The reflection phase on the AMC plane varies continuously from -180° to 180° related to the frequency and is zero at the resonance frequency. The useful bandwidth of AMC performance is defined in the range from +90° to -90°, since in this range, the phase values would not cause destructive interference between direct and reflected waves (Sievenpiper, 1999; Sievenpiper et al, 1999). The surface impedance of an AMC is very high in its bandwidth of AMC performance, so they are also known as High Impedance Surfaces (HIS).

A commonly used technique for AMCs implementation consists in using two-dimensional periodic metallic lattices patterned on a conductor-backed dielectric surface, known as PEC-baked metallo-dielectric Frequency Selective Surfaces (FSSs) and also called Electromagnetic band-gap (EBG) surfaces, as they have one or multiple frequency band-gaps in which no substrate mode can exist. However, in the absence of via holes, the AMC and EBG frequency bands do not always coincide (Goussetis et al 2006). Their unique properties have been applied to design antennas with a better gain and efficiency, lower sidelobes and backlobe level (Mosallaei & Sarabandi, 2004; Feresidis et al, 2005; Mantash et al, 2010a, 2010b). Several narrow band antennas, such as Microstrip patches and dipoles have been mounted on these periodic structures in previous works (McVay et al, 2004; Liang & Yang, 2007; Zhu & Langley, 2009). With the aim of obtaining AMC designs that can be easily integrated in low profile antennas and microwave and millimeterwave circuits, recent research efforts focus on the development of planar unilayer EBGs (in contrast to the use of multilayered FSS (Monorchio et al, 2002)) that do not need vias (Yang et al, 1999; Zhang et al, 2002; McVay et al, 2004; Kern et al, 2005). The main drawback of using unilayer FSSs over a metallic ground plane is the very narrow AMC operation bandwidth, due to EBGs' inherent resonant nature. In addition, designing compact AMCs for frequencies below 1GHz as those required in UHF RFID applications is by itself quite challenging and specially when intended to be used for RFID tags due to their size and thickness restrictions.

Each AMC unit-cell can be seen as implementing a distributed parallel LC network having one or more resonant frequencies. The resonance frequency is where the high impedance and AMC conditions occur and for a parallel LC circuit is equal to $1/\left(2\pi\sqrt{LC}\right)$, while in-phase reflection bandwidth is proportional to $\sqrt{L/C}$ .The resonance frequency and the bandwidth of an AMC depend on the unit-cell geometry together with substrate's relative dielectric permittivity and thickness. So, it is necessary to increase L and reduce C in order to obtain a wider AMC operation bandwidth. Lower frequency applications require higher L and/or C values. L can be increased using a thicker dielectric substrate and also including in the geometry narrow and long strips (lines). C can be reduced by reducing substrate's relative dielectric permittivity ε_r and increasing the gap between the metallization edge and the unit-cell edge (and so the gap between adjacent unit-cells). In order to obtain both compact size and broad AMC operation bandwidth a trade-off solution regarding ε_r and substrate thickness has to be adopted.

With the aim of searching the frequency band in which the periodic structure behaves as an AMC, its reflection coefficient for a uniform incident plane wave is simulated, using Finite Element Method (FEM) together with the Bloch-Floquet theory, modelling a single cell of the structure with periodic boundary conditions (PBC) on its sides, resembling the modelling of an infinite structure (Sievenpiper et al, 1999; Yang & Rahmat-Samii, 2003). The periodic surface is chosen as the phase reference plane. Normal plane waves are launched to illuminate the periodic surface using a waveport positioned a half-wavelength above it. The phase of the reflection coefficient of the AMC plane is compared to that of a PEC plane taken as reference, in the same way as in (Sievenpiper et al, 1999).

The aim of this section is to show an AMC structure design proper to be used for European UHF RFID frequency band tags and for 2.4GHz and 5.8GHz SHF RFID frequency band tags, using the same geometry for the AMC unit-cells and just changing the dielectric substrate and/or the unit-cell size. AMC structures for other UHF RFID bands can be easily obtained just by scaling the unit-cell metallization from the European UHF unit-cell design, and/or slightly scaling the whole unit-cell.

Unit cell size W (mm)	Thickness h (mm)	ε_r	BW (%)	Reso. freq (GHz)
16.93 ($\lambda/20$)	2.54 ($\lambda/136$)	25.0	4.63	0.864
16.93 ($\lambda/7$)	1.27 ($\lambda/98$)	10.2	5.24	2.480
11.52($\lambda/5$)	0.81 ($\lambda/64$)	3.38	7.20	5.820

Table 1. AMC Unit-cell design parameters and resulting resonance frequencies and bandwidths of AMC performance.

Table 1 shows the unit-cell dimensions and the dielectric substrate parameters to achieve the indicated resonance frequencies and bandwidths of AMC performance. The three AMC designs use commercial dielectric substrates: Transtech MCT-25 with relative dielectric permittivity ε_r=25 and loss tangent less than 0.001, Rogers RO3010 with ε_r=10.2 and loss tangent 0.0035 and RO4003C with ε_r=3.38 and loss tangent less than 0.0027.

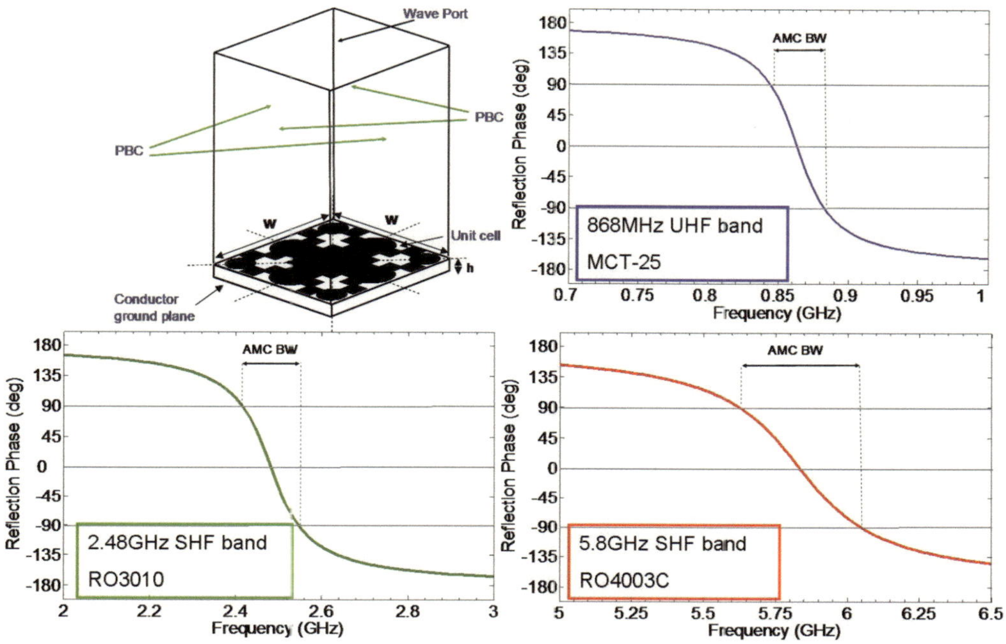

Fig. 1. Simulation model and Reflection phase of the simulated AMC prototypes

The simulated reflection phase of normally incident plane wave (field strength 1 V/m) on the AMC surface versus frequency for the designed unit cell geometry with the three different dielectric substrates is shown in Fig. 1. The bandwidth of AMC performance increases with the thickness of the dielectric substrate but decreases as the relative dielectric permittivity gets higher values. The three presented designs (see Fig.1 and Table 1) show broad bandwidth using neither via-holes nor multilayered structures, which simplifies manufacturing process and reduces the costs.

It is remarkable that the broad AMC operation bandwidth of this specific unit cell geometry makes possible its combination with an antenna without significantly reducing the antenna bandwidth, which is the common drawback pointed out when dealing with AMC structures due to their inherent narrow bandwidth.

Another major concern on AMCs operation is related to their angular stability (Simovski et al, 2005). This can be analyzed from two different points of view: the first analysis is performed with regards to AMC operation under normal incidence condition when the polarization of the incident field is varied. The second analysis is focused on the AMC performance under oblique incidence. Both of them are very important because when combining the AMC with the antenna, the angular stability of the AMC will influence the antenna radiation performance and this will have direct impact on the angular reading range of the final RFID tag depending on the position of the reader with respect to the tagged object. Following this, an AMC design with as higher angular stability as possible is desirable.

As pointed out in section 1, the negative effects of metallic objects in RFID tags are increased at higher frequencies and so the following discussions are going to be focused on an AMC to be used on 5.8GHz SHF RFID frequency band tags.
The reflection phase of the designed AMC surface has been simulated for different incident field (E_{inc}) polarization angles (φ). The unit cell design symmetry makes possible the AMC to operate identically for any polarization of the incident field (assuming normal incidence), as shown in Fig. 2. This also means that reflection phase of both TE and TM polarizations of the incident wave will be identical for normal incidence.
Regarding AMC operation under oblique incidence, it can be extracted from Fig. 3 that resonance conditions are met within an angular margin of $\theta_{inc}= \pm58^o$ (due to the unit cell design symmetry) for TE polarization. In this range the deviation of the resonance frequency is less than 1%. For higher incident angles the resonance frequency shifts to another band. It is also remarkable that the AMC operation bandwidth decreases from 7.20% to 3.39% as the incident angle θ_{inc} is increased from 0° to $+58^o$. However, the 5.8GHz frequency of interest is within the AMC operation bandwidth for all the incident angles in the $\theta_{inc}= \pm58^o$ angular margin. This means that there is almost a 120° angular margin in which the structure performs as an AMC at 5.8GHz. For TM polarization, the angular margin reduces to $\theta_{inc}= \pm40^o$, the deviation of the resonance frequency is 6.83% and the AMC operation bandwidth is preserved. So there is a 80° angular margin in which the structure performs as an AMC at 5.8GHz for both Te and TM polarizations of the incident wave, which can be considered as a very stable AMC structure.

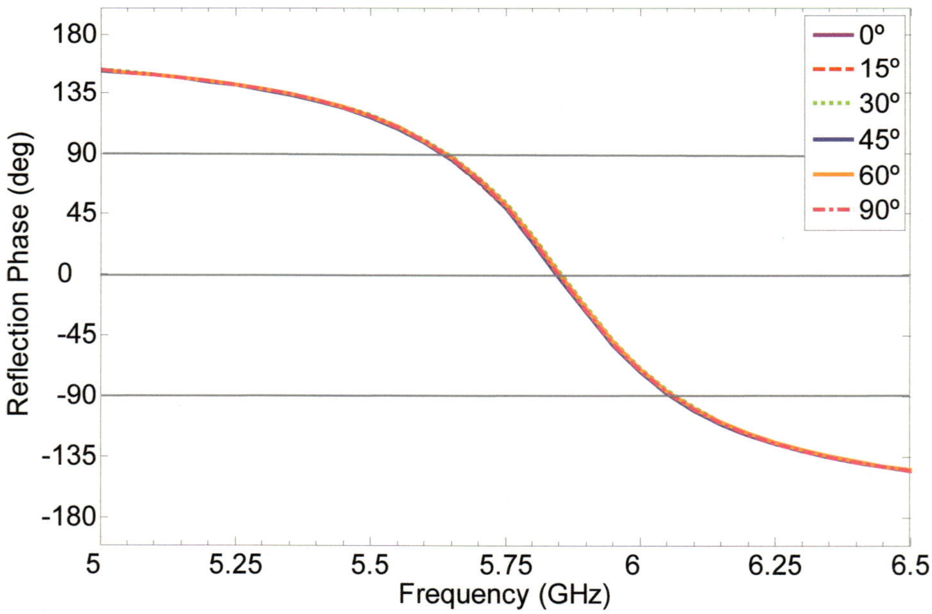

Fig. 2. Simulated Reflection phase of the AMC surface for different incident field (E_{inc}) polarization angles $\varphi=0^o, 15^o, 30^o, 45^o, 60^o$ and 90°.

It is important to point out that angular stability under oblique incidence depends not only on the unit cell design geometry but also on the thickness of the dielectric substrate and on the unit cell size (periodicity) compared to the dielectric substrate thickness (Hosseini et al, 2006; Simovski et al, 2005).

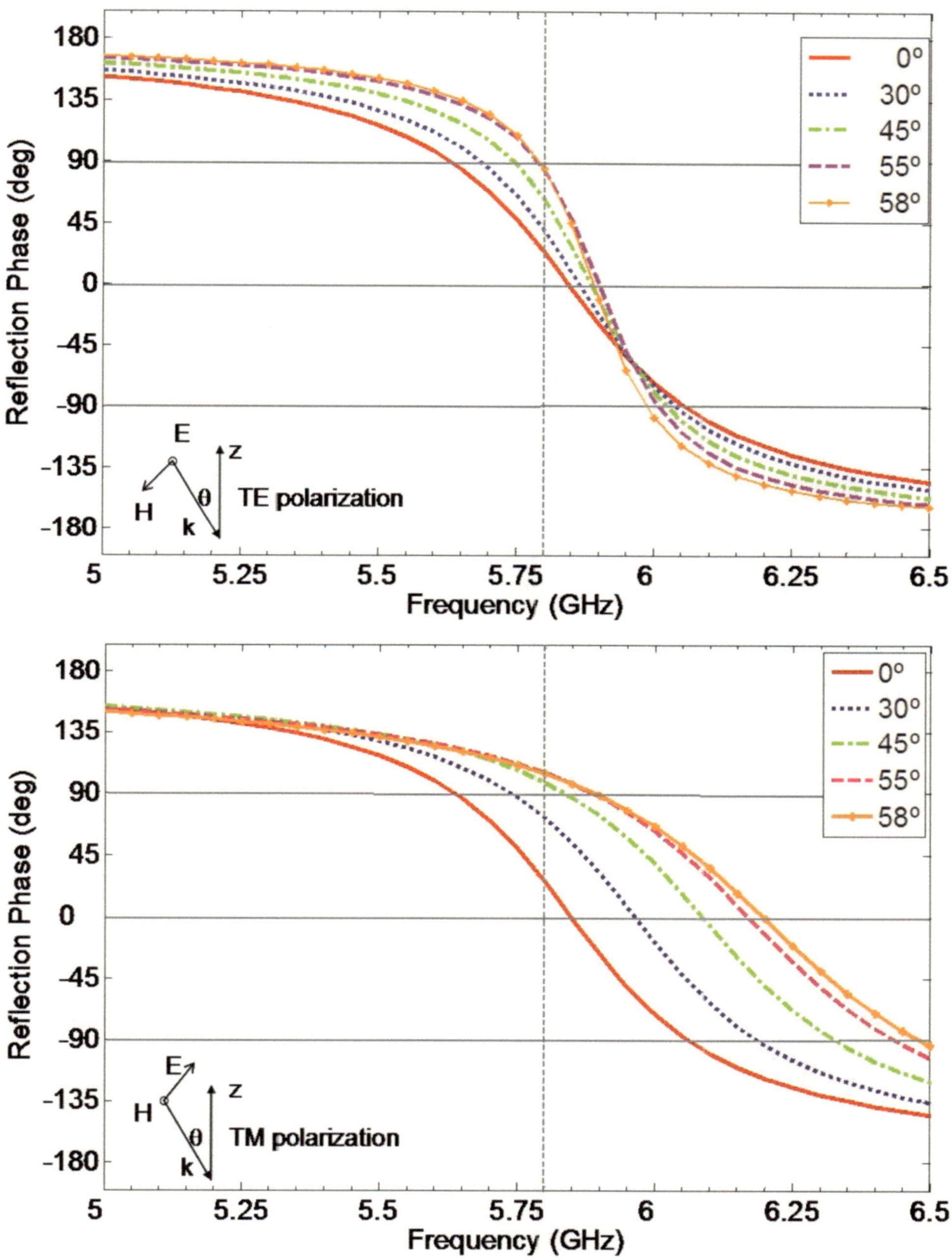

Fig. 3. Simulated Reflection phase of the AMC surface for TE (up) and TM (down) polarizations for different incident angles θ_{inc}=0°, °, 30°, 45°, 55° and 58°.

3. Antenna on AMC to be used in 5.8GHz SHF RFID tags over metallic objects

Firstly, a miniature printed CPW-fed slot antenna (Lin et al, 2005) for operating in the 5.8GHz frequency band has been designed (see Fig.4) using RO4003C, with ε_r=3.38, loss tangent less than 0.0027 and 0.813mm thickness, as dielectric substrate. A slot antenna has been chosen because it will provide wider bandwidth making easier the combination with the narrower bandwidth of AMC performance. There is no metallic layer under the antenna dielectric substrate. This antenna has a simple structure with only one layer of dielectric substrate and metallization.

The antenna dimensions together with simulated return losses for the antenna are shown in Fig. 4. The simulated operating bandwidth of the antenna (range of frequencies with S11≤-10dB) is 1.48GHz (22.0%).

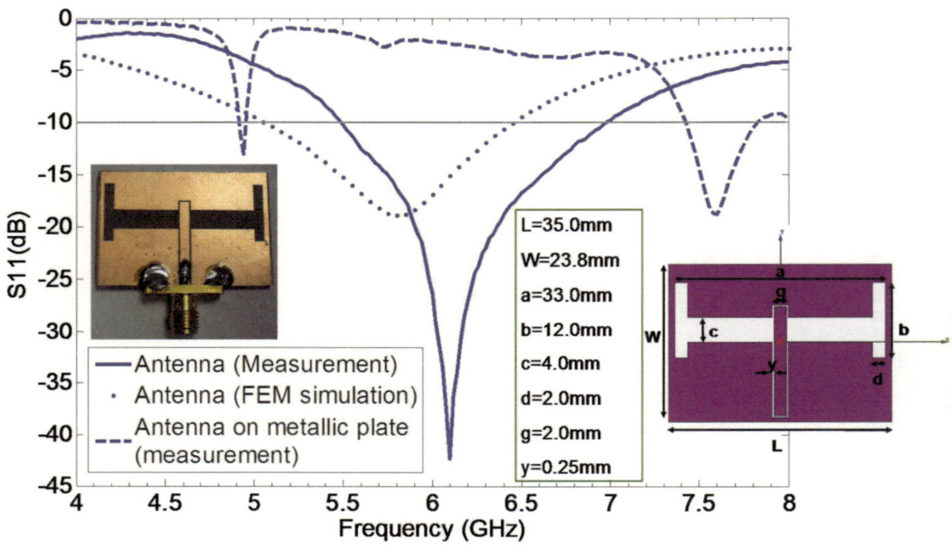

Fig. 4. Return loss of the antenna; Geometry and dimensions; Manufactured prototype.

The simulated antenna gain at 5.8GHz is 5.0 dB with very small variation along the antenna bandwidth (see Fig 5). The simulated E and H-plane radiation pattern in polar form for the antenna at 5.8GHz are shown in Fig.5. Both the CP and the XP components are represented. The E-plane radiation pattern is broadside and bidirectional. The H-plane radiation pattern is almost omnidirectional.

Regarding the AMC arrangement with respect to the antenna, several ideas have been considered. The first one is that the AMC used as antenna ground plane would electromagnetically insulate the antenna from the metallic object, without disturbing the antenna performance. The second is to minimize the size of the final prototype and to facilitate manufacturing process.

Two AMC arrangements having respectively 5x5 and 5x4 AMC unit cells have been combined with the CPW-fed slot antenna and the resulting prototypes (see Fig. 6) have been tested in terms of return loss. In both cases the antenna is fixed to the AMC structure by a 0.1mm double sided non-conducting adhesive tape.

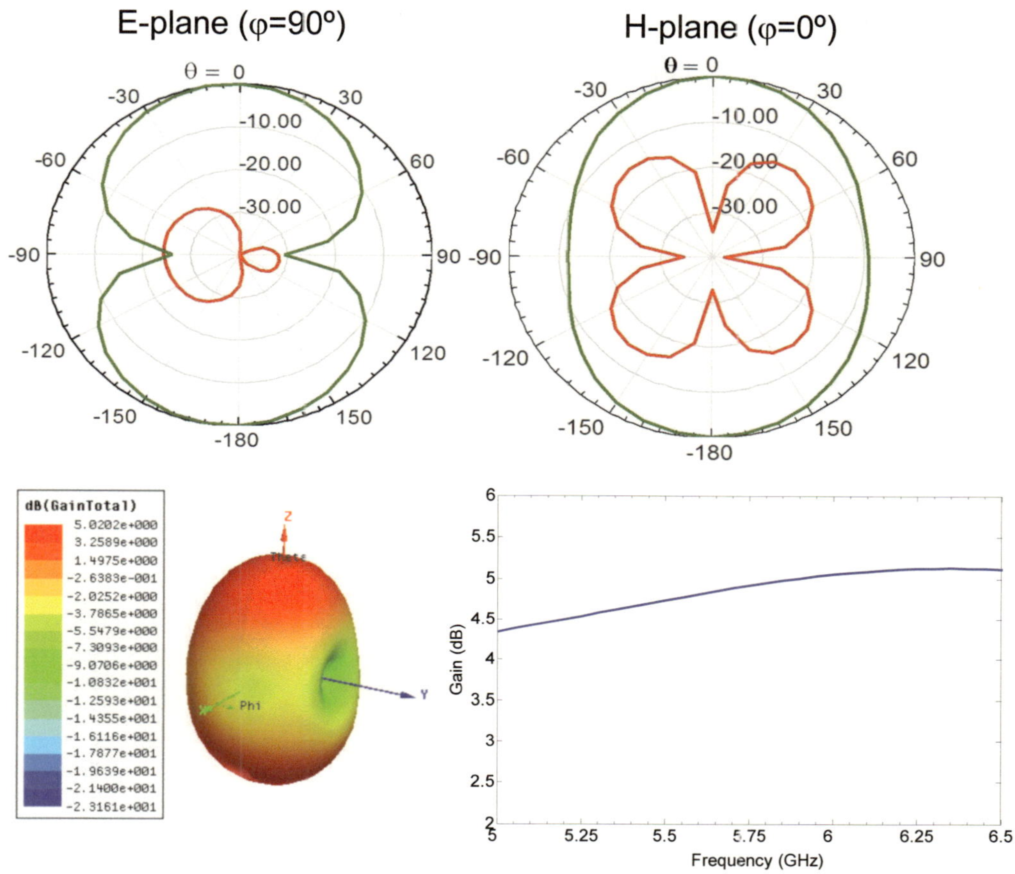

Fig. 5. CPW-fed slot antenna simulated radiation pattern (normalized, in dB) CP (green) and XP (red) components for E plane (up, left) and H plane (up, right). Three-dimensional simulated radiation pattern (down, left). Simulated antenna gain (down, right)

Prototypes of the antenna and the antenna on AMC have been manufactured using laser micromachining. The return losses of each manufactured prototype have been measured. As it can be observed in Fig.4, the measured operating bandwidth of the slot antenna is 1.5GHz (24.0%), which is wider than the 1.48GHz (22.0%) obtained by simulation. The difference in bandwidth and the frequency shift could be due to manufacturing tolerances.

From the measurements results shown in Fig. 6 it can be concluded that although the antenna on 5x5 AMC shows better return loss results than the antenna on 5x4 AMC at some frequencies, both prototypes have the same operating bandwidth and the return loss of the antenna on 5x5 AMC is also proper. So the increase of the prototype size due to the use of 5x5 unit cells is not profitable from the performance point of view. Taking this into account, the 5x4 AMC has been selected to be combined with the CPW-fed slot antenna.

The selected AMC arrangement in terms of a trade-off between performance and size is the one shown in Fig.7. The dimensions of the final structure, antenna on AMC (Fig. 7)), are Lp=57.60mm and Wp=46.08mm. The thickness is 1.626mm in the part corresponding to the antenna on the AMC and 0.813mm in the part corresponding only to AMC unit-cells.

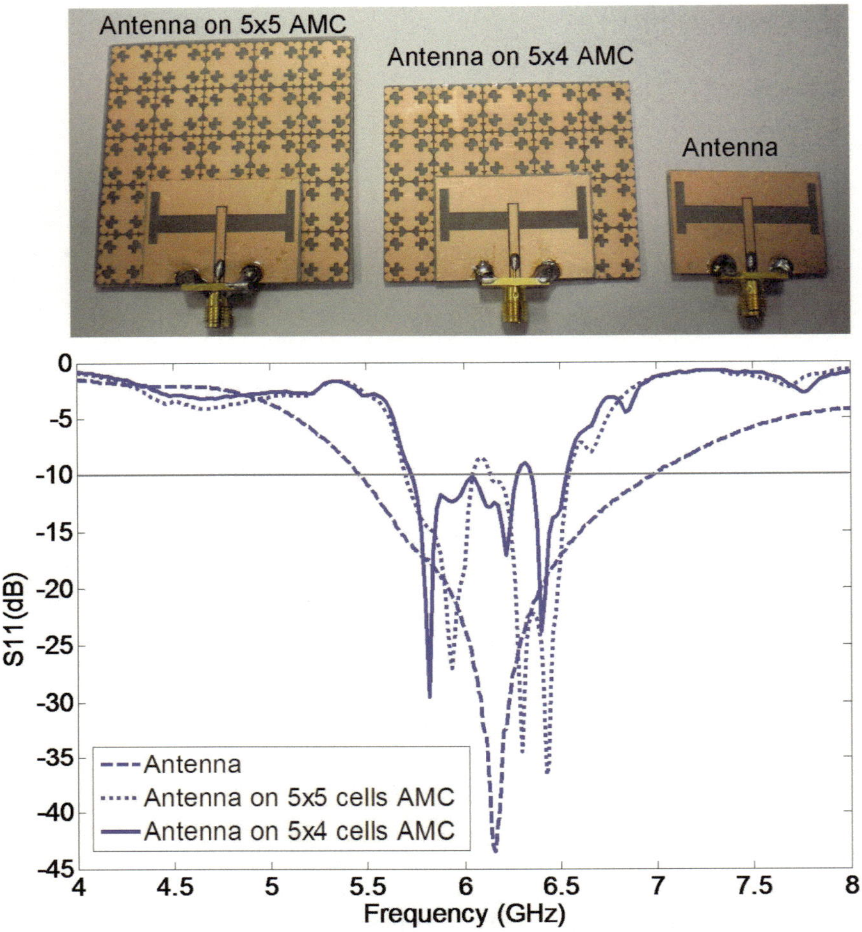

Fig. 6. Manufactured prototypes of the antenna on 5x5 cells AMC, the antenna on 5x4 cells and the antenna (up). Return loss of the Antenna, the antenna on 5x5 cells AMC and the antenna on 5x4 cells (down).

As it could be expected, when placed on a metallic plate the antenna resonance frequency has been shifted out of the SHF RFID band leading to its total malfunctioning (see Fig.4.). However, from Fig.7, it can be extracted that the antenna on AMC combination keeps the antenna operating properly in the whole antenna bandwidth, even when placed on a metallic plate, as the AMC electromagnetically insulates the antenna from the metallic plate. The measured input return loss for the antenna on AMC prototype shows two resonances: the first one is due to the joint operation of the antenna and the AMC, since the AMC operation bandwidth starts at 5.625GHz (See Fig.1). Whereas the second resonance is due to an antenna resonance out of the AMC operation bandwidth, since there is an additional RO4003C metal-backed layer below the original antenna.

According to the measurements, metallic plates do not affect the resonance frequency of the antenna on AMC. In addition, the metallic plates do not degrade the bandwidth of the antenna on AMC.

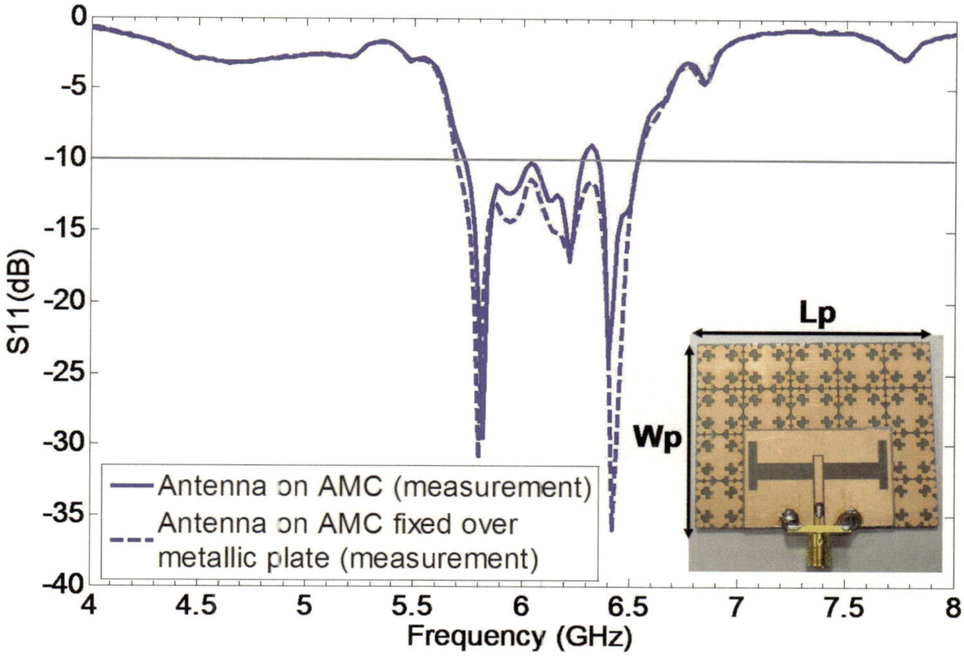

Fig. 7. Measured input return loss for the antenna and the antenna on AMC on a metallic plate; manufactured prototype.

The radiation pattern of the manufactured prototypes has been measured in anechoic chamber (see Fig. 8). The prototypes are placed in the XY plane. The measured antenna radiation pattern is in very good agreement with the simulated one, as can be concluded by comparing Fig 5 and Fig 10.

As can be observed in Fig 9, when the antenna is placed on the AMC, the maximum of the radiation pattern is displaced (the direction of maximum radiation changes). However when the antenna on AMC prototype is fixed over a metallic plate, this maximum is preserved with respect to the antenna on AMC prototype. As could be expected, the back radiation of the antenna on AMC is reduced with respect to the antenna prototype due to the in phase reflection properties of the AMC. So despite the small AMC structure, the antenna on AMC has a relatively low back radiation. Radiation pattern properties of the Antenna on AMC for RFID application are still preserved even when placed on a metallic plate.

Prototype	Gain (dB)	Pattern directivity (dB)	Efficiency (%)
Antenna	4.2	6.3	59.8
Antenna on AMC	2.2	7.0	32.0
Antenna on AMC over metallic plate	3.8	10.0	22.7

Table 2. Measured gain, directivity and radiation efficiency of the manufactured prototypes.

Fig. 8. Measurement set-up in anechoic chamber. Antenna measurement (left) and antenna on AMC over metallic plate measurement (right).

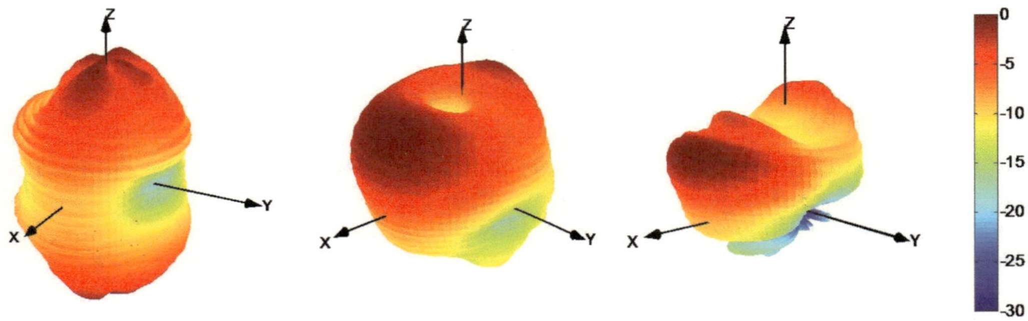

Fig. 9. Three-dimensional representation of the normalized measured radiation pattern for the three manufactured prototypes: antenna (left), antenna on AMC (center) and antenna on AMC over metallic plate (right)

In addition, the gain of the antenna on AMC fixed over a metallic plate almost preserves with respect to the gain of the antenna alone as it is shown in table 2, which represents a significant achievement.

In general, when placing an antenna on an AMC radiation properties such as gain and radiation efficiency are enhanced with respect to the antenna alone. This is due to the fact of using the AMC as a ground plane for the antenna substituting a conventional metallic ground plane i.e. antenna topologies that already have a metallic ground plane under the antenna metallization, such as microstrip patch antennas. As pointed out in section 1, these antennas can perform well with metallic objects but have narrow bandwidth and not negligible thickness. Other approaches combining antennas without metallic layer under the dielectric substrate (such as CPW-fed antennas) with AMCs for gain enhancement purposes, separate the antenna from the AMC by using an additional layer of foam. This also increases the antenna thickness which is not convenient in RFID applications. However, the slot

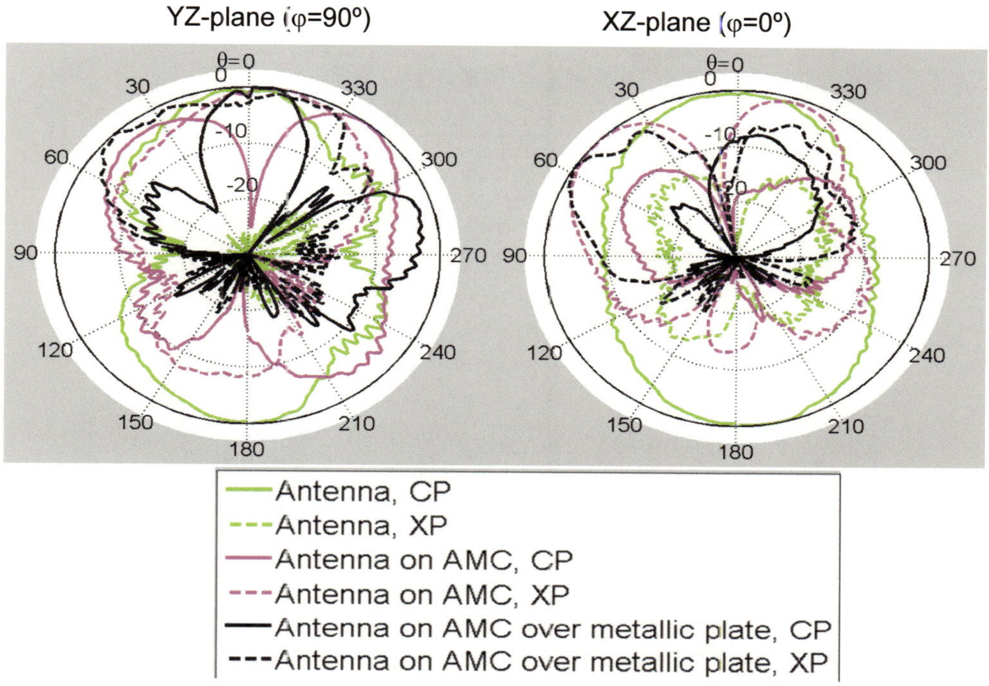

Fig. 10. Measured radiation pattern (normalized, in dB) of antenna, antenna on AMC and antenna on AMC over metallic plate. Planes φ = 90° (YZ-plane) and φ = 0° (XZ-plane).

antenna presented here has no metallic layer under the dielectric substrate and so when placing the AMC directly under the antenna to electromagnetically insulate the antenna from the metallic object, the antenna performance is slightly disturbed to some extent in terms of gain and radiation efficiency (see table II), whereas the obtained prototype exhibit proper operation over metallic objects and the gain is almost preserved compared to the antenna operating alone. All these properties are suitable for RFID application.
Another possibility tested to try to obtain enhanced efficiency (or at least preserved it) is to remove the AMC's unit cells below the antenna but this significantly reduces the antenna bandwidth as can be seen in Fig. 11 and the resonance at 5.8GHz disappears. For this reason the use of this arrangement has been declined. The only resonance that appears is at 6.45GHz and it is due to antenna having and additional dielectric substrate layer and a metallic ground plane bellow this dielectric substrate layer.
Also the antenna could be centred on the AMC arrangement which might preserve and/or enhance the radiation properties of the antenna, but this would require changing the antenna feeding increasing the complexity of the prototype and also its cost. The aim of this chapter it is to show that it is possible to obtain a compact, low profile and low cost antenna on AMC combination proper to be used over metallic objects.

4. Conclusion

A novel CPW-fed-slot antenna on AMC combination prototype suitable to be used in 5.8 GHz RFID tags on metallic objects has been presented. It has been shown that metallic plates

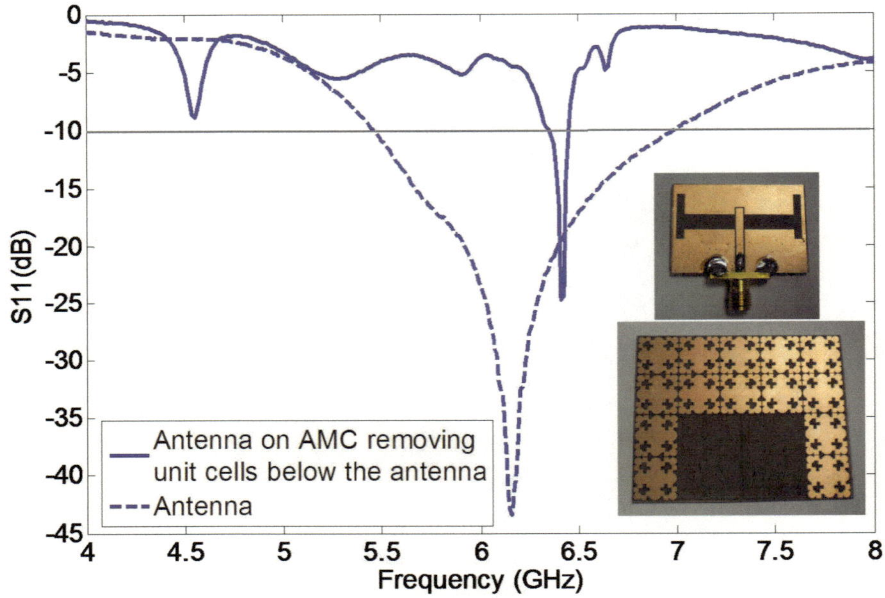

Fig. 11. Measured input return loss for the antenna and the antenna on AMC when the unit cells under the antenna are removed.

do not affect the resonance frequency of the antenna on AMC. In addition, the metallic plates do not degrade the bandwidth of the antenna on AMC.

As a reward for the AMC addition, the manufactured prototype, using a thin and low dielectric permittivity commercial substrate, exhibits proper operation both alone and when placed on a metallic plate.

The presented CPW-fed-slot antenna on AMC combination meets most of the RFID tag antennas requirements pointed out in section 1. Further research is being carried out to obtain a prototype in a bendable dielectric substrate.

By using the other presented AMC designs for UHF and 2.4GHz SHF with antennas operating at those frequency bands, problems related to RFID tags operation with metallic objects can be overcome.

5. Acknowledgment

Authors thanks Ramona C. Hadarig and Dr Yuri Álvarez for their comments and useful discussions. This work has been supported by the "Ministerio de Ciencia e Innovación" of Spain /FEDER" under projects TEC2008-01638/TEC (INVEMTA) and CONSOLIDER CSD2008-00068 (TERASENSE), by PCTI Asturias under project, PEST08-02 (MATID) and by the Principado de Asturias/FEDER Project IB09-081 (CAMSILOC).

6. References

Clarke R., Twede D., Tazelaar J., Boyer K., Radio frequency identification (RFID) performance: the effect of tag orientation and package contents, Packaging Technology and Science vol. 19, no. 1, 2006, pp. 45-54.

Dobkin D. M., Weigand S. M., Environmental effects on RFID tag antennas, IEEE Microwave Symposium Digest, Jun. 2005, pp. 135-138.

Feresidis, A. P., Goussetis G.,Wang S., and Vardaxoglou J. C., Artificial magnetic conductor surfaces and their application to low profiles high-gain planar antennas. IEEE Trans. Antennas Propag., Vol.53, no.1, 209-215, 2005.

Finkenzeller K., RFID Handbook: Radio-Frequency Identification Fundamentals and Applications, 2nd ed.: Wiley, 2004.

Foster P.R. and Burberry R.A., Antenna problems in RFID systems, IEE Colloquium on RFID Technology, pp. 3/1-3/5, October 1999

Geyi, W., Derivation of Equivalent Circuits for Receiving Antenna, IEEE Transactions on Antennas and Propagation, vol. 52, no. 6, pp. 1620-1623, June 2004.

Goussetis G., Feresidis A. P. and Vardaxoglou J. C., Tailoring the AMC and EBG Characteristics of Periodic Metallic Arrays Printed on Grounded Dielectric Substrate, IEEE Trans. Antennas Propag., Vol.54, no.1, pp. 82-89, Jan 2006.

Hirkonen M., Pursula, P. Jaakola K., and Laukkanen K., Planar inverted-F antenna for radio frequency identification, Electronic. Letters., vol. 40, pp. 848-849, July 2004.

Hosseini M., Pirhadi A. and Hakkak M., A Novel AMC with Little Sensitivity to the angle of Incidence Using 2-layer Jerusalem Cross FSS, Progress In electromagnetics Research, PIER 64, 43-51, 2006.

Kern, D. J., Werner, D. H., Monorchio, A., Lanuza, L. and Wilhelm, M. J., The design synthesis of multiband artificial magnetic conductors using high impedance frequency selective surfaces, IEEE Trans. on Antennas and Propag., Vol.53, No. 1, Jan. 2005.

Kwon H. and Lee B., Compact slotted planar inverted-F RFID tag mounted on metallic objects, Electronic Letters, vol. 41, pp. 1091-1092, Nov. 2005.

Liang J. and Yang H.-Y. D., Radiation Characteristics of a Microstrip Patch Over an Electromagnetic Bandgap Surface. IEEE Trans. on Antennas and Propag., Vol. 55, No 6, June 2007

Lin Y.F, Liao P.C., Cheng P.S., Chen H.M., Song C.T.S. And Hall P.S., CPW-fed capacitive H-shaped narrow slot antenna, Electronic Letters, vol.41, No.17, 2005

Mantash M., Tarot A.C., Collardey S. and Mahdjoubi K., Dual-band antenna for WLAN application with EBG. Fourth International Congress on Advanced Electromagnetic Materials in Microwaves and Optics. Metamaterials 2010. pp. 794-796, Sep.2010.

Mantash M., Tarot A.C., Collardey S. and Mahdjoubi, K., Dual-band CPW-fed G-antenna using an EBG structure, Antennas and Propagation Conference (LAPC), 2010 Loughborough. pp. 453 – 456.

McVay J., Engheta N. and Hoofar A., High impedance metamaterials surfaces using Hilbert-curve inclusions, IEEE Microw. Wire. Comp. Lett., vol.14, no.3, 130-132, 2004.

McVay J., Hoofar A. and Engheta N., Small dipole-antenna near Peano high-impedance surfaces, IEEE AP-S Int. Symp.,vol 1., 305-308, 2004.

Monorchio A., Manara G., and Lanuzza L., "Synthesis of artificial magnetic conductors by usisng multilayered frequency selective surfaces", IEEE Ant. Wireless Propag, Lett.,vol1, pp.196-199, 2002.

Mosallaei H. and Sarabandi K.,Antenna Miniaturization and Bandwidth Enhancement Using a Reactive Impedance Substrate, IEEE Trans. on Antennas and Propag, Vol.52, No.9, September 2004.

Pozar, D., Scattered and Absorbed Powers in Receiving Antennas, IEEE Antennas and Propagation Magazine, vol. 46, no. 1, pp.144-145, February 2004.

Rahmat-Samii, Y.; Mosallaei, H.,"Electromagnetic band-gap structures: clasiffication, characterization, and applications", Eleventh International Conference on Antennnas and Propagation, 2001. (IEE Conf. Publ.No.480) vol. 2, 2001. Page(s): 560-564;

Rao, K. V. S., An overview of Backscatter Radio Frequency Identification System (RFID), IEEE Asia Pacific Microwave Conference, vol.3, pp. 746-749, November-December 1999.

Rao K. V. S., Nikitin Pavel V., Lam Sander F. Antenna Design for UHF RFID Tags: A Review and a Practical Application, IEEE Transactions on Antennas and Propagation, vol. 53, no.12, pp. 3870-3876, December 2005.

Rao K. V. S., Nikitin Pavel V., Lam Sander F., Impedance Matching Concepts in RFID Transponder Design, Proceedings of IEEE Workshop on Automatic Identification Technologies, Oct. 2005, pp. 39-42.

Sievenpiper D., High-impedance electromagnetic surfaces, Ph.D. Thesis. University of California. Los ángeles, 1999.

Sievenpiper, D., L. Zhang, R. F. J. Broas, N. G. Alexopolous, and E. Yablonovitch, High-impedance electromagnetic surfaces with a forbidden frequency band, IEEE Trans. Microwave Theory and Techniques. vol.47, no.11, pp.2059-2074, Nov.1999

Simovski, C. R., de Maagt, P. and Melchakova, I. V., High-Impedance Surfaces Having Stable Resonance With Respect to Polarization and Incidence Angle, IEEE Trans. Microwave Theory and Techniques. Vol.53, No. 3, pp. 908-914, March 2005.

Stockman H., Communication by means of reflected power, Proc. IRE, pp 1196-1204, Oct.1948.

Ukkonen L. et al., Operability of Folded Microstrip Pach-Type Tag Antenna in the UHF RFID Bands within 865-928 MHz, IEEE Antennas and wireless Propagation Letters, vol. 5, 2006.

VanBladel, J., On the Equivalent Circuit of a Receiving Antenna, IEEE Antennas and Propagation Magazine, vol. 44, no. 1, pp. 164-165, February 2002.

Yang, F. and Rahmat-Samii Y., Reflection phase characterizations of the EBG ground plane for low profile wire antenna applications, IEEE Trans. Antennas Propag., Vol. 51, No. 10, 2691-2701, 2003.

Yang, F. R., Ma K. P., Qian Y., and Itoh T., A uniplanar compact photonic-bandgap (UC-PBG) structure and its applications for microwave circuit, IEEE Trans. Microwave Theory Tech., vol 47, no.8, 1509-1514, 1999.

Yang F. and Rahmat-Samii Y., Electromagnetic band-gap structures in Antenna Engineering (The Cambridge RF and Microwave Engineering Series). Cambridge University Press. 2008.

Zhang Y., J. von Hagen, and W. Wiesbeck, Patch array as artificial magnetic conductors for antenna gain improvement, Microw. Opt. Technol. Lett., vol.35, no. 3, 172-175, 2002.

Zhu S. and R. Langley, Dual-Band Wearable Textile Antenna on an EBG Substrate, IEEE Transaction On Antennas and Propag., Vol.57, No. 4, April 2009.

10

High Performance UHF RFID Tags for Item-Level Tracing Systems in Critical Supply Chains

Luca Catarinucci, Riccardo Colella, Mario De Blasi,
Luigi Patrono and Luciano Tarricone
University of Salento
Italy

1. Introduction

The need of a traceability system implemented at item level is becoming more and more essential in many business processes and, among the different potential enabling technologies, passive Radio Frequency Identification (RFID) (Finkenzeller, 2003) is undeniably the most adequate candidate. Indeed, its simplicity of use as well as its very attractive cost-benefit ratio, give a strong appeal to RFID.

Among the many application sectors, the pharmaceutical supply chain, with millions of medicines moving around the world and needing to be traced at item level, represents a very interesting test-case. Furthermore, the growing counterfeiting problem raises a significant threat within the supply chain system. Moreover, several international institutions (e.g. Food and Drug Administration, European Medicines Agency, European Federation of Pharmaceutical Industries and Associations) are encouraging the use of innovative solutions in healthcare and pharmaceutics, to improve patient safety and enhance the efficiency of the pharmaceutical supply chain.

In order to select the most adequate hardware solution, though, several aspects must be compulsory taken into account, including the working frequency, the near or far field empowering methods, but also the differences among the various RFID-based checkpoints of a generic supply chain (De Blasi et al., 2009; Uysal et al., 2008).

The choice between the two main RFID solutions, High Frequency (HF) or Ultra High Frequency (UHF), can be aided by several recent works, which highlight how passive UHF RFID systems provide better performance than passive HF ones, see for example (Uysal et al., 2008). Hence, UHF seems to be the most promising technology for item-level traceability on the whole supply chain. The success of UHF can be mainly attributed to the assertion of the EPCglobal (Thiesse et al., 2009) international standard. Furthermore, UHF has several advantages over HF and LF technologies: the capability to enable multiple simultaneous readings of tags, the capacity to offer very high read rates, in addition to the much longer reading distance.

Unfortunately, performance of UHF systems depends on several parameters (Bertocco et al., 2009), which are strongly related to environment, design and setup choices. For example, it is well known that a supply chain is composed of several steps that have different characteristics in terms of traceability procedures (e.g., distance between reader antenna and tag antenna, speed of moving objects, quantity of tags to be read, etc.). In such scenarios, the choice of an RFID tag solution, able to guarantee high performance in each step of the supply chain and in any operating condition, is certainly a hard challenge.

Some approaches proposed in literature, are based on the use of general-purpose Far Field (FF) UHF tags (Rao et al., 2005; Catarinucci et al., 2010) applied on the secondary package of the product. Several studies, in fact, have shown that the use of FF UHF tags guarantees better performance than Near Field (NF) ones in every step of the supply chain. Indeed, as most of FF UHF tags are provided with an inner loop that short-circuits the tag chip technology (hybrid tags), they exhibit good performance even in near field conditions. In fact, this strategy allows an efficient coupling with the magnetic field generated by NF reader antennas (Catarinucci et al., 2010).

In addition to the RFID checkpoints peculiarities, another important aspect is the effect on the tag performance of the platform where the tag is attached. Unfortunately, commercial FF UHF tags still suffer of many drawbacks (Nikitin & Rao, 2006). First of all, they suffer of performance degradation in presence of electromagnetically hostile materials, such as metals and liquids (Catarinucci et al., 2010; De Blasi et al., 2010). Another issue regards the strong dependence of the system performance on the mutual position between reader antenna and tag antenna, which may vary randomly for each item. Consequently, from the electromagnetic (EM) point of view, very strict requirements must be satisfied by the tag antenna.

The sum of the requirements to be met by a single tag, functioning properly in every step of the supply chain, will be extended in the next sections.

Consequently, the first part of this chapter describes the main features of the pharmaceutical scenario, mainly focusing on item-level tracing systems, RFID devices performance, related works and experimental measurement campaigns of commercial UHF RFID tags.

Taking into account the analysis of such aspects, the main causes of performance degradation are individuated and a guideline for the design of a new kind of RFID tag, working properly in each step of the pharmaceutical supply chain and regardless of the kind of traced product, has been drawn in the second part of this chapter. Moreover, a new enhanced tag has been realized by following the guideline, tested, and finally results have been discussed.

2. Related works

The current vision of the RFID market shows, in addition to UHF FF tags already widely used, also the presence of UHF NF tags. These last are based on inductive rather then radiative coupling and usually are energized by specific NF reader antennas, appositely designed to minimize the radiated field. The tag antenna is usually a simple loop, whose diameter is calculated in order to guarantee the resonance at the desired frequency —a few centimeters in the UHF band—, but also more complicated shapes do exist. The short range of both NF reader antennas and small loop antennas restrict the NF tags reading range to only a few centimeters. Nevertheless, the smaller size of NF tags and the higher tolerance to the scenario —inductive field penetrates through liquids and dielectrics— make them useful in some applications where the size is crucial and marked items are electromagnetically complex.

It is important for the scientific community to understand the capabilities and limitations of the emerging passive UHF technology, and just as importantly, to understand where researchers may contribute to face problems and challenges that currently are limiting a large-scale deployment of this technology. Main barriers are: (i) hardware technology current weaknesses (Catarinucci et al., 2010) (e.g. data reliability, read rate in critical conditions, lack of unified standard for interoperability), (ii) software weakness (Barchetti et al., 2010) (e.g. scalability, single-point of failure, integration with information systems), (iii) relatively high costs of tags, software customization and systems integration, (iv) security issues (Staake et al., 2005; Mirowski et al., 2009), (v) lack of scientific literature on the evaluation of potential effects of RFID exposure on molecular structure and potency of drugs (Acierno et al., 2010).

There is a rich literature about developing and evaluating UHF RFID solutions.

(Aroor & Deavours, 2007), for instance, evaluates performance of several commercial passive UHF tags under critical operating conditions (e.g. presence of liquids and/or metals) by using an experimental approach. In order to simplify both measurements and result analyses, an end-user metric has been chosen. In particular, performance of tags is measured in terms of maximum reading distance in a given environment. The tests have demonstrated that no commercial FF UHF tag can be properly read when it is directly applied to metal. Further results have shown that the water presence degrades the tag performance significantly. The tests have also demonstrated that larger tags guarantee better performance. In the same work, a series of experiments has also been carried out by using some NF UHF tags. The results have clearly demonstrated that NF UHF tags do not solve the problem associated with the presence of neither the metal nor the water. On the contrary, it has been highlight that the presence of metal or water has even much more drawbacks in NF rather than FF UHF tags.

(Bertocco et al, 2010) experimentally investigates the relationships between the EM field levels at the tag antenna and the overall performance of a UHF RFID system. The results have underlined the importance of preliminary measurements in the setup of the system, in the evaluation of the maximum distances between tags and reader antenna, and in the estimation of a correction factor to be used in theoretical analyses.

(Ramakrishnan & Deavours, 2006) describes a benchmark suite useful to give good indications about how well UHF solutions work in real world scenarios. These benchmarks are able to compare the reading performance of different tags in terms of distance, quality, and real rates in various situations.

(Fuschini et al., 2010) is another work that aims at investigating the main benefits and performance of NF UHF tags in item-level tagging systems. This study exploits an electromagnetic analysis based on both theoretical evaluations and measurements carried out on real UHF RFID devices. Four different commercial tags (i.e. Alien Squiggle, Texas Instruments, Impinj Button, and Impinj Satellite) have been tested mainly in terms of the system Path Gain, defined as the ratio between the power absorbed by the tag and the available power at the reader. The results have demonstrated no particular electromagnetic benefits in performance in favour to NF UHF tags.

(Tae-Wan Koo, et al., 2010) is a very interesting work focused on the need to improve the performance when an UHF tag is applied on a metallic object.

(Bertocco et al., 2009) highlights the importance to evaluate the performance of UHF RFID systems in real-world conditions by using suitable test bed to perform the experiments. In particular, the system efficiency is considered. This work asserts that there are many parameters that should be known and tuned to maximize the efficiency even in critical

conditions. Some measurements have demonstrated that the deployment of multiple antennas might be totally useless. On the contrary, better results can be obtained using reflecting surfaces, or deploying reading-paths, avoiding reading-gates.
(De Blasi et al., 2010) is focused on the use of passive UHF tags, in order to analyze a performance comparison between near field and far field UHF RFID systems in every of the pharmaceutical supply chain. Some different commercial passive UHF tags (i.e. Impinj Thin Propeller, Impinj Paper Clip, and RSI Cube2) have been tested in an item-level system, simulating each step of the pharmaceutical supply chain in a controlled test environment. Results allow to analyze the advantages and disadvantages of using NF and FF UHF tags for item-level tracing in each step of the pharmaceutical supply chain. Experimental results show that the use of passive FF UHF tags represent a well suitable solution to guarantee both high performance and item level tagging in the whole supply chain. This work highlights also that the pharmaceutical supply chain is characterized by very critical operating conditions where tag improvements are strongly needed in order to guarantee acceptable performance.

3. Test environment to emulate a pharmaceutical supply chain

3.1 Reference scenario

The pharmaceutical supply chain, shown in Fig. 1, is a complex scenario with millions of pharmaceutical products moving around the world each year. Three are the most significant actors of such a supply chain: (i) the *manufacturer* who produces the package of pharmaceuticals, (ii) the *wholesaler* who buys and resells big quantities of medicinal products, and (iii) the *retailer*, which in general is a pharmacy or hospital.

Fig. 1. An abstract vision of the pharmaceutical supply chain.

The item-level traceability of drugs starts just after the packages are filled during the manufacturing process. In this step, each tagged product is individually scanned on the conveyor belt and then cased to be sent to the wholesalers. The wholesalers separate the products according to their identifiers and place them onto the shelves. Wholesalers receive orders from retailers. Such orders often refer to small quantities of many products; they may contain a large number of items. The products in the orders of the retailers are picked and put into some large envelope bags that are scanned and confirmed before their distribution. Upon receipt, the pharmacy retailer scans the contents of each bag without opening it.
In order to select the most adequate RFID hardware solution, though, several aspects must be compulsory taken into account, including the working frequency, the near or far field empowering methods, but also the differences among the various RFID-based checkpoints of a generic supply chain.
In fact, depending on the considered step of the supply chain, at least three different RFID checkpoints are commonly used. They differ each other in terms of interrogation distance,

number of items to be read, reader antenna typology and scanning speed. It is worth pointing out that the tag marking an item must work properly in all checkpoints.

More specifically, one of the possible checkpoints is given by the so-called items line, where the tagged product must be singularly scanned by using NF reader antennas. Whatever tag is used for the item-level traceability, it should guarantee good performance even in near field conditions.

A second kind of checkpoint is given by the so-called cases line, where a case containing a number of homogeneous items packed together, passes through a NF tunnel in order to read all the items in one shot. Consequently, the RFID tags used to assure reliable item-level tracing systems should work correctly even at medium distance from the interrogator antennas. Moreover, the problem of the multiple readings of tags and of the tag overlapping should be considered.

A third kind of checkpoint is given by the so-called border gate. When a pharmacy retailer is restocked it becomes necessary to simultaneously read all the different tagged items contained in a box or in a plastic bag. The border gate, equipped with FF reader antennas, is designed for such a purpose.

Besides the RFID checkpoints peculiarities, another important aspect is the effect on the tag performance of the platform where the tag is attached. UHF tags, more than HF ones, are influenced by the presence of electromagnetically hostile materials, such as liquids and metals; this aspect is crucial because in several scenarios, as the considered pharmaceutical one, metals and liquids are massively present.

3.2 Test bed components

The controlled test environment, shown in Fig. 2, has been realized in order to simulate the main steps of the pharmaceutical supply chain. Such a test environment, in fact, makes it possible to carry out effective experimental campaigns to evaluate the performance of UHF RFID-based tracing systems, even in particularly stressed operating conditions.

The test environment, based on the three main components above described (items line, cases line and border gate), makes possible unbiased and repeatable comparisons among technologies.

More specifically, the items line consists of a conveyor belt whose speed can be tuned in the range from 0.00 to 0.66 m/s, in order to guarantee real requirements to be met by pharmaceutical manufacturing processes. The conveyor belt has a double containment edge to keep products in the same position along the belt. In the middle of each containment edge, a near field reader antenna has been installed and connected to a high performance UHF RFID reader compatible with the EPC Class1 Gen2 standard. The following devices have been used: two Impinj Mini-Guardrail reader antennas and one Impinj Speedway UHF reader.

Similarly, the cases line consists of a conveyor belt, equipped with a line speed regulator in the range from 0.00 to 0.66 m/s, a double containment edge to keep cases in the same position along the belt, one Impinj Speedway reader, and two roller conveyors. In the middle of the line, four small near field reader antennas (Impinj Brickyard) have been placed inside a metallic tunnel. Each reader antenna is in the centre of each tunnel side. The width of the tunnel is equal to 0.6 m. Further characteristics are: 50 Ω of impedance, 6 dBi as maximum far field gain and -15 dB as Return Loss.

Finally, the border gate uses a single UHF RFID reader (Impinj Speedway) and four far field UHF reader antennas.

Fig. 2. Test environment composed of an items line (left), a cases line (middle), and a border gate (top right).

In order to effectively simulate the pharmaceutical supply chain, it is very important to take into account real heterogeneous drugs, so to significantly represent the global market of drugs, that is characterized by a wide heterogeneity of products, which differ for several factors as, for instance, medicine state (i.e. solid, liquid, gas, etc.) and material of the primary package (e.g. glass, metal, plastic, etc.). A complete taxonomy of most popular drugs may be done by considering these factors.

The first classification, which takes into account only the medicine state, splits all pharmaceutical items into four main categories:

- Solid: tablets capsules, granules, etc.
- Semi-liquid: creams, suppositories, etc.
- Liquid products: syrups, oral liquids, solutions, etc.
- Gas: pressurized gasses.

Another useful classification can be done in terms of material of the primary package. Plastic is the most widely diffused material because of the large use of bottles, blister packs, and film layers. Nevertheless, even the use of metal is fairly common: aluminium blister packs and sachets are possible examples. Another common material for pharmaceutical products is glass that is very valuable especially for the liquid products. Classical applications of glass packaging are bottles for liquids, ampoules, and vials.

Based on the above information and discussions, Table 1 summarizes a simple taxonomy of pharmaceutical products according to their physical properties.

It is worth observing that this classification is very important to perform significative tests because different materials interact with RF waves differently. In particular, liquids cause the RF waves attenuation by absorbing their energy, whereas metals do not let RF waves pass through by reflecting them. Moreover, in both cases, the impact on the radiating properties of RFID tag antennas is relevant. The reported taxonomy, hence, becomes a compulsory instrument to select the most adequate drug for the specific laboratory test, so as to evaluate the impact of hostile materials on the performance of the RFID systems in the UHF band.

<table>
<tr><th colspan="2" rowspan="2">Product Type</th><th colspan="3">Package Material</th></tr>
<tr><th>Metal</th><th>Glass</th><th>Plastic</th></tr>
<tr><td rowspan="4">Solid</td><td>Tablets in Blister</td><td>X</td><td></td><td></td></tr>
<tr><td>Tablets in a Bottle</td><td></td><td></td><td>X</td></tr>
<tr><td>Granules in Sachets</td><td>X</td><td></td><td></td></tr>
<tr><td>Powders in a Bottle</td><td></td><td>X</td><td></td></tr>
<tr><td>Semi-liquid</td><td>Cream</td><td>X</td><td></td><td>X</td></tr>
<tr><td rowspan="5">Liquid</td><td>Syrup</td><td></td><td>X</td><td></td></tr>
<tr><td>Single injectable solution in syringe</td><td></td><td>X</td><td></td></tr>
<tr><td>Multiple injectable solution in syringes</td><td></td><td>X</td><td></td></tr>
<tr><td>Oral solution</td><td></td><td></td><td>X</td></tr>
<tr><td>Ophthalmic solution</td><td colspan="2">X</td><td></td></tr>
<tr><td>Gas</td><td>Bomb Spray</td><td>X</td><td></td><td></td></tr>
</table>

Table 1. Classification of pharmaceutical products

3.3 Description of the working conditions

An effective evaluation of RFID reliability in a pharmaceutical supply chain cannot neglect the effects on the performance caused by hostile factors such as: the potential misalignments between tag antenna plane and reader antenna plane, multiple reading of tags, distance between tag antenna and reader antenna.

The misalignment problem is mostly relevant in the items line. To test such a misalignment impact, four different operating conditions should be tested. They are characterized by a mutual orientation between the plane where the tag antenna lies and the plane where the reader antennas lie: 0°, +90°, -90° and 180° are considered. In particular, this last represents the worst case and allows the performance evaluation under unfavourable conditions. Vice versa, the 0° case is the ideal condition. Finally, the -90° case is characterized by the contact between tag and conveyor belt. Instead, in the +90 ° case the tag is attached to the up-side of the item, so that the potential interference with the conveyor belt is avoided but the distance with the reader antennas depends on the size of the item.

Another problem to be analyzed deals with the collisions among tags, impacting both the cases line and the border gate. For the cases line both homogeneous cases (consisting of a single product type) and heterogeneous cases (containing products of different types) should be tested. Moreover, also the configuration of the cases plays an important role. In order to simulate realistic conditions, three different configurations have been adopted for each case:

- Configuration I: the case was prepared placing the items with their tag antenna oriented toward the reader antennas and avoiding the overlapping of tag antennas.
- Configuration II: the case was prepared placing the items with their tag antenna oriented toward the center of tunnel (i.e. opposite to the reader antennas) and trying to obtain the overlapping of tag antennas.
- Configuration III: the case was prepared considering four different random dispositions of items. These dispositions were alternated in progress during the test bed.

In order to better clarify the compositions of cases, in Fig. 3, the configurations I, II and III are schematically reported. The overlapping of two different tags is represented by a double "x".

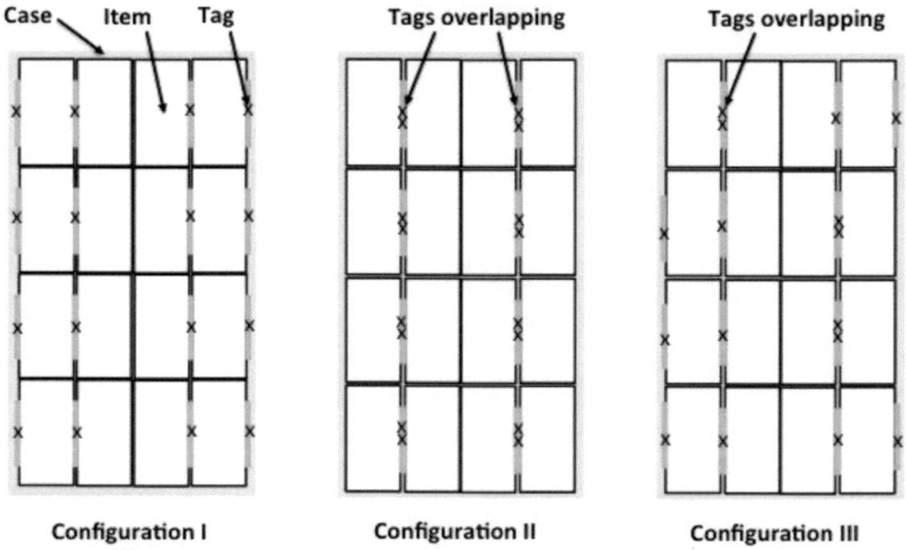

Fig. 3. Some examples of case compositions.

Further tests related to cases line and border gate should be carried out also on sets of heterogeneous drugs. More specifically, cases filled with a mix of 50 items in the cases line and mix of 200 items in the border gate will be considered in out measurements. Three different combinations of items into the case (cases line test) and into the plastic bag (border gate test) have been prepared considering the following percentages:

- MIX1: 40% solids, 40% liquids, and 20% semi-liquids;
- MIX2: 50% solids, 30% liquids, and 20% semi-liquids;
- MIX3: 60% solids, 20% liquids, and 20% semi-liquids.

4. Experimental results of commercial RFID UHF tags

4.1 Commercial RFID tags

In order to carry out an effective performance comparison among commercial RFID tags, able to evaluate the current limits in item level tracing systems in the whole supply chain, a preliminary technological scouting is very important. Note that for item-level tagging applications, the choice of the tags is affected by different requirements as: small size of the tag itself, compatibility with EPCglobal standard, high scanning speed, low cost, and high stress of tag label during product life cycle. As already stated, particular attention is focused on passive UHF tags that can be split into two sub-sets: NF and FF tags. In this chapter, experimental results derived by testing eight different types of passive UHF tags, six FF and two NF, are reported. All preselected tags are characterized by the same memory (96 bit), operating frequency (860-960 MHz) and compliance with the standard EPC Class1Gen2. On the contrary, the main differences are on antenna geometry and on the size of tag. The layout antenna of the eight preselected RFID tags is reported in Fig. 4.

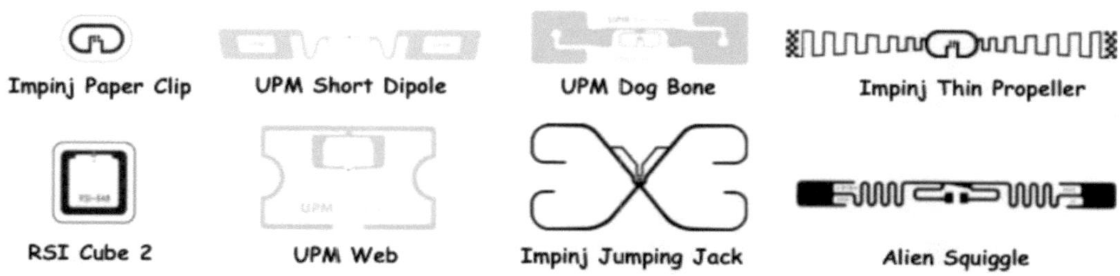

Fig. 4. Layout of the eight preselected commercial passive UHF RFID tags.

More in detail, the following types have considered:

- RSI Cube 2: it is a small near-field tag, whose size is 25.4 x 25.4 mm, with the NXP Ucode G2XL chip, designed for pharmaceutical and applications where small form factor is required;
- Impinj Paper Clip: it is a small near-field tag with the Impinj Monza3 chip, whose size is 19.0 x 12.7 mm, designed for pharmaceutical and applications where small form factor is required;
- Impinj Thin Propeller: it is a far field tag with the Impinj Monza3 chip and with an antenna, whose size is 8.0 x 95.0 mm, that is a high-performance dipole configuration. It guarantees large working bandwidth and is designed for warehouse, logistics, case, carton, and garment applications.
- Impinj Jumping Jack: it is equipped with the Impinj Monza3 chip, a high-performance FF antenna and a NF antenna. Its size is 44.5 x 88.9 mm and it is designed for long-range, multi-orientation warehouse, logistics, carton, baggage, and garment applications.
- UPM Dog Bone: it is a high performance tag for a wide range of RFID Supply Chain Management RFID Apparel and RFID Transportation applications. It is equipped with the Impinj's Monza3 chip. Its size is 27 x 97 mm.
- UPM Web: it is a high performance tag for RFID item-level use, whose size is 34 x 54 mm. Reliable reads/writes when tags are in close proximity to each other. It is equipped with the NXP U-Code G2XL chip.
- UPM Short Dipole: it is equipped with the NXP U-Code G2XL/G2XM chip. It is designed for a wide range of RFID Supply Chain Management. Its size is 15 x 97 mm.
- Alien Squiggle: it is equipped with the Alien Higgs-3 chip. It is designed for a wide range of RFID Supply Chain Management. Its size is 12.3 x 98.2 mm.

4.2 Results in ideal conditions

The first part of the experimental campaigns is focused on a performance comparison between NF and FF commercial UHF tags applied on different drug types in each step of the supply chain. The following RFID tags have been used: RSI Cube 2 and Impinj Paper Clip as NF tags, while, Impinj Thin Propeller, Impinj Jumping Jack, and UPM Dog Bone as FF tags.
In the first test, the most favourable working conditions in the items line are reproduced. Indeed, the mutual orientation between the plane where the tag antenna lies and the plane where the reader antennas lie is 0°. Results reported in Fig. 5 show that the NF tag Cube 2 ensures optimal performance for every drug type, even in presence of critical materials such as liquid and metal. Likewise, also the two FF tags have given good results, even though a slight effect due to the presence of metal and liquid, above all in the bomb spray and ophthalmic solution cases, is observed.

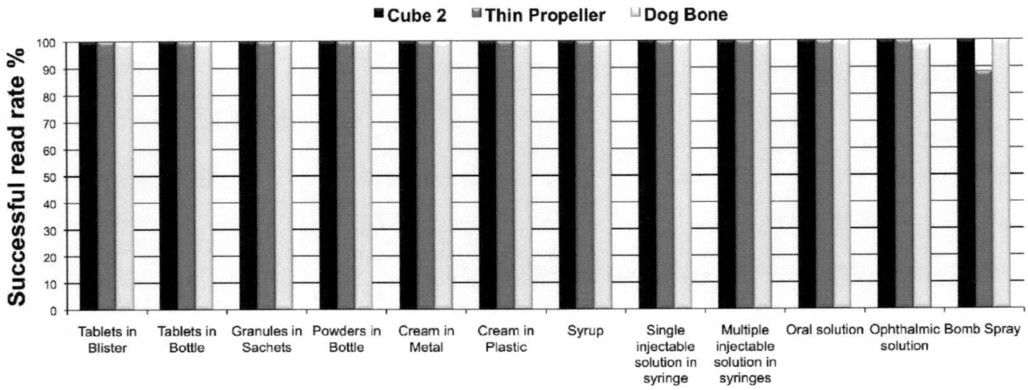

Fig. 5. A performance comparison on the items line by varying the drug type and the tag type considering normal operating conditions (mutual orientation 0°).

In a second test, the worst conditions are reproduced: a mutual orientation between the tag plane and reader antenna plane of +180° has been considered. This has been done by switching off one of the two reader antennas and by applying the RFID tag on the package face opposed of the active reader antenna. The experimental results, reported in Fig. 6, clearly show that FF UHF tags perform better than NF ones. The results where even the Dog Bone tag is not characterized by a successful read rate of 100%, have been obtained mainly in presence of granules in aluminum sachets, ophthalmic solution in aluminum sachets, syrup, and bomb spray. For these cases, though, the Cube 2 tag has shown very poor performance, reaching values of successful read rates near to 0%.
In order to evaluate the reliability in presence of multiple reading of tags at the same time, several experimental campaigns have been carried out. More specifically, eight different types of product, chosen among the four categories previously defined, have been used to perform tests on homogenous (single type of product) and heterogeneous (mix of products) cases (see section 3.3).
Homogenous cases have been analyzed considering a random disposition of items within the case. In particular, the type of case composition is the Configuration III previously described. Fig. 7 shows the experimental results obtained by testing again the NF Cube 2 tag and the two FF, Thin Propeller and Dog Bone, tags. The histogram clearly demonstrates that

the performance of FF UHF tags is better than the NF Cube 2 tag. Furthermore, this test shows that all the analyzed tags are not able to guarantee good performance in presence of materials that are hostile to electromagnetic propagation. In particular, the experimental results show very poor successful read rate when the homogenous case is composed of items such as bomb spray or ophthalmic solution.

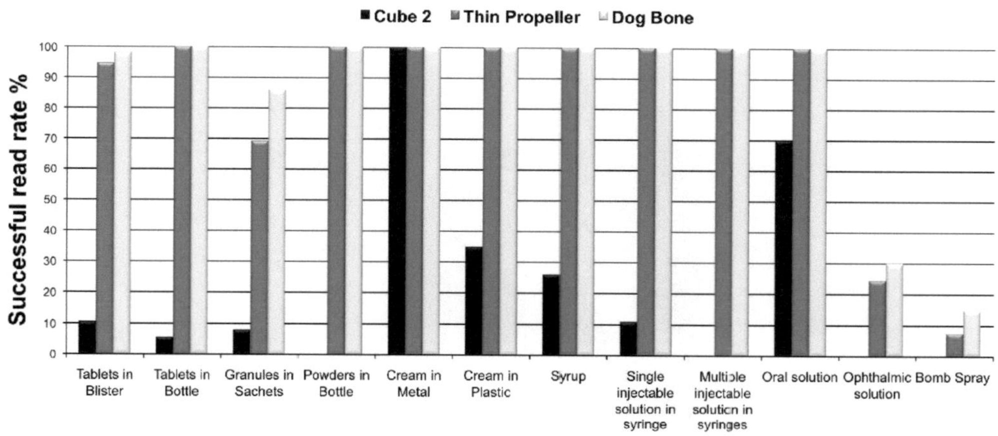

Fig. 6. A performance comparison on the items line by varying the drug type and the tag type and considering extreme operating conditions (mutual orientation +180°).

Further tests allowed the evaluation of the performance, also in this case in presence of multiple reading of tags, but considering the heterogeneity of drugs. In particular, Fig. 8 shows a performance comparison in the cases line by varying the configuration of the case in terms of drug types. The three different and previously described realistic compositions have been considered (MIX1, MIX2 and MIX3). The experimental results clearly show that optimal performance (100% of successful read rate) can be obtained using the two types of commercial UHF tag.

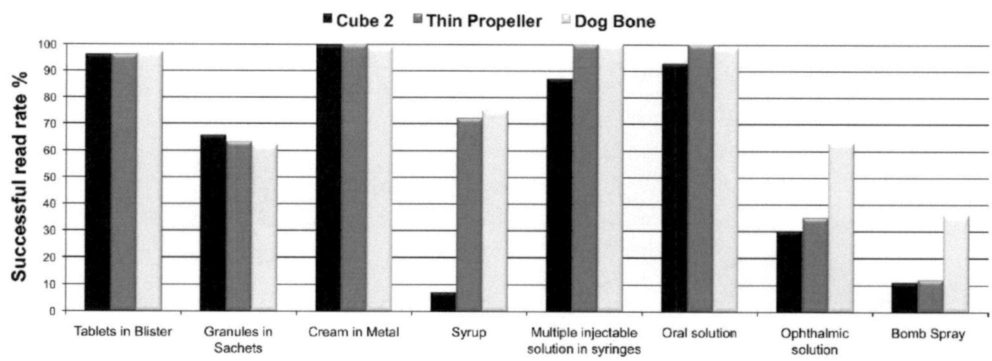

Fig. 7. A performance comparison on the cases line by varying the drug type and the tag type considering homogeneous cases and a random disposition of the items within the case (Configuration III).

Finally, the last test for the first part of the experimental campaigns is aimed to evaluate the reliability of commercial UHF tags in the border gate. An adequate number of 200 heterogeneous items contained in a big plastic bag has been considered. For this test, the same three previous different compositions, in terms of drugs type, have been considered. The experimental results of Fig. 9 show that a NF UHF solution is definitely not suitable for item-level tracing systems in this step of the supply chain.

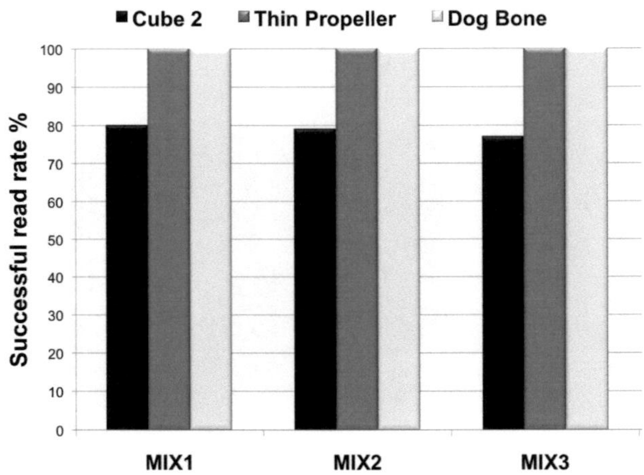

Fig. 8. A performance comparison on the cases line by varying the tag type and the composition of the case containing heterogeneous drugs.

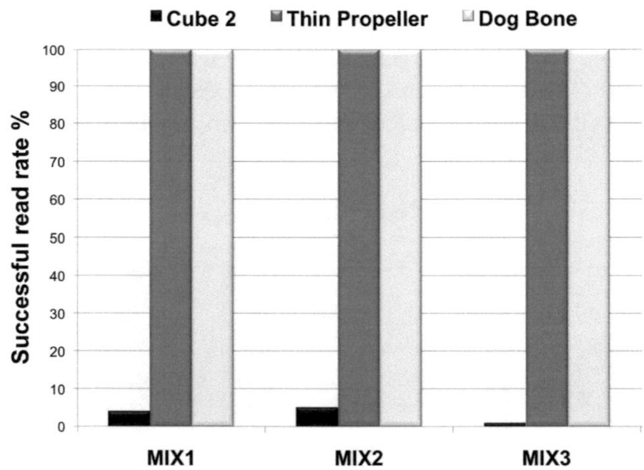

Fig. 9. A performance comparison on the border gate by varying the tag type and the composition of the plastic bag containing heterogeneous drugs.

The first part of the experimental campaigns has demonstrated that the use of commercial UHF RFID tags represents a well suitable solution for the traceability at item level at least in presence of non-critical operating conditions.

4.3 Results in critical conditions

As above stated, when the operating conditions do not present particular criticalities, all the selected FF tags work properly. Nevertheless, the performance degradation observed when the tags have been tested on products containing relevant quantities of liquid or metal is evident. For such a reason, this section focuses on the tag performance evaluation under more severe operating conditions. The products resulted as the most critical in the previously conducted tests, the bomb spray and the ophthalmic solution, have been considered. As previously stated, in fact, when the item to be traced is composed of metal and/or liquid, the tag performance could decrease considerably. This is substantially due to the fact that the horizontal-plane radiation pattern of the tag antenna, that is almost omnidirectional in free space conditions, varies significantly because of the presence of such materials.

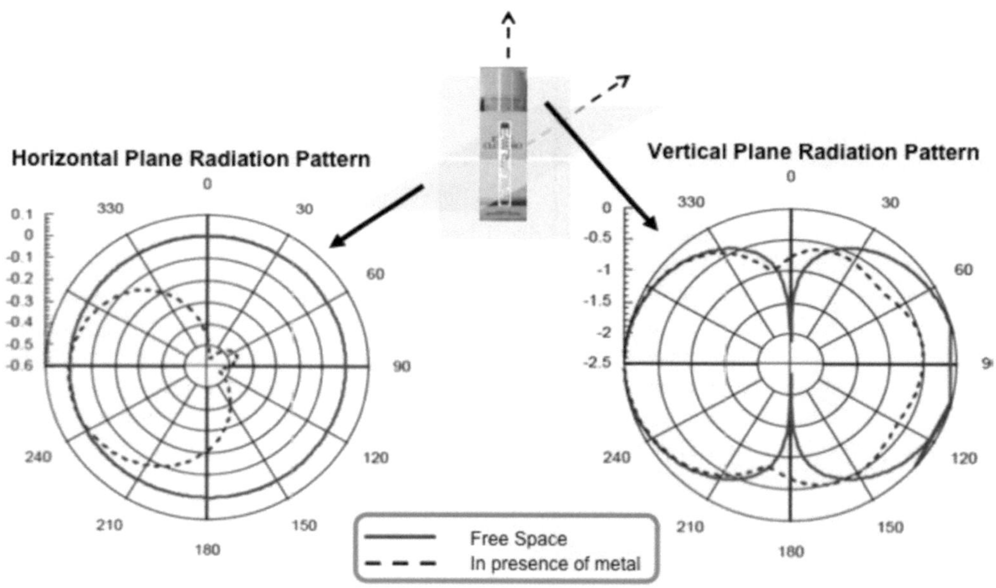

Fig. 10. Horizontal radiation pattern in free space condition (continuous line) and in presence of metal (dot line) of the Alien Squiggle Tag.

In Fig. 10, for instance, the horizontal plane radiation pattern of the Alien Squiggle RFID tag both in free-space conditions and in presence of metal are reported. It can be observed that, for many directions, the potential link between tag and reader is not possible anymore when the tag is attached on the bomb spray case. This effect becomes relevant when there is a 90° f misalignment between the planes individuated by the reader antenna and the tag respectively.

The results reported in Fig. 11 and in Fig. 12, for instance, are referred to the successful read rate measured in the items line respectively for the ophthalmic solution case and the bomb spray case. In order to stress the misalignment problem, tests have been carried out by considering the three different operating conditions previously described and characterized by a mutual orientation between tag antenna and reader antenna equal to 0°, +90° and -90° respectively.

Results are quite relevant: despite the very good values obtained in case of alignment, in general it emerges that no tag guarantees satisfactory performance levels in both cases and for each orientation.

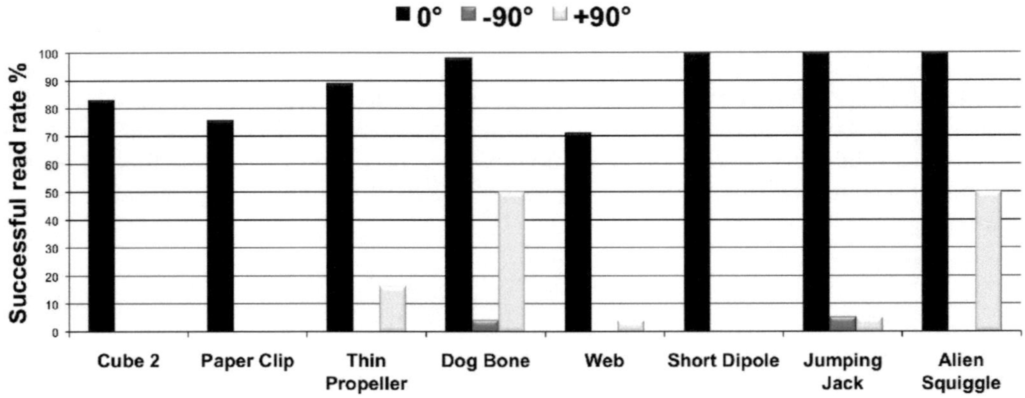

Fig. 11. A performance comparison on the items line by varying tag type and the tag-reader antenna misalignment in presence of liquids and metals (ophthalmic solution).

Even more interesting are the results obtained by testing the tags in the cases line. As an example, Fig. 13 and Fig. 14 reports the successful read rate evaluated by packing together 36 secondary packages of ophthalmic solution and 14 of bomb spray respectively, in the three configuration previously described and named respectively Configuration I, Configuration II, and Configuration III. It can be observed that the strong presence of metal and liquid substantially inhibits the communication between reader and the NF tags. Moreover, also when FF tags are considered, very low performance are obtained in each configuration, demonstrating once more that general purpose commercial tags are not appropriate for the implementation of complex item level tracing systems. It is substantially due to the fact that such tags have been designed not taking into account the peculiarities of the scenario where they must be utilized.

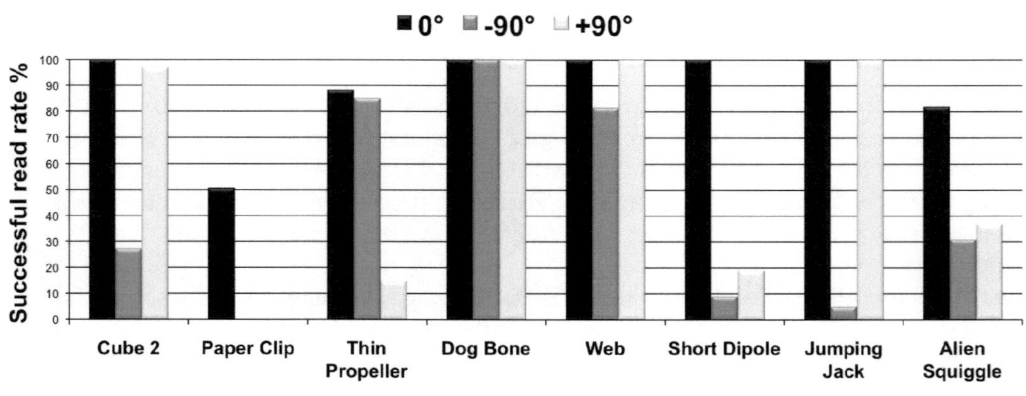

Fig. 12. A performance comparison on the items line by varying tag type and the tag-reader antenna misalignment in presence of metals (Bomb spray).

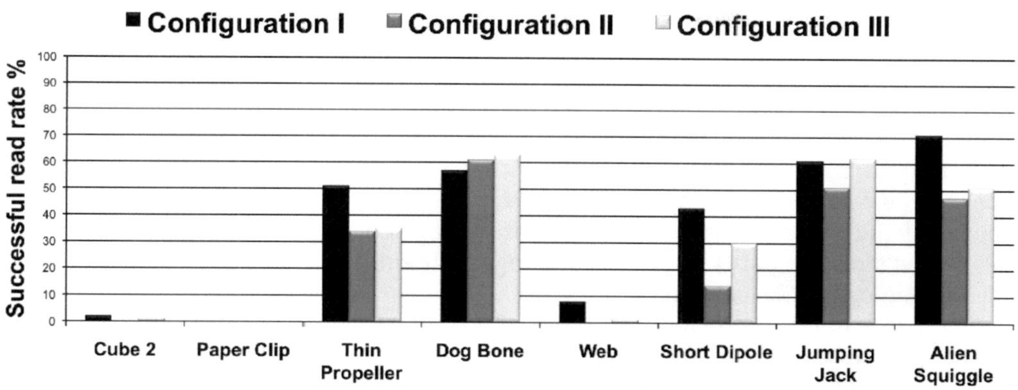

Fig. 13. A performance comparison on the cases line by varying tag type and the homogenous case composition in presence of liquids and metals (ophthalmic solution).

In the next section, a requirement analysis for a tag appositely designed to work in a complex supply chain, such as the pharmaceutical one, are individuated and described and are used to drive the development of a new high performance UHF RFID tag.

5. Requirements and guidelines in tags design

On the basis of the results shown in the previous sections, the realization of a tag designed ad-hoc for the specific supply chain scenario is a must. Consequently, this section focuses on the analysis of the pharmaceutical supply chain peculiarities and on the individuation of the properties that a tag should own in order to guarantee high performance in all supply chain steps, even when used to track items containing electromagnetically critical materials, such as liquids and metals. It is worth observing, though, that the pharmaceutical sector is only one of the many scenarios where a similar study could be of interest.

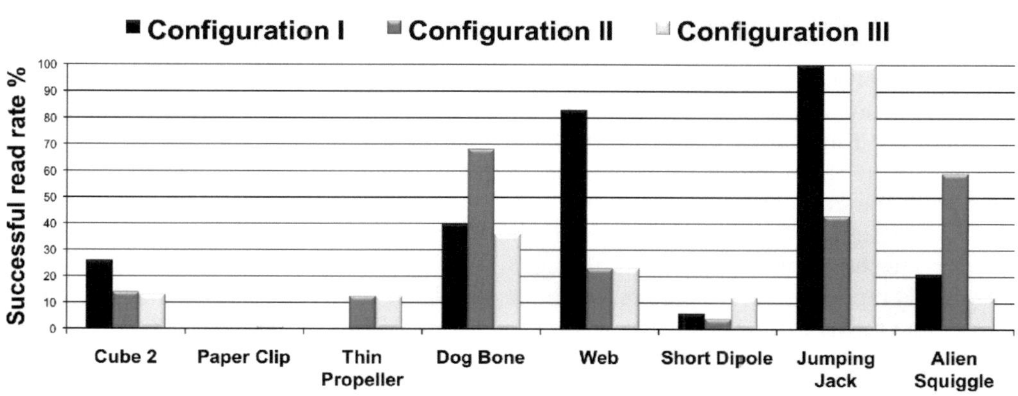

Fig. 14. A performance comparison on the cases line by varying tag type and the homogenous case composition in presence of metals (Bomb spray).

One of the sources of performance degradation in the items line is given by the potential misalignment between the NF reader antenna and the tag attached to the secondary package

of an item. In fact, by means of a conveyor belt, the tagged item passes through two NF RFID antennas.
Nevertheless, it is possible that the item surface on which the tag lies and the plane on which the reader antennas lie are mutually orthogonal. Now, as depicted in Fig. 10, despite the almost omnidirectional tag radiation pattern in free space condition, the presence of metal or liquid inside the item strongly modifies the radiating properties, even inhibiting, in some cases, the communication with the reader.
It can be deduced, hence, that a well performing tag should guarantee at least two main lobes on the radiation pattern in every working condition, above all when it is used to trace items containing hostile materials.
Another reason of reading-failure in the items line is due to the use of a FF tag antenna with a NF reader antenna. Although NF reader antennas are used in the items line, NF UHF tags cannot be used because they would not work properly in the subsequent supply chain steps, where FF reader antennas are adopted. Therefore, a well performing tag should exhibit good performance both in the NF and the FF.
On the items line step, the packages are read one by one and no multiple-readings related problem arises; on the contrary, they will occur in the cases line and in the border gate. In such cases, shielding effects due to the presence of plenty of items as well as the potential overlapping of tags, could lead to a strong performance collapse. Furthermore, also in these cases, problems due to a potential misalignment of tag and reader antennas can arise. Consequently, a well performing tag should take into account such issues. Therefore, the tag should be designed in order to avoid the complete tag overlapping and, moreover, it should guarantee (also in this case) multiple radiation pattern lobes.

6. Design of new passive RFID UHF tags: the prototypal enhanced tag

The designed and realized Enhanced tag (patent pending number TO2010A000493) is substantially based on a dual-lobe (collapsing in a particularly oriented one-lobe) conformal label-type antenna, adaptable to the different shapes of the various item packages and easy to be integrated in them. The shape of the antenna has been modeled in order to make the complete tag overlapping highly improbable. Moreover, the common design solution, based on the use of an inner loop around the microchip, has been adopted in order to guarantee good performance also in NF condition. The antenna has been realized in copper tape. Cost and size are comparable with canonical general-purpose UHF tags. Unfortunately, because of the patent-pending status, no details can be given on the shape and on the electromagnetic solutions adopted in order to reach the prefixed goal. Nevertheless, this is not even fundamental because the primary purpose of this work is, on the contrary, to demonstrate that an ad-hoc design of tags is able to effectively solve many of the performance degradation problems affecting general-purpose UHF tags.
In Fig. 15 is reported the comparison, in terms simulated horizontal plane radiation pattern, between the Enhanced tag (Fig. 15a) and the commercial Thin Propeller tag (Fig. 15b), when the tags are attached to a cardboard-made secondary package containing a metallic cylinder.
It can be observed that the radiative behavior of the two devices is radically different.
In the Thin Propeller tag case, the radiation pattern is not omnidirectional anymore and the link with the reader is possible only if the reader antenna is faced with the tag itself.
On the contrary, in the Enhanced tag case, an almost 45° oriented radiation pattern is found, resulting from the combination of two mutually orthogonal lobes. This way, also reader antennas orthogonal to the tag-plane can communicate with the tag.

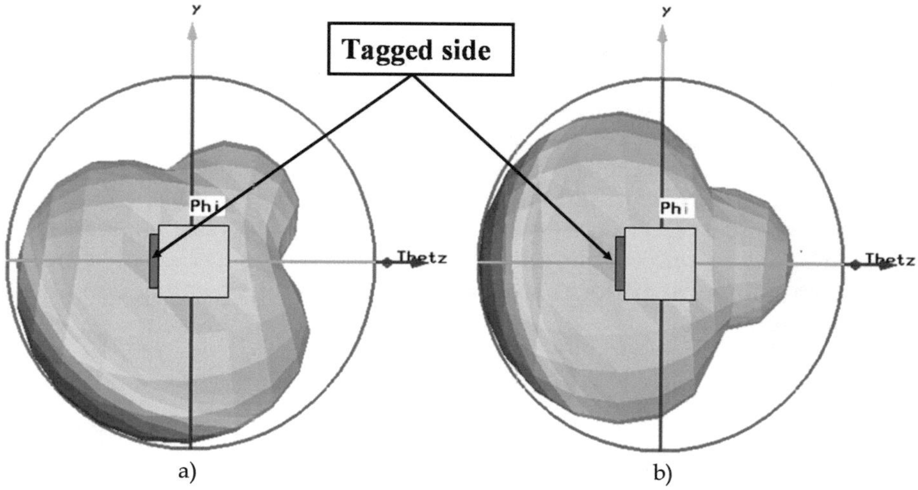

Fig. 15. Simulated horizontal radiation pattern of the a) Enhanced tag, and b) Thin Propeller tag, when they are applied on a cardboard package containing a metal block.

7. Experimental results of the enhanced RFID UHF tag

In order to evaluate the effectiveness of the designed Enhanced tag in the pharmaceutical supply chain, several experimental campaigns have been performed. In particular, a performance comparison of the Enhanced tag with some of the above described commercial UHF tags has been carried out in terms of successful read rate. In particular, taking into account the performance analysis carried out on commercial FF and NF UHF tags, the following four tag types with higher performance have been chosen in the comparison with the Enhanced tag: Jumping Jack, Cube2, Dog Bone, and Thin Propeller. One for the NF group and the others for the FF group.

Experimental campaigns have been mainly focused on particular operating conditions of two steps of the pharmaceutical supply chain: the items line and the cases line. As previously reported, these steps are particularly adequate to carry out an effective validation of novel RFID tags.

In all tests, the speed of the conveyor belt has been set to 0.66 m/s and 0.33 m/s respectively for the items line and cases line. The transmission power of the reader RFID has been set to 1W. Furthermore, the RFID tag is applied on the secondary package (made of cardboard) of the medicine product. Two different types of products have been used: ophthalmic solution in aluminum sachets and metallic bomb-spray.

The first part of the experimental campaign has been carried out on the items line. In this test, the misalignment problem has been stressed. In particular, the three different operating conditions (i.e. 0°, +90°, and -90°), previously described, have been considered.

The second part of the experimental campaign has been focused on the cases line. In such a test, each case was composed off homogeneous items. In particular, the bomb-spray case was prepared with 14 items on one layer, whereas the ophthalmic solution case was prepared with 36 items on three layers.

All the results, reported in this paper, are characterized by a confidence level equal to 95% with maximum relative error of 5%.

Fig. 16 presents the performance comparison when a single item of ophthalmic solution, enclosed in aluminum sachets, (i.e. liquid and metal) is scanned on the items line. The graph clearly shows that the Enhanced tag is able to reach the optimal performance, i.e. a successful read rate equal to 100%, in every critical operating conditions. More in detail, the graph shows that although the performance of all tested tags are comparable under optimal conditions (orientation equals to 0°), in critical conditions (orientation equal to -90° and +90°) the performance of commercial tags decreases so abruptly to achieve in most cases a percentage of successful read rate equal to 0%. Instead, the Enhanced tag reaches, also in these conditions, 100% of successful readings. The results clearly show also that the NF UHF tags are not able to solve performance problems in critical operating conditions (e.g. presence of misalignment).

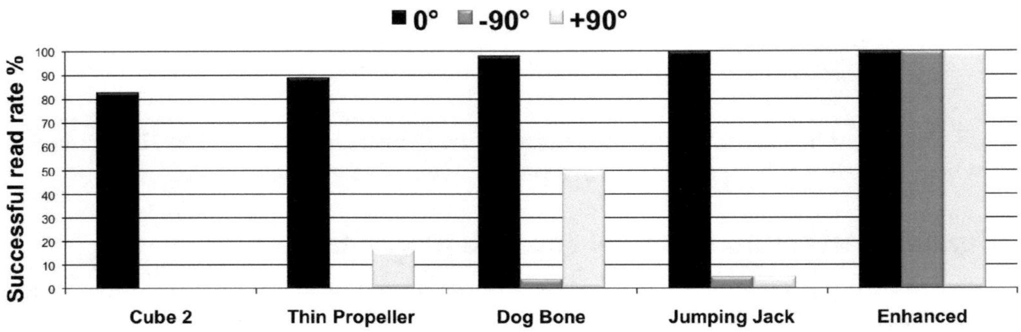

Fig. 16. A performance comparison between high-performance commercial tags and the Enhanced tag on the items line by varying the tag-reader antenna misalignment in presence of liquids and metals (ophthalmic solution).

In Fig. 17, the same performance comparison, using metallic bomb-spray, is shown. The graph confirms the excellent performance achieved by the Enhanced tag in all operating conditions on the items line. In this case, however, the performances obtained by some commercial tags are comparable to those reached by the realized tag (100% of successful read rate).

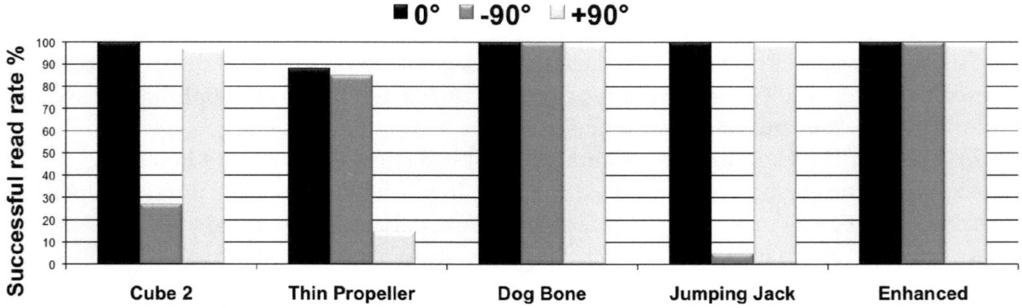

Fig. 17. A performance comparison between high-performance commercial tags and the Enhanced tag by varying tag type and the tag-reader antenna misalignment in presence of metals (Bomb spray).

Vice versa, the second part of the tests is aimed at comparing the tags performance in another challenging step of the supply chain: the cases line. Fig. 18 shows the performance comparison, in terms of successful read rate, of the Enhanced tag with the four commercial tags (i.e. one for NF and three for FF) by varying the composition of the ophthalmic solution case (i.e. Configuration I, Configuration II and Configuration III). It is worth noting that, commercial tags have never reached successful read rate higher than 70%, while in all the configurations the Enhanced tag has achieved the maximum performance. The results have also demonstrated the very poor performance of the NF UHF tags when used in a cases line. Finally, Fig. 19 shows the performance comparison when the case is composed of 14 items of bomb-spray. In this case, only one commercial FF UHF tag (i.e. Jumping Jack) presents good performance especially in Configurations I and III. On the contrary, other commercial tags have shown very low performance. This permits to assert that, also in this case, the Enhanced tag guarantees successful read rates better than the other tags.

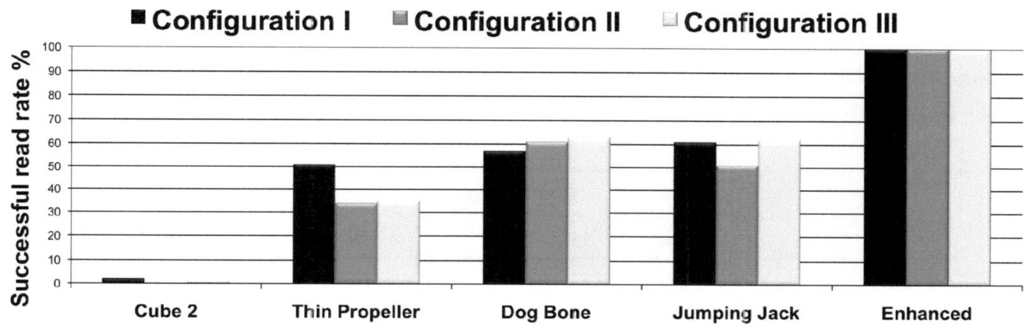

Fig. 18. A performance comparison on the cases line between high-performance tags and the Ehnanced tag by varying the homogenous case composition in presence of liquids and metals (ophthalmic solution).

In order to further emphasize the Enhanced tag robustness also in even more critical applications, an additional test has been performed. In particular, packages of milk have been considered. They are characterized by an external package made in Tetra Pak, where the percentage of metal is relevant, and by the presence of liquid. To test the effectiveness of the Enhanced tag, a performance comparison with one of the most powerful commercial tags (i.e. Dog Bone tag) has been carried out.

Also in this case, the measurement campaign has been carried out by considering both the items line (configurations 0°, +90° and -90°) and the cases line (only the configuration I with a 3 x 3 disposition of the single milk items).

Table 2 summarizes in detail the performance comparison between Enhanced tag end Dog Bone tag in the items line and in the cases line steps considering the Tetra Pak milk package. Also in this case the results are impressive: in the items line the Enhanced tag exhibits always 100% of successful read rate regardless of the package orientation. The commercial Dog Bone tag, instead, shows good results only in the optimal condition. In all other cases it cannot be read.

Even in the cases line the Enhanced tag is much more robust than Dog Bone. In fact, as can be observed in the same Table 2, the Dog Bone is never read, whereas the Enhanced tag achieves a successful read rate higher than 60%. This clearly demonstrates the qualities in terms of robustness and reliability of the proposed Enhanced tag even in contexts different from those the tag has been designed for.

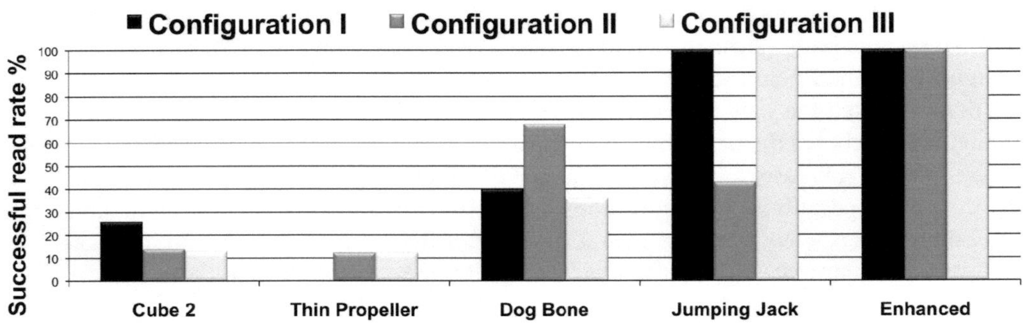

Fig. 19. A performance comparison on the cases line between high-performance commercial tags and the Enhanced tag by varying the homogenous case composition in presence of metals (Bomb spray).

	Items line			Cases line
	0°	+90°	-90°	Conf. I
Enhanced tag	100%	100%	100%	61%
Dog Bone tag	93%	0%	0%	0%

Table 2. Performance comparison between Enhanced tag and Dog Bone tag applied on Tetra Pak package.

8. Conclusion

In this chapter, the problem of the effective RFID-based traceability performed at item level has been addressed. The pharmaceutical supply chain has been considered and its criticalities, in terms of kinds of goods to trace and peculiarities of the checkpoints RFID, have been individuated and discussed. The inadequateness of the use of commercial general purpose tags has been proved through an exhaustive performance evaluation campaign, aimed at evaluating the successful read rate in each step of the supply chain for numerous tagged products. Six different commercial Far Field UHF tags and two Near Field UHF tags have been tested; these last are the less reliable, but also the Far Field ones exhibit strong limits when used to trace products containing metals or liquids. Consequently, by taking into account the traceability scenario, the requirements that a tag should own in order to overcome such limits have been individuated and, on such basis, a new enhanced tag has been realized. Its performance has been rigorously evaluated and the obtained impressive results demonstrate that, if the tag is designed considering the peculiarities of the specific tracing system, a successful read rate of 100% can be obtained, regardless of the supply chain step, the composition of the traced product, and the operating conditions. Finally, a very severe test has been carried out, aimed at evaluating the performance of our Enhanced tag on Tetra Pak packets containing milk. This application is one of the most challenging because of the very massive presence of both metal and liquids without any air in the middle. Very surprisingly, the performance are quite good also in this case, undoubtedly demonstrating once more that when a tag is designed by taking into account the peculiarities of the tracing systems, high performance can be obtained even in particularly critical conditions.

9. Acknowledgment

The authors wish to thank Dr. Vincenzo Mighali and Dr. Maria Laura Stefanizzi, that collaborate with the IDA Lab of the Department of Innovation Engineering of the University of Salento (Lecce, Italy), without whose assistance this study would not have been successful.

10. References

Acierno, R.; De Riccardis, L.; Maffia, M.; Mainetti, L.; Patrono, L.; Urso, E (2010). Exposure to Electromagnetic Fields in UHF Band of an Insulin Preparation: Biological Effects, Proceeding of IEEE Biomedical Circuits and Systems Conference, Paphos, Cipro, November 2010

Aroor, S.R.; Deavours, D.D. (2007). Evaluation of the State of Passive UHF RFID: An Experimental Approach. IEEE Systems Journal, vol.1, no.2, (December 2007), pp.168-176, ISSN : 1932-8184

Barchetti, U.; Bucciero, A.; De Blasi, M.; Mainetti, L.; Patrono, L. (2010). RFID, EPC and B2B convergence towards an item-level traceability in the pharmaceutical supply chain, Proceeding of IEEE International Conference on RFID-Technology and Applications, Guangzhou, China, June 2010

Bertocco, M.; Dalla Chiara, A.D.; Sona, A. (2010). Performance evaluation and optimization of UHF RFID systems, Proceeding of Instrumentation and Measurement Technology Conference, (3-6 May 2010), pp.1175-1180, ISSN: 1091-5281

Bertocco, M.; Dalla Chiara, A.; Gamba, G.; Sona, A. (2009). Experimental analysis of UHF RFID impairments and performance, Proceeding of IEEE International Instrumentation and Measurement Technology Conference, ISBN: 978-1-4244-3353-7, Singapore, May 2009

Catarinucci, L.; Colella, R.; De Blasi, M.; Patrono, L.; Tarricone, L. (2010). Improving Item-Level Tracing Systems through Ad Hoc UHF RFID Tags, Proceeding of IEEE Radio and Wireless Symposium, New Orleans, LA (USA), January 2010

De Blasi, M.; Mighali, V.; Patrono, L. & Stefanizzi, M. L. (2010). Performance Evaluation of UHF RFID tags in the Pharmaceutical Supply Chain, Paper presented at The Internet of Things - 20th Tyrrhenian International Workshop on Digital Communications, Pula, Sardinia, Italy, September 2009.

Finkenzeller, K. (2003). RFID Handbook, Fundamentals and Applications in Contact-less Smart Cards and Identification, Wiley & Sons, ISBN 978-0-470-84402-1

Fuschini F.; Piersanti, C.; Sydanheimo, L.; Ukkonen, L.; Falciasecca, G. (2010). Electromagnetic Analyses of Near Field UHF RFID Systems, IEEE Transaction on Antennas and Propagation, Vol. 58, No. 5, (May 2010) pp. 1759-1770

Mirowski L. et al. (2009). An RFID Attacker Behavior Taxonomy. IEEE Pervasive Computing Magazine, (October-December 2009), pp.79-84, ISSN: 1536-1268

Nikitin, P. V. and Rao, K.V.S. (2006). Performance Limitations of Passive UHF RFID Systems, Proceeding of IEEE Antennas and Propagation Society International Symposium, Albuquerque, NM, July 2006

Rao, K.V.S.; Nikitin, P.V.; Lam, S.F. (2005). Antenna design for UHF RFID tags: a review and a practical application. IEEE Transactions on Antennas and Propagation, vol.53, no.12, pp. 3870- 3876, (December 2005), ISSN: 0018-926X

Ramakrishnan, K. M. and Deavours, D.D. (2006). Performance Benchmarks for Passive UHF RFID Tags, Proceeding of 13th GI/ITG Conference on Measurement, Modeling, and Evaluation of Computer and Communication Systems, Nurenberg, Germany, March 2006

Staake, T.; Thiesse, F.; Fleisch, E. (2005). Extending the EPC network: the potential of RFID in anti-counterfeiting, Proceeding of ACM symposium on Applied computing, ACM Press New York, NY, USA, 2005

Koo, T.W.; Kim, D.; Ryu, J.I.; Kim, J.K.; Yook, J.G.; Kim, J.C. (2010). Design and Implementation of Label-type UHF RFID Tags for the Metallic Object Application, Proceeding of IEEE Antennas and Propagation Society International Symposium, Toronto, Canada, July 2010

Thiesse, F.; Floerkemeier, C.; Harrison, M.; Michahelles, F.; Roduner, C. Technology, Standards, and Real-World Deployments of the EPC Network. IEEE Internet Computing, vol.13, no.2, pp.36-43, (March-April 2009), ISSN: 1089-7801

Uysal, D. D.; Emond, J. P.; Engels, D. W. (2008). Evaluation of RFID performance for a pharmaceutical distribution chain: HF vs. UHF, Proceedings of IEEE international conference on RFID, Las Vegas, Nevada, USA, April 2008.

Part 3

Readers

11

Design and Implementation of Reader Baseband Receiver Structure in a Passive RFID Environment

Ji-Hoon Bae[1], Kyung-Tae Kim[2], WonKyu Choi[1] and Chan-Won Park[1]
[1]*Electronics and Telecommunication Research Institute*
[2]*Pohang University of Science and Technology (POSTECH)*
Republic of Korea

1. Introduction

In this chapter, we present a demodulation structure suitable for a reader baseband receiver in a passive Radio Frequency IDentification (RFID) environment. RFID refers to a technology which uses radio communications to contactlessly identify a tagged physical object [1-2]. An RFID system may include a plurality of electronic tags on objects, animals, and other things having unique identification information and a reader for reading or writing information from or to the tags. RFID systems can be variously classified into the inductively coupled and electromagnetic schemes according to the communication method employed between an RFID reader and a tag, into an active type and a passive type according to whether the tag operates using its own power or not, and into long wave, medium wave, shortwave, ultrashort wave, and microwave depending on the frequency of the electric waves used for the communication [1-2]. Essentially, a passive RFID system consists of a reader and a passive tag without a battery. The International standard, ISO 18000-6C, defines the communication protocol and Ultra High Frequency (UHF) band between the reader and the passive tag [3]. Many studies have been conducted in the field of UHF RFID, as described in [4-14]. In the case of passive UHF RFID technology, the reader must provide the tag with continuous radio power, while the tag sends its information to the reader via a backscatter modulation. The tag encodes the backscattered signal as either FM0 (bi-phase space) or Miller modulation of subcarrier at the given data rate [3]. Recently, UHF Passive RFID has a trend of extending its domain to the application of an item-level-tagging (ILT) from that of a conventional pallet/case-level-tagging. In the ILT RFID environment, tags can be attached on the objects composed partially of a metal or liquid and can be placed at a nearby complicated surrounding in which the metallic objects exist. As a result, if undesired large signal reflected from the complicated surrounding is received at the reader receiver during receiving a desired backscattered tag signal, the performance of the identification for the reader can be easily degraded due to the reflected large signal which can leak to the reader receiver (Fig. 1(a)). In addition, if insufficient isolation is guaranteed between the transmitter and receiver, the transmission power (Tx power) created by the reader transmitter can leak to the receiver (Fig. 1(a)) [2]. A reflected power larger than the backscattered tag signal which is generated by the return loss (S11) of the antenna can also leak to the receiver via the circulator (Fig. 1(a)). Because of these unwanted leakage components in the reader receiver, the DC-offset phenomenon can occur in the baseband of the reader receiver.

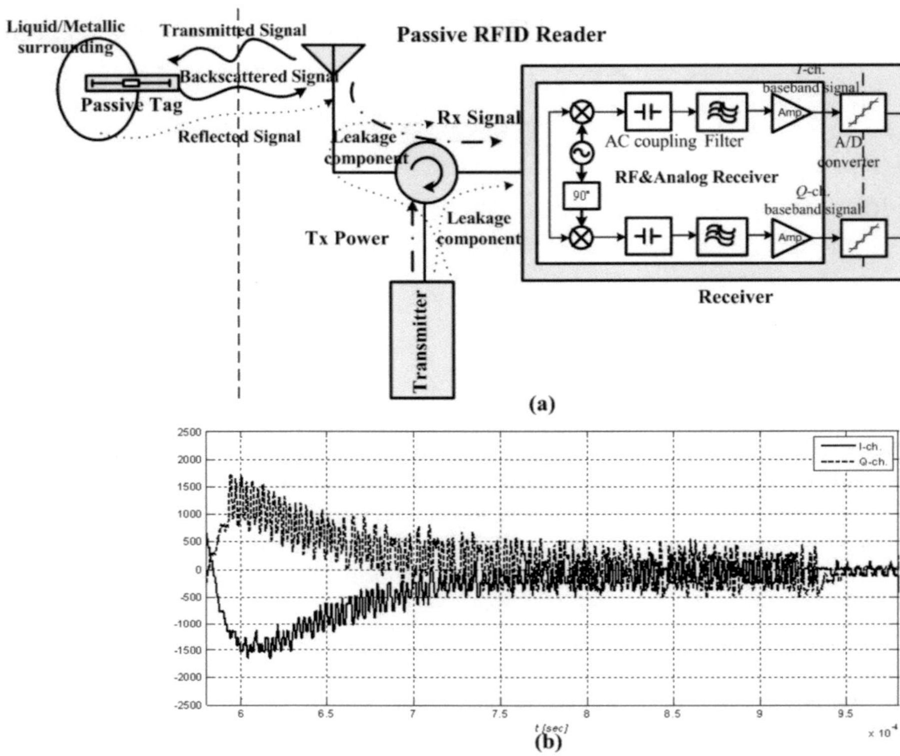

Fig. 1. Description of leakage components (a) and the corresponding DC-offset phenomenon (b) in a passive RFID communication environment

As a result, the received baseband signal can be corrupted by the DC-offset phenomenon (Fig 1(b)). For example, Fig. 1(b) shows the Miller subcarrier signal highly affected by the DC offset phenomenon in our reader receiver measured using an Agilent Logic Analyzer. Due to the unwanted DC-offset phenomenon, the reader baseband receiver may not determine the valid bit data with sufficient reliability. There have been several researches to reduce the originally generated leakage components in advance, as reported in [4-6]. However, it may be difficult to perfectly and adaptively eliminate the leakage components in the ILT RFID field, in which the performance of the reader receiver can be adversely affected by the unwanted large reflected signals. Therefore, although the received baseband signal is contaminated with the DC-offset phenomenon, we attempt to further remove the DC-offset phenomenon from the distorted received signal in the baseband receiver. In the earlier study, we proposed a demodulation structure composed of an edge signal generator, an edge extractor and a signal reconstruction block for the FM0 signal distorted by the DC-offset phenomenon [14]. In this chapter, a similar concept is also applied to the reliable reconstruction of the Miller subcarrier signal not suffering from the DC-offset phenomenon by using the phase inversion information instead of the amplitude information. In order to accomplish this, we detect the valid information from the corrupted signals by making use of a demodulation structure composed of a peak signal generator, a peak detector, and a signal reconstruction block, in order to successfully decode the received baseband signal distorted by the DC-offset phenomenon. According to the proposed demodulation method, the peak signal is created at the position of phase inversion within the Miller subcarrier signal sequence using the phase inversion information. Therefore, although a certain amount of DC-offset noise can be appeared in the baseband of the reader receiver, the proposed method is allowed to supplementally deal with the DC-offset phenomenon once more in the baseband receiver.

This chapter is organized as follows. In Section 2, we describe in detail the demodulation structure and method used to extract meaningful information from the distorted Miller subcarrier signal suffering from the DC-offset phenomenon. In Section 3, we show the simulation and implementation results. Finally, we draw our conclusions in Section 4.

2. Demodulation algorithm

In this section, we introduce the demodulation structure and algorithm suitable for the reconstruction of the Miller subcarrier signal distorted by DC-offset noise.

2.1 Demodulation of the Miller subcarrier-encoded signal

The Miller modulated sequence contains exactly two, four, or eight subcarrier cycles per bit, depending on the M value (M = 2,4,8) specified in [3]. Namely, if M has a value of 2, the Miller basis signal is multiplied by a square-wave at 2 times the symbol rate ($1/(M \cdot T_b)$), resulting in a Miller subcarrier signal with M = 2, as shown in Fig. 2. For the reliable reconstruction of the Miller subcarrier signal under the DC-offset environmnet, Fig. 3 shows the proposed demodulation architecture, which includes a peak signal generator, a peak extractor, and a signal reconstruction block, similar to the FM0 demodulation structure [14]. As shown in Fig. 3, the received signal, r(t) is composed of an in-phase channel (I-channel) signal, $r_I(t)$, and a quadrature-phase channel (Q-channel) signal, $r_Q(t)$ including DC-offset noise, n_{dc}, and the complex additive noise, $n(t)$, which is a sample function of a white Gaussian process with power spectrum $N_0/2$ watts/hertz. At this point, the DC-offset noise (n_{dc}) can be expressed as follows [14]:

$$n_{dc}(t) = A_{dc} \cdot e^{-B \cdot t} e^{j(w_d \cdot t + \psi)} \tag{1}$$

where A_{dc} is the initial DC-offset value, $e^{-B \cdot t}$ and $e^{jw_d \cdot t}$ represent the damping term and oscillation term of the DC-offset noise, respectively, and $e^{j\psi}$ is the initial phase of the DC-offset noise. By adjusting the parametrs A_{dc}, B, and w_d related to the DC-offset noise to proper values, any kind of DC-offset phenomenon in the area of passive RFID can be established.

In order to generate the peak signal with respect to the received baseband signal r(t) which is sampled at a sampling rate of $1/T_s$, the initial peak signal $r_{p1}(t)$ is designed using the predefined $s_{m0}(t)$ and $s_{m1}(t)$ as follows:

$$r_{p1}(t) = \left[LPF\{r_1(t) - r_0(t)\}\right]_{t=(k \cdot T_s), k=0,1,2,\ldots} \tag{2}$$

where $r_0(t)$ is the output signal using $s_{m0}(t)$ via the I and Q channels and is defined as follows:

$$r_0(t) = \frac{1}{A}\left(\underbrace{\left| r_I(t) \otimes s_{m0}(t) \right|}_{I-channel} + \underbrace{\left| r_Q(t) \otimes s_{m0}(t) \right|}_{Q-channel} \right), \tag{3}$$

where, A is the normalized gain, and $r_1(t)$ is the output signal using $s_{m1}(t)$ via the I and Q channels and is defined as follows:

$$r_1(t) = \frac{1}{A}\left(\underbrace{\left| r_I(t) \otimes s_{m1}(t) \right|}_{I-channel} + \underbrace{\left| r_Q(t) \otimes s_{m1}(t) \right|}_{Q-channel} \right) \tag{4}$$

$s_{m0}(t)$ has the same form as data-0 of the Miller subcarrier and $s_{m1}(t)$ has the same form as data-1 of the Miller subcarrier [3]. If M is four, then $s_{m0}(t)$ and $s_{m1}(t)$ can be described by Fig. 4. In our method, the low pass filtering of the difference signal between $r_1(t)$ and $r_0(t)$ is required for the generation of the desired peak signal.

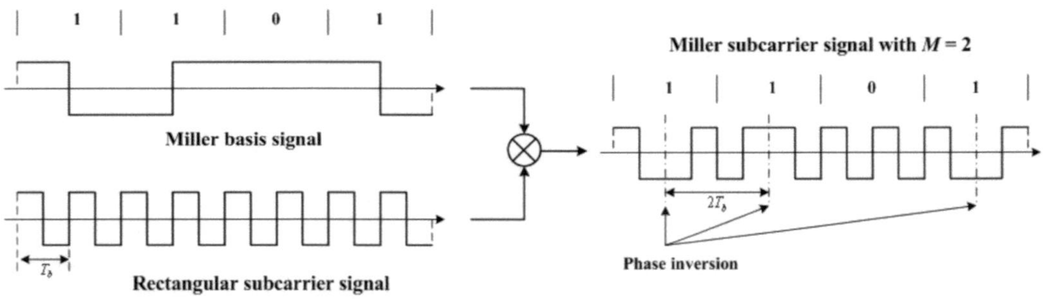

Fig. 2. Configuration of the Miller subcarrier signal with $M = 2$

In the second step, the created initial peak signal $r_{p1}(t)$ is reconstructed by removing the low level noise included in the specific level of the initial peak signal. This operation is implemented in the level decision block by using a reference level r_{ref} (Fig. 3). The reference level for the level decision is fixed at a value of 0. This is because the two orthogonal basis functions $s_{m0}(t)$ and $s_{m1}(t)$ participate in building the peak signal through Eq. (2), while only one basis function is used for the generation of the edge signal for the case of the FM0 signal [14]. Therefore, the demodulation method of the Miller subcarrier signal has the advantage that there is no need to find the optimal decision level, unlike the adaptive level decision method in the case of the FM0 signal. Then, the final peak signal can be obtained after the level decision by using the fixed reference level and the initial peak signal as follows:

$$r_{p2}(t) = \begin{cases} r_{p1}(t), & r_{p1} \geq r_{ref}(=0) \\ 0 \quad , & r_{p1} < r_{ref}(=0) \end{cases} \tag{5}$$

In the next step, from the final peak signal, $r_{p2}(t)$, the peak extractor (Fig. 3) finds the positions of the peaks using a peak detection algorithm which is identical to that of the edge extractor in Fig. 5 [14]. Finally, in the signal reconstruction block (Fig. 3), the basedband signal without DC-offset noise is regenerated by a state diagram which is also identical to that of the signal reconstruction block in Fig. 6 [14]. Therefore, the procedure for the proposed Miller subcarrier demodulation algorithm can be summarized as shown in Fig. 5

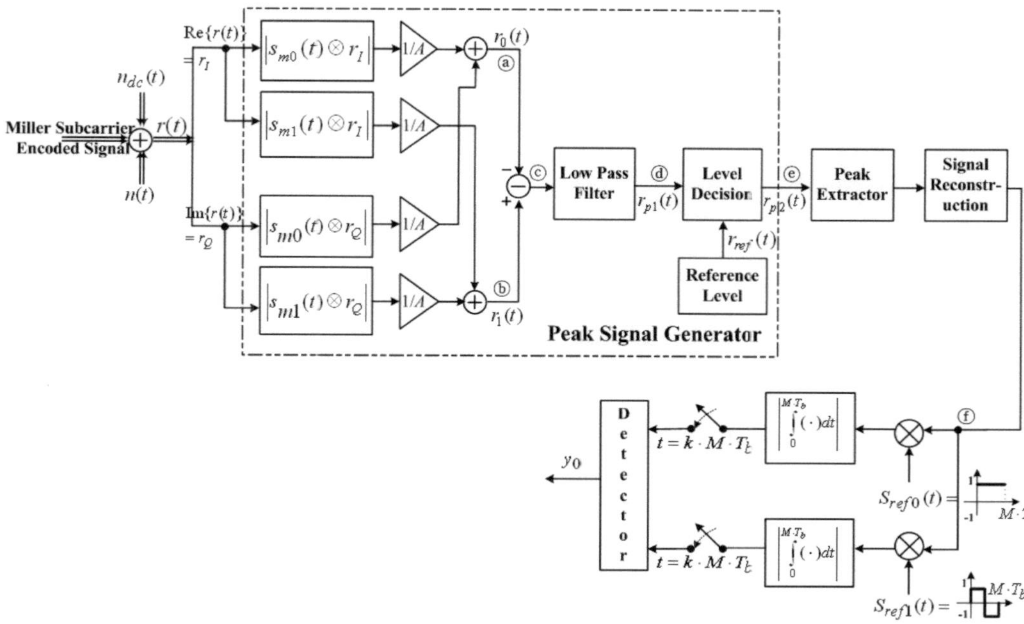

Fig. 3. Proposed demodulation structure for the purpose of reconstructing the Miller subcarrier signal distorted by DC-offset noise

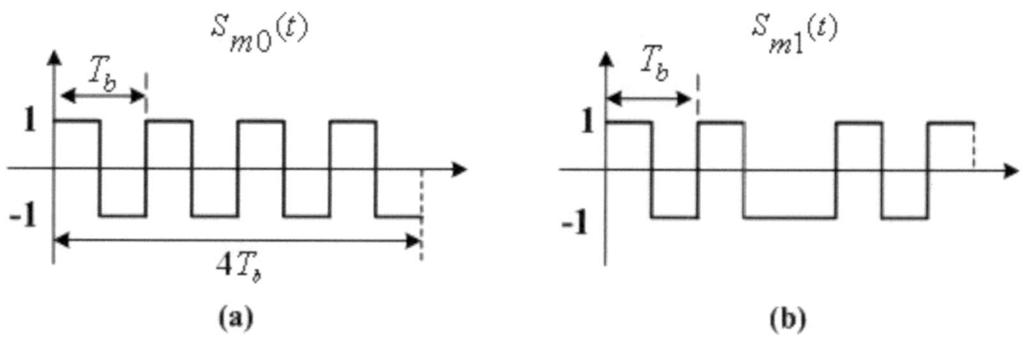

Fig. 4. Description of the two orthogonal basis functions, $s_{m0}(t)$ and $s_{m1}(t)$

2.2 Determination of the low pass filter specification

In our method for the demodulation of the Miller subcarrier signal, the difference signal between $r_1(t)$ and $r_0(t)$ must be reformed using a low pass filter (LPF), as mentioned in

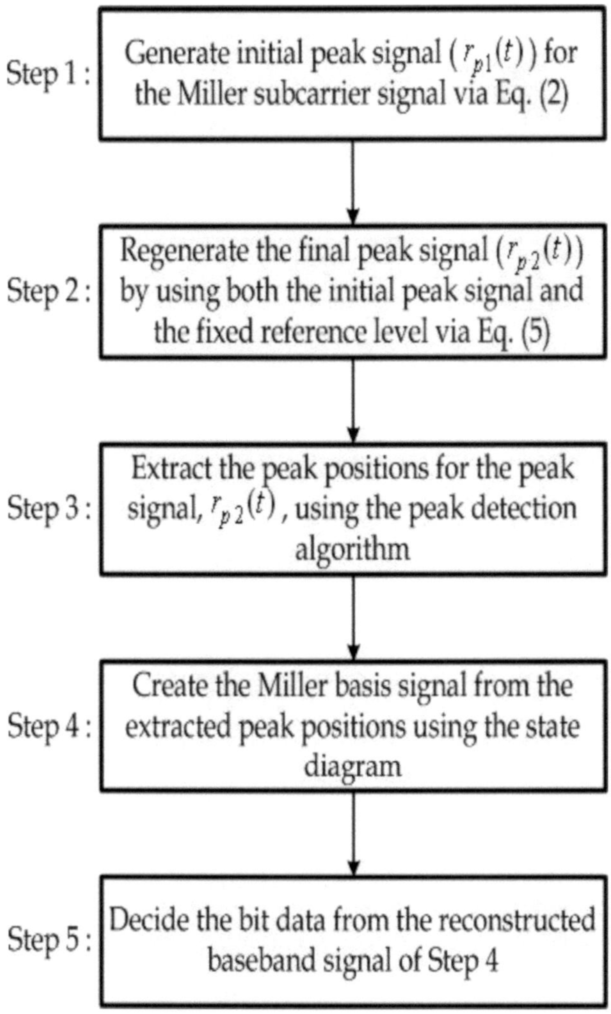

Fig. 5. Flowchart illustrating the Miller subcarrier demodulation method

Section 2.1. Fig. 6(a) shows the spectral response of the difference signal of the Miller subcarrier signal with $M = 4$ before the LPF, and Fig. 6(b) represents the spectral response of the initial peak signal $r_{p1}(t)$ after the LPF. To obtain the peak signal from the difference signal before the LPF, the second harmonic component of the difference signal (Fig. 6(a)) must be attenuated below a certain level, as shown in Fig. 6(b). We observe in Fig. 7 that, as the attenuation level Att_{dB} is increased, the error rate performance (p_e) for the Miller subcarrier signal is improved. However, the computational complexity for the design of the filter is also increased. This is because the order of the designed filter is increased to obtain the high level of the Att_{dB}. Meanwhile, when the Att_{dB} increase to a value larger than about 70dB, there is no noticeable improvement of the error rate performance. From the result shown in Fig. 7, we found that a level of attenuation Att_{dB} ranged from 70dB to 80dB can provide a suitable tradeoff between the computational complexity and the error rate performance for the demodulation algorithm. The result in Fig. 6(b) was obtained using a 39th order LPF with an attenuation level of 70dB and the magnitude response of the designed LPF is shown in Fig. 8

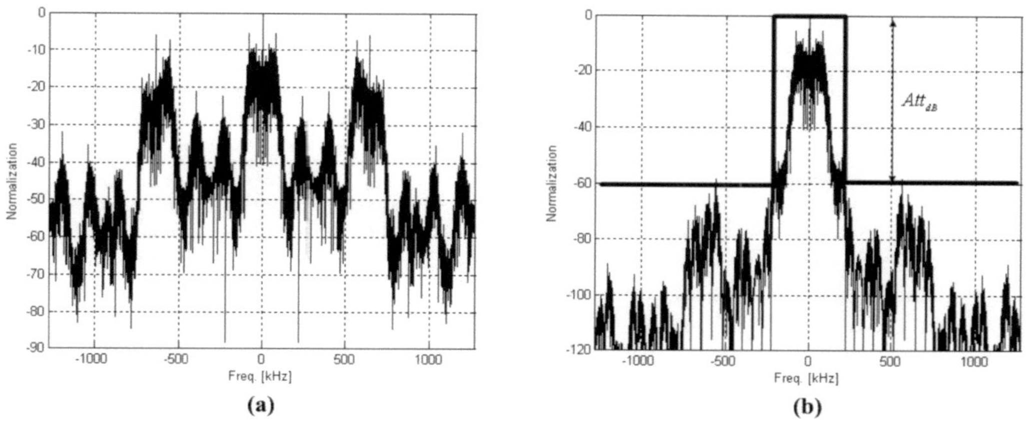

Fig. 6. Spectral responses of the difference signal before LPF (a) and the initial peak signal after LPF (b) (320kHz-Miller subcarrier signal with $M = 4$)

3. Simulation & experimental results

For the first example, we consider the operation of the proposed demodulation structure when the received baseband signal is the distorted Miller subcarrier-encoded signal with DC-offset noise ($A_{dc} = 5$, $B = T_s / T_b \cdot 100$, and $w_d = 1/(T_b \cdot 100)$). Fig. 9 shows the demodulation result of the Miller subcarrier signal with $M = 2$. Case 1 of Fig. 9(a) and Case 2 of Fig. 9(b) represent the output signals $r_0(t)$ in Eq. (3) and $r_1(t)$ in Eq. (4), respectively. From the results, we observe that both output signals, $r_0(t)$ and $r_1(t)$, are robust to the variation of the amplitude, although the I- and Q-channel received baseband signals are severely distorted by DC-offset noise. Fig. 9(c) represents the difference signal between $r_1(t)$ and $r_0(t)$ and the corresponding initial peak signal after the LPF. A 54th order LPF is designed for the generation of the initial peak signal $r_{p1}(t)$, in order to attenuate the amplitude response of the filter to about 80dB. The spectral responses of the difference signal before the LPF and the initial peak signal after the LPF, and the magnitude response of the designed 54th order LPF are shown in Fig. 10. Fig. 9(d) represents the final peak signal $r_{p2}(t)$ using the fixed reference level of 0 and the initial peak signal $r_{p1}(t)$. Note that the generated peaks of $r_{p2}(t)$ are placed at every position at which a phase inversion occurs within the Miller subcarrier signal sequence. Finally, as shown in Fig. 9(e), the reconstructed baseband signal without the DC-offset noise is obtained using the peak extraction algorithm of the peak extractor and the state diagram of the signal reconstruction block. The resulting signal in Fig. 9(e) has the same form as the Miller basis signal, which is obtained by removing the rectangular subcarrier signal from the Miller subcarrier signal, as shown in Fig. 2. This is because the peak signal $r_{p2}(t)$, which appears at a position having phase inversion, is used to reconstruct the Miller baseband signal.

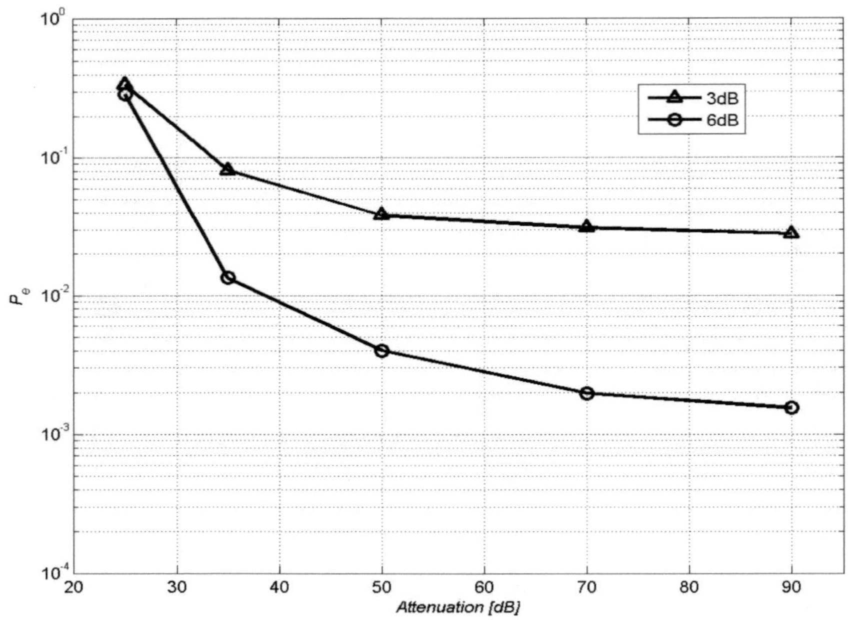

Fig. 7. The error rate performances for several values of Att_{dB}

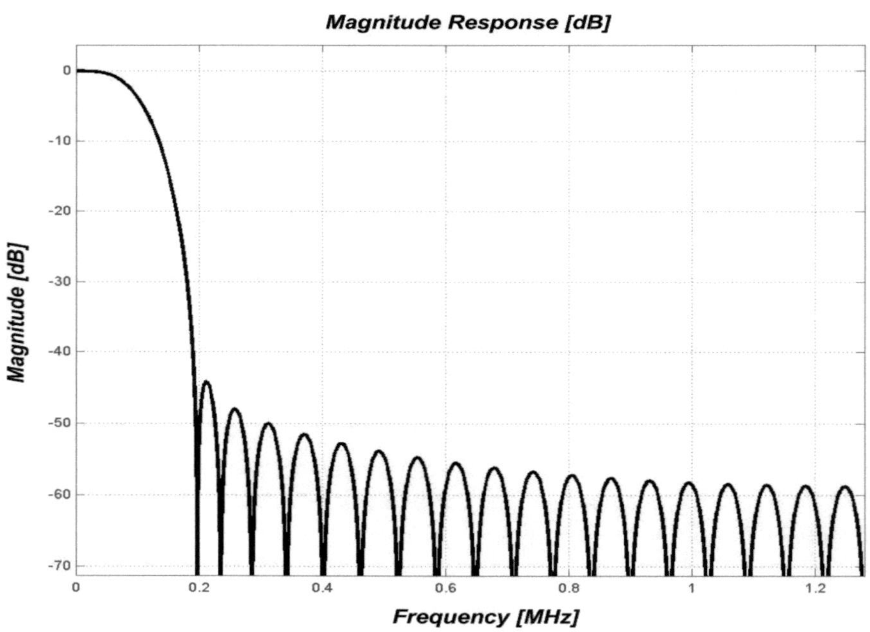

Fig. 8. The magnitude response of the designed 39th order low-pass filter for the generation of the initial peak signal

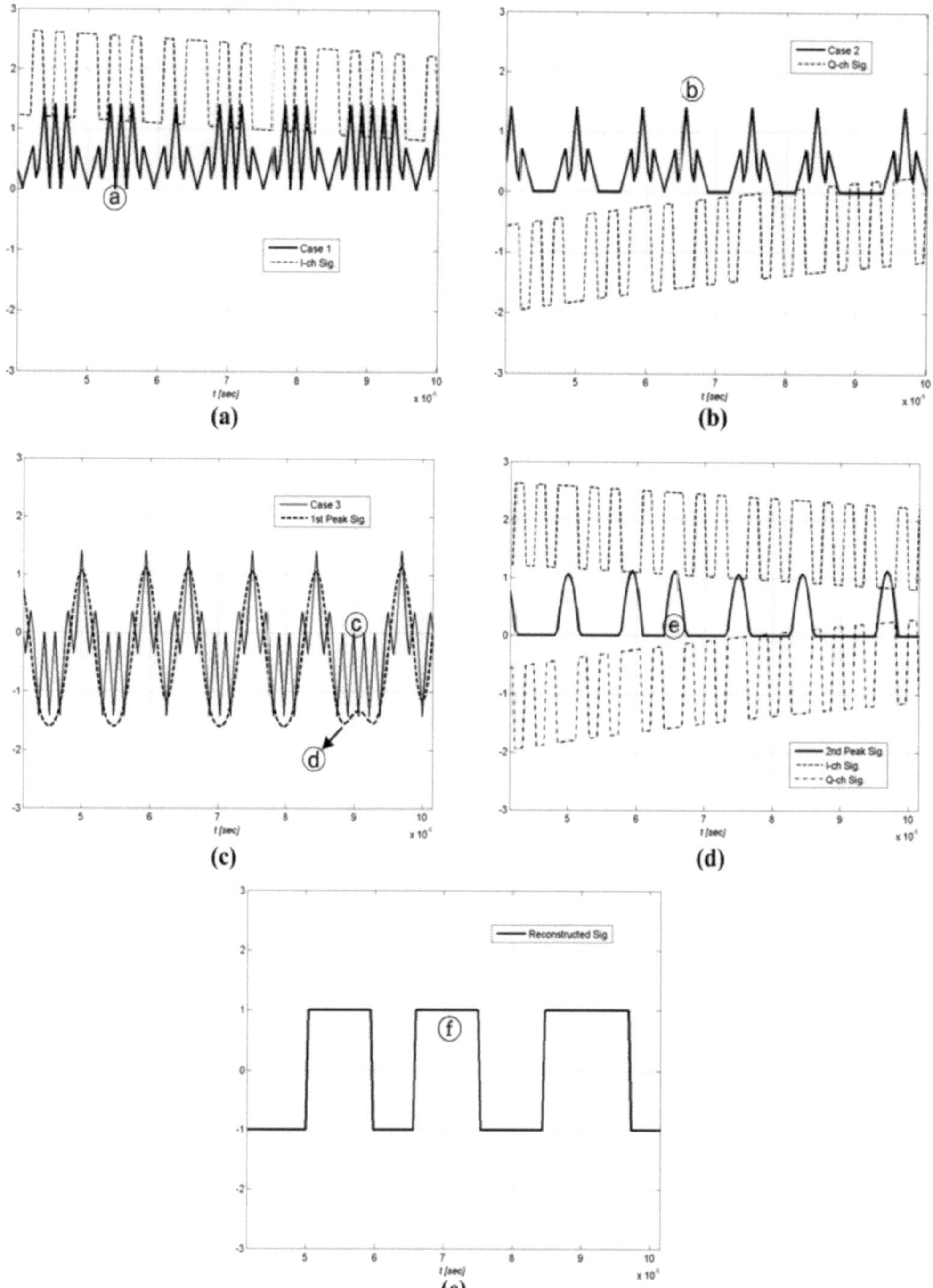

Fig. 9. Operation results of the proposed demodulation structure shown in Figure 3 (320kHz-Miller subcarrier signal with $M = 2$)

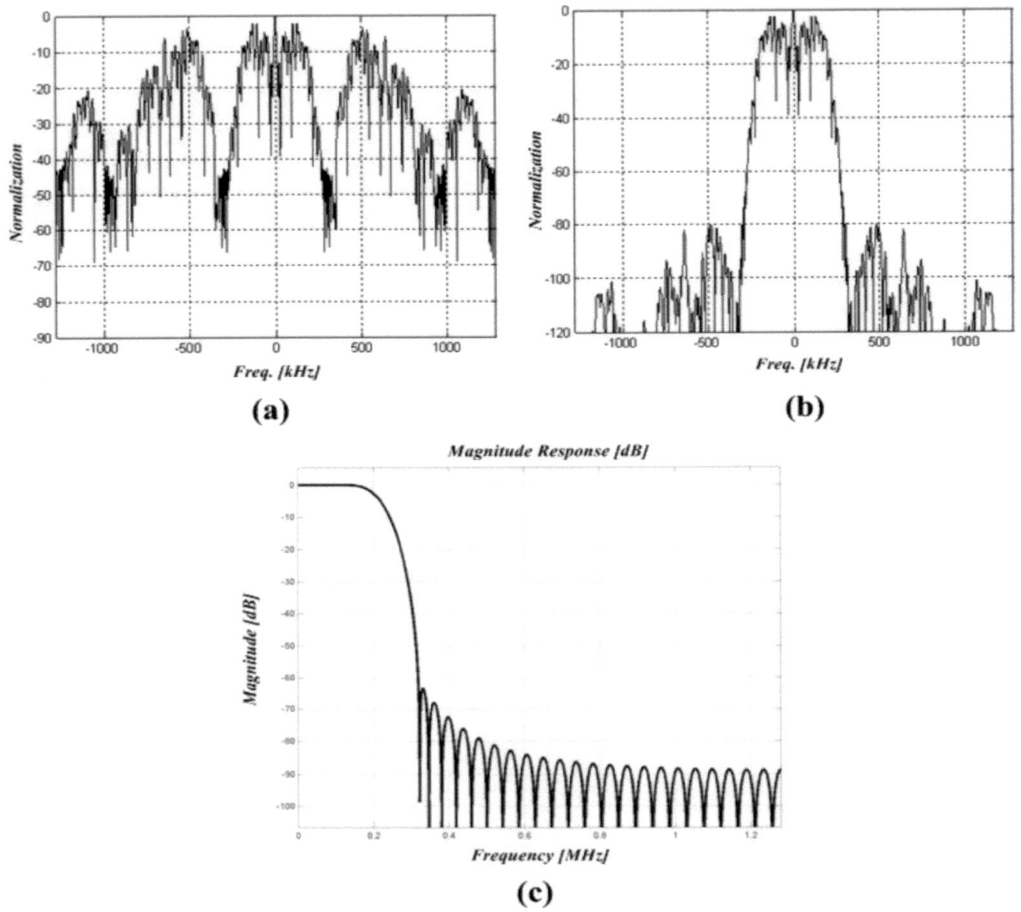

Fig. 10. Spectral responses of the difference signal before LPF (a) and the initial peak signal after LPF (b), and the magnitude response of the designed 54th order LPF (c) (320kHz-Miller subcarrier signal with $M = 2$)

For the second example, the Miller subcarrier signal with $M = 4$ is considered for the signal reconstruction, as shown in Fig. 11. The descriptions of Fig. 11 are explained as follows:

Case 1: the output signal $r_0(t)$ in Eq. (3) using $s_{m0}(t)$

Case 2: the output signal $x_1(t)$ in Eq. (4) using $s_{m1}(t)$

Case 3: the difference signal between $r_1(t)$ and $r_0(t)$ before the 39th order LPF having an attenuation level of 70dB

Case 4: initial peak signal $r_{p1}(t)$ after the 39th order LPF

Case 5: final peak signal $r_{p2}(t)$ using the fixed reference level and the initial peak signal $r_{p1}(t)$

Case 6: Reconstructed Miller baseband signal without DC-offset noise

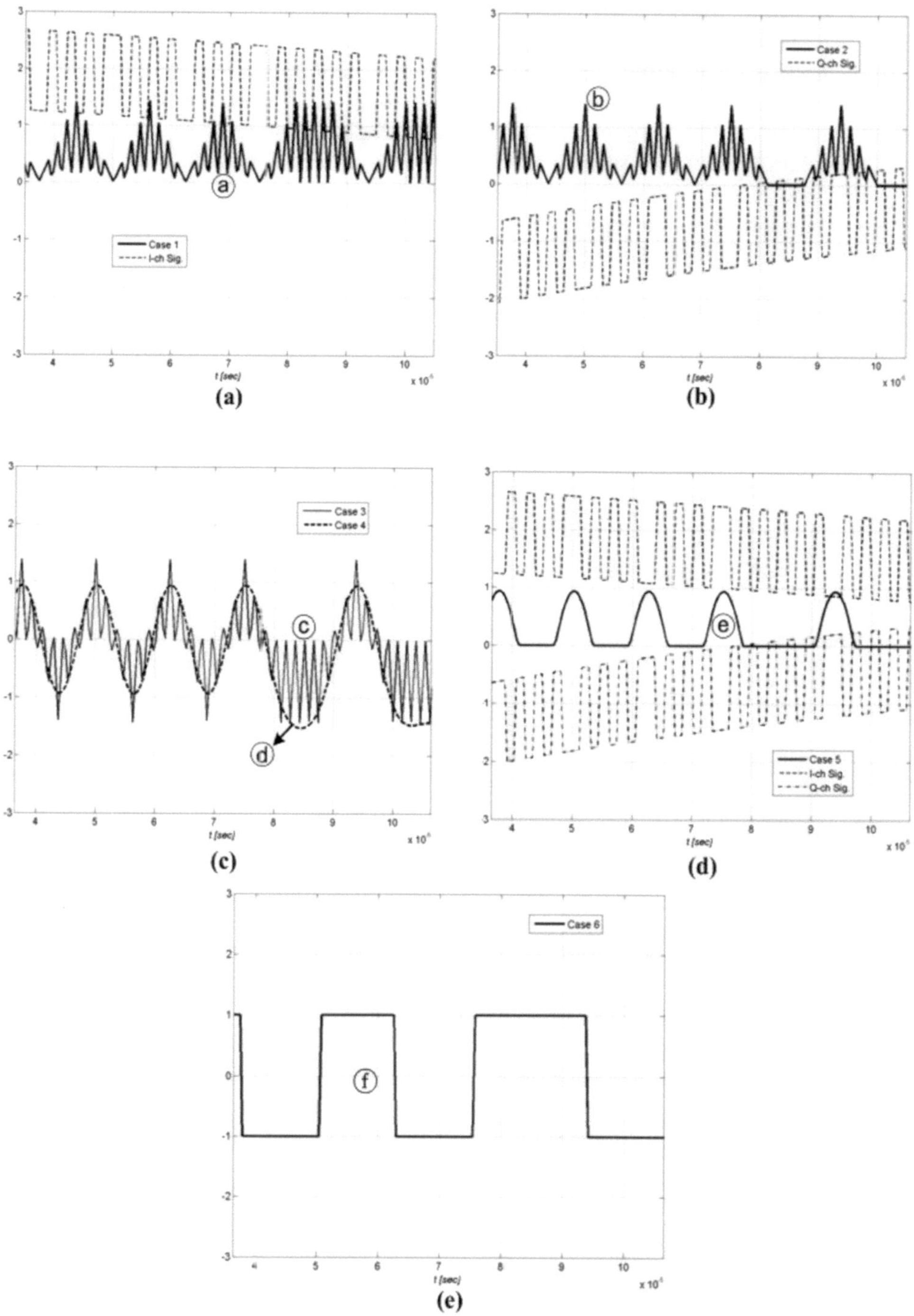

Fig. 11. Operation results o: the proposed demodulation structure shown in Figure 3 (320kHz-Miller subcarrier Eignal with $M = 4$)

In the next example, we implemented the proposed demodulation structure as a hardware device FPGA (Field Programmable Gate Array) and then the operation of the demodulation structure is observed using the commercial DSP design tool, Xilinx System Generator which provides hardware co-simulation, making it possible to incorporate the demodulation design running in an FPGA directly into a MATLAB Simulink simulation [15]. Fig. 12 shows the designed hardware co-simulation model of the proposed demodulation structure using a Black Box and JTAG Co-Sim library block provided by the System Generator. The Black Box library block allows a designed HDL (hardware description language), such as VHDL and Verilog, to be brought into the Simulink design model and enables us to easily observe the corresponding simulation behaviour in MATLAB Simulink.

In order to execute the designed Simulink model of the demodulation structure in Fig. 12, the following hardware co-simulation environment should be considered as shown in Table 1.

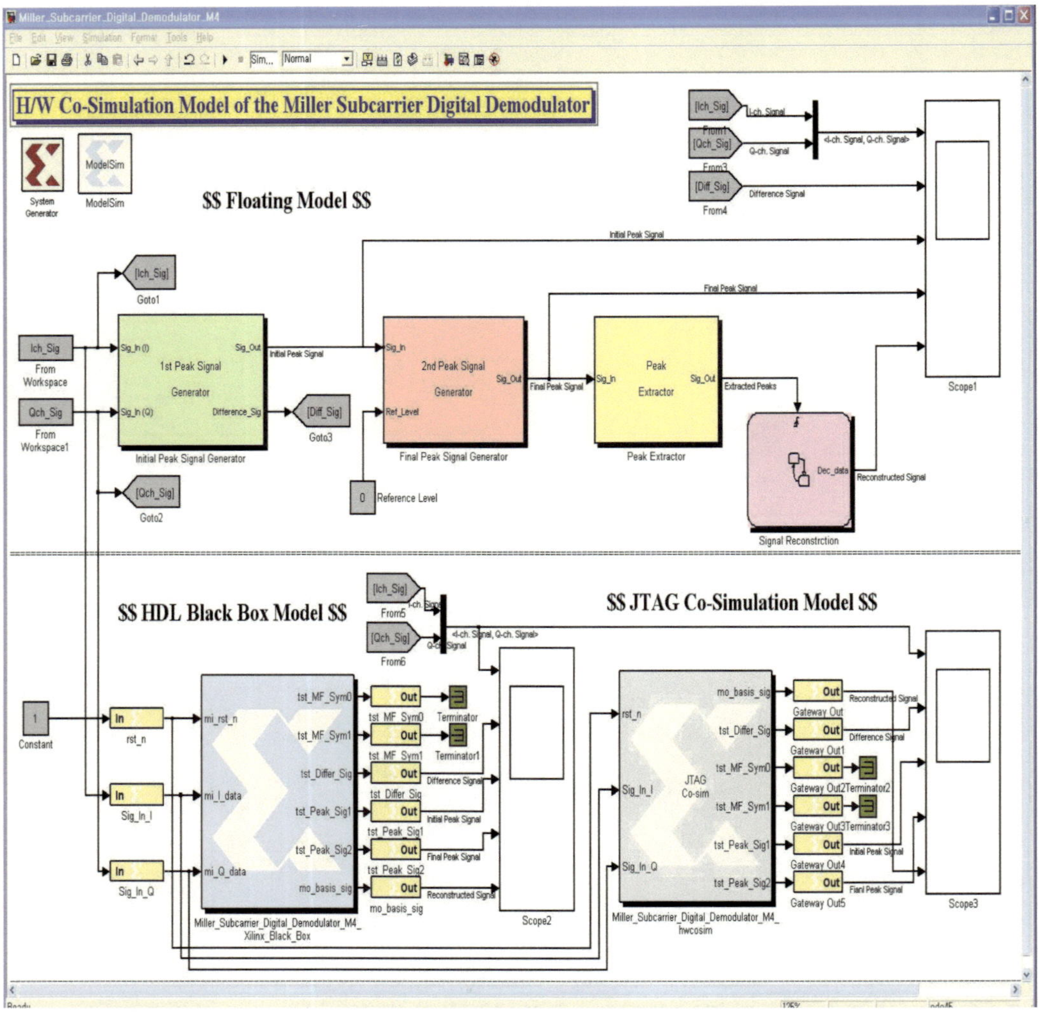

Fig. 12. Hardware co-simulation model with Black Box and JTAG Co-Sim library block

Items	Descriptions
MATLAB	MATALB 2008a
System Generator	Xilinx System Generator 10.1
HDL design tool	Xilinx ISE 10.1
HDL simulation tool	ModelSim SE 6.2b
FPGA (Digital hardware board)	Xilinx SPARTAN-3 XC3S4000FGG676-5G
JTAG Cable	Xilinx USB cable
Simulink system period [sec]	3.90625e-7 (1/2.56MHz)

Table 1. Hardware co-simulation configuration for the verification of the demodulation structure

Meanwhile, the Black Box HDL can be co-simulated with MATLAB Simulink using the System Generator interface to either ISE simulator or the ModelSim simulation software from Model Technology, Inc. Fig. 13 shows the operation result of the demodulation structure using the latter method through ModelSim when the measured Miller subcarrier signal (Fig. 2(b)) is considered. The operation result (Scope 2) using the former method through ISE simulator is also shown in Fig. 14

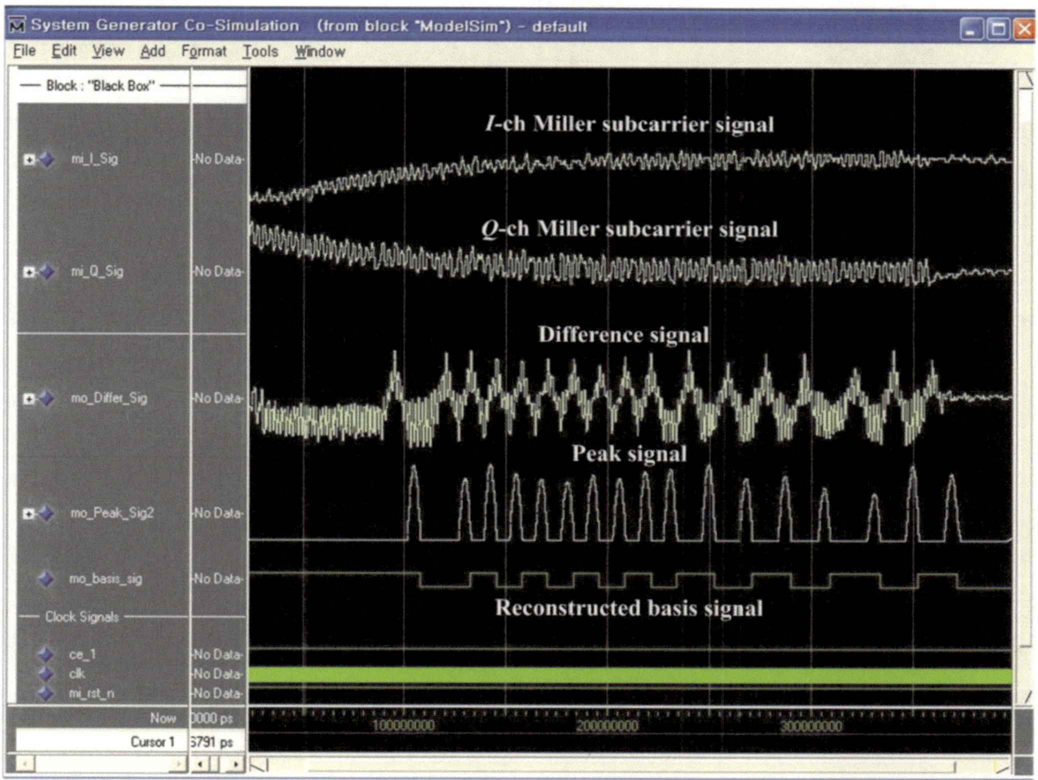

Fig. 13. Simulation result showing Black Box output through ModelSim for the proposed demodulation structure

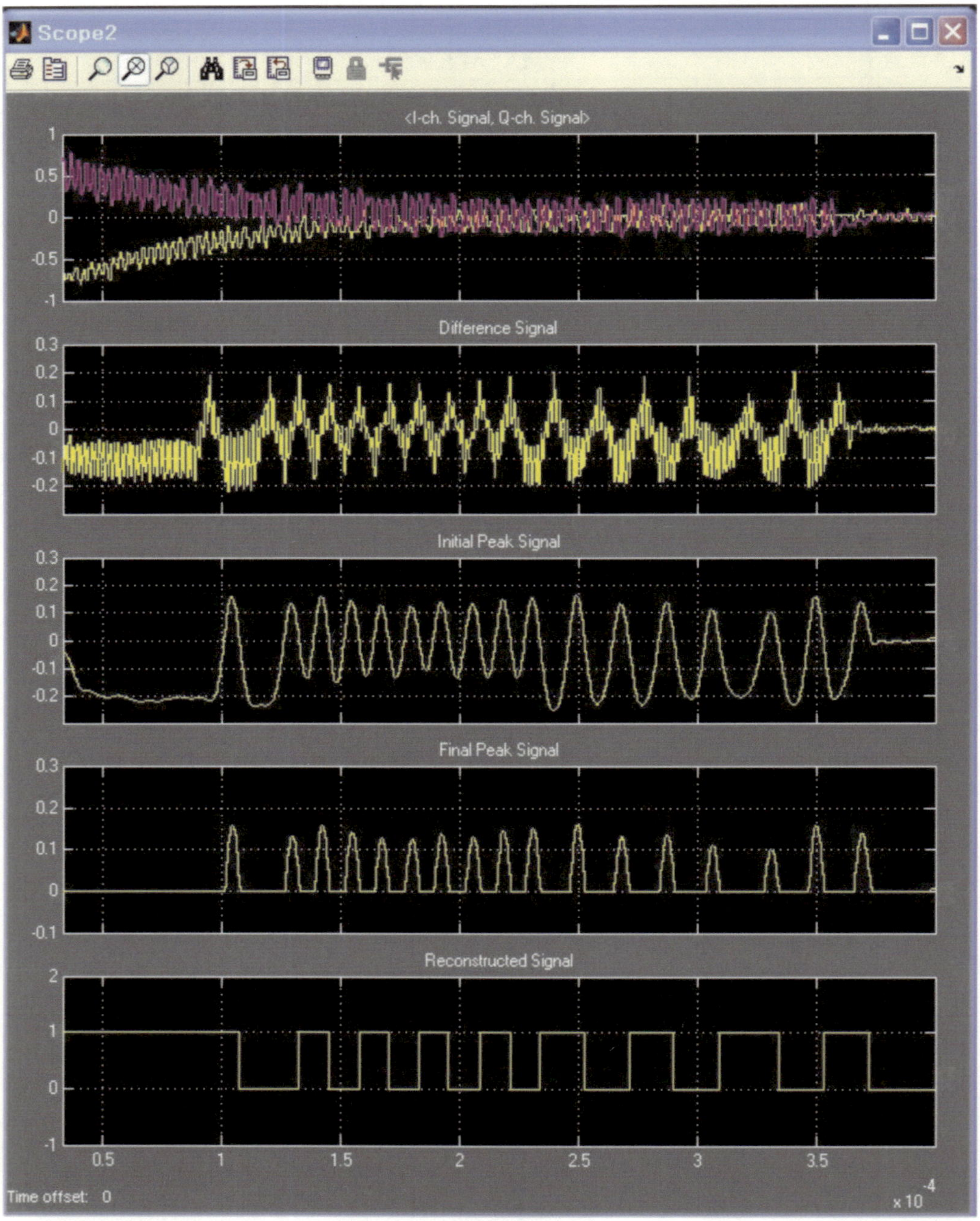

Fig. 14. Simulation result showing Black Box output through ISE simulator for the proposed demodulation structure

Next, Fig. 15 shows the hardware co-simulation result in Scope 3 when the same measured Miller subcarrier signal (Fig. 2(b)) is considered.

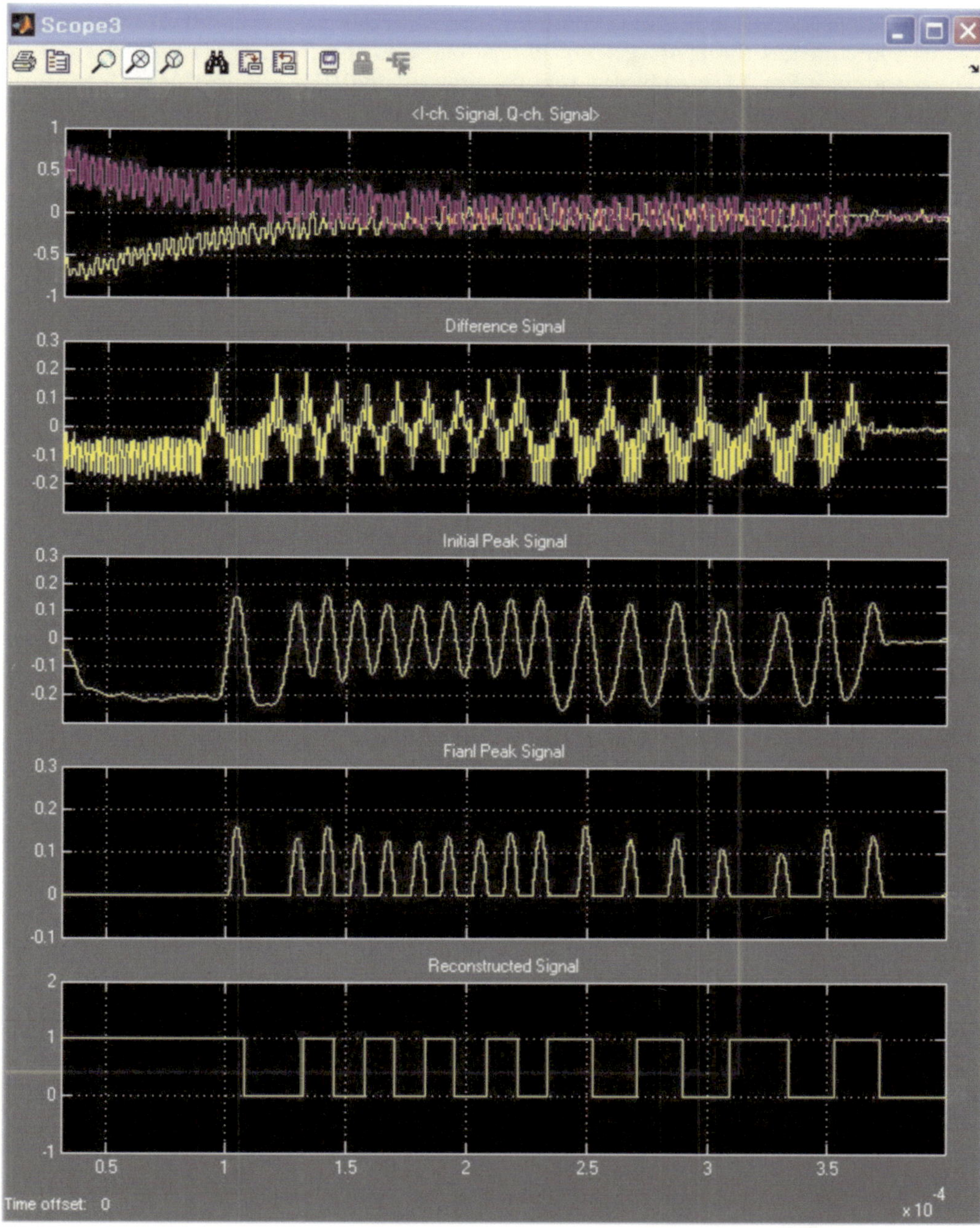

Fig. 15. Simulation result showing JTAG Co-Sim block output for the proposed demodulation structure

To obtain the JTAG Co-Sim library block (Fig. 12) for the demodulation structure, the hardware and software required to run the hardware co-simulation on the FPGA board (Table 1) should be installed and set up in advance [15]. Therefore, once the System

Generator has successfully finished compiling the HDL design into the FPGA bitstream, it automatically creates the JTAG Co-Sim library block as shown in Fig. 12. As a result, in comparison with Fig. 14, we observe that the timing simulation result of Fig. 15 is nearly identical to the functional HDL simulation result of Fig. 14.

Finally, the measured operation results obtained using Agilent Logics Analyzer equipment are described in Fig. 16. From the results of Figs. 13, 14, 15, and 16, we observe that the proposed method can successfully reconstruct the Miller baseband signal, even though the received Miller subcarrier signal is distorted by the DC-offset phenomenon. Although we do not show the corresponding demodulation results, the demodulation structure and algorithm described in Section 2 can be directly used to reconstruct a distorted Miller subcarrier signal with $M = 8$.

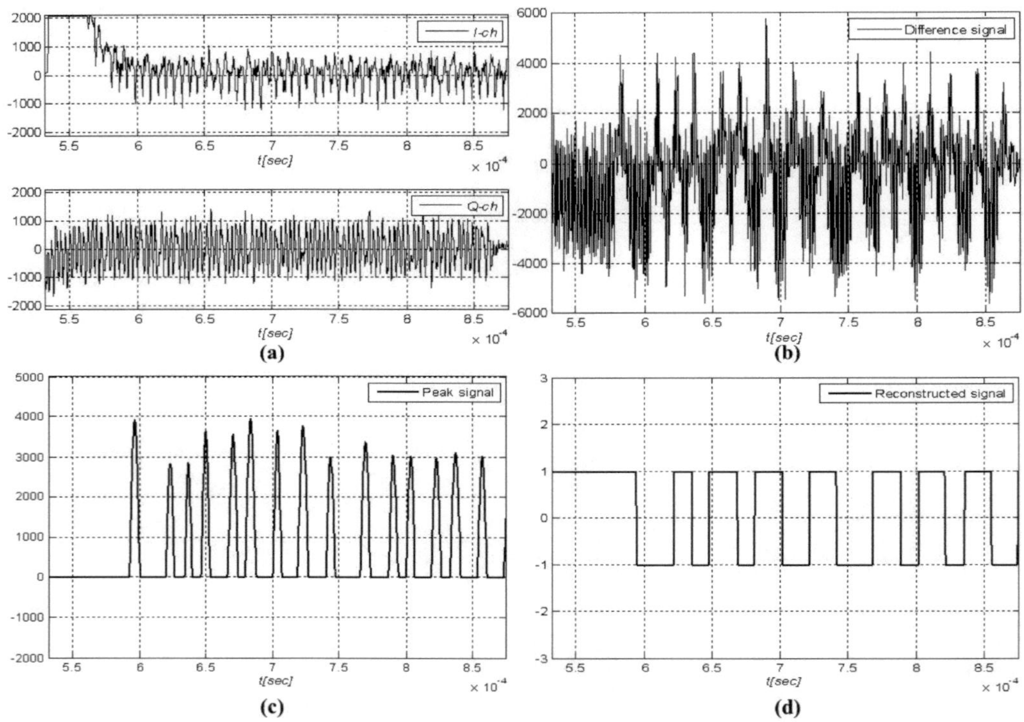

Fig. 16. Experimental results of the proposed Miller subcarrier demodulation structure (320kHz-Miller subcarrier signal with $M = 4$)

3. Conclusion

In this study, we propose a reader baseband receiver structure for the demodulation of the Miller subcarrier signal described in the international UHF RFID standard, 18000-6 Type C, under the DC-offset phenomenon. In order to perfectly remove the DC-offset noise caused by the leakage components in a passive RFID environment, the proposed structure for the passive RFID reader baseband receiver primarily includes the peak signal generator, the peak extractor, and the signal reconstruction block for demodulating Miller subcarrier signals. The phase inversion information is used to generate the peak signal and the resulting reconstructed signal is the Miller baseband signal removed both the DC-offset noise and the rectangular subcarrier signal simultaneously. To verify the operation and functionality of the proposed demodulation structure, we implemented both the HDL co-simulation using the Black Box library block and the hardware co-simulation using the JTAG Co-Sim one simultaneously. The results show that the proposed demodulation method can successfully extracts the valid information from the signal corrupted by the DC-offset noise which can occur in a passive RFID configuration.

4. Acknowledgment

This work was supported by the IT R&D program of MKE/IITA [Development of Next Generation RFID Technology for Item Level Applications], Rep. of Korea.

5. References

[1] Finkenzeller, K., "RFID Handbook - Fundamentals and Applications in Contactless Smart Cards and Identification", John Wiley & Sons Ltd, 2003.
[2] Daniel M. Dobkin, "The RF in RFID", Elsevier Inc., 2008.
[3] "Radio-Frequency Identification for Item Management - Part 6: Parameters for Air Interface Communications at 860MHz to 960MHz", ISO/IEC 18000-6 Type C, 2009.
[4] H.-W. Son, J.-N. Lee, and G.-Y. Choi, "Design of Compact RFID Reader Antenna with High Transmit/Receive Isolation", Microwave and Optical Technology Letters, Vol. 48, No. 12, Dec., 2006.
[5] Scott Chiu, et al.,"A 900MHz UHF RFID Reader Transceiver IC", IEEE Journal of Solid-State Circuits, Vol.42, No.12, December, 2007.
[6] Jeiyoung Lee, et al., "A UHF Mobile RFID Reader IC with Self-Leakage Canceller", IEEE Radio Frequency Integrated Circuits Symposium, pp.273~pp.276, 2007.
[7] Fan Z. G., S. Qiao, J.T. Huangfu, and L.X.Ran, "Signal descriptions and formulations for long range UHF RFID readers", Progress In Electromagnetics Research, Vol. 71, pp.109-127, 2007.
[8] Harrison B. Chung, et al.,"An advanced RFID system to avoid collision of RFID reader, using channel holder and dual sensitivities", Microwave and Optical Technology Letters, Vol. 49, No. 11, Aug., 2007.
[9] WonKyu Choi, et al., "RFID Tag Antenna Coupled by Shorted Microstrip Line for Metallic Surfaces", ETRI Journal, Vol. 30, No. 4, August 2008.
[10] Ickjin Kwon, "A Single-Chip CMOS Transceiver for UHF Mobile RFID Reader", IEEE Journal of Sold-State Circuits, Vol. 43, No. 3, March 2008.
[11] Ki Yong Jeon, and Sung Ho Cho, "A RFID EPC C1 Gen2 System with Channel Coding Capability in AWGN Noise Environments", IEICE Trasn. Commun., Vol. E92-B, No.2, Feb. 2009.
[12] Ah. Wasif Reza, and T. K. Geok, "Objects tracking in a dense reader environment utilizing grids of RFID antenna positioning", International Journal of Electronics, vol.96, no. 12, pp.1281~pp.1307, 2009.

[13] Iker Mayordomo, et al.,"Design and Implementation of a Long-Range RFID Reader for Passive transponders", IEEE Trans. On Microwave theory and tech. Vol. 57, No. 5, May, 2009.

[14] J.-H. Bae, et al.,"Study on the demodulation structure of a reader receiver in a passive RFID environment", Progress In Electromagnetics Research, Vol. 91, pp.243-258, 2009.

[15] "System Generator for DSP, User Guide", Xilinx Inc., Sept., 2008.

12

RFID Readers for the HDX Protocol - A Designer's Perspective

Dan Tudor Vuza[1] and Reinhold Frosch[2]
[1]*Institute of Mathematics of the Romanian Academy,*
[2]*Frosch Electronics OEG, Graz,*
[1]*Romania,*
[2]*Austria*

1. Introduction

Previous work (Gelinotte et al., 2006; Vuza et al., 2007; Vuza et al., 2009) presented the contribution of the present authors to the development of readers for communication with tags according to the FDX protocol. Readers produced so far by Frosch Electronics were classified as voltage-driven and current-driven (Vuza et al., 2009), according to which circuit variable is controlled by the reader and which is controlled by the tag (and sensed by the reader). A voltage-driven reader powers the antenna with an AC voltage of constant amplitude. The FDX tag transmits data by load modulation, which causes the variation of the voltage at the tap point (the junction between the antenna coil and the tuning capacitor). The reader senses the latter voltage and extracts the baseband signal that contains the data. A current-driven reader powers the antenna with an AC current of constant amplitude. Again, the FDX tag transmits data by load modulation, which this time modulates the voltage across the whole antenna circuit. The reader extracts the data from the latter voltage, the tap point connection being not needed in this case. Recently, Frosch Electronics decided to add a new feature to the existing readers, providing them with the possibility of communicating with tags that use the HDX protocol, of interest in applications such as animal identification, so that the same reader could cover a larger variety of applications. In FDX, the tag is continuously powered by the reader. In HDX, the tag is first charged by an RF pulse of limited duration from the reader, and then it transmits the data using the energy stored during the first step, by driving its coil with an AC voltage whose frequency toggles between two values f_C = 134.2 KHz and f_{LOW} = 123.7 KHz prescribed by the HDX standard.
After a brief reminder on the two types of FDX readers in section 2, we present in sections 3 and 6 circuit topologies achieving the HDX extension for each of the mentioned classes of readers. The topologies are different for the two classes according to whether the tap point is accessible or not. For voltage-driven readers, the schematic is based on the usage of the TMS3705 circuit as a bit decoder. For current-driven readers, we consider the option of bit decoding by either a dedicated IC or by building a custom decoder from discrete hardware components supplemented by a software component, which could in principle offer more control over the decoding process.
In section 4 we discuss the important issue of transients, which has to be taken into account when striving for data integrity, and hence reliability. Transient effects have been discussed

in (Vuza et al., 2009) for FDX load modulation and have to be discussed again in the HDX setting, since transients manifest themselves when the tag changes the frequency and may have deleterious effects on data integrity if their duration is too long. The results obtained here are compared with those previously obtained for FDX and recommendations for reader design are drawn.

In section 5 we expose the principle of low coupling approximation that allows, in the case of low coupling between tag and reader antenna which is usually the case in real situations, to replace the tag with a voltage source in series with the reader antenna for the purpose of circuit analysis. We will make use of this principle in the analysis of transients and of the procedure of bit equalization.

Because the reader antenna circuit is tuned to the nominal frequency f_C, the two signaling frequencies used by the tag may induce voltages in the reader circuits whose amplitudes differ in a significant way. Such an inequality in amplification may increase the probability of bit error, especially at higher reading distances when the signal is weak. We present in section 7 a method for equalizing the bit amplification based on the one-pole model of the opamp and the related gain-bandwidth product, which does not require any additional component in order to achieve the required effect.

The material discussed so far has emphasized the importance of the correct choice of the components in the antenna and amplifier circuits in order to ensure that the duration of transients agrees with the bit time and that equalization of bit amplification is achieved as much as possible. The choice is to be made in the design phase and fine-tuning will be needed in the test phase. Both mentioned phenomena are connected to the transitions between the two signaling frequencies employed by the tag. One needs therefore means for generating such transitions in a reproducible and convenient way. Using real tags for testing does not provide the most convenient way. Observing the frequency transition is not easy on a scope, as the frequency difference is rather small. The transition is gradual because of transients, making difficult to estimate when the transition actually started. For this reason it is preferably to rely on simulators. In section 8 we propose a hardware tag simulator for tuning and testing. In order to be able to estimate the parameters of transients, it is necessary to know precisely the moment of transition onset, which cannot be deduced from the gradual system response. The simulator provides the means for generating transitions together with a signal for the transition onset that can be used as a trigger for the scope on which the system response is recorded. The transient is hidden in the signal and only its negative effects on the latter are immediately visible. Displaying the transient itself require an indirect method. We propose in section 9 two such methods aiming at providing a graphical display of transients, allowing thus to estimate their parameters such as duration and magnitude and to assess their effects on the received signal: a software simulation procedure based on PSpice, which can be used in the design phase, and a method based on the usage of the simulator that can be used in the testing and tuning phases.

2. Voltage-driven and current-driven readers for FDX tags

A voltage-driven reader (figure 1) powers the antenna with an AC voltage of constant amplitude at a carrier frequency f_C of 125 KHz or 134.2 KHz. The FDX tag transmits data by opening and closing the switch *SW*, which, due to the magnetic coupling *M*, modulates the current through the antenna. The variation of the current antenna causes the variation of the voltage V_{TAP} at the tap point (the junction between the antenna coil and the tuning capacitor). The reader senses the latter voltage and extracts the baseband signal that contains the data.

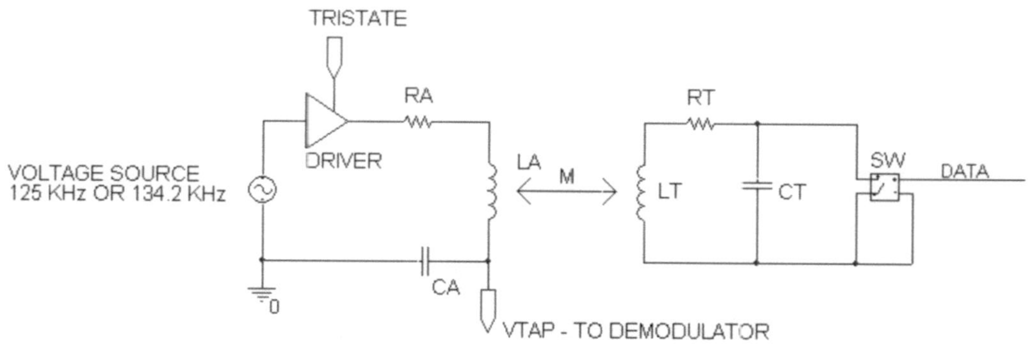

Fig. 1.Voltage-driven reader

A current-driven reader (figure 2) powers the antenna with an AC current of constant amplitude. Again, the FDX tag transmits data by opening and closing the switch *SW*, which this time modulates the voltage across the whole antenna circuit. The reader extracts the data from the latter voltage, the tap point connection being not needed in this case.

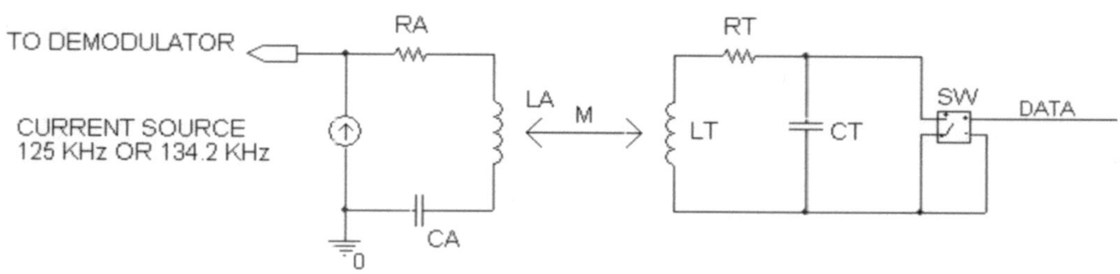

Fig. 2. Current-driven reader

It is to be observed that for the voltage-driven reader, the drivers that provide the amplified voltage to the antenna can be set into high Z mode via the tristate input during the interval when the antenna is not driven. This will be of importance for the extension to HDX tags. The high Z mode is implicit for the current-driven reader, as the (near) ideal current source presents high impedance to the antenna.

The formulas to be presented in the next sections are derived from the following general circuit model of the interaction between reader and tag.

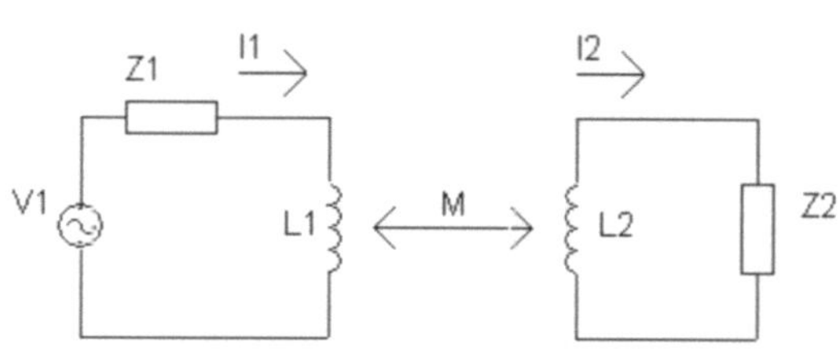

Fig. 3. Model of coupling reader-tag

Consider the circuit of figure 3, in which the two coils are linked by the magnetic coupling $M = k\sqrt{L_1L_2}$. Let I_1 be the current sourced by voltage source V_1 and let I_2 be the current flowing into impedance Z_2. Elementary circuit analysis gives the results below, in which s denotes the Laplace variable.

$$I_1(s) = \frac{(L_2s + Z_2)V_1(s)}{(L_1s + Z_1)(L_2s + Z_2) - k^2L_1L_2s^2}, \tag{1}$$

$$I_2(s) = \frac{k\sqrt{L_1L_2}sV(s)}{(L_1s + Z_1)(L_2s + Z_2) - k^2L_1L_2s^2}. \tag{2}$$

3. Adding the HDX protocol to the FDX voltage-driven reader

In FDX, the tag is continuously powered by the reader and transmits data by load modulation. In HDX, the tag is first charged by an RF pulse of limited duration from the reader, and then it transmits the data using the energy stored during the first step. The tag drives its coil with an AC voltage whose frequency toggles between two values: according to the standard (International Organization for Standardization, 2007), each data bit comprises 16 cycles of the AC voltage, the nominal frequency f_C = 134.2 KHz being used for a zero bit and the frequency f_{LOW} = 123.7 KHz for a one bit.

For the voltage-driven reader (figures 4, 5) we consider the usage of a dedicated integrated circuit (IC) such as TMS3705 produced by Texas Instruments (Texas Instruments, 2003). The manufacturer provided the IC with its own antenna drivers so that a minimal design of an HDX reader could consist of only the IC and a micro-controller. However, in our design we continue to use the drivers of the existing reader in order to keep the FDX functionality. In

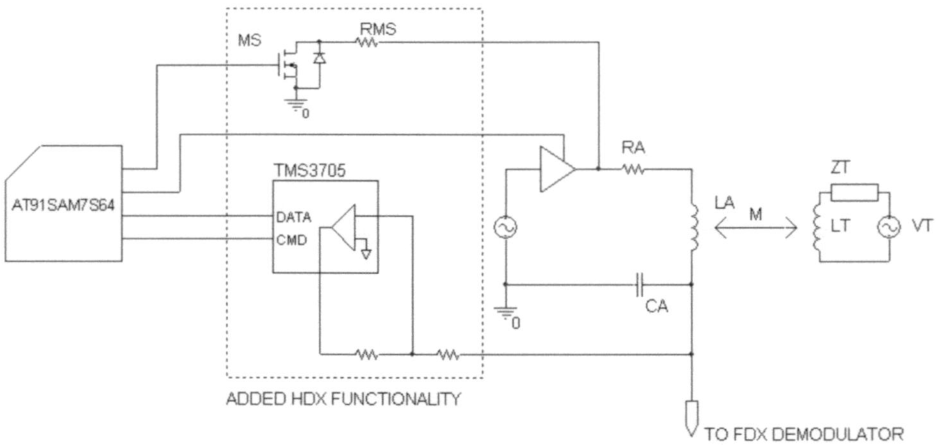

Fig. 4. Adding the HDX protocol to the voltage-driven reader

the schematic of figure 4, we first observe the MOS transistor M_S with low on-resistance that is used as a switch. When the reader is used in FDX mode, M_S is cut off allowing the antenna to be powered by the reader drivers. The same is true during the charge phase of the communication with an HDX tag. After the charge phase, the reader stops driving the antenna and the drivers are tristated. The reader micro-controller (uC) then turns on M_S, establishing thus a low resistance path through which the antenna circuit is closed. The resistor R_A includes the AC resistance of the antenna as well as any additional resistor added in order to limit the antenna current and to damp the transients during transmission/reception; more on this topic in the next section. There is a resistor R_{MS} in series with M_S, the role of which will also be explained later. It is to be observed that only positive voltages are present at the drain of M_S when cut off, which avoids any unwanted conduction through the parasitic diode of the transistor, represented here explicitly in parallel with the latter.

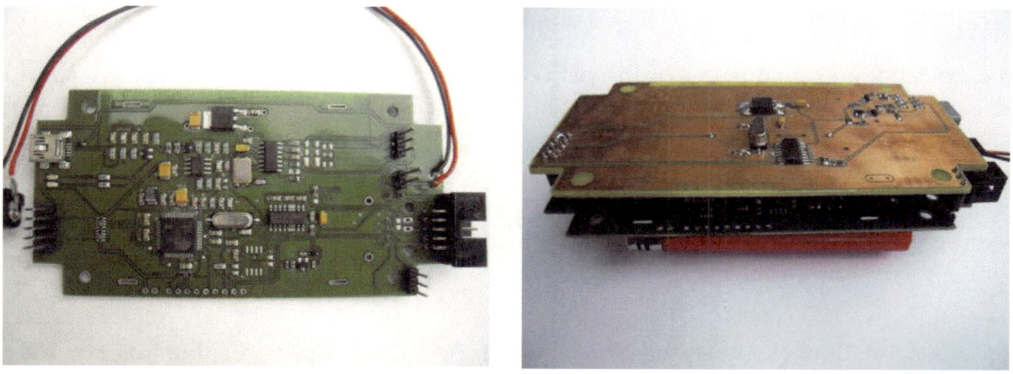

Fig. 5. The voltage-driven FDX reader produced by Frosch Electronics (left) and the reader with the plug-in for the HDX extension (right).

The tag starts the transmission a short delay after the interruption of the power flow from the reader. Meanwhile the uC has informed the decoder IC via the command line that a new decoding cycle is to begin. In our schematic, the tag is represented as a voltage source V_T with output impedance Z_T that drives the tag coil L_T. The voltage source produces an AC voltage of constant amplitude whose frequency toggles between the nominal frequency f_C to which the reader antenna is tuned and the frequency f_{LOW}. The current in the tag coil induces a frequency-modulated voltage in the reader antenna circuit that is sensed at the tap point by the decoder IC. The tap voltage is amplified by an opamp internal to the IC, which is part of an inverting amplifier configuration together with two external resistors provided by the user. The IC extracts the bit information from the frequency modulation and transmits it serially to uC via the data line.

4. Effect of transients on data reception

The effect of transients for the FDX protocol has been discussed in (Vuza et al., 2009). A similar analysis may be carried for the HDX protocol. Consider a circuit described by the linear system

$$\frac{dX(t)}{dt} = SX(t) + Y(t) \tag{3}$$

where $X(t)$ is the state vector and $Y(t)$ is a periodic input. In most cases we may assume that Y is continuous but we may also allow for a discontinuous input such as a square wave. In the latter case we shall assume that Y is integrable on each finite interval, that X is continuous and almost everywhere derivable, and that (3) holds almost everywhere; the periodicity of Y will be understood in the sense that there is $T > 0$ such that $Y(t + T) = Y(t)$ almost everywhere in t, each such number T being called a period of Y. Assume that the circuit is stable, that is, the characteristic roots of matrix S have strictly negative real parts. There is a unique periodic solution $X_P(t)$ for (3), which we shall call the periodic solution for input Y. The general solution of (3) is the sum between X_P and a solution of the homogeneous system

$$\frac{dX(t)}{dt} = SX(t). \tag{4}$$

The existence and uniqueness of the periodic solution are readily established. We consider here only the case when Y is not constant, the proof being easily adapted to the other case. Since Y is periodic and not constant, it has a smallest period T such that any other of its periods is a multiple of T. Let X be any solution of (3); such a solution always exists, for instance the one given by $X(t) = \exp(St)\int_0^t \exp(-S\tau)Y(\tau)d\tau$. The matrix $\exp(ST) - I$ is invertible as S is stable (I being the identity matrix). The function

$$X_p(t) = X(t) + \exp(St)(\exp(ST) - I)^{-1}(X(0) - X(T))$$

is also a solution of (3) satisfying $X_P(0) = X_P(T)$. As Y has the period T, the function $X_2(t) = X_P(t+T)$ is again a solution of (3). Hence $X_3(t) = X_2(t) - X_P(t)$ is a solution of (4) that vanishes at $t = 0$. But such a solution must vanish everywhere; hence X_P must admit T as a period. Let now X_{P2} be another periodic solution of (3) and let T_2 be its period. Since T_2 is also a period for the derivative of X_{P2}, it follows from (3) that it is a period for Y; hence T_2 must be a multiple of T and therefore a period for X_P. Consequently $X_{P2}(t) - X_P(t)$ is a solution of (4) with period T_2. But since S is stable, all solutions of (4) must approach 0 as t goes to infinity, implying that the mentioned periodic solution must vanish identically and hence $X_{P2} = X_P$.

Consider now two periodic inputs Y_1, Y_2 (possibly with different periods) and let X_{P1}, X_{P2} be the respective periodic solutions. Suppose that up to moment t_0, the circuit received input Y_1 and its state vector evolved according X_{P1}. At t_0, the input changes from Y_1 to Y_2. How the state vector will change? After t_0, the state vector can be written as the sum of the periodic part $X_{P2}(t)$ and a transient part $TR(t)$ that is a solution of (4) uniquely determined by its initial value at t_0. The latter value is in turn determined by imposing the continuity of the state vector at t_0, expressed by the equality $X_{P1}(t_0) = X_{P2}(t_0) + TR(t_0)$. Since, because of stability, every solution of (4) tends to 0 for large values of t, it follows that as times goes past t_0, the state vector will approach the periodic solution X_{P2} for input Y_2. Thus, the change of input at moment t_0 results in changing the evolution of the system from one periodic solution to another, but has also the side effect that a transient solution will manifest itself for some time after the change. The time constants of these transients are determined by the characteristic roots of S. As well known from Laplace transform theory, if one is interested in the time constants of the transients that affect an output of the system, one has to look for the roots of the denominator of the transfer function from the driving input to that output and take the inverses of the real parts of those roots, provided that the degree of the denominator equals the order of the system.

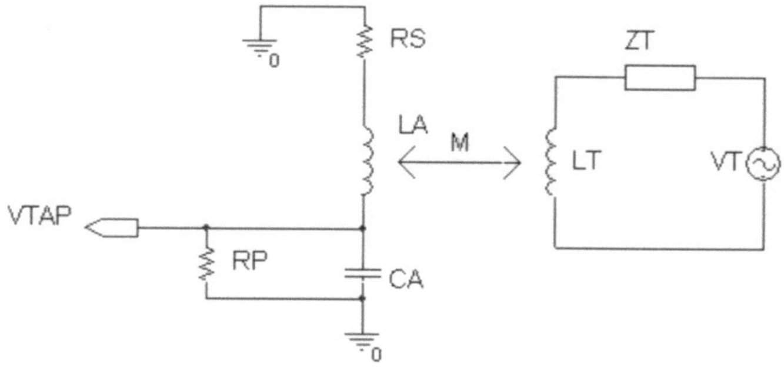

Fig. 6. Model for studying the effect of transients

We apply the above remarks to the case of the HDX reader of section 3. The inverting input of the opamp internal to the decoder IC is a virtual ground. Hence one may use the simplified schematic of figure 6 for analyzing the transients that are induced whenever the tag switches from a frequency to another during data transmission to reader. In this schematic, R_S is the total resistance in series with the antenna, which in this case is the series combination of R_A and R_{MS} in figure 4. Let Z_A be the impedance seen by the reader antenna. According to (2), the antenna current is given by

$$I_A(s) = \frac{k\sqrt{L_A L_T}\, s V_T(s)}{(L_A s + Z_A)(L_T s + Z_T) - k^2 L_A L_T s^2}. \tag{5}$$

We consider the case of weak coupling, as in real situations values around 0.01 for k are common. It is therefore reasonable to approximate the above formula by

$$I_A(s) = \frac{k\sqrt{L_A L_T}\, s V_T(s)}{(L_A s + Z_A)(L_T s + Z_T)}. \tag{6}$$

The tap voltage equals the above current multiplied by the parallel impedance of C_A and R_P. Define the series quality factor $Q_S = L_A\omega_C/R_S$ and the parallel quality factor $Q_P = R_P C_A \omega_C$, where $\omega_C = 2\pi f_C$ and f_C is the nominal frequency to which the antenna is tuned. Introducing also the normalized Laplace variable $x = s/\omega_C$, we have for the tap voltage

$$V_{TAP}(s) = \frac{k\sqrt{L_A L_T}\, s V_T(s)}{P_A(s/\omega_C)(L_T s + Z_T(s))}, \tag{7}$$

where $P_A(x) = x^2 + (Q_P^{-1} + Q_S^{-1})x + Q_P^{-1}Q_S^{-1} + 1$.

When the tag changes frequency, V_{TAP} will be affected by transients whose time constants are computed by finding the roots of the denominator of the transfer function in (7). Specifically, for any such root s_0, $-1/\operatorname{Re} s_0$ will be the time constant for a transient. In the limit of weak coupling, the denominator is the product of two factors, one of them depending exclusively on the tag and the other depending only on the reader antenna circuit. The reader designer has no control over the first factor and may only assume that the time constants related to it have been taken care of in the adequate way by the tag producer. The reader designer shall therefore take care of the time constants related to $P_A(x)$ and ensure that the corresponding transients will be short enough in order not to disturb the data decoding. Provided that $\left|Q_P^{-1} - Q_S^{-1}\right| \le 2$, which is usually the case, the roots of $P_A(x)$ will be complex conjugated and will produce the time constant $2(Q_P^{-1} + Q_S^{-1})^{-1} / \omega_C$. It is reasonable to ask that the 90% - 10% decrease time of the corresponding transient, equal to 2.2 times its time constant, should be less than half of the shortest duration T_B of a bit. It results that the following inequality should be imposed on the quality factors:

$$Q_P^{-1} + Q_S^{-1} \ge \frac{4.4}{\pi f_C T_B}. \tag{8}$$

During the charge phase, the opamp of the decoder IC will be saturated because of the high voltage at the tap point and its inverting input will no longer function as a virtual ground. Protection diodes at the inverting input prevent the opamp to be damaged by the high voltage. In order not to exceed the current rating of the diodes, it is advisable to choose a high value for R_P, resulting in a high Q_P. Inequality (8) will then be satisfied if we impose $\pi f_C T_B/4.4$ as an upper bound for Q_S. In the case of HDX protocol, T_B equals $16/f_C$ so 11.4 is an upper bound for Q_S.

Let us compare the above situation with the case of the reader in figure 4 working in FDX mode. Now the voltage source V_R is on the reader side as in figure 1 and the tag transmits data by modulating the load Z_T. The voltage at the tap point is obtained with the aid of (1):

$$V_{TAP}(s) = \frac{(L_T s + Z_T(s))V_R(s)}{P_A(s/\omega_C)(L_T s + Z_T(s)) - k^2 L_T \omega_C^{-1} s^2 (Q_P^{-1} + s/\omega_C)} \tag{9}$$

where $P_A(x)$ is as above. In the limit of weak coupling, the denominator is again approximated by the product of two factors, one determined by the tag and the other by the reader. Transients occur when the tag changes the value of Z_T. Similar considerations as above lead to the upper bound $\pi f_C T_B/4.4$ for Q_S, where this time T_B is the shortest bit duration for the FDX protocol. The latter is in general two times larger than the bit duration for HDX, resulting in a two times higher upper bound for Q_S.

The current for a tuned antenna circuit is given by

$$I_A = \frac{V_R}{R_S} = \frac{Q_S V_R}{L_A \omega_C}.$$

A higher antenna current means that the tag can be at a larger distance from the antenna and still receive the amount of power required for the activation of its internal circuits. Higher Q_S means a higher antenna current. Since the upper bound on Q_S is higher for FDX compared with HDX, it makes sense to use a lower R_S for FDX. This is the reason for using the resistor R_{MS} in figure 4. When the reader works in FDX mode, transistor M_S is cut off, R_{MS} does not play any role and Q_S is determined by R_A, adjusted to fulfill the upper bound for Q_S in the FDX case. In the charge phase of HDX, M_S is also cut off and the current is again determined by R_A. Choosing the minimal allowed value for the latter would ensure the largest possible activation distance for the HDX tag. Finally, during reception of HDX data, M_S is turned on and R_{MS} is now in series with R_A, lowering thus Q_S in order to agree with the upper bound for HDX. A mean for increasing the antenna current without exceeding the upper bound for Q_S is to decrease L_A, with simultaneous decrease of R_A (to maintain the same Q_S) and increase of C_A (to maintain the tuning). However, the reader designer should be aware that, as shown by (7), decreasing L_A while maintaining the quality factors constant would decrease the tap voltage and hence reduce the signal received by the decoder. It is to be observed that in the FDX case, the modification in question does not change the tap voltage and the signal received from the tag at all, as proved by (9).

5. The principle of low coupling approximation

We have seen above in passing from (5) to (6) that, in the limit of low coupling k, the transfer functions conveniently factor into a product of three terms, namely a transfer function that depends only on tag parameters, a transfer function that depends only on reader parameters, and the constant $k\sqrt{L_A L_T}$. This is in fact a consequence of a general principle that we state and derive in this section. In section 7 we shall have another opportunity to apply it.
Consider the interaction between the reader antenna and an HDX tag as represented in the upper left side of figure 7. The principle of low coupling approximation states that *in the limit of low coupling k, the tag may be replaced with a voltage source in series with the reader antenna coil, the Laplace transform of the voltage produced by that source being given by*

$$\frac{k\sqrt{L_A L_T} s V_T(s)}{L_T s + Z_T}. \tag{10}$$

For the derivation we start by replacing the coupled coils L_A and L_T by the equivalent circuit consisting of the leakage inductance $(1 - k^2)L_A$, the magnetizing inductance $k^2 L_A$ and the ideal transformer with voltage ratio $k\sqrt{L_A / L_T} : 1$.

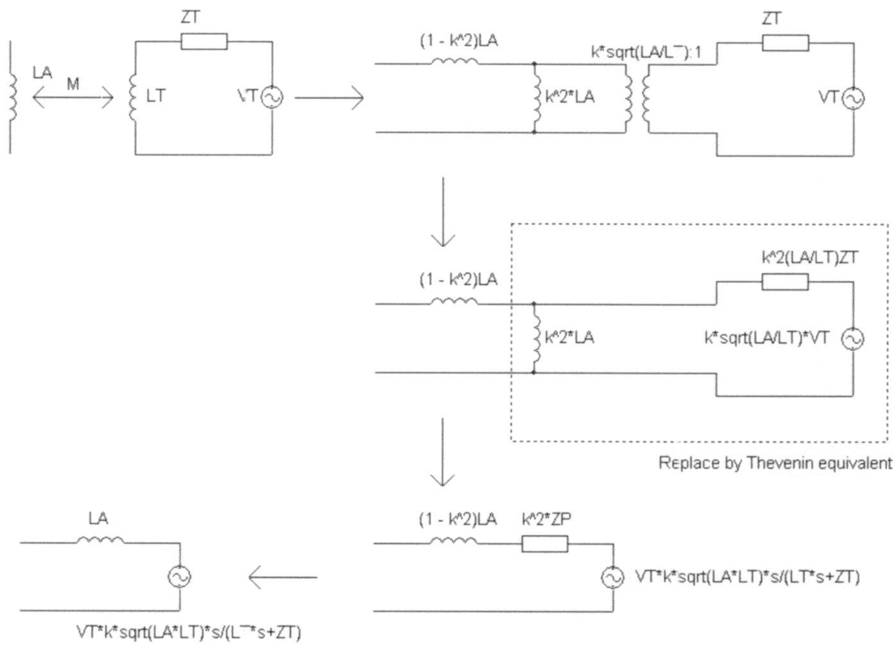

Fig. 7. Steps in deriving the principle of low coupling approximation

In the second step we reflect to the left of the transformer everything found to its right. In this way the voltage source V_T gets multiplied by the transformer voltage ratio, the impedance Z_T gets multiplied by the square of the latter ratio, and we get rid of the transformer. In the third step we replace that part of the circuit enclosed in the rectangle by its Thevenin equivalent, consisting of a voltage source in series with an output impedance. In the original circuit we had a voltage source in series with a voltage divider formed by two impedances k^2L_A and $k^2(L_A/L_T)Z_T$. The new voltage source produces the voltage at the open-circuited output of the voltage divider, while the new output impedance is the parallel combination of the impedances forming the divider, and hence equals k^2 times the parallel combination Z_P of L_A and $(L_A/L_T)Z_T$.

All transformations so far were equivalent transformations and no approximation was made. The low coupling approximation comes at this final step, and consists in replacing, for low k, $(1 - k^2)L_A$ by L_A and ignoring k^2Z_P. In this way we arrive at the approximate circuit in the lower left side of figure 7.

6. Adding the HDX protocol to the FDX current-driven reader

As already mentioned, the tap point connection is no longer available in the current-driven reader. The voltage-driven reader is connected via a three-wire cable to the end points and to the tap point of the antenna circuit, while the current-driven reader is connected via a two-wire cable only to the end points of the antenna circuit. Consequently, a different HDX topology is needed for the current-driven reader, which is presented in figure 8.

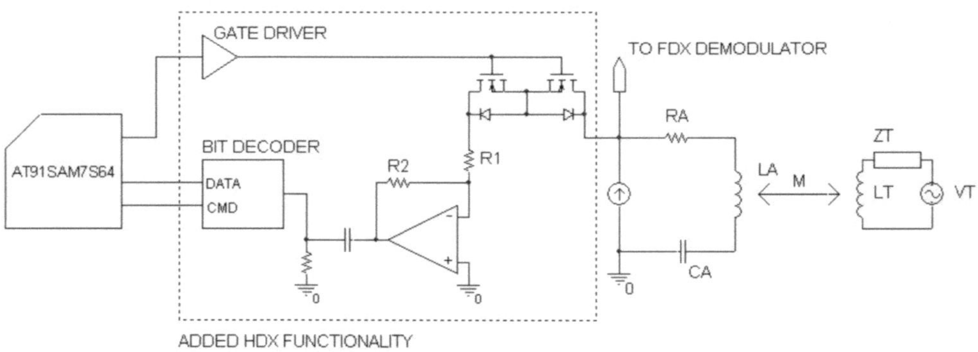

Fig. 8. Adding the HDX protocol to the current-driven reader

One remarks first that the newly added part of the schematics is connected to the existing part via two MOS transistors with low on-resistance. The transistors have their sources tied together with their parasitic diodes back-to-back so that the unwanted conduction through them is eliminated. The reader is powered from a positive source VCC and a negative source VSS. The voltage present on the antenna, which is sensed by the reader for decoding the data sent by the tag, is confined to the range from VSS to VCC. Therefore, in order to cut off both transistors, it is enough to apply the most negative voltage VSS to their gates tied together. For this reason, unlike to the voltage-driven reader where the gate of the MOS switch can be driven directly by uC, a gate driver is needed here to provide the positive voltage for turn on and the negative voltage for cut off. When the reader works in FDX mode, the transistors are cut off so that the HDX part of the schematic is isolated and plays

no part. The transistors are also cut off during the charge phase of the HDX protocol, when the reader drives the constant amplitude current at the nominal frequency f_C through the antenna. At the end of the charge phase, the reader stops driving the antenna and turns on the MOS transistors; since the current source presents high impedance to the antenna circuit, the latter is now closed through the transistors. The voltage induced by the tag on the antenna is amplified by the opamp connected in the inverting configuration, with a much higher gain than in the voltage-driven case since now we lack the amplification that was provided by the tap point. There is a high pass filter at the output of the opamp, with the purpose of eliminating any DC component in the signal; such a DC component may occur because the high gain that is used may amplify any non-ideal characteristic of the opamp such as input offset voltage.
There are now two options for decoding the amplified and filtered signal. One of them is to use the same decoder IC as in figure 4.

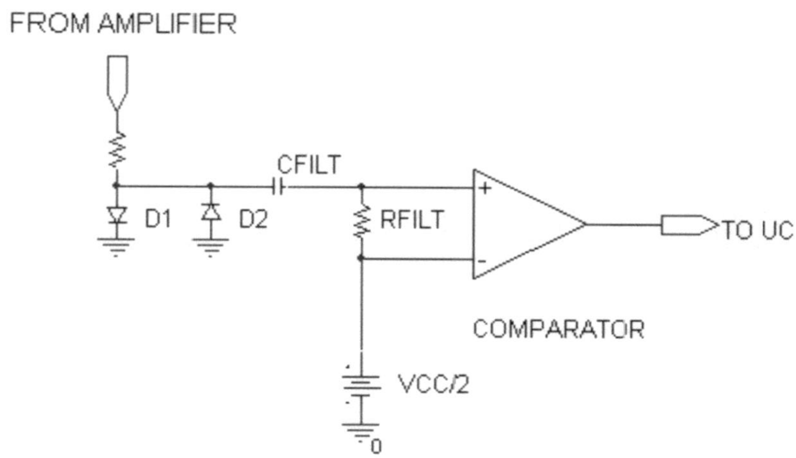

Fig. 9. Analog to digital interface for a bit decoder

Another option is to build a custom decoder that splits the task of data retrieving between a hardware part, built with discrete components as in figure 9, and a software part, included in the uC program. The input is limited by diodes D1 and D2 and then shifted by the high pass filter formed by RFILT and CFILT to an AC voltage with a DC component equal to the reference provided by voltage source VCC/2. The output of the filter together with the reference voltage is applied to the comparator. Shifting the AC voltage is necessary since the comparator admits only positive voltages at its inputs. The output of the comparator is a square wave whose frequency toggles between two values, as determined by the tag. This signal goes to an input line of uC, which is connected to an internal timer. The timer is programmed to run at a certain frequency, 24 MHz in our case. Each raising transition on the input line causes the value of the running counter of the timer be stored in a register and then the counter be reset. At the same time, the transition triggers an interrupt to uC. The uC interrupt routine reads the value of the register and stores it in memory. After the whole record is stored, the uC uses the stored values as estimates of the period of the signal coming from the tag and divides the record into intervals of high, respectively low frequency, according to whether the values are below, respectively above a certain threshold. Ideally, an interval of high frequency containing N values should correspond to a sequence of exactly N/16 zero bits in the tag response. In practice, there are errors caused by noise, so that correction algorithms should be used. The performance of these algorithms is one of the factors on which the reading distance depends. This is one reason for preferring the custom-built decoder to the decoder IC: the latter is a black box to the reader designer and one has no control over its internal decoding algorithms.

7. Using the gain-bandwidth product in the equalization of HDX bit amplification

Because the reader antenna circuit is tuned to the resonant frequency f_C, the two signaling frequencies used by the tag may induce voltages whose amplitudes differ in a significant way. Consider the transition between a zero bit and a one bit. The zero bit is transmitted at the resonant frequency f_C of the antenna circuit and hence the resulted signal at the reader is of high amplitude. The tag then shifts to the lower frequency f_{LOW} that is outside resonance, resulting in a signal of lower amplitude. The transients that are triggered by the transition have a frequency close to f_C and in general start with an amplitude close to that of the signal before the transition. If the signal after the transition has significantly lower amplitude, the transients will have a greater chance to disturb the decoding of the latter signal (figure 12); this effect is especially manifest at higher reading distances when the whole signal is weak, imposing thus a limitation on the reading distance if not taken care of properly.
We present a method for equalizing the bit amplification based on the one-pole model of the opamp and the related gain-bandwidth product (Gray & Meyer, 1993). The one-pole model assumes that the transfer function between the differential voltage at the input and the voltage at the output of the opamp is given by

$$A(s) = \frac{A_0}{1 + \frac{s}{p_1}}. \tag{11}$$

By definition, the gain-bandwidth product is the product between the DC gain A_0 and the 3 dB frequency $p_1/2\pi$. Consider the opamp in the inverting configuration as in figure 10.

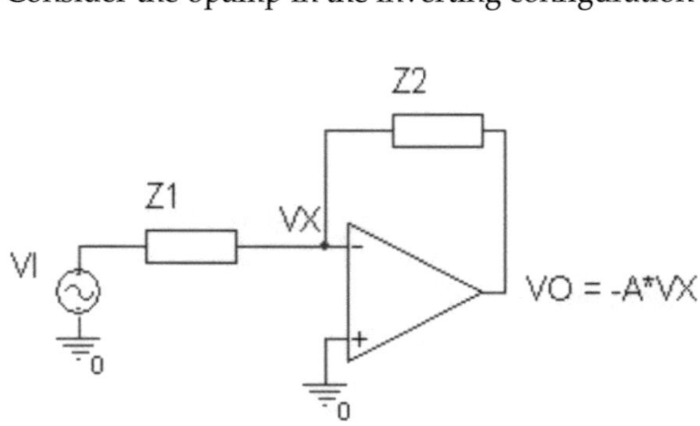

Fig. 10. Inverting amplifier

Assuming that there is no current into the inverting input, the current law gives $(V_I - V_X)/Z1 = (V_X + A(s)V_X)/Z_2$. Solving for $V_O = -A(s)V_X$ gives, taking into account (11),

$$V_O = -\frac{V_I}{\frac{1}{A_0} + \frac{s}{A_0 p_1} + \frac{Z_1}{Z_2}\left(1 + \frac{1}{A_0} + \frac{s}{A_0 p_1}\right)}.$$

Because A_0 is in general high, we may neglect $1/A_0$ in the above formula. Using the notation ω_{GB} for $A_0 p_1$, that is, 2π times the gain-bandwidth product, we obtain

$$V_O = -\frac{V_I}{\frac{s}{\omega_{GB}} + \frac{Z_1}{Z_2}\left(1 + \frac{s}{\omega_{GB}}\right)}. \quad (12)$$

Let us again consider interaction between reader and tag represented in the left side of figure 11 in the limit of weak coupling, in which situation we may apply the approximation principle of section 5 and replace the tag by a voltage source with Laplace function (10) in series with the reader antenna, as in the right side of figure 11. We may then use (12) in which we set $Z_1 = L_A s + R_S + 1/C_A s$ and $Z_2 = R_2$, where R_S denotes the total resistance in series with the antenna, that is, R_A in series with R_1 in figure 8.

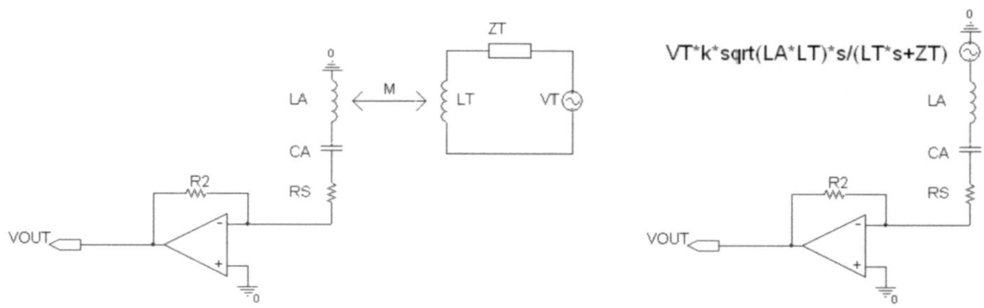

Fig. 11. Replacing the tag by the equivalent source in the limit of weak coupling

The output voltage V_{OUT} can be written as the product between the voltage V_T of the source in the tag and the gain functions G_T and G_R, with the remark that the dependence of $s = j\omega$ had been moved from the numerator of (10) to the numerator of G_R:

$$V_{OUT} = G_R G_T V_T,$$

$$G_T(j\omega) = \frac{k\sqrt{L_A L_T}\,\omega_C}{L_T j\omega + Z_T},$$

$$G_R(j\omega) = -\frac{R_2 j\omega / \omega_C}{\left(L_A + \frac{L_A j\omega}{\omega_{GB}} + \frac{R_S + R_2}{\omega_{GB}}\right) j\omega + R_S + \frac{1}{C_A \omega_{GB}} + \frac{1}{C_A j\omega}}.$$

We want V_{OUT} to have the same amplitude for $\omega = \omega_C$ and $\omega = \omega_{LOW}$ ($= 2\pi f_{LOW}$), which translates into the equality of absolute values $|V_{OUT}(\omega_C)| = |V_{OUT}(\omega_{LOW})|$. We assume that V_T keeps constant its amplitude when switching between ω_C and ω_{LOW}, hence $|V_T(\omega_C)| = |V_T(\omega_{LOW})|$. We also assume that by design, the quality factor of the tag is low enough to neglect the variation of the absolute value of G_T when ω varies around ω_C; however, we still have to consider the variation with frequency of the factor $s = j\omega$ in the numerator of (10) whose presence accounts for the magnetic coupling and for this reason we have moved it to the numerator of G_R. We now make the following approximations for G_R. First, since ω takes values around ω_C and we shall assume ω_{GB} much larger than ω_C, we may neglect the term $L_A j\omega/\omega_{GB}$ in comparison with L_A. Second, the required high gain asks for a resistance R_2 much higher than R_S, so that we may neglect R_S in the sum $R_S + R_2$. We arrive at following approximation of the gain G_R

$$G_R = -\frac{R_2 j\omega / \omega_C}{\left(L_A + \frac{R_2}{\omega_{GB}}\right) j\omega + R_S + \frac{1}{C_A \omega_{GB}} + \frac{1}{C_A j\omega}} \tag{13}$$

in which the inductance L_A appears as augmented by the quantity R_2/ω_{GB}, R_S as augmented by $1/C_A\omega_{GB}$ while the capacitive term $1/C_A j\omega$ is not changed. Consequently, the resonant frequency of the compound circuit antenna plus amplifier appears as diminished with respect to the nominal resonant frequency f_C of the antenna circuit. We now have to determine R_2 so that the two signaling frequencies f_C and f_{LOW} employed by the tag are equally amplified by the above transfer function. This brings us to the general problem that given a transfer function of the form $j\omega/Z(j\omega)$, where $Z(j\omega) = j(L\omega - 1/C\omega) + R$ is the impedance of a series LRC circuit, find the condition for two frequencies ω_1, ω_2 to be equally amplified by the function, that is, $|\omega_1/Z(j\omega_1)| = |\omega_2/Z(j\omega_2)|$. If we had not $j\omega$ in the numerator, the condition would be, as well-known, $\omega_1\omega_2 = \omega_r = 1/LC$, ω_r being the resonant frequency of the LRC circuit. However, because of that numerator, the condition is here different and to find it we start by squaring the moduli and inverting the fractions, which leads us to

$$\left(L - \frac{1}{C\omega_1^2}\right)^2 + \frac{R^2}{\omega_1^2} = \left(L - \frac{1}{C\omega_2^2}\right)^2 + \frac{R^2}{\omega_2^2}.$$

Then some straightforward algebra gives the required condition as

$$\frac{1}{2}\left(\frac{1}{\omega_1^2} + \frac{1}{\omega_2^2}\right) = \frac{1}{\omega_r^2}\left(1 - \frac{1}{2Q^2}\right) = LC - \frac{R^2C^2}{2}$$

where $Q = L\omega_r/R$ is the quality factor. Applying the above condition to (13) yields for the choice of R_2

$$R_2 = \frac{L_A \omega_{GB}}{2}\left(\left(\frac{1}{Q_S} + \frac{\omega_C}{\omega_{GB}}\right)^2 + \left(\frac{f_C}{f_{LOW}}\right)^2 - 1\right) \tag{14}$$

where $Q_S = L_A\omega_C/R_S$ is the quality factor of the antenna circuit. For the present choice, the amplifier gain is reduced from its maximal value of R_2/R_S corresponding to an infinite gain-bandwidth product, to the value

$$\frac{R_2}{R'_S}\left(1 + \left(\frac{R_2}{R'_S}\frac{\omega_C}{\omega_{GB}}\right)^2\right)^{-1/2}$$

where $R'_S = R_S + 1/C_A\omega_{G3}$. In our design we use the LT1224 opamp for which a gain-bandwidth product of 45 MHz is specified. For L_A = 1 mH and Q_S = 21, (14) gives a resistance of 25.4 KOhms and an amplification of 294. The results in figure 12, based on a simulation to be described in section 9.1, make use of these values and confirm the theoretical prediction; truly the employed Q_S is in excess of that recommended by (8) but it was nevertheless used in order to clearly display the effect of inequal bit amplification that is magnified by a higher Q_S.

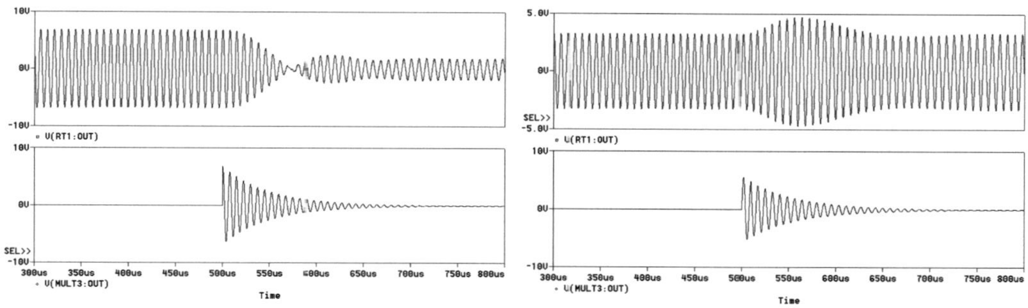

Fig. 12. Left: unequal amplification of bits. Right: equalization of bit amplification. Upper traces show voltages V_{OUT}, lower traces show transients. Frequency transition at 500 us.

8. A simulator for FDX and HDX tags

Why do we need simulators? Because, during the development of a reader, we may need to generate in a systematic and reproducible way situations that with real transponders occur only randomly and unpredictably. Such a need may arise in connection with the following tasks: testing the system response (antenna plus reader) to signals from tags; testing the behavior of demodulation hardware and decoding software of the reader; generating test data for the information system in which the reader is to be integrated.

The first author's work on simulators started in collaboration with Frosch Electronics (Vuza & Frosch, 2008; Vuza et al., 2009) and responded to the need of simulating a forthcoming tag not yet available by the time when a reader had to be developed. It continued with the work (Vuza et al., 2010a) that presented the general principles of a multifunction simulator intended for both FDX and HDX tags and realized as a stand-alone PC-configurable device. The simulator covered the case of "transponder talks first" (TTF) tags, meaning tags that transmit data as soon as they are powered by the reader, which is opposed to the "reader talks first" mode, where the tag transmits only in response to a command from the reader. The simulator described here was presented in (Vuza et al., 2010b) as a further elaboration of the preceding one. It is based on the AT91SAM7S64 micro-controller (uC), which provides the signal and data processing capabilities for the communication both with the reader to which it simulates the tag, and with a standard PC for the purpose of configuration. In our application, the software programmed into uC addresses the simulation of tags compatible with the FDX transponder EM4102 (EM Microelectronic-Marin SA, 2005) and the HDX transponder TIRIS (Texas Instruments, 2003). Of course, many other cases can be addressed by programming the adequate software. We start by describing the functioning of the analog part. With reference to figure 13, FDX/HDX, FREQMOD and LOADMOD are inputs from uC while CLOCK is an output to uC. As it will be indicated below, the antenna circuit should be tuned to the nominal FDX frequency in order to achieve the maximal amplitude of the baseband signal decoded by the reader. One sees that the resonance capacitor CS is not connected directly to ground but to inverters INV1 and INV2. Their role will be explained in section 8.2 on simulation of HDX transponders.

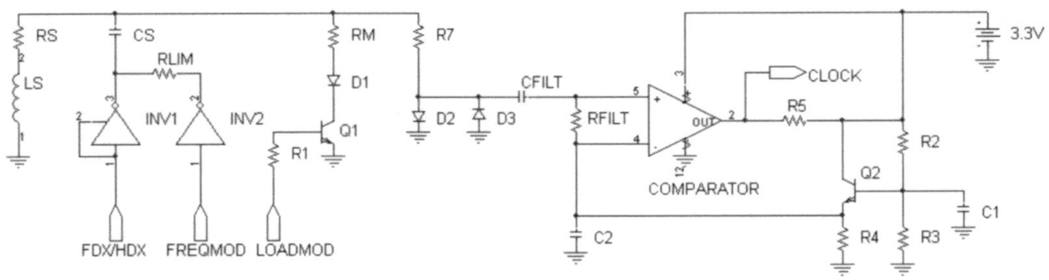

Fig. 13. Schematic of the analog part of the simulator

The output of INV1 is tristateable and the input and tristate pins are connected together. For load modulation, RM is switched in and out by transistor Q1. D1 prevents inverse current through Q1. Attached to the antenna circuit is the circuit that converts the RF signal from the reader into a digital clock. When the reader antenna is powered, an RF voltage is induced in the simulator antenna circuit. This voltage, which has a zero DC component, is limited by diodes D2 and D3 and then shifted by the high pass filter formed by RFILT and CFILT to an RF voltage with a DC component equal to the reference voltage provided by R2, R3 and Q2. The output of the filter together with the reference voltage is applied to the comparator. Shifting the RF voltage is necessary in order to use a single power supply: if the original voltage was fed to the comparator, the latter would have needed a positive and a negative supply. The comparator converts the shifted RF voltage into a square wave, which is fed to an internal counter of uC; R5 is a pull-up resistor needed by the comparator. An internal timer based on the uC clock generator is used for measuring the frequency of the square wave. If the latter matches, with a certain tolerance, the frequency imposed by the standard (either 125 KHz or 134.2 KHz), an optical indicator is activated for signaling the presence of RF power from the reader. The square wave is also used by uC as a clock for synchronizing the data transmission with the reader RF signal, as described in the next section.

8.1 Simulation of FDX tags

When simulating FDX tags, the lines FDX/HDX and FREQMOD are driven high by uC. In this situation, the output of INV1 is active and, through it, the pin of the resonance capacitor is connected to ground. The uC waits for the RF signal from the reader that is supposed to power the tag. As soon as this signal is detected by the procedure explained above, the simulator starts the data transmission, which lasts as long as RF power from reader is maintained.

Transmission is achieved with the aid of load modulation and uC can be programmed to use one of several bit-encoding schemes, among of which Manchester and Biphase (figure 14). As an example, let us explain how data is transmitted using Manchester encoding. A bit consists of 64 cycles of the reader RF signal. As we have seen, the latter is converted to a digital signal that clocks an internal counter of uC. The counter is programmed to reset automatically after each 64 clocks. The hardware is also programmed to do two things when the counter reaches the 32-th clock after each reset. First, it toggles the LOADMOD line, creating thus the transition in the middle of the Manchester bit. Second, it triggers an interrupt to the uC program. The interrupt routine will program the hardware to either set or reset the LOADMOD line by the time when the counter would reach the 64-th clock, according to whether the next bit to be sent is one or zero.

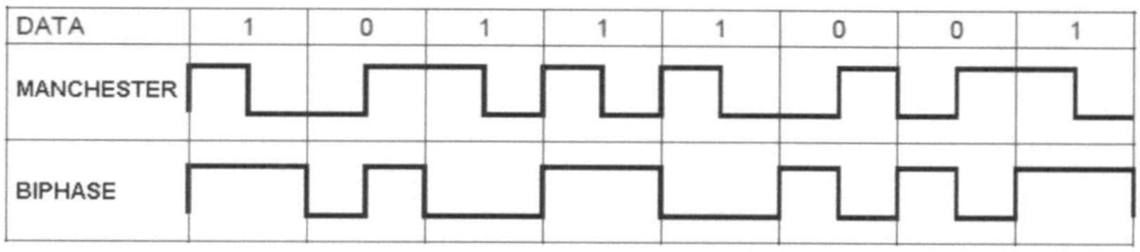

Fig. 14. Methods of bit encoding. Traces show the digital signal on the LOADMOD line.

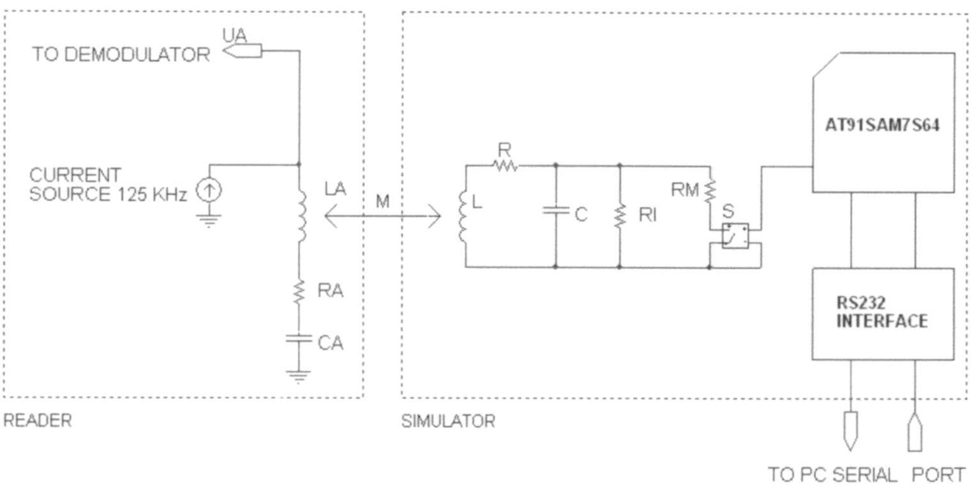

Fig. 15. Interaction between current-driven reader and simulator

We indicate now why it is necessary to tune the antenna circuit to the frequency f_C. In figure 15 we show in a simplified way the interaction between a current-driven reader and the simulator in FDX mode. Load modulation is achieved by switching in and out resistor R_M. L, R and C are the parameters of the antenna circuit of the simulator, M is the magnetic coupling between the reader and simulator coils and R_I denotes the lumped input resistance of other circuits attached to the antenna circuit. It was shown in (Vuza et al., 2010a) that the maximal signal amplitude that can be achieved at the reader by synchronous demodulation of the load modulation is given by

$$\frac{I_A M^2 \omega_C^2 |R_2 - R_1|}{\sqrt{(R_1\Delta + R)^2 + \omega_C^2 (L + RR_1C)^2} \sqrt{(R_2\Delta + R)^2 + \omega_C^2 (L + RR_2C)^2}} \tag{15}$$

where we have set $\Delta = 1 - LC\omega_C^2$, $R_1 = R_I$ and $R_2 = R_I || R_M$. We see that a higher amplitude is achieved when $\Delta = 0$, that is, when the antenna circuit of the simulator is tuned at f_C.
Figure 16 shows the baseband signal decoded by the reader and representing a sequence of Manchester encoded bits sent by the simulator. For comparison the baseband signal received from a real tag is shown. One may remark the similarity between them.

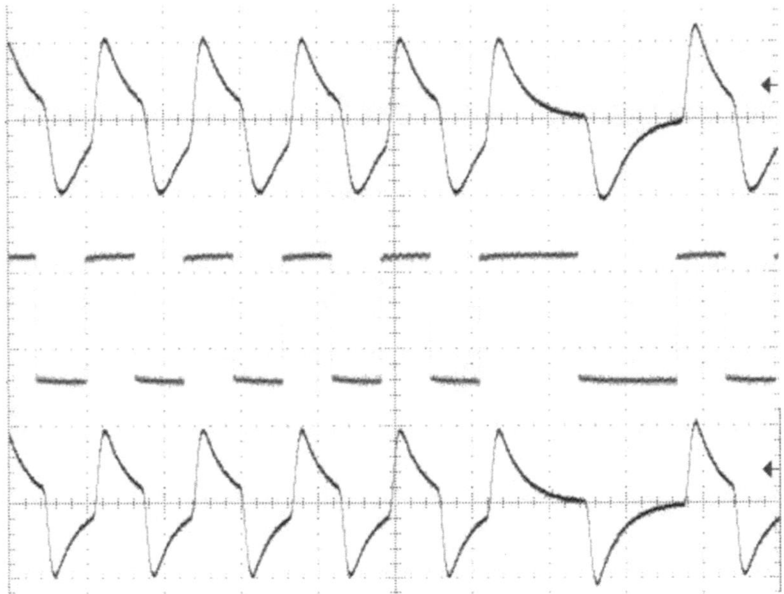

Fig. 16. Upper: baseband signal from simulator in FDX mode retrieved by reader. Middle: digital signal on simulator LOADMOD line that generated the upper trace. Lower: baseband signal retrieved from an FDX tag.

8.2 Simulation of HDX tags

For HDX tags the simulator has to reproduce the two steps of the process: charge and transmission. For the charge step, the simulator only has to detect the start and the end of the reader RF pulse, as it has its own supply and does not need to store energy from the reader. The detection is accomplished with the procedure described in the introduction to section 8. During this procedure, the FDX/HDX and FREQMOD lines are set as in section 8.1. As soon as the end of the charging pulse is detected, the FDX/HDX line is driven low by uC. This has the effect of putting the output of INV1 in the high Z state. The antenna resonant circuit is now driven by the output of INV2 and becomes a transmission circuit. RLIM has the role of limiting the current supplied by INV2. The value of RLIM is typically ten times that of RS. This explains why INV1 had to be used: if current limitation would be achieved with RS instead of RLIM, then the amplitude of the baseband signal decoded by a reader when receiving from the simulator in FDX mode would be substantially reduced as one may see from (15). Hence RLIM is shorted out by INV1 when the simulator works in FDX mode and the only resistance left in the antenna circuit is represented by the resistance of the antenna coil together with RS. The latter is added in order to damp the transients that otherwise could have deleterious effects on the data decoding at the reader, as discussed in

(Vuza et al., 2009). Data is transmitted with the aid of frequency modulation. The FREQMOD line is driven by an internal uC timer that generates a digital signal of programmable frequency. Besides driving the FREQMOD line, the timer is also programmed to clock a uC counter. The latter is set to trigger an interrupt every 16 clocks. The interrupt routine programs the frequency (f_C or f_{LOW}) of the timer that will be in effect during the next 16 clocks, according to the value (0 or 1) of the next bit to be sent. In agreement with the description in (Texas Instruments, 2003), the uC has to use the following data format in order to simulate a HDX tag of TIRIS type: 16 leading zero bits, a start byte equal to 0x7F, 64 data bits, 16 CRC bits, a stop byte equal to 0x7F, 16 trailing zero bits.

8.3 Connectivity

The data to be transmitted to the reader is stored in the internal non-volatile memory of uC. Therefore the simulator is a stand-alone device. However, for the purpose of configuration, the simulator can be connected to a PC. The configuration process allows the modification of the data to be sent to the reader, the choice of protocol (FDX or HDX) and, in the FDX case, the choice of bit encoding (Manchester or Biphase). The communication between the simulator and the PC is achieved either via the RS232 serial link or the USB link. The latter takes advantage of the USB transceiver embedded into the uC. The simulator may be powered from a battery or from the USB port when connected to a PC.

8.4 Applications

The simulator is a useful device for the process of customization and tuning the RFID hardware and software as it allows doing things that would be difficult or even not possible with real tags.

The two kinds of tags considered here, FDX and HDX, are typically used in access control and animal identification. They transmit to the reader data consisting of several fields that will be used as keys in databases containing information about the identified subject. The simulator offers a quick way to test the functioning of the database system for arbitrary values of the data fields, without the need of disposing of large collections of pre-programmed tags.

Another application is simulating anomalous tag behavior. During the realization of their joint work, the authors of (Vuza & Frosch, 2008) observed that some FDX tags have the tendency to skip some cycles of the reader RF signal during data transmission. A Manchester bit would then appear to the reader as containing, for instance, 65 RF cycles instead of the nominal 64. A well-designed decoding algorithm in the reader should be able to handle this situation. The simulator may be programmed to skip cycles on purpose in order to test the behavior of the reader decoding algorithms.

Finally there are the important applications of the simulator to the study of transient effects and of equality of bit amplification, to which we dedicate the next section.

9. Usage of simulators in studying the effects of transients in the HDX protocol

In testing a HDX system, it is important to find the behavior of the combined system reader plus antenna in response to the transition from one frequency to another. Consider the interaction between reader and HDX tags as presented in figure 11. The schematic is that of a linear system whose input is the voltage source V_T internal to the tag and the output is the analog signal VOUT that is to be taken by the reader for further processing. Assume that up to moment t_0, V_T produced a square wave of frequency f_1, to which the system responded with a steady-state periodic signal of the same frequency at the output. At t_0, V_T switches to a new frequency f_2. Recalling the discussion in section 4, the output will be the sum of two parts after the transition: the new steady-state response corresponding to the new input and the transients induced by the frequency change. The transients vanish gradually so that the output is evolving towards the steady-state response. The problem for the reader designer is to ensure that the transients would vanish quickly enough in order not to disturb the bit decoding. Observing the frequency transition is not easy on a scope, as the frequency difference is rather small compared to the nominal frequency. Generated transients make the transition gradual and because of this it is difficult to estimate when the transition actually started; not having access to the interior of the tag implies not knowing the moment when the tag changed the frequency. For these reasons it is more convenient to use simulators rather than tags in assessing the effects of transients in the reader design. The method we propose for visualizing the transient relies on the following considerations. Let VIN_{12} be the input consisting of a square wave of frequency f_1 before t_0 and of a square wave of frequency f_2 and of same amplitude after t_0. Let VIN_2 be the input consisting of a square wave of frequency f_2 and of same amplitude as VIN_{12} and let $VOUT_{12}$, $VOUT_2$ be the respective outputs of the system corresponding to the defined inputs. Then the transient in the system response induced by the frequency change can be obtained as the difference between $VOUT_{12}$ and $VOUT_2$, provided one condition holds: VIN_{12} and VIN_2 must be aligned so that they overlap after t_0, that is, $VIN_{12}(t) = VIN_2(t)$ for $t \geq t_0$ (figure 17).

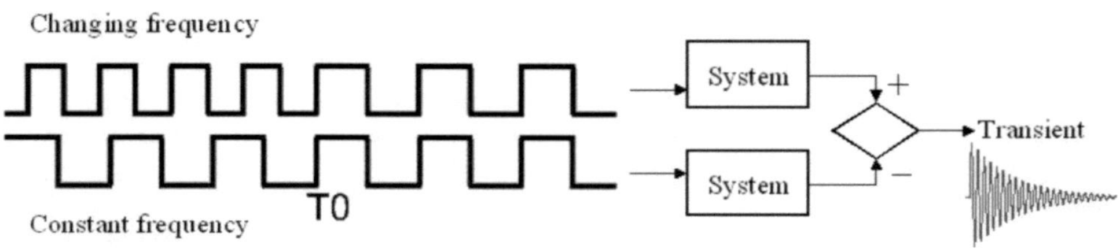

Fig. 17. Alignment of input signals VIN_{12} (upper) and VIN_2 (lower) fed simultaneously to identical copies of the system

9.1 Watching transients with the aid of a PSpice simulation

We may dispose of two identical copies of the system, which are fed simultaneously with the inputs VIN_{12} and VIN_2. This is the principle on which relies the PSpice simulation that we propose as a CAD tool to be used during reader design. Its aim is to provide a graphical display of transients, allowing thus to estimate their duration and magnitude and to assess their effects on the received signal. The two copies of the system are produced with the aid of PSpice hierarchical blocks in order to avoid duplication of the schematic: any modification to the schematic is automatically reflected in both copies. The blocks are fed with the inputs VIN_{12} and VIN_2. The outputs go into a difference block that isolates the transient from the output $VOUT_{12}$ by subtracting the steady-state response $VOUT_2$. We illustrate the above method with the simulation that was used for producing the results on equalization of amplification of HDX bits presented in section 7. Figure 18 shows the schematic of the composite system reader plus tag. The tag is represented on the right side as a tuned antenna circuit driven by the voltage present at the input port. A current-driven reader is represented on the left side and consists of the tuned antenna circuit and the amplifier, which produces the signal available at the output port.

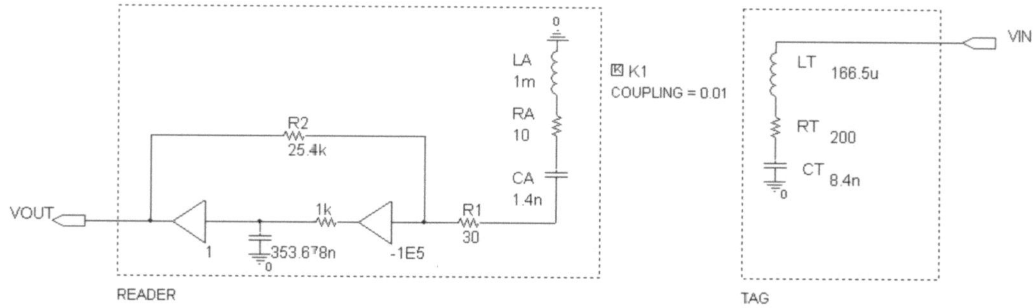

Fig. 18. Schematic of the composite system reader-tag used in simulation

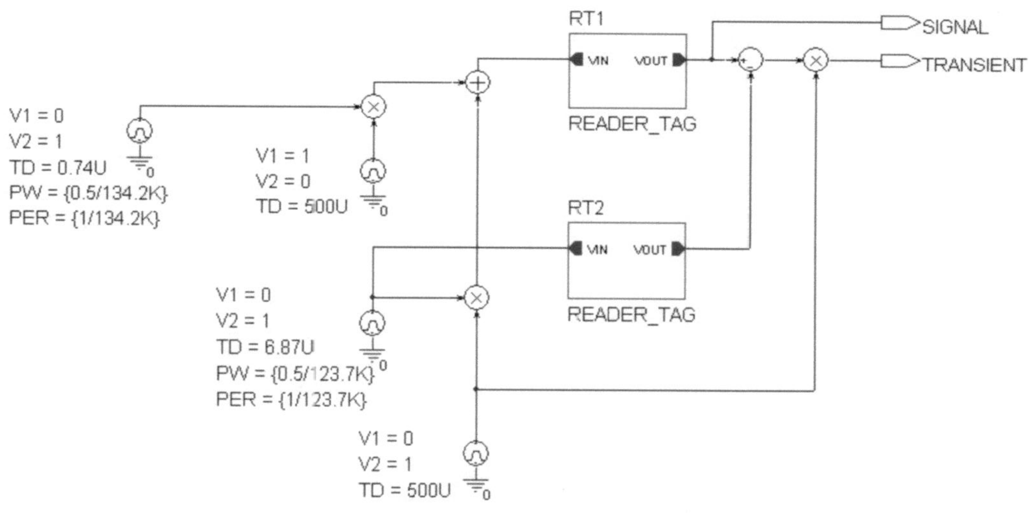

Fig. 19. Simulation of transients in the composite system

The amplifier is based on an opamp connected in the inverting configuration. Two gain blocks and an RC low-pass filter are used for simulating an opamp with a DC gain of 100000 and a gain-bandwidth product of 45 MHz. The reader and tag antennas are magnetically coupled, with a coupling constant $k = 0.01$. Figure 19 shows the schematic of the simulation. The two copies of the system are represented by the hierarchical blocks RT1 and RT2. The input to RT1 consists of a square wave of frequency f_C up to time t_0 and of a square wave of frequency f_{LOW} after t_0; the two square waves are combined into a single signal with a summing block. The input to RT2 consists of a square wave of constant frequency f_{LOW}. Delays TD are used in order to properly align inputs RT1 and RT2 as in figure 17. The difference block used for isolating the transient is followed by a multiplication block. The purpose of the latter is to eliminate the part of the graphical display of the transient that precedes the transition time t_0, as it has no meaning for the simulation. The equal bit amplification seen in figure 12 is achieved by choosing R2 according to formula (14). If the RC filter is removed from the opamp schematic, the gain of the amplifier does no longer depend on frequency and the frequency dependence of the overall gain is set by the antenna circuit. In this situation one obtain the unequal amplification seen in figure 12.

9.2 Watching transients with the aid of the tag simulator

In practice we may not always dispose of copies of the system and much less of identical copies. We may however successively feed the inputs VIN_{12} and VIN_2 to the same system and make use of time invariance. Suppose that we first feed VIN_{12} that was described above and with the aid of a recording device such a scope, we take a record of $VOUT_{12}(t)$ in the interval from $t_0 - a$ to $t_0 + b$. Then at a later time we feed VIN_2 and we take a record of $VOUT_2(t)$ in the interval from $t_1 - a$ to $t_1 + b$. We do not assume that the alignment condition of figure 17 holds, which was meaningful for the case of inputs fed at the same time to identical systems. Instead, we assume the equality $VIN_{12}(t) = VIN_2(t + t_1 - t_0)$ is satisfied for each $t \geq t_0$ (figure 20). If we define the time displaced input $VIN_{2D}(t) = VIN_2(t + t_1 - t_0)$, then VIN_{12} and VIN_{2D} satisfy the alignment condition of figure 17 and hence the difference of the corresponding outputs $VOUT_{12}$ and $VOUT_{2D}$ would produce the transient we look for. By time invariance, $VOUT_{2D}(t) = VOUT_2(t + t_1 - t_0)$. Consequently, the transient is obtained as the difference $VOUT_{12}(t) - VOUT_2(t + t_1 - t_0)$ between the records taken by the recording device.

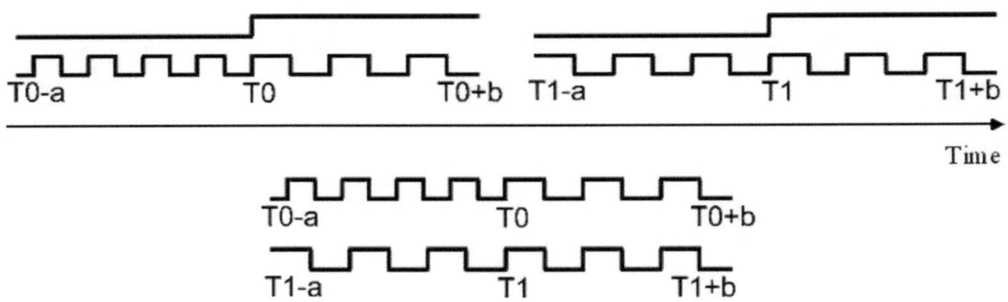

Fig. 20. Upper: input signals VIN12 and VIN2 fed one after the other to the same system. The traces above the signals show the trigger provided by the simulator. Lower: signals superimposed for displaying overlap condition

One has still to ensure that the device would use the recording intervals ($t_0 - a$, $t_0 + b$) and ($t_1 - a$, $t_1 + b$) which are properly aligned with respect to t_0 and t_1. For this purpose, our simulator provides a separate output line that may be used as a trigger by the recording device. In the first step of the recording process, the simulator produces the signal VIN_{12} together with a raising transition on the trigger line at the moment t_0 when the frequency changes. In the second step, the simulator produces the signal VIN_2 together with a raising transition on the trigger line at some moment t_1 corresponding to a raising edge in VIN_2. If the recording device allows computations with stored waveforms, one may use it for displaying the transient as the difference between the records of $VOUT_{12}$ and $VOUT_2$. Or one may transfer the records on a PC and use CAD tools such as PSpice for displaying the difference. Assuming that one uses a scope with memory and arithmetic capabilities and that one wishes to visualize the transient at the transition between f_C and f_{LOW}, the algorithm for displaying the transient would be the following.

- Set up the scope for displaying the difference between channel 1 and memory record on the math channel.
- Set up the scope in single sequence (one shot) mode with trigger on channel 2 that records the trigger signal provided by the simulator.
- Generate with the simulator a signal of constant frequency f_{LOW}, record the reader response and store it to scope memory.
- Generate with the simulator a signal of that changes frequency from f_C to f_{LOW} and record the reader response on channel 1.
- The transient shows on the math channel.

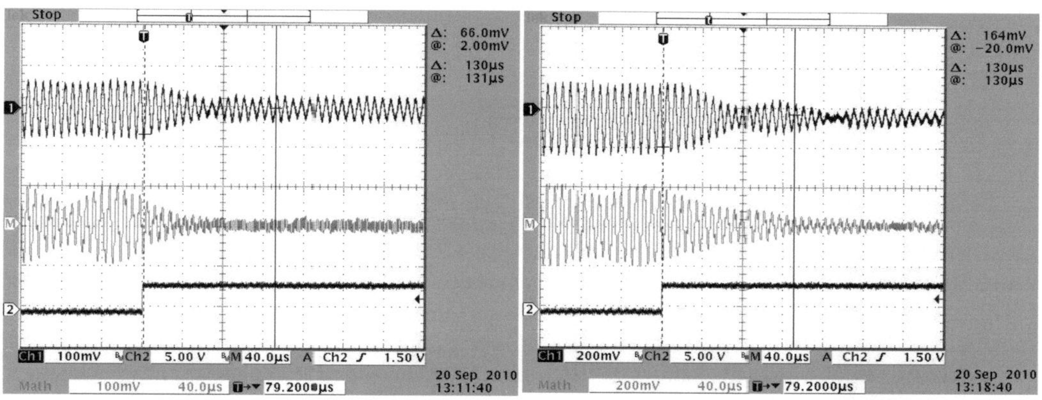

Fig. 21. Effect on transients on bit decoding. Upper traces: signal amplified by reader. Middle traces: transient induced by transition (note that only the part that follows the transition represents the transient). Lower traces: trigger at transition provided by simulator.

In figure 21 we show the result of the application of the described algorithm to the study of the effects of transients on data decoding. The onset of the frequency change is marked by the raising transition on the trigger line provided by the simulator. Knowing the start time of the new bit, one may precisely demarcate the bit interval which here is shown enclosed between the vertical cursor lines. In the left side was recorded a transient of normal duration and the bit was correctly decoded by the decoder IC. The right side shows a transient of abnormally long duration produced by a reader antenna with a too high Q, which resulted into incorrect decoding by the bit decoder IC. In figure 22 we show how the simulator may be used for assessing the amount of equalization of bit amplification by the procedure described in section 7.

9.3 A Low cost alternative for the tag simulator

In the case of readers that achieve bit decoding with a dedicated IC, a low-cost alternative for the simulator is available, that may be used for testing system response and bit decoding. The only hardware of the simulator consists of just a resonant antenna circuit to be plugged in an output port of the reader (figure 23). In this case, the AT91SAM7S64 uC already existent in the reader provides the software component (program) and hardware

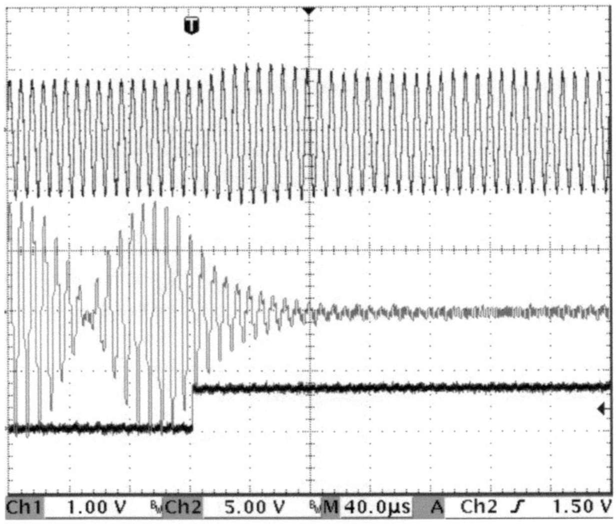

Fig. 22. Scope visualization of frequency transition generated with the tag simulator, after application of equalization of bit amplification. Traces have same meaning as in figure 21.

components (timers and interrupts) needed by the simulator simultaneously with the function of receiving the data from an IC specialized in decoding the answer of an HDX tag. In other words, the reader is receiving the data simulated by itself, which saves the cost of a stand-alone board for the simulator with its own controller, power and communication components. Other hardware components such as the carrier detector are no longer needed, as the reader knows of course the moment when the charge phase ends. In fact the only purpose of such a "charge phase" is to inform the decoder IC that a new decoding phase is to be started. Subsequently the reader uC starts driving the simulator antenna with a preloaded bit pattern. As the simulator and reader antennas are magnetically coupled, the bit pattern transmitted by the reader uC is received by the decoder IC, which sends the decoded bits back to the reader uC. Thus, two tasks are simultaneously performed by the reader uC - driving the simulator antenna and receiving the decoded bits, which is possible by using the system of prioritized interrupts.

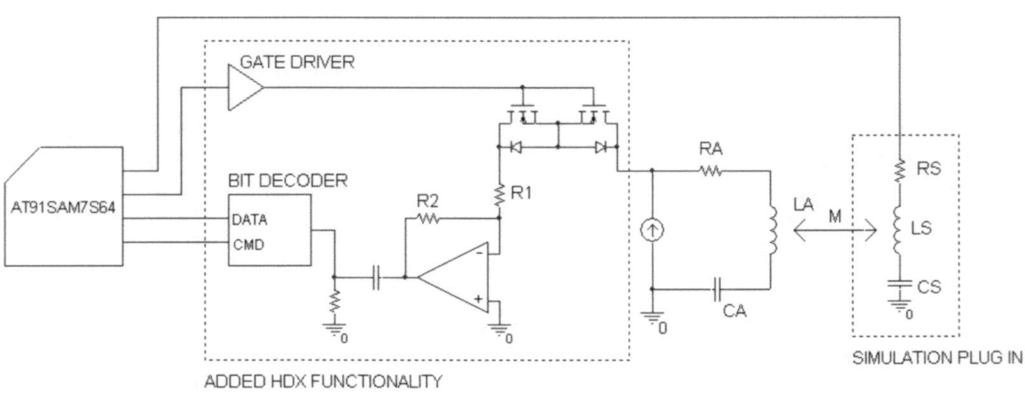

Fig. 23. Simulation plug-in added for test purposes to the current-driven reader

10. Conclusion

We presented two procedures for adding HDX functionality to an existing FDX reader, together with some design issues that influence the reader performance. All these originated in our joint work of developing and producing new readers. We applied the proposed design procedures and tools to the development of an expanded version of the portable voltage-driven proximity reader that is now able to read HDX tags up to 16 cm and of an expanded version of the current-driven long-range reader that can read HDX tags up to 60 cm. In both cases, it was the tag activation, not the reception, which limited the reading distance. The simulator here described, intended to assist the reader developer and the system integrator, allowed us to conveniently perform test and tuning procedures that would have been difficult or nearly impossible with real transponders.

11. References

EM Microelectronic-Marin SA (2005). Read Only Contactless Identification Device. Available from www.emmicroelectronic.com

Gelinotte, E., Frosch, R., Vuza, D.T. & Pascu, L. (2006). An RFID Reader Based on the Atmel AT91SAM7S64 Micro-Controller, *Proceedings of the 1st Electronics Systemintegration Technology Conference*, pp. 1158-1165, ISBN 1-4244-0552-1, Dresden, Germany, September 2006

Gray, P. R. & Meyer, R. G. (1993). *Analysis and Design of Analog Integrated Circuits*, 3rd ed. John Wiley & Sons Ltd, ISBN 0-471-57495-3, New York, USA

International Organization for Standardization (2007). Radio Frequency Identification of Animals, ISO/DIS 14223-1, Part 1: Air Interface

Texas Instruments (January 2003). TMS3705A Transponder Base Station IC, Rev. 1.1. Available from: www.ti.com

Vuza, D.T., Frosch, R. & Koeberl, H. (2007). A Long Range RFID Reader Based on the Atmel AT91SAM7S64 Micro-Controller, *30th ISSE 2007 Conference Proceedings*, pp. 445-450, ISBN 1-4244-1218-8, Cluj, Romania, May 2007

Vuza, D.T. & Frosch, R. (2008). Simulation of Multiple ISO/IEC 18000-2: 2004 Transponders with the AT91SAM7S64 Controller, *SIITME 2008 Conference Proceedings*, pp. 41-45, ISSN 1843-5122, Predeal, Romania, September 2008

Vuza, D.T., Frosch, R., Koeberl, H. & Boissat, D. (2009). A Low Cost Anticollision Reader, In: *Development and Implementation of RFID Technology*, C. Turcu, (Ed.), pp. 201-216, I-Tech, ISBN 978-3-902613-54-7, Vienna, Austria

Vuza, D.T., Chițu, S. & Svasta, P. (2010a). An RFID Tag Simulator Based on the Atmel AT91SAM7S64 Micro-Controller, *33rd ISSE Conference Proceedings*, pp. 229-234, ISBN 978-83-7207-874-2, Warsaw, Poland, May 2010

Vuza, D.T., Chițu, S. & Svasta, P. (2010b). An RFID Tag Simulator for the FDX and HDX Protocols, *16th SIITME 2010 Conference Proceedings*, pp. 53-58, ISBN 978-60-6551-013-5, Pitești, Romania, September 2010

Vuza, D.T. & Frosch, R. (2010). RFID Readers for the HDX Protocol - Design, Simulation and Testing, *16th SIITME 2010 Conference Proceedings*, pp. 47-52, ISBN 978-60-6551-013-5, Piteşti, Romania, September 2010

Part 4

Protocols and Algorithms

13

F-HB$^+$: A Scalable Authentication Protocol for Low-Cost RFID Systems

Xiaolin Cao and Máire P. O'Neill
Centre for Secure Information Technology (CSIT), Queen's University Belfast, Northern Ireland

1. Introduction

RFID technology has received much attention both in industry and academia in recent years and it is seen as the leading ubiquitous computing technology. A typical RFID system consists of a reader, $\mathcal{R}$, and a set of tags, $(\mathcal{T}_i)_{1\le i\le N}$. The reader $\mathcal{R}$ is composed of a set of transceivers and a powerful backend database. Each tag $\mathcal{T}_i$ is a passive transponder identified by a unique ID. However, the fact that RFID tags can be read without line-of-sight results in security risks, especially in relation to the privacy of tag users. Therefore, developing privacy preserving authentication protocols for low-cost RFID tags is a major security challenge that needs to be addressed if RFID systems are to be widely deployed in the coming years.

In previous research in this area the majority of authentication protocols use challenge-response mutual authentication based on symmetric-key ciphers. In order to preserve privacy, on receiving a challenge from a reader, a tag uses pseudonyms, which are the result of using symmetric-key ciphers to process the secret key or ID, as authenticators to the reader. The reason symmetric-key ciphers are used is that the hardware cost of existing asymmetric-key ciphers is too expensive for low-cost tags. For example, ECC and RSA require more than 40,000 gates, which are too large for low-cost tags in which only 200 - 2,000 gates out of 1,000 - 10,000 gates are available for security features (Juels, 2006). A lightweight algorithm known as the learning parity with noise (LPN) problem was first introduced in the HB protocol for human authentication by Hopper & Blum (2001). Juels & Weis (2005) first employed the LPN problem in the HB$^+$ protocol for RFID authentication. The simplicity and novelty of the HB$^+$ protocol has led to the proposal of other HB-related protocols (Jr *et al.*, 2010). Gilbert *et al.* (2008) introduced a simple but effective man-in-the-middle attack against these types of protocols, in which the adversary can derive the secret key of the LPN problem through modifying the tag's response messages. This attack is known as a GRS-MIM attack. In the Trusted-HB protocol (Bringer & Chabanne, 2008), a universal hashing based message authentication code (MAC) is introduced to effectively resist GRS-MIM attacks. Although cryptographic attacks to the Trusted-HB protocol have been reported, they are impractical as they are too complex to implement (Frumkin & Shamir, 2009). Meanwhile, in the F-HB protocol (Cao & O'Neill, 2011), the LNP problem is first introduced to protect the forward privacy of low-cost tags. The operations in the LPN problem involve the calculation of inner products of binary vectors and Bernoulli noise bit generation. Computing the binary inner product only requires bitwise AND and OR operations that can be computed on the fly. Therefore the LPN problem is a hardware-friendly primitive, and very attractive for low-cost RFID security. The most recent progress in LPN-based protocols was reported by Kiltz *et al.* (2011). They introduced an authentication protocol based on a variant of the LPN problem, known as the Subspace LPN problem, and also proposed an efficient MAC construction based on the LPN problem.

The rigorous definition and modelling of privacy in RFID systems has also been investigated in previous research (Avoine, 2005; Juels & Weis, 2007; Vaudenay, 2007; Ha *et al.*, 2008). This research differs in how they treat the adversary's ability to corrupt tags and their different privacy notions for corrupted tags. Compared to the general privacy notion that only considers adversaries that are unable to corrupt a tag, forward privacy is a stronger privacy notion because it also considers the privacy of a corrupted tag. Ma *et al.* (2009) prove that the unpredictable privacy notion (Ha *et al.*, 2008) is stronger than the indistinguishable privacy notion (Juels & Weis, 2007), and that the unpredictable privacy notion is equivalent to a pseudo random function (PRF). It can be observed that the majority of existing forward privacy schemes (Ohkubo *et al.*, 2003; Berbain *et al.*, 2009; Billet *et al.*, 2010) are based on the indistinguishable privacy notion, and the F-HB protocol is based on the unpredictable privacy notion.

Scalability must also be considered in forward private protocols based on symmetric-key ciphers. In order to protect a tag's privacy, before the tag is authenticated by the reader, it must not reveal its identity (its secret key) to the reader. As a result, in order to locate the identity of a tag, the reader must perform a brute-force search of all the tags to check all the keys in its database. As the number of tags increases, this brute-force search will inevitably lead to scalability problems. Existing research into scalability protocols are composed of three categories. The first category comprises protocols that perform a brute-force search of all the tags in the database (Weis *et al.*, 2003; Ohkubo *et al.*, 2003), the time complexity of which is $O(N)$, where N is the number of tags in the system. This method is only suitable for systems with a small number of tags. The second category involves tree-based protocols (Molnar and Wagner, 2004; Molnar *et al.*, 2005), with a time complexity of $O(\log_b N)$ where b represents the branch factor of the tree. These protocols consider each tag as a leaf in a balanced tree, and each tag needs to store $\lceil \log_b N \rceil$ secrets corresponding to the path from the root to the tag leaf. The disadvantage of this method is that because this approach requires that each tag stores correlated keys, the system privacy is weakened when an adversary is able to corrupt at least one tag. The more tags that are corrupted, the more the privacy of this system is compromised. The advantage of this method is that it supports dynamic scalability, so that new tag entries can be easily added without affecting the operation of the protocol. The third category of scalable protocols are hash-table based protocols (Henrici and Muller, 2004; Dimitriou, 2005; Tsudik, 2006; Lim and Kwon, 2006; Le *et al.*, 2007; Song, 2009; Alomair *et al.*, 2010; Cao & O'Neill, 2011). These protocols require only constant-time, $O(1)$, running time to identify a tag. These protocols need to store pre-computed hash-tables in the database associated with the reader. The reader uses pseudonyms from a tag as the indices of the hash-table to match a value, realizing constant-time tag identification. Compared to the tree-based protocols, hash-table based protocols need smaller storage on a tag and maintain a constant response time even when the number of tags increases. The disadvantage of these protocols is that the backend database needs a large storage to build a hash-table. Although, it is assumed that in RFID systems the database possesses infinite computational ability, from a practical viewpoint, all previously proposed protocols in this category require unrealistic large storage, and lack dynamic scalability (Avoine *et al.*, 2010).

In this chapter, building on previous work in this area, a novel scalable and forward private authentication protocol, F-HB^{+}, suitable for low-cost RFID applications is proposed. The contributions are as follows. Firstly, similar to the F-HB protocol, the proposed protocol uses an LPN problem and a pseudo random number generator (PRNG); however, a hardware counter is introduced to the tag to enhance its desynchronization resistance, and the MAC code generation based on the proposal of Kiltz *et al.* (2011) is more efficient than in the F-HB protocol. Secondly, a new Re-Hash technique is presented to effectively reduce the storage requirement of the hash-table over previous protocols. The Re-Hash technique is adapted to support dynamic scalability and it is used to construct the hash-table required in the F-HB^{+} protocol. Thirdly, the security proof of the F-HB^{+} protocol is derived under the standard model. Overall, the proposed protocol features: (i) from the tag's perspective, low-cost implementation and forward privacy; (ii) from the reader's perspective, constant-time scalability, small hash-table storage and dynamic scalability.

The rest of the chapter is organized as follows. In section 2, the mathematical definitions and previous related work are introduced. In section 3, the Re-Hash technique is presented, and how it can be adapted to include dynamic scalability is discussed. The proposed F-HB^{+} scheme with the Re-Hash technique is described in section 4. The unpredictable forward privacy framework and security proof are derived in section 5. Section 6 presents a performance evaluation and comparison results, while Section 7 concludes the chapter.

2. Preliminary

2.1 Mathematical definitions

Definition 1. LPN Problem (Hopper & Blum, 2001). Let Ber_η denote the Bernoulli distribution with parameter $\eta \in (0,1/2)$. A bit $v \leftarrow \mathrm{Ber}_\eta$ is such that $\Pr[v=1]=\eta$ and $\Pr[v=0]=1-\eta$, while an l-bit vector $v \leftarrow \mathrm{Ber}_{l,\eta}$ is such that each bit of v is independently drawn according to Ber_η. Let $\mathrm{Hwt}(v)$ denote the hamming weight of vector v. Let T be a random $(l \times n)$ binary matrix, let x be a random n-bit vector, let $\eta \in (0,1/2)$ be a noise parameter, and let v be a random l-bit vector according to $\mathrm{Ber}_{l,\eta}$, such that $\mathrm{Hwt}(v) \le \eta l$. Given T, η and $z=(T\cdot x)\oplus v$, find an n-bit vector y such that $\mathrm{Hwt}((T\cdot y)\oplus z) \le \eta l$.

For a fixed n-bit string, k, let $\pi_{k,\eta}$ denote the oracle returning an independent $(n+1)$-bit string according to the LPN problem:

$$\{(a,(k\cdot a)\oplus v) | a \in_R \{0,1\}^n, v \leftarrow \mathrm{Ber}_\eta\}. \tag{1}$$

The following Lemma 1 upper-bounds the probability that an adversary predicts the secret n-bit string k given some instances of oracle $\pi_{k,\eta}$, which implies that the two oracles, $\pi_{k,\eta}$ and U_{n+1}, are computationally indistinguishable, where U_{n+1} denotes an oracle that returns an independent uniformly random $n+1$-bit string.

Lemma 1. Indistinguishability of LPN Problem (Katz & Shin, 2006). Assume there exists an algorithm A making q oracle queries, running in time t, and such that

$$|\Pr[A^{\pi_{k,\eta}}(1^n)=1]-\Pr[A^{U_{n+1}}(1^n)=1]| \ge \epsilon. \tag{2}$$

Then there is an algorithm B making $O(q\cdot\epsilon^{-2}\log n)$ oracle queries, running in time $O(t\cdot n\epsilon^{-2}\log n)$, and such that

$$\Pr[B^{\pi_{k,\eta}}(1^n)=k | k \in_R \{0,1\}^n] \ge \epsilon/4. \tag{3}$$

Definition 2. PRNG (Goldreich, 2001). A PRNG is a function $g: \{0,1\}^m \rightarrow \{0,1\}^n$ that takes as input an m-bit hidden seed and returns an n-bit string, where $n > m$. The output of the PRNG is called a pseudo random number, which appears to be random. A (t, ϵ_g)-secure PRNG represents that the output of this PRNG cannot be discriminated with a true random string in time t with advantage at most ϵ_g.

The PRNG can be implemented using stream ciphers such as those proposed in the STREAM project (Cid & Robshaw, 2009) and a secure stream cipher is seen as a PRF (Billet *et al.*, 2010).

Definition 3. Universal Hash Functions (Wegman & Carter, 1981). A family of functions $\{h_u: \{0,1\}^l \rightarrow \{0,1\}^m\}_{u \in U}$ is called a strongly universal hash family if $\forall x \in \{0,1\}^l$, $\forall y \in \{0,1\}^m$:

$$\Pr[h_u(x) = y] = 2^{-m}, \tag{4}$$

and $\forall x_1 \neq x_2 \in \{0,1\}^l$, $\forall y_1, y_2 \in \{0,1\}^m$:

$$\Pr[h_u(x_2) = y_2 \,\&\, h_u(x_1) = y_1] = 2^{-2m} \tag{5}$$

where any hash function is easily selected by $u \in U$.

An $(l \times m)$-bit Toeplitz matrix is a matrix for which the entries on every upper-left to lower-left diagonal have the same value. Since the diagonal values of a Toeplitz matrix are fixed, the entire matrix is specified by the top row and the first column. Thus a Toeplitz matrix can be stored in $(l + m - 1)$ bits rather than the $(l \times m)$ bits required for a truly random matrix. For any $(l + m - 1)$-bit vector u, let T_u denote the Toeplitz matrix whose top row and first column are represented by u.

Definition 4. Toeplitz based Universal Hash Function (Krawczyk, 1994). Let $\{T_u\}_{u \in U}$ be the family of Toeplitz matrices where the $(l + m - 1)$-bit vector u is chosen at random, and z is a random m-bit vector. Then the following is a strongly universal hash function family:

$$\{h_u(x) = (T_u \cdot x) \oplus z: \{0,1\}^l \rightarrow \{0,1\}^m\}_{u \in U}. \tag{6}$$

Meanwhile, according to the property in (5), the Toeplitz based universal hash function is also a pairwise independent hash function (Naor & Reingold, 1997).

Definition 5. LPN based MAC (Kiltz *et al.*, 2011). Let $h_u: \{0,1\}^l \rightarrow \{0,1\}^m$ be a pairwise independent hash function, $\rho(\cdot)$ be a pairwise independent permutation on $\{0,1\}^{l \times n + n + w}$, $v \leftarrow \mathrm{Ber}_{n,\eta}$, $s_i \in_R \{0,1\}^l$, $r \in_R \{0,1\}^w$, and $T \in_R \{0,1\}^{l \times n}$. Given a secret key $\left(\{s_i\}_{(0 \le i \le m)}, h_u, \pi\right)$ and a message x, the LPN based MAC for the message, x, can be defined as:

$$\mathrm{MAC}_{(s,h,\pi)}(x) = \rho(T, T^{\mathrm{T}} \cdot s(y) \oplus v, r), \tag{7}$$

where $y = h_u(x, r)$ and $s(y) = s_0 \bigoplus_{i:y[i]=1} s_i (0 \le i \le m)$.

The verification steps of the LPN based MAC are as follows. Firstly, use $\rho^{-1}(\cdot)$ to obtain (T, z, r); if $\mathrm{rank}(T) \neq n$, then reject. Secondly, use $h_u(x, r)$ to obtain y and $s(y)$. Thirdly, if $\mathrm{Hwt}\left(z \oplus T^{\mathrm{T}} \cdot s(y)\right) \le n\left(\frac{1}{4} + \frac{\eta}{2}\right)$, accept the MAC, otherwise reject.

One disadvantage of this MAC is that if the standard pairwise independent permutation $\rho(x) = a \times x + b$ (where a and b are random strings) is used, the computation for the multiplier will be a bottleneck for the LPN based MAC (Kiltz *et al.*, 2011). But it can be observed that the function of $\rho(\cdot)$ prevents the adversary from directly choosing the input of a MAC. The protocol proposed in this chapter solves this limitation by using a simplified pairwise independent permutation, $\rho(x) = x + b$, where $a = 1$. Another disadvantage is that the key $(\{s_i\}_{(0 \le i \le m)}, h_u, \pi)$ requires a large storage cost. The proposed protocol solves this by using a PRNG that is able to generate successive random strings.

2.2 Related work

In this section, a brief introduction and analysis of previous research is presented. The most relevant work for comparison is the hash-table based scalable and forward private protocols. These protocols can be divided into two classes according to their methods for generating pseudonyms. In the remainder of the chapter, the word "pseudonyms" is taken to mean indices used to look up a hash-table.

In the first class of protocols, each tag stores a unique key, which can be used as the tag's authenticator to the reader. The pseudonyms are derived from this secret key, and the pseudonym update method on the tag depends on a one-way secure hash function without interference from the reader. In the first hash-table based protocol proposed by Weis *et al.* (2003), on any query from a reader, a tag always replies with the fixed pseudonym of its unique secret key. Therefore, it is vulnerable to tracking attacks and tag impersonation. In the protocols proposed by Henrici and Muller (2004) and Dimitriou (2005), the tag's response comprises a pseudonym and an authenticator. Due to the fixed pseudonym used between successful mutual authentications, these protocols fail to resist tag tracking. The protocols proposed by Lim and Kwon (2006) and Tsudik (2006) also use a response pair. But the pseudonyms in these protocols will recycle in a brute-force desynchronization attack, so they fail to provide forward privacy.

In the second class of protocols, each tag needs to store two secrets, where one secret is used as the tag's final authenticator key and the other one is used to generate the pseudonym chain. These protocols possess the advantage that pseudonyms are unrelated to the secret key, but they use more non-volatile memory on the tag. The O-FRAP protocol was proposed by Le *et al.*, (2007) for RFID authentication under a universally composable framework and provides forward privacy. It updates pseudonyms using the same method as in the first class of protocols. The O-FRAP protocol constructs a hash-table using the output of a PRF implemented by a PRNG. But it is difficult to validate that the output of a PRF possesses the collision-free property. Two further protocols in this class (Song, 2009; Alomair *et al.*, 2010) require the help of the reader to update pseudonyms and send the updated pseudonyms to tags, which does not relieve the burden on the tag and adds to the risk of desynchronization.

The desynchronization threats in the above protocols can be alleviated by using more than one pseudonym for a secret key. There are two methods to achieve this purpose. One method is based on the time-stamp concept (Tsudik, 2006), and involves adding a hardware timer to the tag, inevitably increasing the cost of the tag. This technique is unsuitable for low-cost tags. Another technique relies on a hardware counter on the tag (Le *et al.*, 2007; Song, 2009; Alomair *et al.*, 2010). This counter is used to limit the maximum number of pseudonyms associated with a secret key. The maximum threshold value of this counter determines the ability to resist desynchronization attacks. Although the hardware counter also increases the cost of the tag, it is more practical than a hardware timer. Another problem of the above protocols is that they utilise cryptographic secure hash functions, the hardware cost of which exceeds the budget of low-cost tags. For example, according to the latest literature reports, the standard algorithm, SHA-1, requires at least 5,000 gates (O'Neill, 2008).

The most recent progress in constant-time scalable protocols is presented by Alomair *et al.* (2010). It also uses a counter with threshold Th to control the number of pseudonyms for each secret key. Compared to the previous proposals, this protocol considers a further step: how to build a hash-table with a reasonable storage in the database. This paper points out that impractically large hash tables are a result of the fact that the bit-length of a pseudonym, L, must be long enough to avoid collision. And in order to directly address the hash-table, the size of the hash-table must be $O(2^L)$ bits, which is unrealistic in practice. In order to reduce the storage requirement, a 2-level hash-table construction method is proposed. The 1st level is a hash-table with the s most significant bits (MSB) of the L-bit pseudonyms as its indices, and that stores the addresses of the 2nd level. The 2nd level is a linear table composed of the remainding $(L-s)$ bits of the L-bit pseudonym, that stores the addresses of the actual information. Assuming that the number of pseudonyms is N', the protocol recommends the use of the following parameters: the 1st level storage is $O(2^s)$ bits, where $s = \lceil \log_2(N' \times Th) \rceil$, and the 2nd level storage is $O(N' \times Th)$ bits. Using these parameters, constant-time authentication can be achieved with the 2-level hash-table. Avoine *et al.* (2010) noted that although this method is very efficient, its total storage requirement for the 2-level structure is still very large and does not support dynamic resizing.

3. Proposed Re-Hash technique

3.1 Basic Re-Hash technique

As mentioned before, in the hash-table based protocols, a tag can be identified in constant-time by its L-bit pseudonyms. The total number of valid pseudonyms for each tag in a synchronized state is controlled by a counter with a maximum threshold, Th. Firstly, let us take an example to show how much storage is required if these pseudonyms are directly used as look-up indices of a hash-table. The total number of tags, N, is assumed to be 2^{30} (greater than 1 billion) and the value of Th is 2^{10}. Therefore 2^{40} $(= N \times Th)$ indices are needed for the hash-table, so the collision-free bit-length of an index should be at least 40 bits. According to Alomair *et al.* (2010), the bit-length of pseudonyms should be large enough to obtain a collision-free 40-bit index of a hash-table. Assuming $L = 60$ bits, the collision-free hash-table needs at least 2^{17} terabytes (TB) of storage with 2^{60} slots ($2^{60} \times 1$ bit, i.e., assume every slot in the hash-table stores 1 bit) to meet the demands of direct addressing. This storage requirement is too large for practical use.

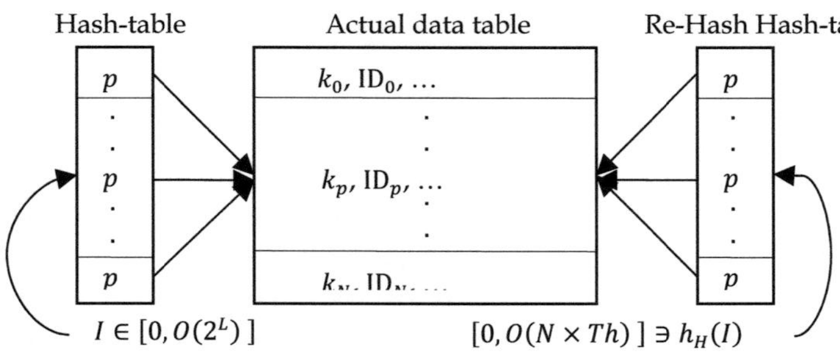

Fig. 1. The traditional Hash-table vs. basic Re-Hash hash-table

It can be observed that in the above example only 2^{40} slots out of the total 2^{60} slots are used in each authentication session, so that the truly useful storage of all the indices during each authentication session is 0.125 TB ($2^{40} \times 1$ bit), which is practical. Therefore, of the total $O(2^L)$ bits of storage, the true requirement is at most $O(N \times Th)$ bits, which causes a huge storage waste.

Therefore, in order to reduce the storage cost, a mathematical mapping is needed, $f: \{0,1\}^{60} \rightarrow \{0,1\}^{40}$, which is the essence of the Re-Hash technique proposed in this chapter. The function $f(\cdot)$ can be implemented as a look-up table hash function $h_H(\cdot)$, which uses the 60-bit pseudonyms of tags as its inputs and outputs 40-bit strings. These 40-bit outputs can then be used as look-up indices of a hash-table. If this technique is used, the storage cost of the directly addressed hash-table in the above example can be reduced to 0.125 TB ($2^{40} \times 1$ bit). Fig. 1 illustrates the difference between the traditional hash-table and the basic Re-Hash hash-table, where I represents the pseudonym of a tag, and p represents the address of the actual information related to the tag.

The Re-Hash technique for hash-table construction can be generalized as follows:

1. Determine the number of pseudonyms required during each authentication session, $N \times Th$, in the RFID system.
2. Determine the collision-free bit-length of a pseudonym, L.
3. Select an appropriate look-up table hash function, $h_H: \{0,1\}^L \rightarrow \{0,1\}^{N \times Th}$, which uses the pseudonyms as its input values.
4. Use the output of h_H as indices to construct the hash-table, in which every slot stores a pointer to the address storing actual tag information.

The important advantage of this technique is the storage cost saving. One possible disadvantage is that the collision probability among hash-table indices may increase, because the number of hash-table indices is equal to the number of pseudonyms in each authentication session. However in section 6.1 analysis shows that if an appropriate Re-Hash hash function is used, constant-time look-up is maintained.

3.2 Dynamic Re-Hash

In this section it is illustrated that it is necessary to build a dynamic hash-table to accommodate frequent database changes, insertions and deletions. Firstly, dynamic table should effectively utilize the storage available. Assume a large-scale supermarket respectively sells and buys 2^{20} (greater than 1 million) items per month, the change in the number of indices for the hash-table is 2^{31} ($2 \times 2^{20} \times 2^{10}$). Thus, the change in storage will be at least 2 gigabytes (GB) ($2^{31} \times 1$ bit). If the hash-table is fixed, then this 2 GB storage may not be fully utilized. Secondly, a dynamic table should be able to process concurrent transactions without affecting the system response time. For example, merchandize is checked out in a supermarket at the same time. This would need many hash-table insertions and deletions at the same time.

Linear-Hashing (Black, 2009) is a dynamically updateable hash-table construction method which implements a hash-table that grows or shrinks one slot at a time through splitting a current slot into two slots. In general, assuming the Linear-Hashing scheme has an initial hash-table with M slots, then it needs a family of look-up table hash functions $h_{H,j}(\cdot) = f(\cdot) \bmod (2^j M)$. At any time, there is a value $j (\geq 0)$ that indicates the current splitting round and the current look-up hash functions; a pointer $p \in [0, \ldots, 2^j M - 1]$ which points to the slot to be split next; a total of $(2^j M + \text{p})$ slots, each of which consists of a primary page and possibly some overflow pages; and two hash functions $h_{H,j}$ and $h_{H,j+1}$. The look-up process works as follows: If $h_{H,j}(\cdot) \geq p$, choose slot $h_{H,j}(\cdot)$ since this slot has not been split yet in the current round; otherwise, choose slot $h_{H,j+1}(\cdot)$, which can either be the slot $h_{H,j}(\cdot)$ or its split image slot $h_{H,j}(\cdot) + 2^j M$.

The final proposed dynamic hash-table construction method, in which the Re-Hash technique is adapted to include the Linear-Hashing technique, can be described as follows:

1. Determine the system capacity, i.e., the maximum tag number N_{MAX} the system can accommodate, and the collision-free bit-length of a pseudonym L.
2. Determine the output range of the Re-Hash hash function, L', such that $L' \geq L/2$.
3. Select an appropriate look-up table hash function, which is used as the Re-Hash hash function, $h_H: \{0,1\}^L \to \{0,1\}^{L'}$.
4. Determine the initial tag number of this RFID system, N, and the initial dynamic hash-table size, M, such that $M \geq N \times Th$.
5. Determine the Linear-Hashing look-up hash function family, $h_{H,j}(\cdot) = h_H(\cdot) \bmod (2^j M)$.
6. Use the outputs of $h_{H,j}(\cdot)$ as indices to construct the dynamic hash-table, in which every slot stores a pointer to the address storing actual tag information.

4. F-HB$^+$ protocol description

4.1 Initialization

The initialization steps involved in the proposed F-HB$^+$ protocol are as follows.

- Tag: Every tag is independently assigned a secret key $k \in_R \{0,1\}^m$, which is shared with the reader. Each tag can compute a PRNG $g(\cdot)$ as in Definition 2, multiple instances of $\pi_{k,\eta}$ at the same time, and an m-bit counter $ct_{\mathcal{T}} \leftarrow 0$ whose maximum threshold value is Th. They also have enough non-volatile memory to store the value of k and $ct_{\mathcal{T}}$.
- Reader: In the database, there is an old key $k_{old} \leftarrow k$, a current key $k_{cur} \leftarrow k$, a counter $ct_{\mathcal{R}} \leftarrow 0$ with threshold Th, and Th hash-table entries $\{h_{H,j}(I_i) \mid 0 \leq i < Th\}$ for every tag, where $I_i = (T_k \cdot i) \oplus r_i$ and r_i is the i-th iteration result of $g(k_{cur})$. The two secret keys are used to resist brute-force desynchronization attacks, and the Th hash-table entries are used to enhance the desynchronization resistance. The variables for Linear Hashing are also initialized: the current splitting round indicator $j \leftarrow 0$ and the current splitting pointer $p_s \leftarrow 0$. All the information is organized into a pre-computed 2-level database structure, which is illustrated in Fig. 2. In addition, the database can compute a look-up hash function family $\{h_{H,j}(\cdot)\}_{j\geq 0}$. The 1st level of the database is the pre-computed

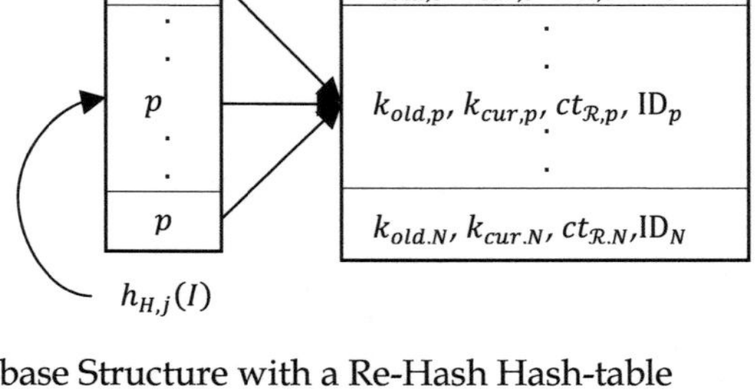

Fig. 2. The 2-level Database Structure with a Re-Hash Hash-table

dynamic hash-table. For every tag, there are Th slots (maybe not successive) in this hash-table, which store the pointers p indicating an address in the 2nd level table. The address of the 1st level hash-table is computed by $h_{H,j}(I_i)$. The 2nd level of the database is a pre-organized linear table. For each tag, there is only 1 slot in this level to store k_{old}, k_{cur}, $ct_{\mathcal{R}}$ and the actual information about each tag.

4.2 Authentication interaction

An overview of the proposed authentication protocol is illustrated in Fig. 3. It is a 3-pass mutual authentication protocol.

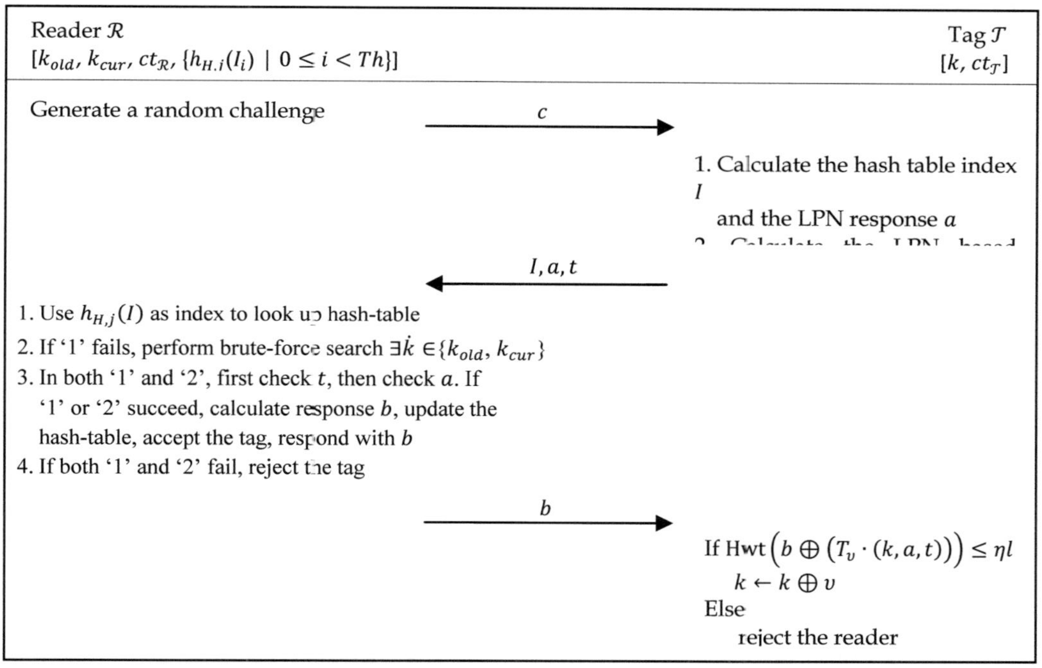

Fig. 3. The Proposed F-HB⁺ Protocol

Fig. 4 illustrates the tag's operation after the tag receives the challenge message c from the reader. It can be observed that the Toeplitz matrix T_k is used in the LPN problem such that $a \leftarrow \left(T_k \cdot (c, I)\right) \oplus v$, and in the strong universal hashing such that $I \leftarrow (T_k \cdot ct_{\mathcal{T}}) \oplus r$ at the same time. Meanwhile, the PRNG g is also used in the strong universal hashing such that $\{r \leftarrow g(k), I \leftarrow (T_k \cdot ct_{\mathcal{T}}) \oplus r\}$. More importantly, the PRNG is in charge of generating all the secret keys of the LPN based MAC, such that $\left(\{s_i\}_{(0 \le i \le m)}, r_1, r_2\right) \leftarrow g(k)$.

Fig. 5 explains the reader's key search method in detail after it receives the authentication message (I, a, t) from the tag. Only if both the MAC code t and authenticator a pass the verification will the reader accept the tag and generates a confirmation message, b. It can be observed that the reader does not use k_{cur} as the secret key for the LPN problem again, but uses the noise vector v' such that $b \leftarrow \left(T_{v'} \cdot (k_{cur}, a, t)\right) \oplus v''$. This is to prevent GRS-MIM attackers from recovering the secret key k_{cur}. The difference between steps 1 and 2 is that (i) step 1 only involves the current key k_{cur} of one tag providing constant-time scalability; but (ii) step 2 involves the secret key pair (k_{old}, k_{cur}) of all the tags, and needs to try all keys.

Step 1:	Step 2:
$v \leftarrow \mathrm{Ber}_{l,\eta}$, $r \leftarrow g(k)$	Generate random r and T, $v \leftarrow \mathrm{Ber}_{n,\eta}$,
If $ct_{\mathcal{T}} < Th$	$\left(\{s_i\}_{(0 \le i \le m)}, r_1, r_2\right) \leftarrow g(k)$,
$\quad I \leftarrow (T_k \cdot ct_{\mathcal{T}}) \oplus r$, $ct_{\mathcal{T}} \leftarrow ct_{\mathcal{T}} + 1$	$y \leftarrow \left(T \cdot (c, a, I, r)\right) \oplus r_1$,
Else	$s(y) = s_0 \oplus_{i:y[i]=1} s_i (0 \le i \le m)$
$\quad I \in_R \{0,1\}^n$, $ct_{\mathcal{T}} \leftarrow ct_{\mathcal{T}}$	$t = (T, T^{\mathrm{T}} \cdot s(y) \oplus v, r) + r_2$,
$a \leftarrow \left(T_k \cdot (c, I)\right) \oplus v$	

Fig. 4. Tag's response operation in the Proposed F-HB+ Protocol

Step 1:	Step 2:
$r \leftarrow g(k_{cur})$, $\left(\{s_i\}_{(0 \le i \le m)}, r_1, r_2\right) \leftarrow g(k_{cur})$	$r \leftarrow g(\dot{k})$, $\left(\{s_i\}_{(0 \le i \le m)}, r_1, r_2\right) \leftarrow g(\dot{k})$
$(T, z, r) \leftarrow t - r_2$, if rank$(T) \ne n$, reject	$(T, z, r) \leftarrow t - r_2$, if rank$(T) \ne n$, reject
$y \leftarrow \left(T \cdot (c, a, I, r)\right) \oplus r_1$	$y \leftarrow \left(T \cdot (c, a, I, r)\right) \oplus r_1$,
$s(y) = s_0 \oplus_{i:y[i]=1} s_i (0 \le i \le m)$	$s(y) = s_0 \oplus_{i:y[i]=1} s_i (0 \le i \le m)$
If $\mathrm{Hwt}(z \oplus T^{\mathrm{T}} \cdot s(y)) \le n\left(\frac{1}{4} + \frac{\eta}{2}\right)$	If $\mathrm{Hwt}(z \oplus T^{\mathrm{T}} \cdot s(y)) \le n\left(\frac{1}{4} + \frac{\eta}{2}\right)$
$\quad v' \leftarrow \left(T_{k_{cur}} \cdot (c, I)\right) \oplus a$	$\quad v' \leftarrow \left(T_k \cdot (c, I)\right) \oplus a$
$\quad$ If $ct_{\mathcal{R}} < Th$ and $\mathrm{Hwt}(v') \le \eta l$	$\quad$ If $ct_{\mathcal{R}} < Th$ and $\mathrm{Hwt}(v') \le \eta l$
$\quad\quad v'' \leftarrow \mathrm{Ber}_{l,\eta}$, $ct_{\mathcal{R}} \leftarrow 0$	$\quad\quad v'' \leftarrow \mathrm{Ber}_{l,\eta}$, $ct_{\mathcal{R}} \leftarrow 0$
$\quad\quad b \leftarrow \left(T_{v'} \cdot (k_{cur}, a, t)\right) \oplus v''$	$\quad\quad b \leftarrow \left(T_{v'} \cdot (\dot{k}, a, t)\right) \oplus v''$
$\quad\quad (k_{old}, k_{cur}) \leftarrow (k_{cur}, k_{cur} \oplus v')$	$\quad\quad (k_{old}, k_{cur}) \leftarrow (\dot{k}, \dot{k} \oplus v')$
$\quad\quad$ update $\{h_{H,j}(I_i) \mid 0 \le i < Th\}$	$\quad\quad$ update $\{h_{H,j}(I_i) \mid 0 \le i < Th\}$
$\quad\quad$ accept the tag	$\quad\quad$ accept the tag

Fig. 5. Reader's authentication operation in the Proposed F-HB+ Protocol

4.3 Hash-table update procedure

This protocol supports dynamic update. The update procedure consists of insertion and deletion. Let us first to describe the insertion procedure. There are two insertion scenarios. One is when a tag is successfully authenticated, the old secret key is updated for this tag, therefore, the associated old Th pseudonyms also need to be updated. The other scenario is when new tags are added into the system, new pseudonyms should also be included. Assuming that there is a new pseudonym called I_{new}, and its corresponding hash-table index is $h_{H,j}(I_{new})$. Therefore, I_{new} is inserted into the slot $h_{H,j}(I_{new})$ as follows:

- If no overflow occurs, its position is within the primary page of this slot. Insertion process is completed.
- Otherwise I_{new} is put into the overflow page of the slot $h_{H,j}(I_{new})$. The pseudonyms in the current splitting slot p_s are split into 2 slots: p_s and $p_s + 2^j M$ using the look-up hash function $h_{H,j+1}(\cdot)$. The splitting pointer p_s moves to the next slot, $p_s \leftarrow p_s + 1$. If $p_s \ge 2^j M$, increment the current splitting round indicator, $j \leftarrow j + 1$, and reset the splitting pointer, $p_s \leftarrow 0$. Insertion process is completed.

Deletion will cause the hash-table to shrink. Slots that have been split can be recombined. The operation of two slots merging together is the reverse of splitting a slot in the insertion process.

Overall, the update procedure can be divided into two stages. The first stage is to insert the new pseudonyms according to the above insertion procedure in an on-line mode, which runs concurrently with other transactions. The second stage is to delete the old pseudonyms according to the deletion procedure, which can be done in an off-line mode, in order to obtain optimal system performance.

5. RFID privacy definition and proof

5.1 Adversary assumptions

In this chapter, an adversary A is assumed to be a probabilistic polynomial algorithm that is allowed to perform oracle queries during attacks. The reader side is assumed to be secure. The tag and wireless communication channel are assumed to be insecure, which means that an adversary can intercept all the wireless communications between the reader and tags, and can corrupt a tag. The reader is assumed to have the ability to handle several authentication exchanges simultaneously, but a tag cannot. In order to model the majority of known attacks against authentication protocols in RFID systems, five oracles are defined as follows.

i. $O_1(\mathcal{R})$: It invokes the reader $\mathcal{R}$ to start a new session of the authentication protocol. This oracle returns the reader's challenge message c.
ii. $O_2(\mathcal{T}_i, c)$: It invokes a tag $\mathcal{T}_i$ to start an authentication session exchange related to challenge message c. The tag $\mathcal{T}_i$ responds with the response message a.
iii. $O_3(\mathcal{T}_i, c, a)$: It returns the unmodified and modified challenge, c, and response, a, related to a tag $\mathcal{T}_i$.
iv. $O_4(\mathcal{T}_i)$: It returns the final authentication result of a tag $\mathcal{T}_i$.
v. $O_5(\mathcal{T}_i)$: It returns the current key and internal state information of a tag $\mathcal{T}_i$, and also updates the key and state information of tag $\mathcal{T}_i$ if necessary.

For example, eavesdropping can be modelled as: first query O_1 to get c, then query O_2 to get a, and finally query O_4 to get authentication results. The message interception can be modelled by O_3. Any key compromised due to tag corruption, or side-channel attacks can be modelled by sending the O_5 query to the tag.

Definition 6. (q, t)-adversary. An adversary whose running time is upper-bounded by t and has the ability to disturb at most q authentication exchanges in this interval is called a (q, t)-adversary. The adversaries are assumed to only be able to attack the RFID system at a specific position and during a limited time period. The term "exposure period" (Vaudenay, 2007) is used to name this specific attack time. During an exposure period, an adversary is able to observe and disturb all interactions involving a target tag $\mathcal{T}_i$ and a legitimate reader $\mathcal{R}$ using oracle $(O_i)_{1 \le i \le 5}$ according to the defined security model. After an exposure period, no adversary is allowed to continue his attack. But attacks do not need to be completed within only one exposure period, and can continue in several successive or discrete exposure periods.

5.2 LPN problem characteristics

From the protocol description, it can be found that in every authentication session, the tag needs to calculate multiple instances of $\pi_{k,\eta}$ at the same time: the secret is a Toeplitz matrix rather than a vector, the noise is a vector rather than a single bit. The usage is the same as in the HB# protocol (Gilbert *et al.*, 2008), but HB# reduces its security proof based on the hardness of the LPN problem. In this chapter, the security proof is based on the computational indistinguishability of the two oracles, $\pi_{k,\eta}$ and U_{n+1}, in Lemma 1.

First of all, a new oracle returning multiple bits of $\pi_{k,\eta}$ at the same time is defined as follows. For a fixed $(l \times n)$ matrix K, let $\Pi_{K,\eta}$ be the oracle returning an independent $(n+l)$-bit string according to:

$$\{(a,(K \cdot a) \oplus v) | a \in_R \{0,1\}^n, v \leftarrow \mathrm{Ber}_{l,\eta}\}. \tag{8}$$

Theorem 1 below upper-bounds the probability that an adversary predicts the secret $(l \times n)$ matrix K given some instances of oracle $\Pi_{K,\eta}$, so it implies that the two oracles, $\Pi_{K,\eta}$ and U_{n+l}, are computationally indistinguishable.

Theorem 1. Assume there exists an algorithm A making q oracle queries, running in time t, and such that

$$|\Pr[A^{\Pi_{K,\eta}}(1^n) = 1] - \Pr[A^{U_{n+l}}(1^n) = 1]| \geq \epsilon. \tag{9}$$

Let t_π be the time taken to calculate a $\pi_{k,\eta}$ instance. Then there is an algorithm B making $O(q)$ oracle queries, running in time $t + \frac{l(l-1)}{2} t_\pi$, and such that

$$|\Pr[B^{\pi_{k,\eta}}(1^n) = 1] - \Pr[B^{U_{n+1}}(1^n) = 1]| \geq \epsilon / l. \tag{10}$$

Proof. A hybrid argument technique is used to prove it. Let K' denote a $(l-j) \times n$ binary matrix. Firstly, define the following hybrid distribution, D_j, with $j \in [0, l]$ as

$$\{(a, r, (K' \cdot a) \oplus v)\}, \tag{11}$$

where $a \in_R \{0,1\}^n$, $r \in_R \{0,1\}^j$ and $v \leftarrow \mathrm{Ber}_{l-j,\eta}$. Upon receiving an $(n+1)$-bit input, B gerneates a random value, $j \in [0, l]$ to construct an $(n+l)$-bit input as A's input. When $j < l$, it also needs to generate a random $(l-j) \times n$ binary matrix K'. It is clear that when B's input complies with U_{n+1}, $j \in [1, l]$; when B's input complies with $\pi_{k,\eta}$, then $j \in [0, l-1]$. The distribution of D_l is the same as U_{n+l}, and D_0 the same as $\Pi_{K,\eta}$. And B uses A's outputs as its outputs. Thus

$$|\Pr[B^{\pi_{k,\eta}}(1^n) = 1] - \Pr[B^{U_{n+1}}(1^n) = 1]|$$

$$= \frac{1}{l} \left| \sum_{j=0}^{l-1} (A^{D_j}(1^n) = 1) - \sum_{j=1}^{l} (A^{D_j}(1^n) = 1) \right|$$

$$= \frac{1}{l} |\Pr[A^{\Pi_{K,\eta}}(1^n) = 1] - \Pr[A^{U_{n+l}}(1^n) = 1]| \geq \frac{\epsilon}{l}. \tag{12}$$

A contradiction with the Lemma 1 is obtained, which concludes the proof.

Defintion 7. Indistinguishability of Oracle $\Pi_{K,\eta}$. The oracle $\Pi_{K,\eta}$ is said to be (q, t, ϵ)-secure if there is no (q, t)-adversary who can distinguish $\Pi_{K,\eta}$ from U_{n+l} with advantage ϵ.

Secondly, due to the fact that Bernoulli random noise may exceed the acceptable threshold, even the legitimate tag may be rejected, which is called a false rejection. This property can also result in an adversary impersonating a tag successfully by simply guessing without any prior knowledge, which is called a false acceptance. According to probability theory, the false rejection probability P_{FR}, and false acceptance probability P_{FA} in every authentication session can be defined as follows:

$$P_{FR} = \sum_{i=\eta l+1}^{l} \binom{l}{i} \eta^i (1-\eta)^{l-i}, \tag{13}$$

$$P_{FA} = \sum_{i=0}^{\eta l} \binom{l}{i} 2^{-l}. \tag{14}$$

Thirdly, in the protocol, the universal hashing MAC code is used to protect the integrity of communication messages. If the adversary uses the GRS-MIM attack and its variants (Gilbert *et al.*, 2008), the check for the universal hashing MAC code will fail, then, the reader will not continue to check the LPN problem as illustrated in Fig. 3. Therefore, the adversary cannot know whether or not his modification is successful according to the authentication result and the GRS-MIM attacks cannot succeed. Therefore, the GRS-MIM attack and its variants will not be considered in the following analysis.

5.3 Security

Experiment $\mathrm{Exp}_A^{\mathrm{Secure}}(\kappa, N, q, t)$
1. Setup a reader $\mathcal{R}$ and a set of tags $\mathcal{T}$, $\lvert\mathcal{T}\rvert = N$
2. $(\mathcal{T}_c, st_0) \leftarrow A^{(O_i)_{1\le i\le 5}}(\mathcal{R}, \mathcal{T})$ //learning stage, q sessions
3. $A(\mathcal{R}, st)$ //guessing phase

Fig. 6. Security Experiment

An RFID authentication protocol is said to be secure if it resists impersonation attacks by any (q,t)-adversary without using relay or corruption attacks. Consider the experiment in Fig. 6. This experiment proceeds in two phases: a learning phase and a guessing phase. In the learning phase, the adversary A is given an RFID system $(\mathcal{R},\mathcal{T})$ as input. During a time interval at most t, A is allowed to launch $(O_i)_{1\le i\le 5}$ oracle queries in every authentication session without exceeding q sessions. At the guessing phase, adversary A only interacts with the reader, and uses the information obtained from the learning phase to impersonate the tag $\mathcal{T}_c$, but can no longer access any oracle. Therefore, the security of an authentication protocol is defined as the successful impersonation probability in the above experiment.

Theorem 2. Let the oracle $\Pi_{K,\eta}$ in the F-HB+ protocol be (q,t,ϵ_Π)-secure. Under the attack of a (q,t)-adversary, the security adversary's advantage of F-HB+ protocol is upper-bounded by:

$$\epsilon_S = P_{FA} + \frac{\epsilon_\Pi}{4l}. \tag{15}$$

Proof. The adversary may use two methods to impersonate a tag: (i) randomly guessing, and (ii) recovering the secret key (Toeplitz matrix). The successful probability of randomly guessing a response is P_{FA} as mentioned before. Let us start to analyse how the adversary can deduce the secret key. There are two ways to obtain useful information about the tag's current key.

The first way is to block the tag's response message, as a result, the tag authentication is unsuccessful, and the current key cannot be updated. So the adversary can obtain valid instances of oracle $\Pi_{K,\eta}$, which can help to reveal the current key. According to Lemma 1 and Theorem 1, the probability of inferring the current key successfully is upper-bounded by $\frac{\epsilon_\Pi}{4l}$.

The second way is to block the reader's acknowledge message, as a result, the tag cannot update its current key. So the adversary can obtain valid instances of oracle $\Pi_{K,\eta}$, which can help to reveal the current key. Once again, the probability of inferring the current key is successfully is upper-bounded by $\frac{\epsilon_\Pi}{4l}$.

It is impossible that the adversary can block the two messages in the same session, because the reader or tag will terminate the session if they do not receive the corresponding message. Therefore, combining the situations above, for a (q, t)-adversary, the security of F-HB can be expressed as $\epsilon_s \leq P_{FA} + \frac{\epsilon_\Pi}{4l}$. This completes the proof.

5.4 Correctness

An authentication protocol exchange involving a legitimate tag and a legitimate reader is said to be undisturbed if all messages sent by both parties are correctly transmitted, received and neither modified nor lost in either direction.

The correctness for RFID authentication protocols implies that the legitimate reader should always accept the legitimate tag for all undisturbed authentications between them. But it is observed that the undisturbed session may happen before or after an attack. Therefore the correctness of an authentication protocol is defined as the acceptable probability of an legitimate tag in an undisturbed authentication session, where the tag may have experienced an impersonation attack.

Theorem 3. Let the oracle $\Pi_{K,\eta}$ in F-HB+ protocol be (q, t, ϵ_Π)-secure. Under the attack of a (q, t)-adversary, the correctness of the F-HB+ protocol is at least:

$$\epsilon_c = (1 - {\epsilon_s}^2)(1 - P_{FR}) + {\epsilon_s}^2 P_{FA}. \quad (16)$$

Proof. According to the flow of the F-HB+ protocol, a reader only rejects a legitimate tag when the tag cannot answer the challenge with a correct response. The reasons are composed of (i) falsely rejecting a tag as mentioned before, and (ii) an adversary successfully impersonating a tag two times in succession such that both the old and current keys are updated, thus, this tag cannot be authenticated again.

In the first situation, the correctness is at most $(1 - P_{FR})$ for a legitimate tag due to the inherent property of Bernoulli random noise, whenever this tag is under a synchronized (look-up table search) or desynchronized (brute-force search) state.

In the second situation, the probability of occurrence is ${\epsilon_s}^2$. Once this situation becomes true, this tag cannot be authenticated like a legitimate tag. But it still could be falsely accepted. So the correctness is ${\epsilon_s}^2 P_{FA}$.

Combining the two rejection situations, the correctness probability can be represented as $\epsilon_c = (1 - {\epsilon_s}^2)(1 - P_{FR}) + {\epsilon_s}^2 P_{FA}$. This concludes the proof.

5.5 Forward privacy

The unpredictable forward privacy experiment $\mathrm{Exp}_A^{\mathrm{UFP}}$ involving a (q, t)-adversary A is illustrated in Fig. 7. During the learning phase, adversary A chooses a random number $r \in_R [0, q]$, and disturbs r protocol sessions between $\mathcal{R}$ and tag set $\mathcal{T}$ with oracle $(O_i)_{1 \leq i \leq 5}$. Then adversary A outputs useful information st_0 and chooses one uncorrupted tag $\mathcal{T}_c$ as its challenge tag. On entering the guessing phase, the experiment chooses a random bit b for adversary A, and b is concealed from A. Then if $b = 1$, A disturbs r' sessions involving $\mathcal{T}_c$ with oracle $(O_i)_{1 \leq i \leq 4}$. These interactions happen during a single (or several) exposure period of each tag such that $r + r' \leq q$. If $b = 0$, A interacts with random strings rather than true protocol messages in r' protocol session exchanges. Then, A is given the internal state, st_3, of $\mathcal{T}_c$ using oracle O_5. After this moment, A is no longer able to access any oracle related to $\mathcal{T}_c$, but A can access any other oracle. Then A outputs useful information st_2. Eventually, A is asked to guess the random bit b by accessing oracle $(O_i)_{1 \leq i \leq 5}$ to the tag set $\mathcal{T}'$.

Experiment $\text{Exp}_A^{\text{UFP}}(\kappa, N, q, t)$

1. Setup a reader $\mathcal{R}$ and a set of tags $\mathcal{T}$, $|\mathcal{T}| = N$
2. A chooses a random $r \in_R [0, q]$
3. $(\mathcal{T}_c, st_0) \leftarrow A^{(O_i)_{1\leq i\leq 5}}(\mathcal{R}, \mathcal{T})$ //learning stage, r sessions
4. Set $\mathcal{T}' = \mathcal{T} - \{\mathcal{T}_c\}$
5. $b \in_R \{0,1\}$ //guessing stage
6. A chooses a random r' such that $r + r' \leq q$
7. If $b = 1$, then $st_1 \leftarrow A^{(O_i)_{1\leq i\leq 4}}(\mathcal{R}, \mathcal{T}_c)$; otherwise A interacts with random strings and outputs st_1 //r' sessions
8. $st_2 \leftarrow A^{O_5}(\mathcal{T}_c)$
9. $b' \leftarrow A(\mathcal{R}, \mathcal{T}', st_0, st_1, st_2)$
10. If $b' = b$ output 1, otherwise output 0

Fig. 7. Unpredictable Forward Privacy experiment

Definition 8. The advantage of (q, t)-adversary A in the experiment $\text{Exp}_A^{\text{UFP}}$ is defined as:

$$\text{Adv}_A^{\text{UFP}} = \left|\Pr\left[\text{Exp}_A^{\text{UFP}}(\kappa, N, q, t) = 1\right] - \frac{1}{2}\right| \tag{17}$$

where the probability is taken over the choice of tag set $\mathcal{T}$ and the coin tosses of the adversary A. An authentication protocol is said to be (q, t, ϵ)-forward-private if there exists no (q, t)-adversary able to break its unpredictable forward privacy with advantage $\text{Adv}_A^{\text{UFP}} \geq \epsilon$.

This unpredictable forward privacy experiment extends and improves upon the basis of the unpredictable privacy notion proposed by Ha *et al.* (2008). Firstly, the previous model is designed for the general privacy notion in 3-pass and reader initiated protocols, but our experiment has no such limitation, can include any number of passes and protocols initiated by tags. Secondly, the security model presented here uses a variable to simulate the possible transition point between the learning phase and guessing phase. The previous model does not have this property.

Theorem 4. Let the oracle $\Pi_{K,\eta}$ in the F-HB$^+$ protocol be (q, t, ϵ_Π)-secure, let g be a (t, ϵ_g)-secure PRNG, and let $\{h_u: \{0,1\}^l \to \{0,1\}^m\}_{u \subset U}$ be a strongly universal hash function family. Under the attack of a (q, t)-adversary, the adversary advantage for the unpredictable forward privacy of the F-HB$^+$ protocol can be upper-bounded by

$$\epsilon_{up} = \begin{cases} \epsilon_\Pi + \epsilon_{up\_p}, & \text{successful mutual authentications} \\ \frac{1}{2} + \left(\epsilon_\Pi - \frac{1}{2}\right) P_{FR} + q\epsilon_s + \epsilon_{up\_p}, & \text{otherwise} \end{cases} \tag{18}$$

where $\epsilon_{up\_p} \leq (3q+2)(2q+1)\epsilon_g + 2Th(m+3)\left(\epsilon_g + 2^{-l} + 2^{-m} + q2^{-2m+2}\right)$.

Proof. The protocol is composed of an LPN problem and a PRNG, so the forward privacy should be preserved for the LPN problem and PRNG at the same time.

Let us first analyse the forward privacy of the LPN problem. The forward privacy proof of the LPN problem is discussed under two situations. The first situation is that the latest mutual authentication session of the F-HB$^+$ protocol before the corruption query in the unpredictable forward privacy experiment is successful. The other one is that the latest session is unsuccessful.

Under the first situation, the tag and the reader can successfully authenticate each other and maintain synchronization. The exchanged messages are random strings and a series of $\Pi_{K,\eta}$ instances, thus, this protocol meets the demands of the unpredictable forward privacy experiment: the exchanged messages cannot be distinguished from random strings. The forward privacy adversary's advantage is upper-bounded by ϵ_{Π} according to Theorem 1.
Under the second situation, the analysis is as follows.

a. If the last tag authentication in the forward privacy experiment is successful, but the adversary uses a desynchronization attack on the reader's acknowledge message, then the reader authentication is unsuccessful. The adversary can obtain the secret and valid LPN instances about this secret, thus he can use this information to check the protocol messages in the previous authentication session. Therefore, the adversary can accurately determine if the previous exchanged messages are random strings.
b. If the last tag authentication in the experiment is unsuccessful, the adversary can obtain the secret and invalid LPN instances about this secret. But these failed instances cannot help him to check the authentication results in previous sessions, because in the LPN problem only the valid instances can help. Therefore, the probability of a correct guess is at most $(1/2 + \epsilon_{\Pi})$ according to Theorem 1.
c. If the adversary can use tag impersonation attacks in the experiment, then the adversary can guess right with probability of 1. The total impersonation probability is at most $q\epsilon_s$.

Therefore, the above situations are combined to illustrate that the forward privacy advantage of the LPN problem is at most

$$\epsilon_{up\_l} \le (1 - P_{FR}) + \left(\frac{1}{2} + \epsilon_{\Pi}\right) P_{FR} + q\epsilon_s - \frac{1}{2}$$

$$\le \frac{1}{2} + \left(\epsilon_{\Pi} - \frac{1}{2}\right) P_{FR} + q\epsilon_s. \quad (19)$$

Then, let us discuss the proof of the PRNG. When the authentication is successful, the secret keys of the PRNG cannot be recovered since the key is updated by adding the noise vector. So it is useless to consider the PRNG in this situation. When the authentication is unsuccessful, the secret key of the PRNG is not updated. The possible search length of the PRNG for each session is limited by Th, and in each session the PRNG needs to generate $m + 3$ strings (1 for the strong universal hashing, and $m + 2$ for the LPN based MAC).
In the PFP protocol (Berbain *et al.*, 2009), a secure PRNG is used to update the key chain, and a strong universal hash function is used to generate the authentication response. This is similar to the look-up index generation in the F-HB$^+$ protocol. The forward privacy of the PFP protocol can be expressed as in the following Lemma 2.
Lemma 2 (Berbain *et al.*, 2009). Let g be a (t, ϵ_g)-secure PRNG, let $\{h_u\}_{u \subset U}$ be a strongly universal hash function family, and let $q < \min(2^{m-1}, \omega/2)$ where ω represents the possible search length of the PRNG. The PFP protocol is (q, t_p, ϵ_p)-forward-private with $\epsilon_p = (3q+2)(2q+1)\epsilon_g + 2\omega\left(\epsilon_g + 2^{-l} + 2^{-m} + q2^{-2m+2}\right)$.
Therefore, according to Lemma 2, the forward privacy advantage of the PRNG in the proposed protocol when authentication fails can be expressed as:

$$\epsilon_{up\_p} \le (3q+2)(2q+1)\epsilon_g + 2Th(m+3)\left(\epsilon_g + 2^{-l} + 2^{-m} + q2^{-2m+2}\right), \quad (20)$$

where $q < \min(2^{m-1}, Th(m+3)/2)$.
Overall, the forward privacy advantage of the proposed protocol can be expressed as:

$$\epsilon_{up} \leq \epsilon_{up\_l} + \epsilon_{up\_p}. \tag{21}$$

Remark. Weak forward privacy in the unsuccessful sessions is as a result of (i) the false rejection probability of the HB related protocols and (ii) desynchronization attacks applied to the reader's acknowledge message in the F-HB+ protocol. However, the false rejection probability P_{FR} can be improved using the parameters proposed by Gilbert *et al.* (2008), and this weak forward privacy is only meaningful to two successive unsuccessful sessions. Therefore, this kind of attack is not very practical.

6. Performance evaluation and comparison

6.1 Re-Hash collision analysis

In the proposed protocol, an appropriate look-up hash function for the Re-Hash feature must be chosen. The strong universal hash functions can be used due to their excellent collision resistant characteristics. The Toeplitz-based strongly universal hash function is used to analyze the collision performance of hash-table indices after Re-Hash is implemented. According to the random oracle model, the output of a cryptographic hash function can be seen as a random number with uniform distribution. Therefore the inputs to the Re-Hash function have uniform distribution. The collision performance for an output $y \in \{0,1\}^M$ can be measured as follows: how many inputs $x \in \{0,1\}^M$ (as described before, the number of truly usable pseudonyms in each authentication session is equal to the output range) are mapped to the output y by the Re-Hash hash function. Let S be the random variable representing the input number for the same output, then the expected number of S is analyzed as follows:

$$\mathrm{E}[S] = \sum_x \Pr[h_u(x) = y] = 1. \tag{22}$$

The above analysis indicates that the average length in every slot of the hash-table is only 1. Therefore, this hash-table can be used to achieve constant-time performance. After every successful mutual authentication, there are at least Th hash-table slots updated, but the total number of true usable pseudonyms still is kept unchanged, 2^M. So the above analysis is still valid.

6.2 Storage case study

The first case that will be examined is a static system with a fixed tag number. The parameters used by Alomair *et al.* (2010) are adopted to illustrate the practical storage of the proposed protocol. It is assumed that the total number of tags N is 10^9 and the value of Th is 10^3. The storage cost of the hash-table is composed of address pointers to the 2nd level database. The storage of pointers is analyzed as follows. The number of elements in the 2nd level is 10^9 $(= N)$, so the bit-length of a pointer in the 1st level is no more than 30 bits $(\geq \lceil \log_2 N \rceil)$. Therefore, the total storage cost of the hash-table is no more than 4 TB $(\geq N \times Th \times \lceil \log_2 N \rceil)$.
The second case considered is a dynamic system where the tag number can change. Assume the maximum system tag number N_{MAX} is 10^{12}, and the value of Th is 10^3. Then the collision-free bit-length of pseudonyms L is 100 bits, and the output range of the Re-Hash

hash function L' is 50 bits. If the initial system tag number N is 10^9, the initial hash-table slot number M is 10^{12}. The storage cost can be obtained as follows: (i) the initial table size is upper-bounded to 7 TB ($M \times \lceil \log_2 N_{MAX} \rceil$); (ii) when a new tag is added, 10^3 slots are added into the dynamic hash-table, and the additional storage is about 7 KB ($Th \times \lceil \log_2 N_{MAX} \rceil$); (iii) when the system number N increases to N_{MAX}, the largest table size is no more than 7,000 TB.

6.3 Implementation on the tag

Firstly, the PRNG $g(\cdot)$ can be implemented using any candidate in the eSTREAM project (Cid & Robshaw, 2009). If $g(\cdot)$ is implemented using the Grain-v1, only 1,294 gates are required to achieve an 80-bit security level. Secondly, from equations (1) and (6), it can be seen that if the LPN problem is implemented using Toeplitz universal hashing, a linear feedback shift register (LFSR) is required for T_u, a 1-bit multiplier plus a 1-bit accumulator is needed for the "·" operator, and an XOR operator is also required. Because the $g(\cdot)$ (Grain-v1) needs an LFSR structure, the LPN problem and $g(\cdot)$ can share the LFSR, so T_u can be derived from the state variable of $g(\cdot)$. The two inputs, x and y of the LPN problem can be derived from the output of $g(\cdot)$. Therefore, the main hardware cost of $g(\cdot)$ and the LPN problem equals the hardware cost of $g(\cdot)$ plus a 1-bit "·" operator and an XOR. Thus, the final estimate for the hardware cost of these functions is no more than 2,000 gates to achieve an 80-bit security level.

Secondly, the overall hardware cost of the proposed protocol on a tag is 2,000 gates, in addition to the cost of a counter and non-volatile memory for storing the secret key and current value.

6.4 Performance comparison

In this section the proposed F-HB$^+$ protocol is compared with previous protocols reported in the literature in terms of their forward privacy properties, the tag resource requirements and the database storage cost. The forward privacy properties are compared in Table 1. Although the proposed protocol cannot protect the forward privacy of failed authentication sessions, it can be observed that it not only supports forward privacy under the unpredictable privacy notion, but also provides a security proof under the standard model.

	Le *et al.*, 2007	**Song, 2009**	**Alomair *et al.*, 2010**	**This work**
Forward Privacy	For successful sessions	For successful sessions	For successful sessions	For successful sessions
Forward Privacy Notion	Universal composable notion	Indistinguishabl e notion	Indistinguishable notion	Unpredictable notion
Forward Privacy Proof	Universal composable model	Random oracle model	Random oracle model	Standard model

Table 1. Forward Privacy Comparison Results

The tag hardware cost and desynchronization resistance are compared in Table 2. Although the protocol proposed by Le *et al.* (2007) does not use a counter, it does not provide any desynchronization resistance because the tag only has one index for a secret key. This work requires only 2,000 gates by using a combination of the LPN problem and a PRNG. And among the three counter-related protocols, the proposed protocol consumes a reasonable non-volatile storage and requires simpler operations in the LPN problem.

	Le *et al.*, 2007	Song, 2009	Alomair *et al.*, 2010	This work
Crypto hardware	1 PRF ≈ 3,000 gates	2 h_C > 5,000 gates	1 h_C > 5,000 gates	1 g + 1 LPN ≈ 2,000 gates
Non-volatile storage	1 key + 1 index	1 key + 1 ct_T	2 key + 1 ct_T	1 key + 1 ct_T
Other hardware	None	1 ct_T	1 ct_T	1 ct_T
Desynchronization attack resistance	None	Th	Th	Th

Table 2. Tag Resource Comparison Results

	Le *et al.*, 2007	Song, 2009	Alomair *et al.*, 2010	This work
Time complexity in synchronization / desynchronization	$O(1)$ / $O(N)$	$O(1)$ / $O(N)$	$O(1)$ / None	$O(1)$ / $O(N)$
Hash-table storage with the example in (Alomair *et al.*, 2010)	None	None	26 TB	4 TB
Dynamic scalability	–	–	–	+

Table 3. Database Performance Comparison Results

The database cost is compared in Table 3. According to the case study for a static system described in section 6.2, the proposed protocol requires storage for the hash-table of no more than 4 TB, but the protocol proposed by Alomair *et al.* (2010) needs about 26 TB. The trade-off in achieving a smaller storage cost is that the proposed protocol needs to compute a look-up table hash function in on-line mode to retrieve the data in the hash-table. The data stored in the hash-table is pre-computed in off-line mode or dynamically inserted in on-line mode. But for the same tag, the look-up procedure and insertion procedure are unlikely to happen at the same time. Because the universal hash function is the fastest hash function in software (Black *et al.*, 1999) and linear hashing is the fastest dynamic hash-table technique, this new look-up hash function will not affect the system performance. Additionally, this proposal is the only to support dynamic scalability.

7. Conclusion

In this chapter, the previous authentication protocols for low-cost RFID applications are introduced. In relation to the characteristics of low-cost tags, three important properties are highlighted: (i) hardware cost must be within 200 ~ 3,000 gates, (ii) forward privacy of a tag must be assured, and (iii) scalability of the entire system cannot be compromised.
Therefore, a novel scalable and forward private authentication protocol, F-HB$^+$, is proposed for low-cost RFID tags. The hardware-friendly LPN problem and PRNG are used to reduce

the protocol cost on the tag, which only requires about 2,000 gates plus a hardware counter and some non-volatile memory. A more efficient MAC code is utilized in comparison to the previous F-HB protocol. In the MAC code implementation implementation, a simplified pairwise independent permutation is used to accelerate the MAC code computation, and a PRNG is used to reduce the storage requirement. A new Re-Hash technique is proposed for hash-table based scalable protocols to effectively reduce the storage requirement. In addition, the Re-Hash technique is adapted to a linear-hashing technique, thus, the proposed protocol possesses dynamic scalability. The security proof of the proposed protocol is given under the standard model. It is proven that F-HB$^+$ achieves unpredictable forward privacy for all its transactions before successful mutual authentication sessions.

Finally, a comparison between the proposed protocol and previous protocols is provided. From a hardware perspective, the proposed protocol is among the smallest and it requires the smallest storage cost for its hash-table in addition to supporting dynamic scalability. It also provides unpredictable forward privacy. Overall, the proposed F-HB$^+$ protocol achieves a new and practical balance between hardware cost, scalability and forward privacy.

8. References

Avoine, G. (2005). Adversary Model for Radio Frequency Identification, *Technical Report LASEC-REPORT-2005-001*, EPFL, Lausanne, Switzerland, September 2005.

Avoine, G. ; Coisel, I. ; & Martin, T. (2010). Time Measurement Threatens Privacy-Friendly RFID Authentication Protocols. In *Workshop on RFID Security (RFIDSec)*, June 2010.

Alomair, B. ; Clark, A. ; Cuellar, J. ; & Poovendran, R. (2010). Scalable RFID Systems: a Privacy-Preserving Protocol with Constant-Time Identification. In *IEEE/IFIP International Conference on Dependable Systems and Networks, (DSN'10)*, June 2010.

Black, J. ; Halevi, S. ; Krawczyk, H. ; Krovetz, T. & Rogaway, P. (1999). UMAC: fast and secure message authentication, *Advances in Cryptology – CRYPTO' 99*, LNCS, Volume 1666/1999, 79, DOI: 10.1007/3-540-48405-1_14.

Bringer, J. & Chabanne, H. (2008). Trusted-HB: A Low-Cost Version of HB$^+$ Secure Against Man-in-the-Middle Attacks, *IEEE Transactions on Information Theory* 54(9): 4339-4342 (2008).

Black, P. E. (2009). "linear hashing", in Dictionary of Algorithms and Data Structures [online], Paul E. Black, ed., *U.S. National Institute of Standards and Technology*. 25 July 2006. Available from: http://xw2k.nist.gov/dads/HTML/linearHashing.html.

Berbain, C. ; Billet, O. ; Etrog J. & Gilbert, H. (2009). An Efficient Forward Private RFID Protocol, *ACM Conference on Computer and Communications Security (CCS)*, November 2009.

Billet, O. ; Etrog, J. & Gilbert, H. (2010). Lightweight Privacy Preserving Authentication for RFID Using a Stream Cipher, *International Workshop on Fast Software Encryption (FSE)*, February 2010.

Cid, C. & Robshaw, M. (2009). The eSTREAM Portfolio 2009 Annual Update. July 2009. Available from http://www.ecrypt.eu.org/stream/.

Cao, X & O'Neill, M. (2011). F-HB: An Efficient Forward Private Protocol. *Workshop on Lightweight Security and Privacy: Devices, Protocols and Applications (Lightsec2011)*, March 14-15, 2011, Istanbul, Turkey.

Dimitriou, T. (2005). A Lightweight RFID Protocol to Protect Against Traceability and Cloning attacks. In *International Conference on Security and Privacy in Communication Networks (SecureComm)*, September 2005.

Frumkin, D. & Shamir, A. (2009). Un-Trusted-HB: Security Vulnerabilities of Trusted-HB, *Cryptology ePrint Archive*. Available from : http://eprint.iacr.org/2009/044.

Goldreich, O. (2001). The foundations of Cryptography, Volume I, Basic Tools, *Cambridge University Press*, 2001.

Gilbert, H. ; Robshaw M. J. B. & Seurin, Y. (2008). HB#: Increasing the Security and Efficiency of HB+, *Annual International Conference on the Theory and Applications of Cryptographic Techniques*, EUROCRYPT 2008: 361-378.

Hopper, N. J. ; & Blum, M. (2001). Secure Human Identification Protocols, *International Conference on the Theory and Application of Cryptology and Information Security*, ASIACRYPT 2001: 52-66.

Henrici, A. & Muller, P. (2004). Hash-based enhancement of location privacy for radiofrequency identification devices using varying identifiers. In R. Sandhu, R. Thomas (Eds.), *International Workshop on Pervasive Computing and Communication Security – PerSec 2004*, IEEE Computer Society, Orlando, Florida, USA, 2004, pp. 149 – 153.

Ha, J. ; Moon, S. ; Zhou J. & Ha, J. (2008). A New Formal Proof Model for RFID Location Privacy, *European Symposium on Research in Computer Security conference (ESORICS)*, October 2008.

Juels, A. & Weis, S. A. (2005). Authenticating Pervasive Devices with Human Protocols, *International Cryptology Conference*, CRYPTO 2005: 293-308.

Juels, A. (2006). RFID Security and Privacy: A research Survey, *IEEE Journal on Selected Areas in Communications*, February 2006.

Juels, A. & Weis, S. (2007). Defining Strong Privacy for RFID, *IEEE Pervasive Computing and Communication (PerCom) conference*, March 2007.

Jr, N.J. *et al.* (2010). Lightweight Cryptographic Algorithms (D.SYM.5) revision 1.0, 1 July 2010. Available from : http://www.ecrypt.eu.org/documents.html.

Krawczyk, H. (1994). LFSR-based hashing and authentication, *International Cryptology Conference, Proc. Crypto'94*, LNCS 839, Y. Desmedt, Ed., Springer-Verlag, 1994, pp. 129-139.

Katz, J. & Shin, J. S. (2006). Parallel and Concurrent Security of the HB and HB+ Protocols, *Annual International Conference on the Theory and Applications of Cryptographic Techniques*, EUROCRYPT 2006: 73-87.

Kiltz, E. ; Pietrzak, K. ; Jain, D. A. & Venturi, D. (2011). Efficient Authentication from Hard Learning Problems. In *Eurocrypt 2011*.

Lim, C. H. & Kwon, T. (2006). Strong and Robust RFID Authentication Enabling Perfect Ownership Transfer. In *International Conference on Information and Communications Security*, December 2006.

Le, T. V. ; Burmester, M. & de Medeiros, B. (2007). Universally Composable and Forward-secure RFID Authentication and Authenticated Key Exchange, *ACM Symposium on InformAtion, Computer and Communications Security (ASIACCS)*, March 2007.

Molnar, D. & Wagner, D. (2004). Privacy and Security in Library RFID: Issues, Practices, and Architectures. In *ACM Conference on Computer and Communications Security (CCS)*, October 2004.

Molnar, D. ; Soppera, A. & Wagner, D. (2005). A scalable, delegatable, pseudonym protocol enabling ownership transfer of RFID tags. In *Ecrypt Workshop*, July-August 2005.

Ma, C. ; Li, Y. ; Deng R. & Li, T. (2009). RFID Privacy: Relation Between Two Notions, Minimal Condition, and Efficient Construction, *ACM Conference on Computer and Communications Security (CCS)*, November 2009.

Naor, M. & Reingold, O. (1997). On the Construction of Pseudorandom Permutations: Luby—Rackoff Revisited. In *Journal of Cryptology*, Volume 12, Number 1, 29-66, DOI: 10.1007/PL00003817.

Ohkubo, M. ; Suzuki, K. & Kinoshita, S. (2003). Cryptographic Approach to Privacy-Friendly Tags. *RFID Privacy Workshop*, November 2003.

O'Neill, M. (2008). Low-Cost SHA-1 Hash Function Architecture for RFID Tags. In *RFID Security Workshop 2008 (RFIDSec'08)*, July 2008.

Song, B. (2009). RFID Authentication Protocols using Symmetric Cryptography. In *PhD thesis*, December 2009. Available from: http://www.avoine.net/rfid/.

Tsudik, G. (2006). YA-TRAP: Yet Another Trivial RFID Authentication Protocol. In *IEEE Pervasive Computing and Communication (PerCom) conference*, March 2006.

Vaudenay, S. (2007). On Privacy Models for RFID, *International Conference on the Theory and Application of Cryptology and Information* Security (ASIACRYPT), December 2007.

Wegman, M.N. & Carter, J.L. (1981). New hash functions and their use in authentication and set equality. In *Journal of Computer and System Sciences*, Vol. 22, No. 3, 1981, pp. 265-279.

Weis, S. ; Sarma, S. ; Rivest, R. & Engels, D. (2003). Security and privacy aspects of low-cost radio frequency identification systems. In *International Conference on Security in Pervasive Computing*, March 2003.

14

RFID Model for Simulating Framed Slotted ALOHA Based Anti-Collision Protocol for Muti-Tag Identification

Zornitza Prodanoff[1] and Seungnam Kang[2]
[1]*University of North Florida*
[2]*National Seoul University*
[1]*USA*
[2]*South Korea*

1. Introduction

Radio Frequency Identification (RFID) networks use radio signal broadcast to automatically identify items with attached RFID tags. A tag consists of a microchip that stores a unique identifier and an antenna. The tag's antenna is attached to the chip and can transmit a unique tag identifier to a reader (also called interrogator). The reader is capable of learning the set of tags within its interrogation range. The process of learning in-range tags is called a census. After an initial census is completed, the reader can answer queries about the presence of specific tag(s) within its range sent to it from other type of devices.

RFID systems have abundant benefits as compared to the barcode and smart card systems. RFID networks use radio frequency as a method of data transmission. Thus, unlike barcode labels, a tag does not need to be placed in a line of sight position from the reader, or even get in contact with a reader as smart cards, in order to be identified successfully. Depending on whether they use low, high, or ultrahigh transmission frequencies, RFID tags are identifiable within 3 meters span in case of a typical far-field reader [Want06] or at even further distances. Therefore, RFID tags are used more flexibly and conveniently than existing barcode and smart card implementations.

Moreover, some commercial implementations of RFID tags can store data in the amount of 16bytes - 64Kbytes [Finkenzeller03]. RFID tags can hold the same amount of data compared to smart cards, and much larger volume than barcodes. In addition, RFID tags are getting less expensive. The cost of RFID chips at the time of this study is less than 10 cents, while back in 1999, for example, was around 2 US dollars. Since tag readers have limits on their operations range imposed by the frequency of the wireless signal used, when RFID networks need to cover large spaces, multiple readers need to be used. The cost of current reader implementations is hundreds of US dollars. As a result, RFID networks may not be yet suitable to track large inventories of inexpensive items, but they are certainly becoming more affordable and can be used to track different types of items, e.g. live stock, pets, and valuable goods. Due to these advantages RFID systems are emerging as one of the alternative technologies of our time.

One of the world biggest supply chains Wal-Mart has required suppliers to implement RFID networks in at least 12 of its 137 distribution centres by the end of 2006. The Proctor & Gamble Co. is the first of about 100 suppliers to conform to Wal-Mart's requirements to tag its products with RFID chips [Computerworld07]. The US Navy finished its pilot of a passive RFID system to support the loading of supplies into cargo containers in May 2004. According to the related final report the RFID process increased the speed and efficiency of the cargo checking process, while less people were needed to support the new RFID based system as compared to the legacy implementation [Weinstein05].

1.1 Physical composition

An RFID system is made up of an application, a reader and tags.

- The application is a program installed on a (proxy) computer which can control readers.
- The reader is a device which runs functions such as reading, writing and authentication. When the reader gathers data from tags it transmits to the computer application.
- The tag is used to identify an object and is located on (or in) the object itself.

A reader is connected to the computer and has a transmitter and receiver, while a tag has a control unit (chip) and a coupling element (antenna).

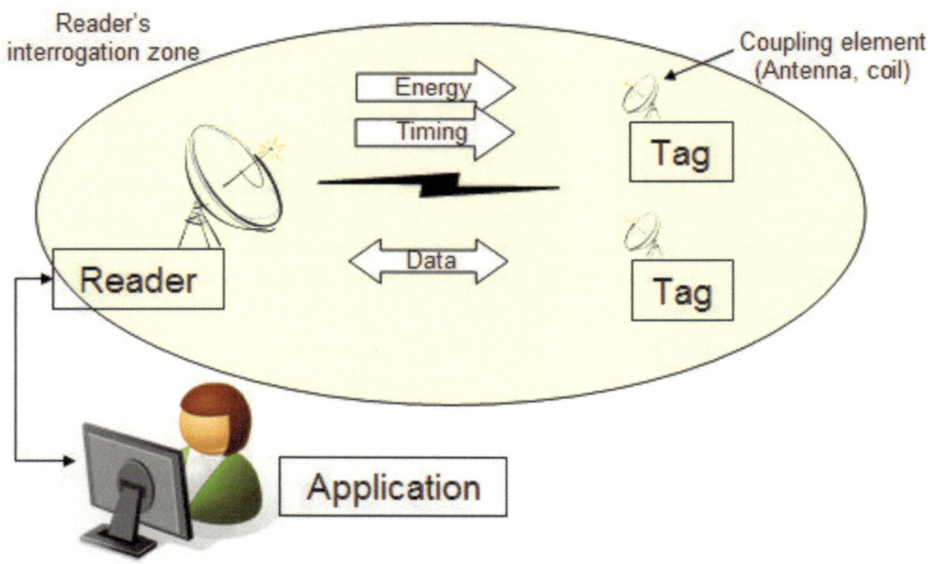

Fig. 1. RFID Physical Composition [Finkenzeller03]

RFID tags can be *passive*, i.e. not having an internal energy source or *active*, internal battery powered. A reader typically charges a set of passive tags within its interrogation zone using *inductive coupling*; the reader broadcasts electromagnetic signal then the tag's antenna absorbs and stores the signal's energy into an on-board capacitor. This technique is called *load modulation* for near-field coupling and *back scattering* for far-field coupling. After charging its battery it can be activated.

1.2 Framed slotted ALOHA anti-collision algorithm

The ALOHA algorithm is a collision resolution algorithm based on Time Division Multiple Access (TDMA). There are three flavors of the original ALOHA algorithm: (Pure) ALOHA, Slotted ALOHA and Frame Slotted ALOHA [Zürich04].

In Figure 2, X and Y axis represents the read cycle and tags respectively. The read cycle is the time interval between neighboring two *REQUEST* commands and it can be repeated until all tags in the interrogation range are identified. Note that there no slots are used in the (Pure) ALOHA algorithm (Figure 2: (a)) while the read cycle is divided into several continuous slots in the Slotted ALOHA (Figure 2: (b)) and Framed Slotted ALOHA algorithm. Furthermore, a frame is comprised of the number of slots in the Framed Slotted ALOHA algorithm (Figure 3: A *slot* is a discrete time intervals synchronized by the reader, sufficiently long in duration to allow a tag to transmit its ID and the ID's 16-bit CRC code. A set of slots are grouped into *frames*. When size is fixed, each consecutively transmitted frame has the same number of slots.

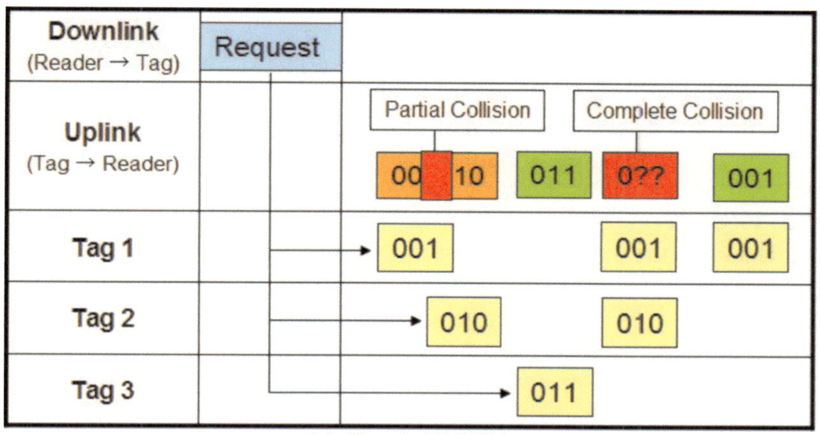

(a) (Pure) ALOHA

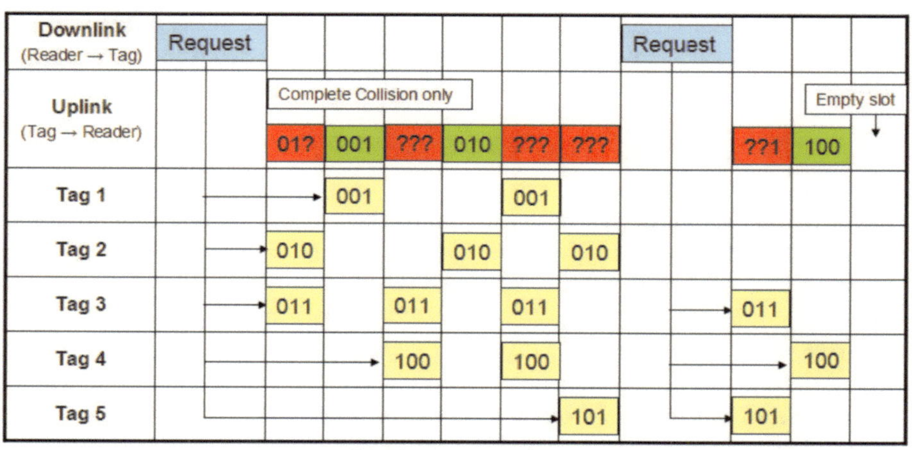

(b) Slotted ALOHA

Fig. 2. Pure and Slotted ALOHA Algorithms

The reader broadcasts the *REQUEST* command to the tags located in the reader's interrogation range during the downlink while the tags transmit their data to the reader during the uplink. As all activated tags share the uplink partial or complete collision can occur in the (Pure) ALOHA algorithm. However, if the data is transmitted using the slot of frame the partial collision can be eliminated. Furthermore, to reduce the fraction of collision occurrence tags send their data no more than once within a frame, which is the Frame Slotted ALOHA algorithm. We next present in more detail the operation of the three ALOHA algorithms introduced above.

1.2.1 (Pure) ALOHA

A tag itself decides the data transmission time randomly as soon as it is activated. The transmission time is not synchronized with both the reader and the other tags at all. When the electricity is charged by the reader's electromagnetic wave tags transmit data after receiving the *REQUEST* command from the reader. If multiple tags transmit data imminently (whether earlier or later) then a complete or partial collision occurs (Fig. 2 (a)). Retransmitting after random delay is the solution for a collision. During the read cycle the reader receives the data and identifies tags sent data without collision. When a read cycle is done then the reader broadcasts the *SELECT* command with the tag's unique identifier received from the tag. Once tags are selected the tags stop responding for the request command i.e. the selected tags keep silence until whether they receive other commands e.g. authenticate, read and write or the tag's power is off by being located out of the reader's power range. When the tag is reentered into the reader's interrogation range it restart transmitting its data to the reader. The advantage of this algorithm is simplicity.

1.2.2 Slotted ALOHA

It is obtained by the addition of a constraint to the (Pure) ALOHA. The read cycle is divided into discrete time intervals called *slot* and which is synchronized with the entire tags by the reader. Thus, tags must choose one of the slots randomly and transmit data within a single slot. Transmission begins right after a slot delimiter (Fig. 2 (b)). This causes that packets either collide completely or don't collide at all i.e. there is no partial collision in the Slotted ALOHA algorithm. This reduces wasting the read cycle relatively as compared with the (Pure) ALOHA algorithm. However, the empty slot can be occurred in the read cycle and the disadvantage is that it requires a synchronization mechanism in order for the slot-begin to occur simultaneously at all tags.

1.2.3 Framed slotted ALOHA

Framed Slotted ALOHA algorithm uses the frame which is the discrete time interval of the read cycle and each frame is divided into the same number of slots. There are multiple frames in a single read cycle and the frame size is decided by the reader (Figure 3: There is a constraint that the tags can transmit data only once in each frame. It may reduce the number of collided slots and it shows the best performance among them.

1.3 Classification of the framed slotted ALOHA protocol

FSA (Framed Slotted Aloha) can be classified into the BFSA (Basic Framed Slotted Aloha) and the DFSA (Dynamic Framed Slotted Aloha) according to whether which uses fixed frame size or variable frame size [Klair04]. If the number of actual tags is unknown DFSA can identify tags efficiently rather than BFSA by changing frame size since BFSA uses fixed frame size. In addition, BFSA and DFSA can be further classified based on whether they support muting or/and early-end features [Klair04]. The muting makes tags remain silent after being identified by the reader while the early-end allows a reader close an idle slot early when no response is detected. Figure 4 is shown for the classification of the FSA.

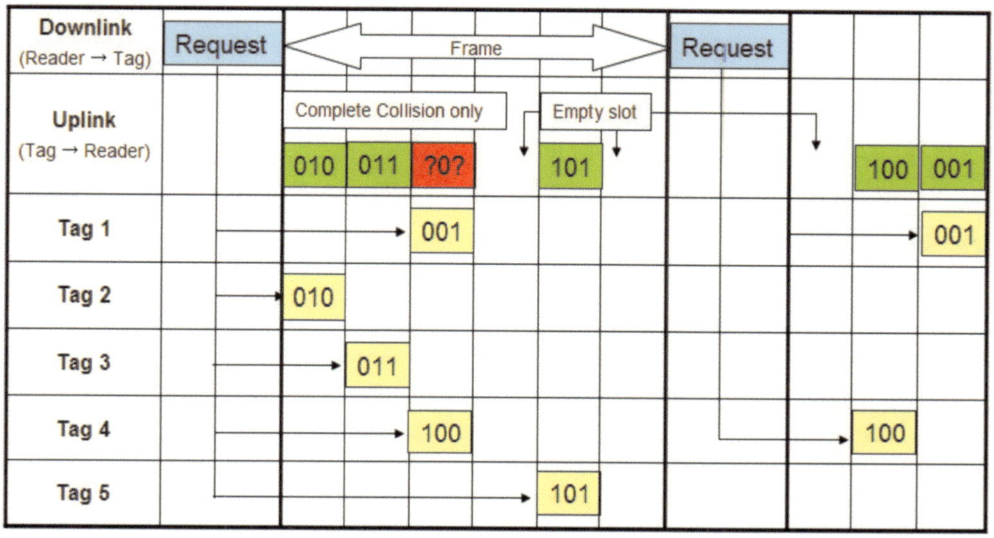

Fig. 3. Framed Slotted ALOHA Algorithms

1) BFSA
- BFSA-Non Muting
- BFSA-Muting
- BFSA-Non muting-early-end
- BFSA-Muting-early-end

2) DFSA
- DFSA-Non Muting
- DFSA-Muting
- DFSA-Non muting-early-end
- DFSA-Muting-early-end

Fig. 4. Classification of FSA

2. RFID network protocol simulation using OPNET

Framed Slotted ALOHA and Binary Tree are the two most widely used multi tags identifying anti-collision protocols. Fabio Cappelletti et al. simulated the Binary Tree protocol of RFID by using the OPNET IT Guru 11.0 in 2005 [Cappelletti06]. In the paper, they measured the network throughput and the census delay through the simulation. And they compared simulation performance and analytical results. What they measure is shown in Figure 5.

Analytical Parameters	Simulation parameters	Unit
Network throughput	Network throughput	(%)
Throughput per node	Throughput per node	(%)
Total census delay Lower bound Upper bound Arithmetic average Heuristic	Total census delay	Number of slots
Time required to detect a single tag Lower bound Upper bound Arithmetic average Heuristic	Time required to detect a single tag	Number of slots
Number of transmitted packets (Total, Average)		Number of packets

Fig. 5. The Measured Parameters for Analysis and Simulation

The network throughput represents the ratio between the number of successfully transmitted packets and the total number of packets sent by the tags while the throughput per node denotes the average number of packets sent by a single tag. The results of the paper showed that the analytical performance was in good agreement with the simulation results.

2.1 Framed slotted ALOHA protocol performance evaluation using Philips I-Code system

2.1.1 The I-Code RFID system

It is commercial product of RFID system which is comprised of the actual tags and a reader connected to the computer. The application is installed in the computer to control the reader collecting data from tags. The built-in Framed Slotted ALOHA protocol is provided by the system. The memory size of an I-Code tag is the total of 64 bytes which are available for 46 bytes application data, 8 bytes serial number and 10 bytes functionalities such as write protection, maintaining quiet state of tag and reset quite state, etc. And, the reader provides the interface for setting configuration parameters such as the serial connection speed and commands for handling communication with tags. Examples of commands are following:

- *Anti-collision/select (ACS)*: After broadcasting this command, tags begins transmitting their serial numbers. Once tags become "selected" status then remain quite in following ACS commands.
- *Read*: This command makes the selected tags transmit their data to the reader.
- *Write*: This command makes the selected tags write data transmitted by the reader.

2.2.2 The procedure of tag identification

The procedure of tag identification is different based on the classification of RFID [Finkenzeller03]. There are two types of procedure based on the tags characteristic in the Frame Slotted ALOHA protocol. One of them switches off when read while the other not switches off but replays transmission. The I-Code system uses the switching off tag. Figure 6 depicts the census procedure used in the I-Code System implementation.

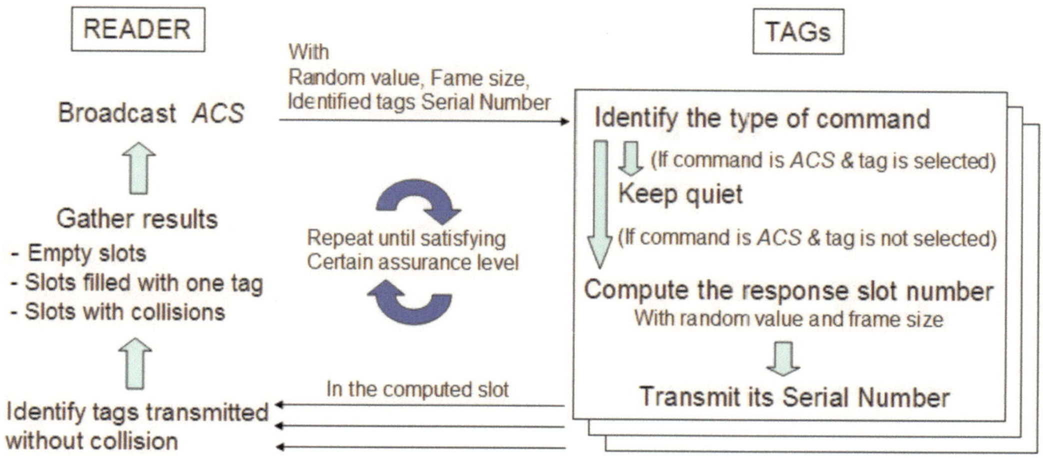

Fig. 6. Tag Identification Procedure of the I-Code System

The reader broadcasts the *ACS* command with a random value, frame size to make tags transmit their serial number to the reader. When the tag receives a command from the reader first of all it identifies the type of command, and if it is *ACS* and the status of tag is not 'selected' then it computes the response slot number using the random value and frame size ($0 \leq$ response slot $<$ frame size) as parameters. Random value is used to prevent the same collisions from occurring repeatedly. The serial number in the computed slot is transmitted to the reader by the tag. During an uplink, data transmission from tags to reader, the reader can identify tags transmitted without collision. The results of a read cycle (the number of empty slots, the number of slots filled with one tag and the number of collided slots) are used for analysis. Once the tag is identified the tag remains quiet until other command e.g. read or write is broadcasted with serial number. And if the tag re-enters the reader's power zone after moving out, that needs re-identification. The read cycle is repeated till reaching the assurance level, the probability of identifying all tags in the reader's range.

2.2.3 Basic framed slotted ALOHA

Through simulation we can measure the total census delay by varying the frame size for given number of tags. Then we can find the optimal frame size which results in the minimum total census delay, for given number of tags (see Figure 7).

The optimal frame size, resulting in minimum census delay, can be determined according to the total number of tags. Figure 8 presents an example of the relationship between total number of tags and frame size. For example, the optimal frame size for 80 active tags is 45, while for 30 passive tags it is 40.

2.2.4 Dynamic slot allocation for dynamic framed slotted ALOHA

To maximize network throughput frame size (the number of allocated slots in the read cycle) should be chosen in accordance with the number of tags since for the same fixed slot size, number of total collisions during a census increases with increase in total number of tags.

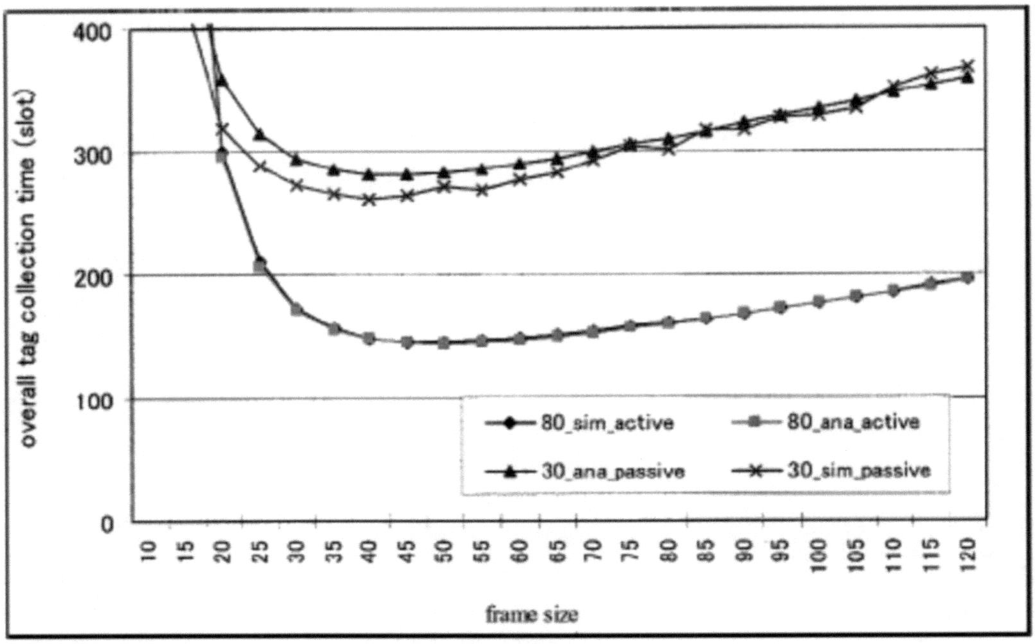

Fig. 7. Example of Total Census Delay (Tag Collection Time) Using Static Frame Size [Bin05]

Due to the nature of the Framed Slotted ALOHA protocol, the read cycle time is divided by the number of slots in a frame and packet data (tag ID number plus CRC) to be transmitted should occupy a single slot. If there are lots of tags and frame size is small then the probability of collisions will be increased and the number of identified tags is decreased, because tags will be competing for a lesser number of slots within a frame. On the contrary, if the reader reads few tags with too big frame size then the probability of collision is decreased , but at the expense of the response time being increased. There is optimal frame size that makes minimum number of read cycles for certain number of tags. The dynamic slot allocation is choosing the optimal frame size (in number of fixed slots) during the tag read cycle. However, the problem is that the number of tags is unknown. So, the reader should estimate the number of tags in each read cycle with the result of the previous read cycle (number of empty slots, number of slots filled with one tag, number of collided slots) and current frame size.

2.3 Developed approaches of the framed slotted ALOHA protocol

The key difference of the developed Framed Slotted ALOHA protocol is how they estimate the number of tags and what they estimate.

2.3.1 H. Vogt's algorithm

In this scheme, they estimate the number of tags using the current frame size and the result of read cycle. And then updates the current frame size using the estimated number of tags and previous frame size. The procedure of this algorithm is shown in Figure 9. Variable 'N', 'N0', 'n_est' and 'stepN' represent the frame size, temporary frame size, the estimated number of tags and the counter for the cycle performed with currently estimated framed size, respectively. Variable 'c', 't' represents the result of a read cycle comprised of three integers; number of empty slots, number of slots filled with one tag and number of collided slots, and variable 't' represents the temporary number of estimated number of tags.

In order to estimate the number of tags two estimation functions are used; lower bound and Chebyshev's inequality. Lower bound simply estimates the number of tags is bigger than the summation of the number of slots filled with one tag and two-times of the number of collided slots:

Number of tags $\geq$ Number of filled slots with one tag $+ 2 \times$ Number of collided slots (1)

When the lower bound is used the real value of number of tags is underestimated.

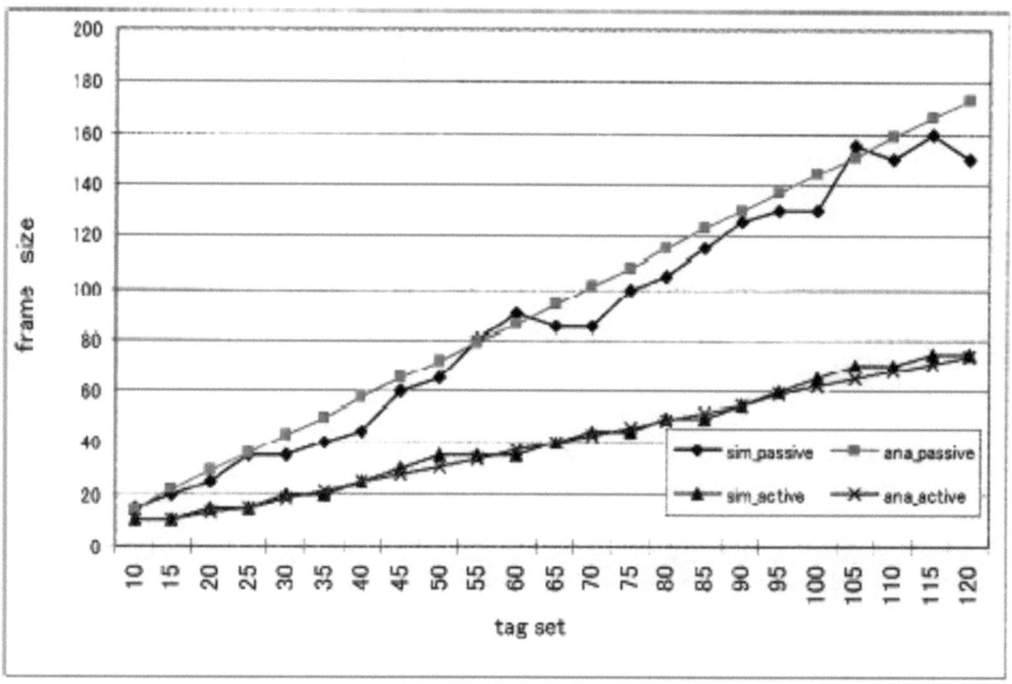

Fig. 8. Example of Optimal Frame Size For Minimum Census Delay [Bin05]

Chebyshev's inequality measures the difference the real values and expected values to estimate the number of tags for which the difference becomes minimal [Vogt02]. The number of tags are estimated using the currently used frame size (N) and the results of previous read cycle $<c_0, c_1, c_k>$ representing the number of empty slots, the number of slots filled with one tag and the number of collided slots respectively. And $< a_0^{N,n}, a_1^{N,n}, a_{\geq 2}^{N,n} >$ denotes the expected number of empty slots, the expected number of slots filled with one tag and the expected number of collided slots respectively where N and n represent the frame size and the number of tag respectively. Lower bound is more accurate for low values of n while Chebyshev's inequality is steadier for a wider range of n [Vogt02].

$$\varepsilon_{vd}\left(N, c_0, c_1, c_k\right) = \min \left\| \begin{pmatrix} a_0^{N,n} \\ a_1^{N,n} \\ a_{\geq 2}^{N,n} \end{pmatrix} - \begin{pmatrix} c_0 \\ c_1 \\ c_k \end{pmatrix} \right\| \tag{2}$$

```
* Function: VogtAlgorithm ()
- Variable: integer : N, N0, n_est, stepN, c, t
- N = minimum frame size
- repeat begin while stepN is smaller than maxStep:
    stepN++;
    Perform a read cycle with N and store the result in c
    Estimate number of tags with N and c, and then store result in t
    If t is bigger than n_est
    then save t in n_est.
        Call adaptFrameSize(N, n_est) and save value in N0.
        If N0 is bigger than N
        then reset stepN with 0. And, save N0 in N.
- repeat end
* Function: adaptFrameSize(N, n_est)
- Return type : integer
- repeat begin while n_est is smaller than low value for choice of N:
    Store N/2 in N.
- repeat end
- repeat begin while n_est is bigger than high value for choice of N:
    Store 2*N in N.
- repeat end
```

Fig. 9. Pseudo Code of H. Vogt's Algorithm

Figure 10 presents the optimal frame size resulting from the execution of function adaptiveFrameSize (*N*, *n*_est) using as an input the estimated number of tags and the value of the current frame size *N*. The optimal frame size is selected for a range of tags, based on a low and high margin.

N slots	1	4	8	16	32	64	128	256
low	–	–	–	1	10	17	51	112
high	–	–	–	9	27	56	129	∞

Fig. 10. Choosing Optimal Frame Size [Vogt02]

2.3.2 Bin ZHEN et al.'s algorithm

This algorithm estimates the number of tags using the *posteriori* probability. The *posteriori* probability of *k* tags for an observed slot is as below:

$$p_k^0(i) = \begin{cases} 0 & \text{if } k = 0, 1 \\ \dfrac{p_k(i)}{1 - p_0(i) - p_1(i)} & \text{if } k \geq 2 \end{cases} \tag{3}$$

The a *posteriori* expected value of the number of tags is respectively, 0 for an empty slot, 1 for a slot filled with one tag, and $\sum_{k=2}^{N} kp_k^0(i)$ tags for a collided slot. Thus, the estimated tag sets from the current read cycle is $p_1(i) + \sum_{k=2}^{N} kp_k^0(i)$. Then, the estimated number of tags comes from the result of the *i*th read cycle is,

$$n_{est}(i+1) = s + Kg \tag{4}$$

Where K = 2.39 is a constant for collided slots [Bin05]. And, the frame size (N) in the (i+1)th read cycle is

$$\begin{aligned} N(i+1) &= H * n_{est}(i+1) \\ &\begin{cases} H = 1 - 1.4 \text{ Passive tag} \\ H = 0.8 - 1 \text{ Active tag} \end{cases} \end{aligned} \tag{5}$$

where H is a constant, which maps the tags to the frame size. The update of the frame size occurs when $n_{est}(i+1) \geq \gamma * n_{est}(i)$. Here, γ is constant value which is 1.15 and it denotes a threshold to handle random jitter of number of estimated tags. The procedure of this algorithm is shown in Figure 11.

```
* Function: BinZHENAlgorithm ()
-Variable: integer : N, i, c, s, nest(i)
 N = minimum frame size
 i = 0
- repeat begin while i is smaller than the number of read cycle for confidence level:
     i ++;
     Perform a read cycle with N and store the result in c for collided slots, s for success
     slots.
     Estimate number of tags with N, c and s, and then store result in nest(i).
     If nest(i) is bigger than γ * nest(i - 1)
     then Call calculateFrameSize(nest(i), tagType) and save value in N.
          i = 0.
- repeat end
* Function: calculateFrameSize(nest(i), tagType)
- Return type : integer
- If tagType is passive
  then N(i) = (random value of 1 – 1.4) * nest(i)
  else N(i) = (random value of 0.8 – 1) * nest(i)
- return N(i)
```

Fig. 11. Pseudo Code of Bin ZHEN et al.'s Algorithm

2.3.3 EDFSA (Enhanced Dynamic Framed Slotted ALOHA)

This algorithm estimates the number of unread tags instead of number of tags to determine the frame size. H. Vogt's algorithm shows poor performance when the number of tags becomes large because the variance of the tag number estimation is increased according to the number of tags increase [Rom90]. Therefore, to handle the poor performance of large number of tag identification EDFSA algorithm restricts the number of responding tags as much as the frame size. Conversely, if the number of tags is too small as compared with the frame size it reduces the frame size. To estimate the number of unread tags equation (2) is used. The procedure of EDFSA algorithm's read cycle is shown in Figure 12.

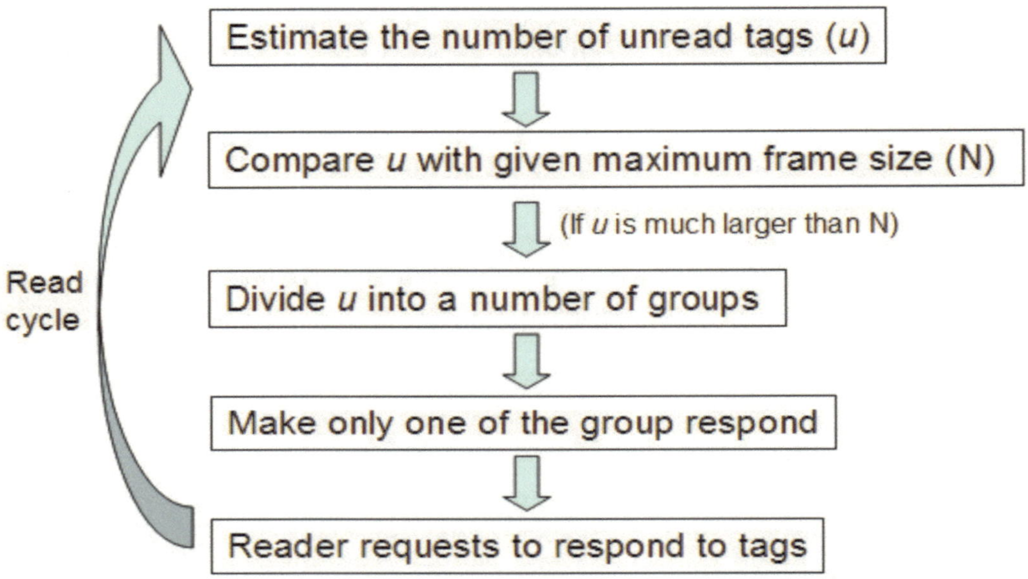

Fig. 12. Read Cycle of EDFSA Algorithm

3.1 Evaluating delays

To evaluate the implementation of the BFSA protocol I first evaluated *the total census delay* of the tag reading process. It is comprised of three different delays; *success delay, collision delay* and *idle delay*. Thus, the total census delay is defined as

$$T[n] = n + C[n] + I[n] \tag{6}$$

where n is success delay, $C[n]$ is collision delay and $I[n]$ is idle delay [Cappelletti06]. The unit of delay can be defined as a slot duration T *(sec)* and it is defined as,

$$T = \frac{\text{ID (bits)}}{\text{data\_rate (bps)}} \tag{7}$$

where *ID (bits)* is the size of the packet containing tag's ID, and *data_rate (bps)* is the data rate from tag to reader.

3.1.1 BFSA-non-muting

It is necessary that evaluating of the read cycles satisfying the confidence level α since it is used to determine total census delay. The assurance level α is the probability of identifying all tags in the reader's interrogation range [Vogt02] e.g. if a = 0.99 which means one or more missing tags, less than 1% of all, are allowed. The probability of r tags responding in a slot in the ith read cycle is given by [Bin05]

$$p_r(i) = \binom{n}{r}\left(\frac{1}{N}\right)^r\left(1-\frac{1}{N}\right)^{n-r} \quad (8)$$

where N is the given frame size (slots) and n is the number of tags to be read in the ith read cycle. From the equation 8, the probability of having one or more idle ($p_o(i)$), successful ($p_1(i)$), and collide ($p_k(i)$) slots in the ith read cycle are defined as:

$$p_0(i) = \left(1-\frac{1}{N}\right)^n \quad (9)$$

$$p_1(i) = \frac{n}{N}\left(1-\frac{1}{N}\right)^{n-1} \quad (10)$$

$$p_k(i) = 1 - p_1(i) - p_0(i) \quad (11)$$

Then the expected number of the successful transmissions in the ith read cycle becomes $Np_1(i)$ since a read cycle has N slots [Bin05]. The probability of having an unread tag after R read cycle is given by [Bin05] [Klair04]

$$p_{miss}(i) = \prod_{i=1}^{R}\left(1-\frac{Np_1(i)}{n}\right) = 1-\alpha \quad (12)$$

R represents the number of required read cycles to identify a set of tags with a confidence level α. As the number of tags n and the frame size N are the same for all read cycle, $p_1(i)$ is constant. That makes equation 13 as,

$$p_{miss}(i) = \left(1-\frac{Np_1}{n}\right)^R = 1-\alpha \quad (13)$$

If we solve the equation 13 for R we can obtain the condition of R as below: [Klair04]

$$R \geq \left\lceil \frac{\log(1-\alpha)}{\log\left(1-\frac{Np_1}{n}\right)} \right\rceil = \left\lceil \frac{\log(1-\alpha)}{\log\left(1-\frac{n\left(1-\frac{1}{N}\right)^{n-1}}{n}\right)} \right\rceil = \left\lceil \frac{\log(1-\alpha)}{\log\left(1-\left(\frac{N-1}{N}\right)^{n-1}\right)} \right\rceil \quad (14)$$

The ceil function is used since R is the integral value. By using R and if the number of tags is known, we can evaluate the theoretical delay of successful (n), idle ($I[n]$), and collision ($C[n]$) transmission as follows [Klair04]:

$$n = Np_1RT \tag{15}$$

$$I(n) = Np_0RT \tag{16}$$

$$C(n) = NRT(1 - p_0 - p_1) \tag{17}$$

where N is a frame size, T is slot duration. The summation of those three delays represents the total census delay.

3.1.2 BFSA muting

Muting decreases the number of tag's responses after every read cycle. Hence, the number of responding tags in the $(i+1)th$ read cycle is less than or equal to those in the ith read cycle. The number of responding tags in the $(i+1)th$ read cycle is evaluated as [Bin05],

$$n(i+1) = n(i) - p_1(i) \times N \tag{18}$$

where $p_1(i) \times N(i)$ represents the number of tags muted in the ith read cycle. And we can calculate the R with the given n and N by using the equation 14. Then the collection rounds to read all tags R is given by solving the following equation [Bin05]

$$p_{miss}(i) = \prod_{i=1}^{R}\left(1 - \frac{Np_1(i)}{n(i)}\right) = 1 - \alpha \tag{19}$$

By using R_{min}, if the number of tags is known, we can evaluate the theoretical minimum delay of successful (n), idle ($I[n]$), and collision ($C[n]$) transmission by using the equation 15, 16, and 17. And, their summation yields the minimum total census delay.

3.2 Evaluating network throughput

Network throughput can be defined as the ratio between the number of successfully transmitted packets (one per tag) and the total number of packets sent by the tags during the census [Cappelletti06]. Suppose that there are n tags to be read. Then, the total number of packets sent by n tags during a census for non-muting BFSA is

$$P[n] = nR \tag{20}$$

where R is the number of required read cycles needed to identify a set of tags with a confidence level a. Since tags can transmit only once in a read cycle. Now we can calculate the network throughput as

$$S[n] = \frac{\alpha n}{P[n]} = \frac{\alpha}{R} \tag{21}$$

where a is assurance level, n is total number of identified tags, and $P[n]$ is the total number of packets sent by the tags during the census.

4.1 Validation of the models

In this project, I implemented two Aloha models; BFSA-Muting and BFSA-Non-Muting. To validate the model I analyzed the log file [appendix A, B] of the models and compared with the pseudo code. For easy comparison I put the figures describing the events of the simulation comes from the log file.

4.1.1 Simulation information

For the simplicity I put a reader and eight tags, and the same given conditions are used between two simulations. The reader and tags being used in the simulation are shown in Figure 13 (a) while the given conditions are shown in Figure 13 (b).

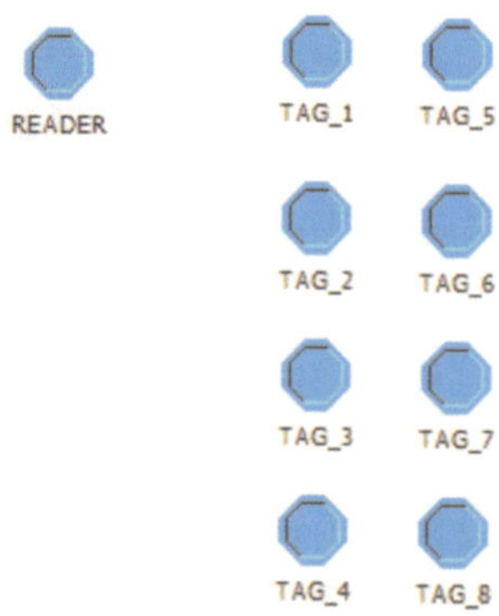

(a) A Reader and Tags

List	Value
Assurance level	0.99
Data rate (reader ↔ tag)	500,000 (bps)
Number of slots in a read cycle	8 (ea)
Duration of a read cycle	0.00128 (sec)
Duration of a single slot	0.00016 (sec)
REQUEST packet size (reader → tag)	88 (bits)
SELECT packet size (reader → tag)	72 (bits)
RESPONSE packet size (tag → reader)	80 (bits)

(b) Given Simulation Conditions

Fig. 13. Simulation Information

The time required for the packet transmission can be calculated by using the given packet size and the data rate among reader and tags. They are shown in Figure 14.

List	Value
REQUEST packet	0.000176 (sec)
SELECT packet	0.000144 (sec)
RESPONSE packet	0.00016 (sec)

Fig. 14. Packet Transmission Time

I assume that the propagation delay is negligible since in case of a typical far-field reader has 3 meters span interrogation range [Want06]. Consider the speed of light is 299,792,458 m/s then the delay of 3 meters will be 1^{-8} seconds. And I also assume the calculation delay of the reader and of the tag is negligible as simplicity is the strong point i.e. it does not need complex calculation both for the reader and for the tag.

4.1.2 BFSA-muting

For the validation of the simulation model we compared the analytical results (obtained based on an algorithm presented in [Klair04] (see Figure 15)) with our simulation results.

When the reader starts a census procedure the number of unread tags is initialized to the number of actual tags in range. While the census is performed to identify unread tags the number of identified tags, collided slots, idle slots, and the current frame size are stored as a running total. If there is no collision from tags the total delay, collision delay, and idle delay are calculated. T represents the duration of a single slot.
The log from the BFSA-Muting simulation is shown in Appendix A. Figure 16 depicts the sequence of events during the BFSA-Muting simulation. Through analyzing the log we can check the correctness of the implementation.

```
BEGIN;
Initialize unread tags = actual number of tags;
while True do
        Perform a read cycle for unread tags;
        Store the number identified tags;
        Store the number slots filled with collisions;
        Store the number of slots filled with idle responses;
        Store current frame size;
        if (No Collisions) then
            Break;
        else
            Unread Tags = actual - identified tags;
        end
end
Total delay = T × ∑ stored frames;
Collision Delay = T × ∑ stored collision slots;
Idle Delay = T × ∑ stored idle slots;
END;
```

Fig. 15. Pseudo Code of the BFSA Muting

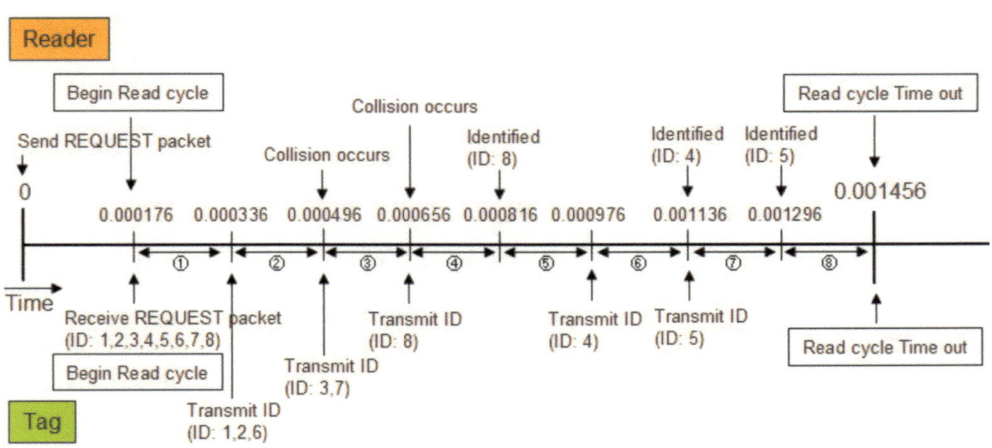

(a) First Read Cycle of the Simulation

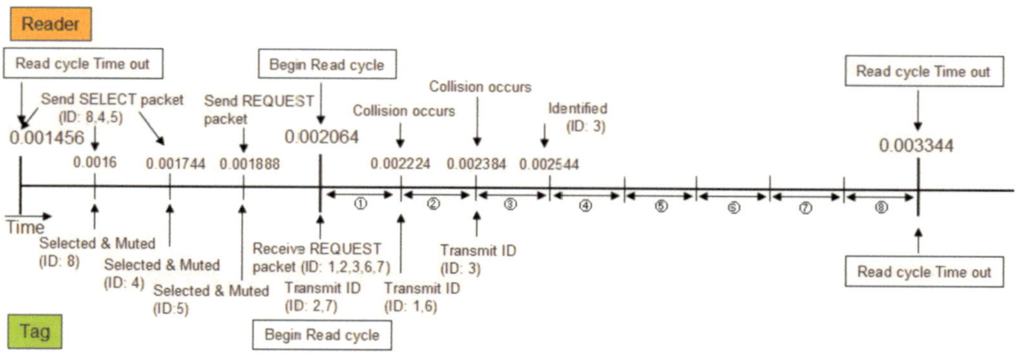

(b) Second Read Cycle of the Simulation

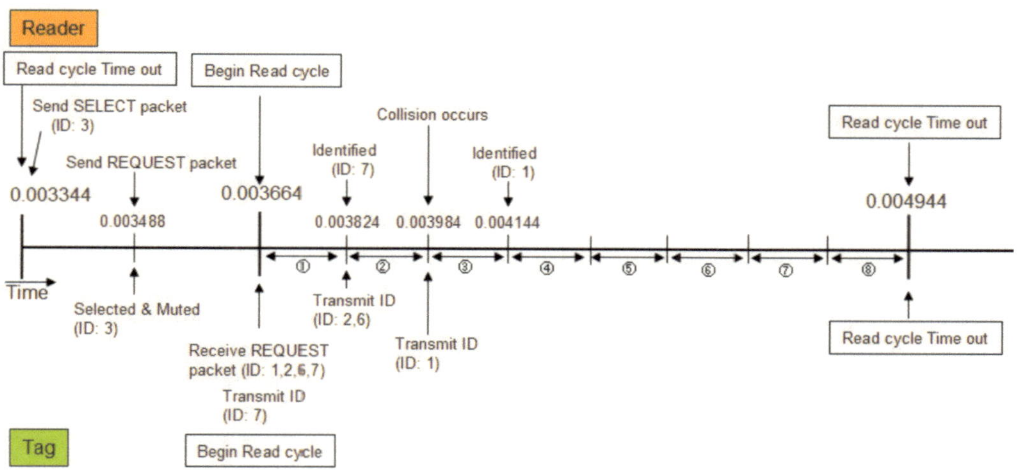

(c) Third Read Cycle of the Simulation

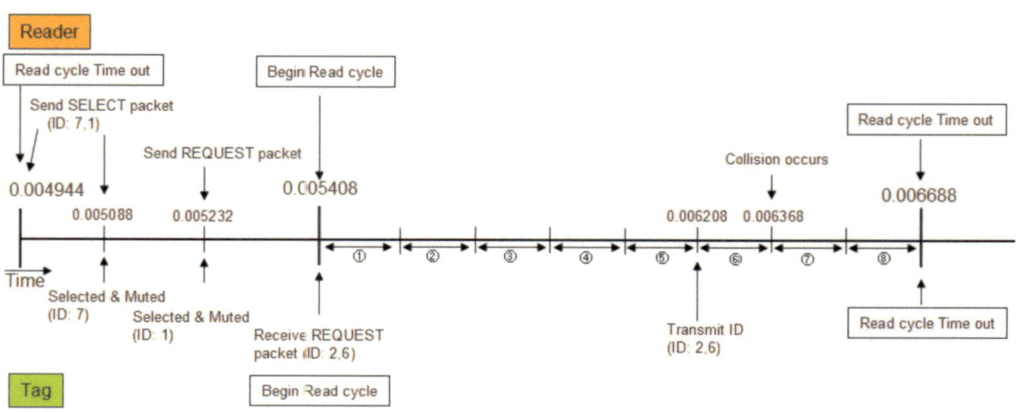

(d) Fourth Read Cycle of the Simulation

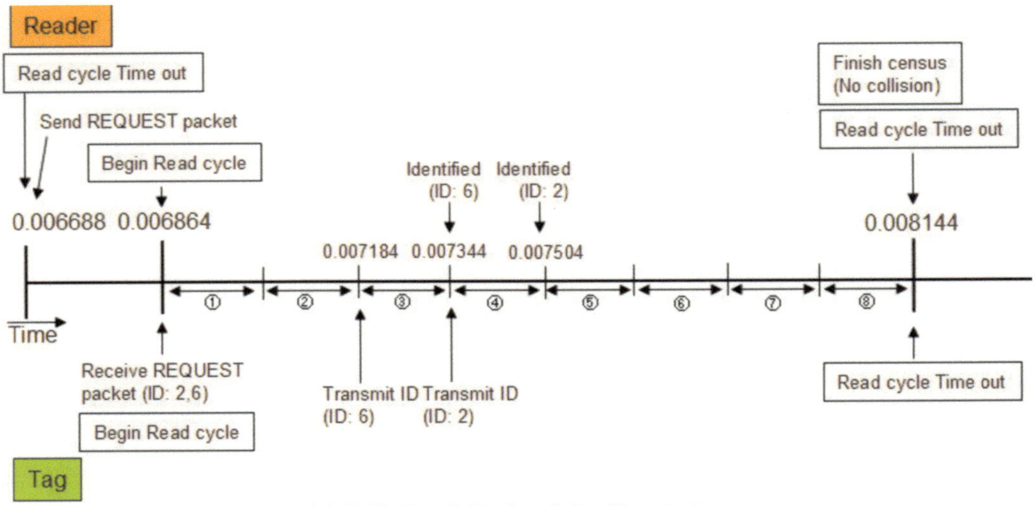

(e) Fifth Read Cycle of the Simulation

Fig. 16. BFSA-Muting Simulation Log

As we can see from Figure 16 (a), when the census begins the reader broadcasts a REQUEST packet to all tags. The transmission delay of a REQUEST packet is 0.000176 seconds since the size of the packet is 88 bits while the data rate is 500,000 bps. We assume propagation and calculation delay are negligible, since events are generated at slot boundaries and propagation delay and computation time will not have an effect on census delay and throughput. As soon as tags receive the REQUEST packet they start their timer to synchronize the read cycle between the reader and tags. Tags can select only one of the slots in the read cycle randomly and transmit a RESPONSE packet which contains tag's ID and CRC to the reader by occupying a single slot, e.g. as we see from Figure 16 (a) each tag send its ID only once in a read cycle based on the definition of the FSA protocol. There are eight slots in a frame in this simulation. And we can see every slot durations in the read cycle is identical. The delay for the transmitting of the RESPONSE packet is the definition of the slot duration. As you see at Figure 13 (b) the size of RESPONSE packet is 80 bits while data rate is 500,000 bps. That makes the transmission time of the REQUEST packet to 0.00016 seconds. When multiple tags transmit their ID to the reader with the same slot it causes a collision then the reader can't identify tag's ID successfully. Two collisions occur in the first read cycle, see Figure 16 (a). Three tags (IDs: 1, 2, and 6) transmits their ID by occupying the second slot and two tags (IDs: 3 and 7) are also transmitting during the third slot. Both of them collide and are being discarded. However, a single tag transmission without collision is identified by the reader successfully as can be seen from the fourth, sixth and seventh slot. The first, fifth and eighth slots are idle slots in the first read cycle (frame). When a read cycle (frame) is finished tags can't transmit their ID until the next read cycle begins and the number of identified tags, collided slots, and idle slots are computed and stored by the reader. If there is no collision during a read cycle the census will be completed.

SELECT packets are transmitted together with the tag's ID identified by the reader as soon as a read cycle has completed (as shown in Figure 16 (b)). The purpose of sending SELECT packet is to mute the already identified tags, i.e. forcing them to stop transmitting their IDs. This reduces collisions.

Three SELECT packets are transmitted as shown in Figure 16 (b) with a tag's ID identified in the previous read cycle. The size of the SELECT packet is 72 bits and because of the data rate being 500,000 bps the transmission delay will be 0.000144 seconds. After transmitting SELECT packets the REQUEST packet is broadcasted to all tags. However, selected tags will disregard this message. Only unread tags will response to the REQUEST packet.
When the REQUEST packet is delivered to all tags the read cycle is started again. The reader can synchronize the start time of the read cycle with tags since reader can calculate the transmission delay of SELECT and REQUEST packet with packets size and data rate. Once the read cycle is started, the procedure of transmit tag's ID, of detecting collision, and of identifying tag's ID is same with ones in the previous read cycle. When the reader detects no collision during a read cycle the census will be finished shown in Figure 16 (e).

4.1.3 BFSA-non-muting

There are two major differences from BFSA-Muting; identified tags are not muted and the assurance level is used for finishing the census. For measuring the assurance level after finishing every read cycle and finishing the census successfully, the line 9 of Figure 15 would be replaced with Figure 17.

```
measure current assurance level
if (Given Assurance level <= Current Assurance level) then
```

Fig. 17. Computing Assurance Level of BFSA-Non-Muting

In the BFSA-Non-Muting, tags are not muted at all. Thus, the probability of collision occurrence is higher than the BFSA-Muting and SELECT packet is not necessary to be transmitted to tags.

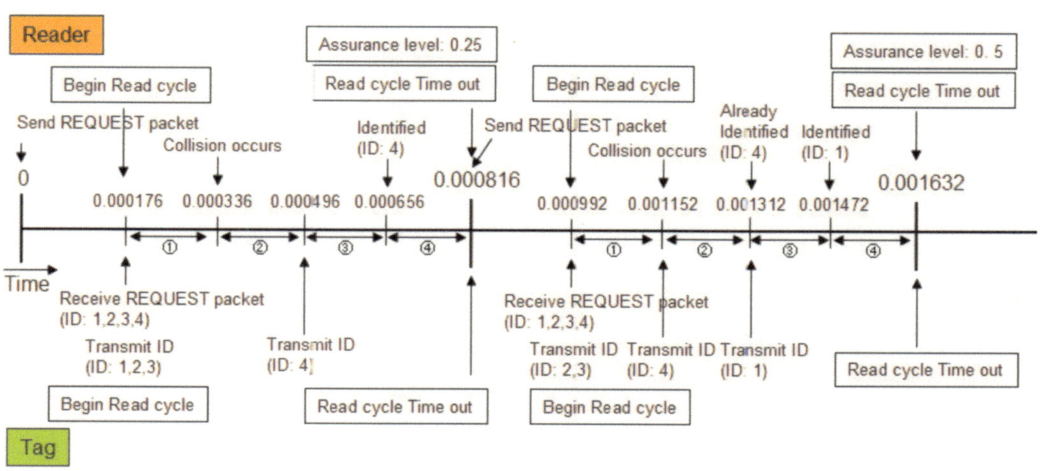

(a) First and Second Read Cycle of the Simulation

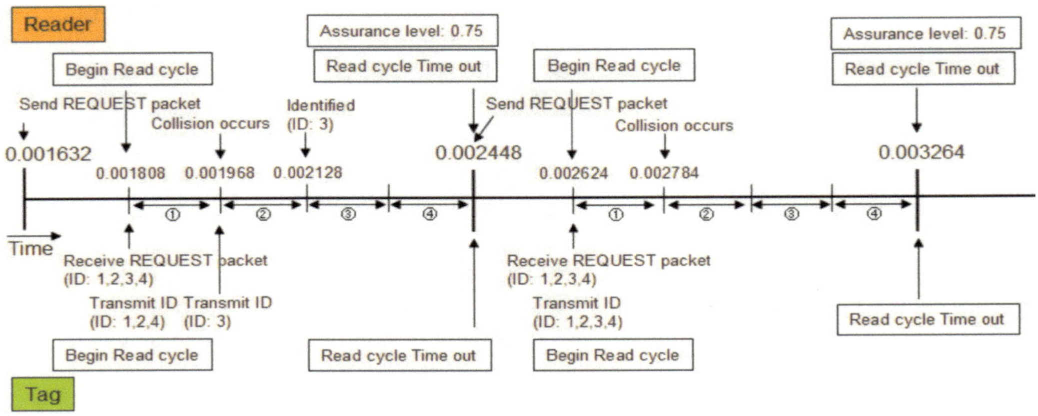

(b) Third and Fourth Read Cycle of the Simulation

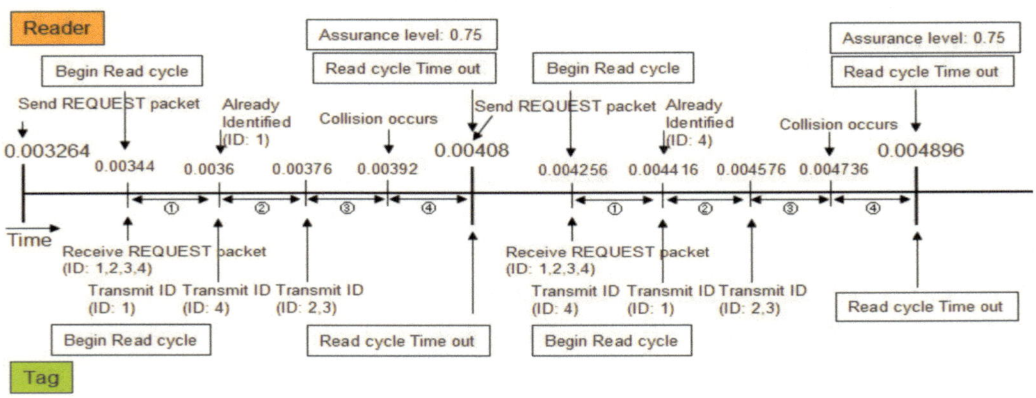

(c) Fifth and Sixth Read Cycle of the Simulation

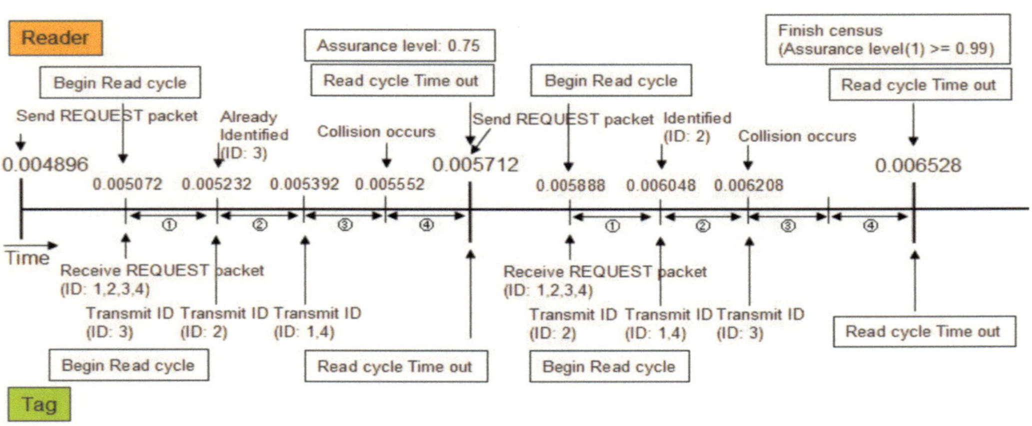

(d) Seventh and Eight Read Cycle of the Simulation

Fig. 18. BFSA-Non-Muting Simulation Log

In BFSA-Non-Muting when the census begins the reader transmits REQUEST packet to all tags and they start transmitting their IDs (once in a read cycle). Tags are never muted, so that all tags continue to transmit for the duration of a census, once every read cycle. Another difference between BFSA-Non-Muting and BFSA-Muting is that they have different behavior at the end of each read cycle. As can be seen from Figure 18, the assurance level is measured at the end of every read cycle and is changed according to the total number of identified tags given the total number of actual tags. As shown in Figure 18 (a) the identified tag (ID: 4) sends its ID again during the next frame. Census completes when the assurance level is satisfied, as shown in Figure 18 (d).

5. Evaluation

In this Section, we evaluate two parameters: total census delay and network throughput. We compare our simulation results with analytical results, computed by using the equations from Section 3.

5.1 Total census delay

Total census delay varies depending on the frame size and the number of actual tags in the BFSA model. If a frame size is either too big or too small as compared to the total number of tags the delay will be longer because of the increased number of idle slots and collision slots respectively, i.e. there is an optimal frame size resulting in the least total census delay for given number of tags. Thus, we first measure the optimal frame size to find the minimal total census delay for given fixed number of tags to be read (identified). Simulation runs were conducted by varying the initial number of tags from 10 to 100 with step of 5 while the given static frame size varies from 10 to 120 with step of 5. 10 census procedures were simulated for each frame size and given a specific number of tags.

The minimal total census delay for given static number of tags is shown in Figure 19. Triangle line represents the analytical result of BFSA-Non-Muting, 'x' line represents simulation result of BFSA-Non-Muting, square line represents analytical result of BFSA-Muting, and '+' line represents simulation result of BFSA-Muting. For computing the analytical result of BFSA-Non-Muting and BFSA-Muting a computing program was developed [appendix C] and equation 6, 7, 9, 10, 11, 14, 15, 16, 17, 18 and 21 in Section 3 are used.

The minimum total census delay was increased linearly with the number of tags and 100 tag set was identified within 0.25 sec using BFSA-Non-Muting with assurance level 0.99 and 500 Kbps data rate. BFSA-muting took less than 0.1 sec with the same given conditions with BFSA-Non-Muting. The simulation result of BFSA-Muting shows approximately 70% shorter minimum total census delay than BFSA-Non-Muting simulation result. Both BFSA simulation results show about less than 15% shorter minimum total census delay than its analytical results in the experiment.

The optimal frame size is acquired from the simulation being used for computing the minimum total census delay in Figure 19. The symbols in Figure 20 are identical with Figure 19. Figure 20 shows us good agreement between the simulation result and the analytical result. The optimal frame size was increased linearly with the number of tags and BFSA-Muting has smaller optimal frame size than the one BFSA-Non-Muting has.

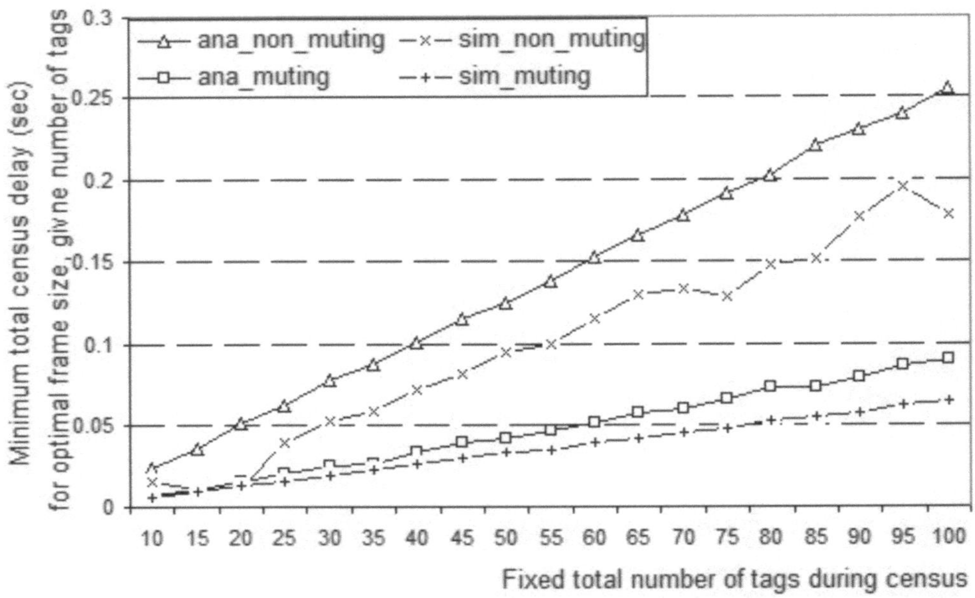

Fig. 19. Minimum Total Census Delay for Given Number of Tags

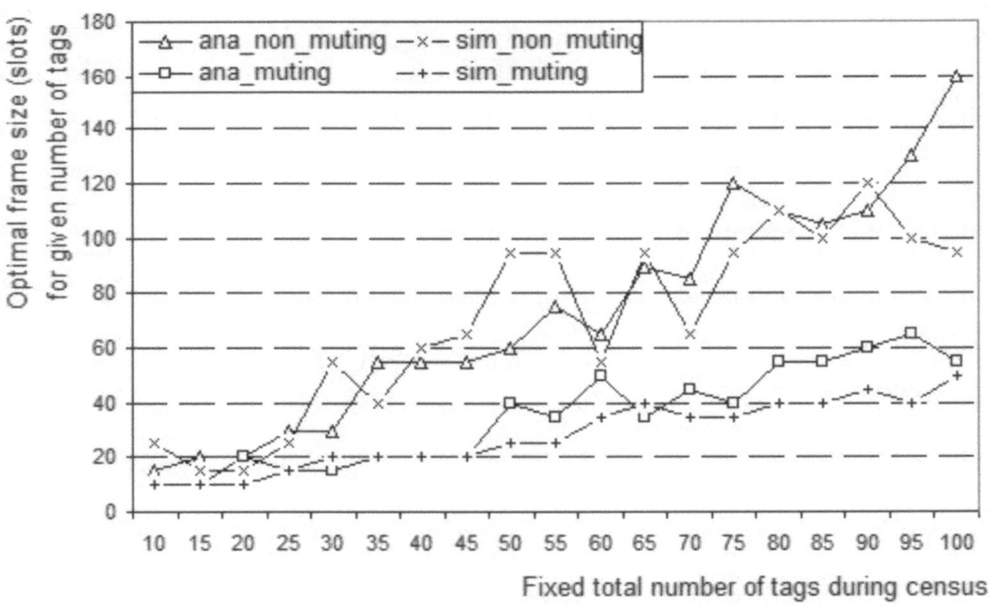

Fig. 20. Optimum Frame Size for Given Number of Tags

5.2 Network throughput

We evaluate three types of network throughput: maximum throughput, minimum throughput, and mean throughput. Network throughput represents the ratio between successfully transmitted number of packets and total number of transmitted packets during census. All of them show good agreement between analytical result and simulation result and they are shown in Figure 21.

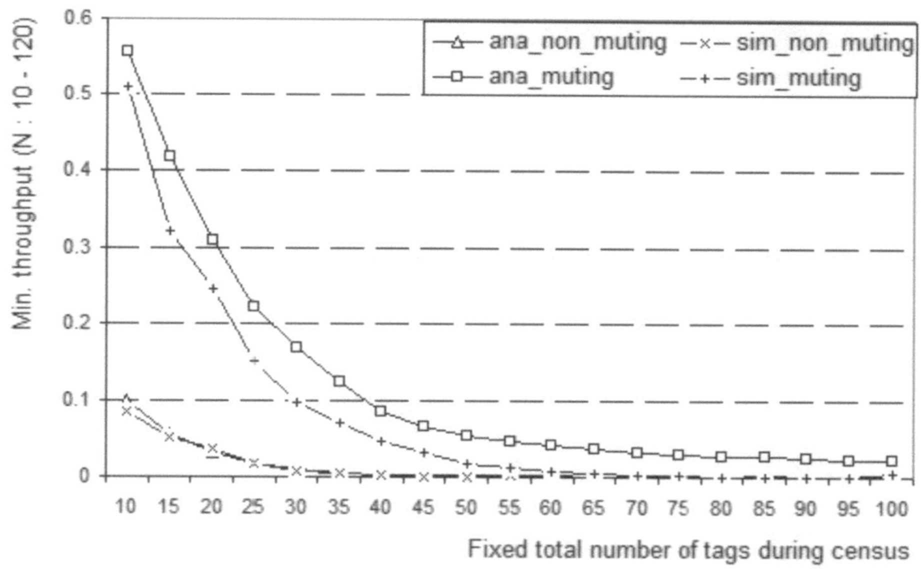

(a) Minimum Network Throughput

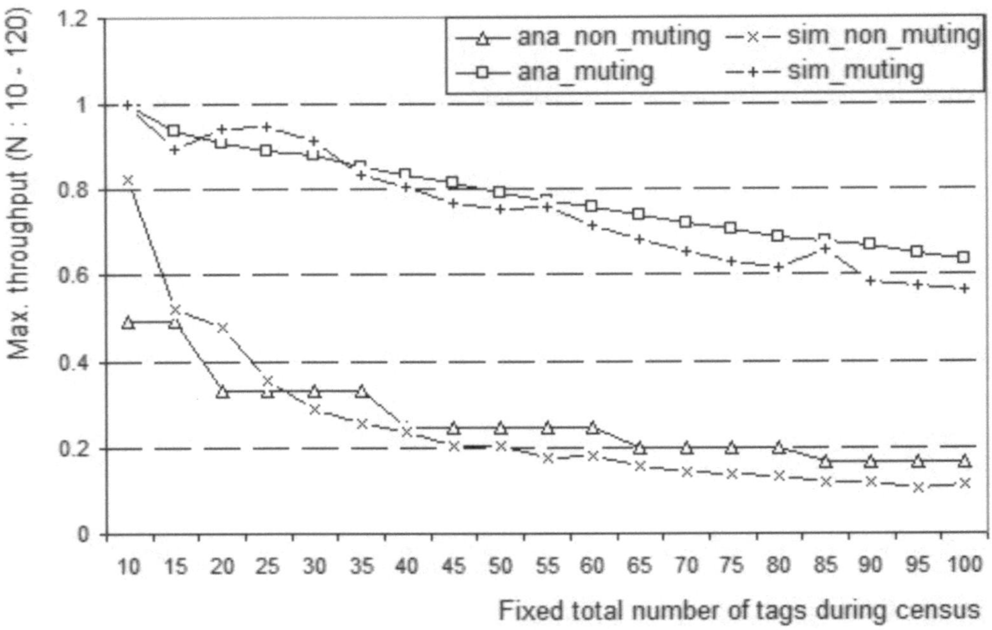

(b) Maximum Network Throughput

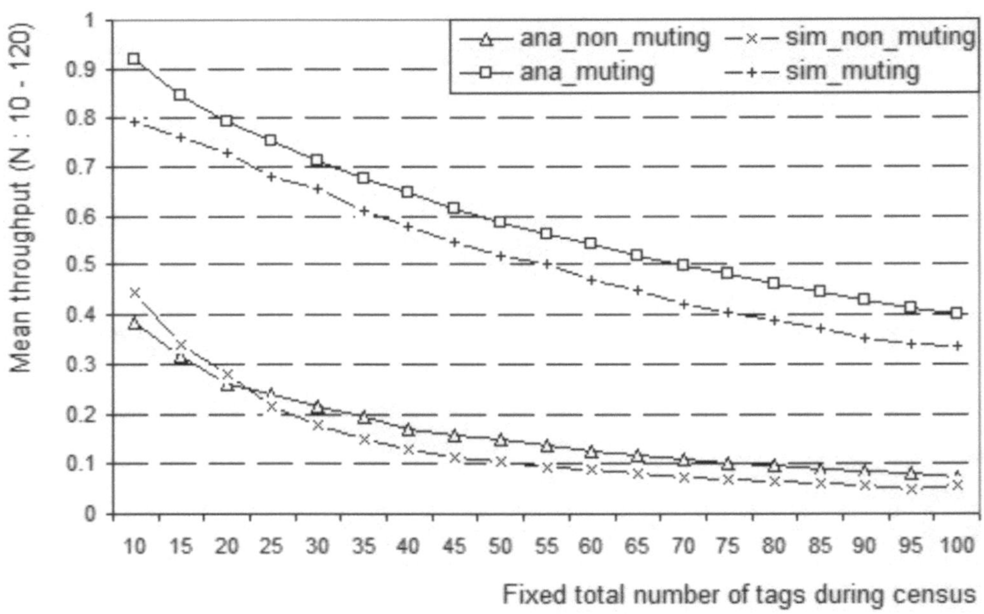

(c) Mean Network Throughput

Fig. 21. Network Throughput

In Figure 21, network throughput shows good agreement between analytical and simulation result and simulation throughput shows slightly lower than analytical one. Figure 21 tell us that the network throughput of the two BFSA models is getting lower according to the increment of the fixed total number of tags. In Figure 21 (c), we can see mean network throughput of BFSA-muting is 200 - 400% greater than the throughput of BFSA-Non-Muting. Since in the BFSA-Muting the identified tags keep silent thus the total number of transmitted packet would be reduced while in the BFSA-Non-Muting the identified tags never stop transmitting its ID. Thus the difference between two RFID models makes network throughput different.

6. Conclusion

To evaluate the performance of RFID protocols we implemented two BFSA models (Muting and Non-muting). We have used the simulation tool, OPNET Modeler 14. The simulation models were validated by analyzing the log in the validation Section. In addition, we compared the simulation results against analytical results, generated by using the equations presented in Section 3.

In Section 4, we evaluated total census delay and network throughput by comparing simulation and analytical results. Our simulation results show good agreement with analytical results both for total census delay and for network throughput. We also could see the performance difference of the two BFSA models in terms of the total census delay and the network throughput. As expected, BFSA-Muting performed better in terms of both network throughput and total census delay as compared to BFSA-Non-Muting due to reduction in the total number of transmitted packets.

7. References

[Finkenzeller03] Finkenzeller, K., "RFID Handbook," 2nd edition, John Wiley & Sons, 2003.

[Klair04] Klair, D. K., Chin, K. W. and Raad, R., "On the Suitability of Framed Slotted Aloha based RFID Anti-collision Protocols for Use in RFID-Enhanced WSNs," Computer Communications and Networks, Proceedings of 16th International Conference (August, 2007), pp. 583-590.

[Rom90] Rom, R., and Sidi, M., "Multiple Access Protocols/Performance and Analysis," Springer-Verlag, March 15, 1990. pp. 47-77.

[Want06] Want, R., "An Introduction to RFID Technology," IEEE CS and IEEE ComSoc, Pervasive computing, 2006, pp. 25-33.

[Weinstein05] Weinstein, R., "RFID: A Technical Overview and Its Application to the Enterprise," IEEE Computer Society, May 2005, pp. 27-33.

Electronic Sources:

[Bin05]Bin, Z., Mamoru, K. and Masashi, S., "Framed Aloha for Multiple RFID Objects Identification," IEICE Trans. Comm., Vol.E88–B, No.3 March 15, 2005.

[Cappelletti06] Cappelletti, F., Ferrari, G., and Raheli, R., "A Simple Performance Analysis of Multiple Access RFID Networks Based on the Binary Tree Protocol", ISCCSP March 15, 2006.

[Computerworld07] www.computerworld.com, "Proctor & Gamble: Wal-Mart RFID Effort Effective,"http://www.computerworld.com/action/article.do?command=viewArticleBasic&articleId=284160, February 26, 2007. [OPNET] OPNET Technologies, http://www.opnet.com.

[Vogt02] Vogt, H., "Efficient object identification with passive RFID tags," Inter. Conf. on Pervasive Computing, LNCS, pp.98–113, Springer-Verlag, March 15, 2002.

[Zürich04] Zürich, E., Burdet, L. A., "RFID Multiple Access Methods," Seminar "Smart Environments", August 15, 2004.

Using CDMA as Anti-Collision Method for RFID - Research & Applications

Andreas Loeffler
Friedrich-Alexander-University of Erlangen-Nuremberg
Germany

1. Introduction

The increasing number of deployed RFID systems and the resulting need for fast recognition of a given amount of RFID tags puts great demand on future RFID readers. Applications requesting for a fast capture of RFID tags are mainly found in logistic and manufacturing processes. Imagine trucks driving through large RFID gates, where each RFID tagged package or even item has to be identified. Also, fast production with tagged units approaching in quick succession would need a fast recognition of RFID tags. Therefore, if several tags are located within the range of a reader, signals from some of these tags will clash. For this very reason anti-collision procedures are widely deployed to prevent tags from broadcasting their information simultaneously. Existing RFID multiple access solutions for the uplink channel are based on Time Division Multiple Access (TDMA). Fig. 1 shows the TDMA method, pointing out that the tags in the reader's field transmit their data at different moments in time (slots) (Finkenzeller, 2003).

As may be imagined, the application of TDMA in RFID systems ensures that each RFID tag in range will be detected on condition that the amount of time available is sufficient. If this condition is false, the system comes to hard decisions, so that, finally, not every tag in range has been recognized successfully. Therefore, the usage of TDMA methods pushes the

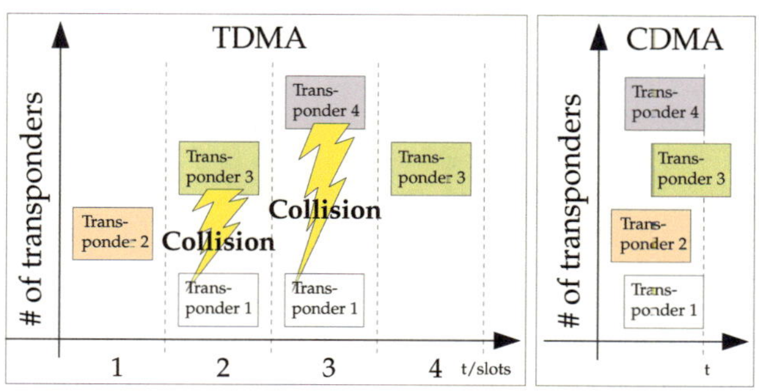

Fig. 1. Comparison of TDMA and CDMA communication channel access techniques for RFID

envelope of the system when a very high number of tags have to be scanned within a given time span.

"Graceful degradation". This quotation from Aein (1964) describes very well the behavior of Code Division Multiple Access (CDMA). In comparison to TDMA with its "hard decisions",

CDMA-based systems take "soft decisions", which means, that within the system each additionally introduced RFID tag decreases the overall probability of detection of all tags. However, for a particular amount of tags, the system may be optimized in such a way that the time needed for detection may be minimized. Therefore, for an RFID system under those certain circumstances, the introduction of CDMA may offer a way out (Fig. 1). The transponders, each equipped with a unique quasi-orthogonal spreading code (e.g., Gold codes (Gold, 1967a)), may use the radio channel whenever the transponders are ready to transmit their data (asynchronous CDMA). The objective is the realization of a DS (direct sequence)-CDMA-based RFID system using semi-passive UHF transponders, with the reader providing the recognition of multiple transponders simultaneously. This means that the transponders are transmitting data within the same time range and frequency band, in contrast to the existing systems based on TDMA.

The realized UHF transponders operate in semi-passive mode, meaning that the digital part of the transponder, i.e., the data generation, has an active power supply, whereas the high frequency (HF) part works in passive mode taking advantage of the backscatter principle.

The attendant RFID reader, though, is separated into two parts. Part one, described as transmitting system, generates a carrier wave at around 867 MHz. Part two, the receiving system, mainly demodulates the incoming backscattered signals of the RFID tags.

This chapter is organized in seven sections. The first section gives a brief introduction to the topic of anti-collision in UHF-RFID-based systems. The following sections introduce CDMA by outlining the advantages over the current used TDMA schemes. After introducing particular problems backscattering RFID systems have to deal with, a concept and an implementation of such a CDMA-based RFID system is shown. The chapter ends with various measurements concerning the system and subsequent results.

2. Anti-collision: EPC class 1 Gen 2

This section outlines some basic issues regarding anti-collision methods within RFID. Basic and state-of-the-art anti-collision methods are shown in Subsection 2.1. Subsection 2.2 presents theoretical performance issues regarding the throughput by comparing state-of-the-art TDMA methods with CDMA anti-collision methods.

2.1 ALOHA and slotted ALOHA

Before elucidating the state-of-the-art anti-collision method for UHF RFID systems, the principle of ALOHA and unslotted ALOHA (Bertsekas & Gallager, 1992) is illustrated, as the principle of ALOHA provides the basis for the modern anti-collision protocols. The ALOHA protocol (or pure ALOHA), first published by Abramson (1970), is a very simple transmission protocol. The transmitter sends its data, no matter if the transmission channel is free or not. This means the transmitter does not care about collisions with other transmitters. The transmitter resends its data later, if the acknowledgment from the receiver is missing. RFID systems based on the principle of pure ALOHA are, e.g., based on the TTF principle, i.e., transponder-talks-first. The IPX protocol from IPICO (2009) is an example for RFID systems using unslotted or pure ALOHA.

An extension of the ALOHA protocol, called slotted ALOHA (Roberts, 1975) introduces time slots in which the transmitter must send its data at the beginning. Therefore, collisions only occur within a full time slot. This extension doubles the maximum throughput of the system. Most current RFID protocols are based on the principle of slotted ALOHA, as is also the very commonly used EPC standard UHF Class-1 Generation-2 air interface protocol

V1.2.0 (ISO 18000-6C), commonly known as "Gen2". Basically, the "Gen2" standard defines, that every communication is triggered by the RFID reader, i.e., RTF (reader-talks-first). An inventory round, i.e., the process of detecting all available transponders, is started with the *Query*-command to acquire all transponders available in the read range. This command inherits a so called *Q*-parameter. Using this *Q*-parameter, every transponder generates a random number *RN* in the range $[0; 2^Q - 1]$ and initializes its internal slot counter with this random number. If, at a given moment, the value of the slot counter of one or more transponders equals 0, the transponders send a 16 bit random number called RN_{16}. After the acknowledgment of the RN_{16} through the reader, the electronic product code (EPC) is transmitted from the transponder to the reader and the transponder will be marked as *inventoried*. All the left-over (non-marked) transponders are prompted to decrement its slot counter by sending a *QueryRep*-command, and the procedure starts all over again. In the case of several transponders initalizing their slot counters with the same random number *RN*, it will come sooner or later to a signal collision as the slot counters will reach zero at the same time slot. If the reader recognizes such a collision, another inventory round will be initiated to identify the left-over transponders. Therefore, a newly value of *Q* will be introduced and new random numbers will be calculated. To sum up, one could say that the choice of *Q* is a typical trade-off. Choosing a high *Q* will lead to a smaller number of collisions, at the expense of an increasing time needed for an inventory round. Indeed, a smaller *Q* will lead to less acquisition time, but to more collisions.

Plenty of work has been done to improve the current EPC standard. Improving the current standard anti-collision method by choosing an appropriate value of *Q*, e.g., dynamically, is described in Maguire & Pappu (2009); Pupunwiwat & Stantic (2010); Wang & Liu (2006). The right choice of *Q* is of great importance for the overall system performance, so that an accurate estimation would improve the time needed for an inventory round. Slightly new algorithms, based on the current EPC "Gen2" standard are outlined, e.g, in Cui & Zhao (2009); Lee et al. (2008). New better performing algorithms for the slotted ALOHA protocol for RFID are described in Bang et al. (2009); Choi et al. (2007); Liu et al. (2009); Makwimanloy et al. (2009). A complete new system with time hopping on the communication link from tag to reader is outlined in Zhang et al. (2010).

2.2 Comparison: CDMA versus TDMA in UHF-RFID

The throughput *S* in dependence of the traffic channel rate *G* describes the performance of a given transmission system regarding how many packets must be transmitted (statistically) until a successful transmission occurs. This statement is given with the term $\frac{G}{S}$ as described by Kleinrock & Tobagi (1975). The reciprocal of this term, i.e., $\frac{S}{G}$ defines accordingly the probability of a successful transmission. The channel capacity is determined by maximizing *S* with respect to *G* (Kleinrock & Tobagi, 1975). According to Abramson (1970) the pure ALOHA transmission has a relation between *S* and *G* of

$$S = Ge^{-G} \tag{1}$$

, whereas the throughput of the slotted ALOHA transmission is defined after Roberts (1975) with

$$S = Ge^{-2G} \tag{2}$$

. Accordingly, the maximum channel capacity is $\frac{1}{2e} \approx 18.4\%$ for pure ALOHA and $\frac{1}{e} \approx 36.8\%$ for slotted ALOHA.

For a fair comparison between CDMA-based systems and ALOHA systems, the total

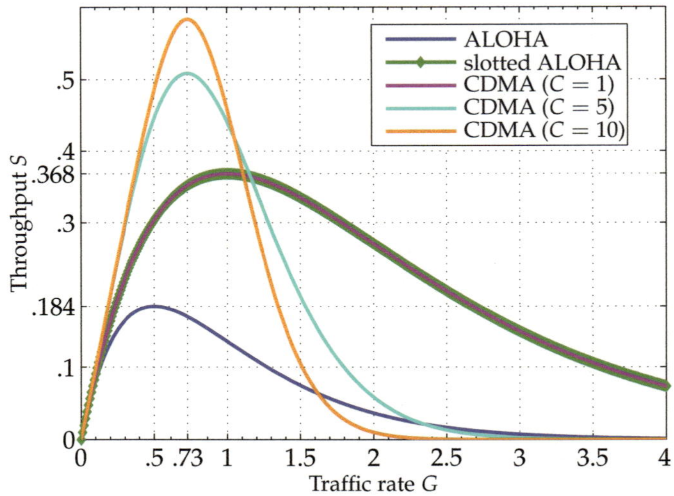

Fig. 2. Various throughputs S over traffic rate G for ALOHA, slotted ALOHA and CDMA

bandwidth has to be maintained the same for both systems. A CDMA system has a so called spreading factor C, which is proportional to the length of the spreading codes respectively the ratio between chip rate and bit rate (i.e., R_{chip}/R_{bit}) used. According to Linnartz & Vvedenskaya (2009) the throughput S and the offered traffic rate G is

$$S = G\, e^{-CG} \sum_{k=0}^{C-1} \frac{(CG)^k}{k!}. \tag{3}$$

Setting $C = 1$ leads to the slotted ALOHA transmission scheme. Figure 2 shows various throughputs S over the traffic rate G. The figure shows the throughputs for ALOHA (channel capacity 18.4%), slotted ALOHA and CDMA with spreading factor $C = 1$ (channel capacity 36.8%), CDMA with $C = 5$ (channel capacity 50.87%) and CDMA with $C = 10$ (channel capacity 58.31%). This graph shows the basic difference between TDMA (ALOHA-based) and CDMA systems. In general, TDMA-based RFID systems can handle much more RFID transponders with a lower overall throughput. CDMA-based system, on the other hand, are able to handle a limited amount of RFID tags with higher overall throughput. For instance, assuming a limited amount of RFID transponders for a traffic rate $G = 0.73$. The throughput of unslotted ALOHA would be $S_{\text{ALOHA}} = 16.95\%$ and the throughput of slotted ALOHA $S_{\text{unslotted ALOHA}} = 35.18\%$. A CDMA-based system with a spreading factor C of 10 would have there its maximum throughput of $S_{\text{CDMA},C=10} = 58.31\%$. This scenario is shown in Figure 2.

Finally, it can be stated that CDMA-based RFID systems may be better for particular applications, in which the number of transponders is limited and the inventory process has to be made very fast, e.g., fast production lines and automation processes.

Particular slotted ALOHA CDMA systems and corresponding performances may be found in Gopalan et al. (2005); Sakata et al. (2007). Other works describe certain CDMA systems with error correction which really outperform the TDMA-based systems. Examples can be found in Liu et al. (2001); Liu & El Zarki (1994); Lo et al. (1996); Sastry (1984); van Nee et al. (1995). Also this list is not complete it gives a short overview of CDMA-based system performances.

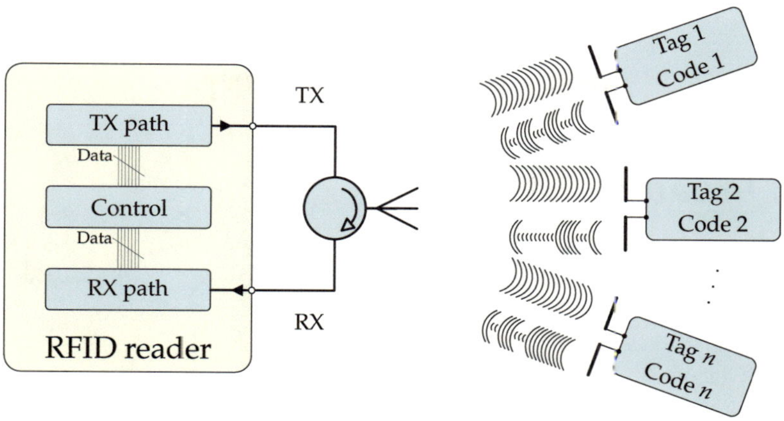

Fig. 3. Basic architecture of RFID system; depicted in monostatic antenna configuration

3. Concept of CDMA-based system

This section presents the basics of the proposed RFID system. Going into more detail, Subsection 4.1 shows the architecture of the Transmitting System (TX system path) followed by Subsection 4.2 presenting the proposed semi-passive RFID transponders and Subsection 4.3 describing the Receiving System (RX system path).

Figure 3 shows the basic architecture of the CDMA-based RFID system. Generally, it consists, as any other RFID system of two major parts. First, the RFID reader itself and second, one or more transponders. The main difference between this system and other current systems is the channel access method in the uplink (transponder to reader communication) layer, in this case based on CDMA; this fact is illustrated in Figure 3 showing each transponder (Transponder 1 to Transponder n) with a unique spreading code (Code 1 to Code n).

The basic working principle is also indicted in Figure 3, showing the RFID reader transmitting a sinusoidal wave over its transmit antenna TX, thus allowing the various transponders in the field to modulate and reflect (principle of backscatter) this incident wave back to the RFID reader. Therefore, the total backscattered signal consists of the additive superposition of n (if multipath is negligible) backscattered transponder signals with each transponder using its own unique spreading code. Receiving this superimposed signal over RX, the reader is, generally, able to separate the various transponder signals from each other (process of despreading) in order to restore the transponders' data.

Figure 3 and Figure 5, respectively, show the concept and the architecture of the realized RFID reader. The following paragraphs will refer to these figures. Fig. 6 shows the setup of the system for directly measuring the backscattered baseband signals.

4. Implementation

Within this section, the implementation of the CDMA-based RFID system is described. By referring to Figure 4 and 5, Subsection 4.1 describes the overall *TX path*, involving the PLL-based RF synthesizer, a power amplifier (PA) and the TX antenna, whereas Subsection 4.3 reveals the fundamentals of the *RX path*, consisting of RX antenna, low-noise amplifier (LNA), a demodulator module, a baseband processing unit, a baseband sampling module (ADC module) and a subsequent DSP module for evaluating the incoming data. However, the basics of the *transponder* are depicted in Subsection 4.2.

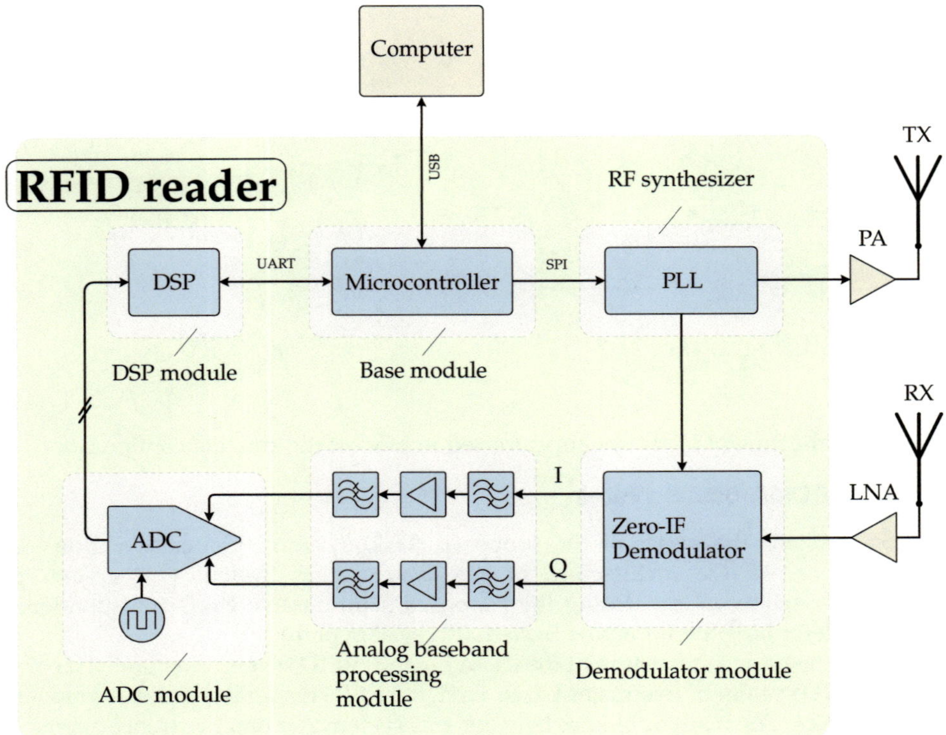

Fig. 4. Basic concept of RFID reader; depicted in bistatic antenna configuration

4.1 TX system path

The proposed semi-passive UHF transponder works in accordance with the principle of backscattering. The incident wave to be backscattered is generated by the *Transmitting System*. Considering the RFID uplink channel (tag to reader), the introduced *Transmitting System* (see Figure 3 and Figure 5) consists of a PLL-based RF synthesizer (Figure 3 and Figure 5) , generating a sine wave (here with $f_{\text{carrier}} = 866.5$ MHz, maximum output power $P_{out} =$ 1 dBm at 50 Ω), an upstream power amplifier (PA, Gain $G_{PA} = 20$ dB, 1 dB compression point = 24 dBm), and a linear polarized 50 Ω antenna (TX, Gain $G_{TX} \approx 7$ dBi). The purpose of the transmitter is to generate an RF wave to be reflected (backscattered) by the UHF transponder whereby the reflected wave is received by the *Receiving System* further discussed in Subsection 4.3.

It has to be mentioned that the RF synthesizer not only generates a sine wave for the transmitting part, but also for the receiving part of the system. Indeed, it is used as local oscillator (LO) source for the downmixing part of the receiver. However, both synthesized RF waves inherit the same frequency as they are both created by the same PLL; the waves only differ in π in phase.

4.2 Transponder

The major tasks of the semi-passive UHF transponders are:

- Generate spreading code
- Create spreaded data

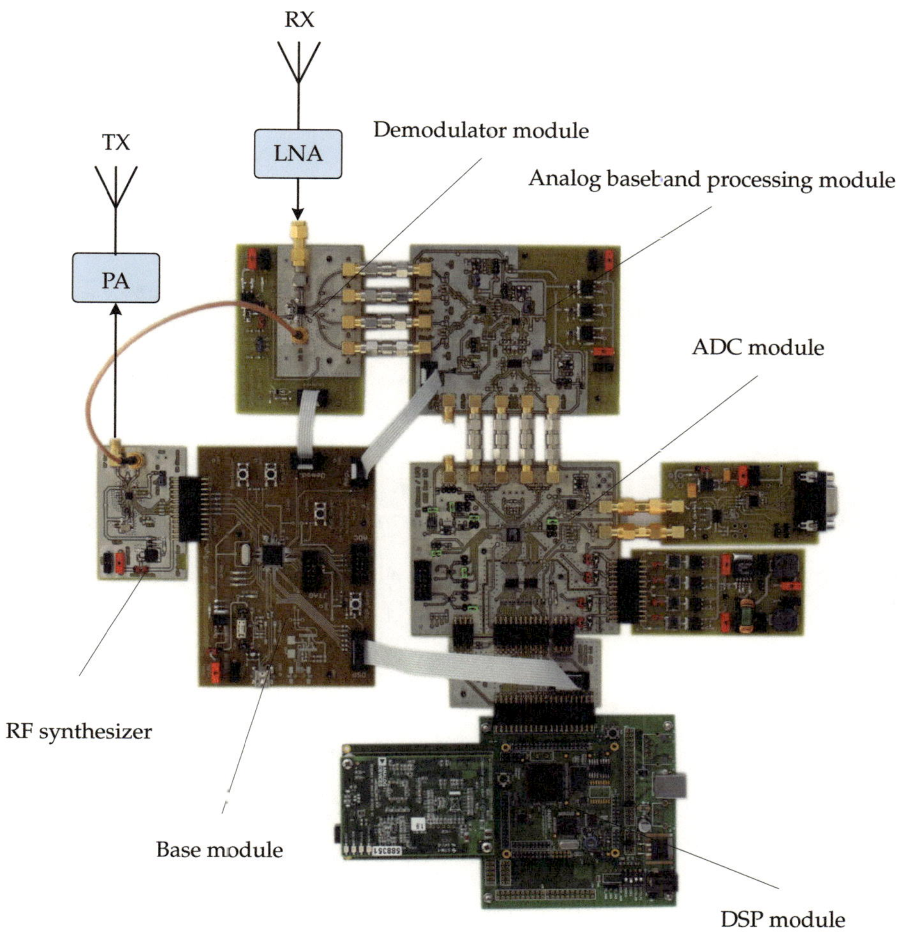

Fig. 5. Architecture of CDMA-based RFID reader

- Modulate and reflect incoming RF signal at $f_{\text{carrier}} = 866.5$ MHz (principle of backscatter)

Figure 7 shows the basic principle of an RFID transponder. An incident RF wave is reflected by the transponder. The phase and amplitude of the reflected wave is affected by three major issues: The first two issues includee structural mode and antenna mode scattering (Hansen, 1989; Penttila et al., 2006), the third issue is the multipath propagation. Multipath effects are a non-changeable fact, so they can be neglected at this point. The structural mode scattering of an antenna is dependent on the structure of the antenna itself (material, antenna geometry, etc.) and cannot be changed - therefore, the structural mode may not be used for a normal data transmission. The antenna mode scattering, on the other hand, describes the receiving and emitting effects of an antenna, which usually depend on the impedances used; particularly the impedance of the antenna Z_{ant} itself and the corresponding load impedance Z_{load} of the following transponder system. Assuming that Z_{load} can adopt two values being Z_1 and Z_2. According to Figure 7 the antenna mode scattering may be changed by altering the load impedance Z_{load} of the transponder's antenna according to the data the transponder

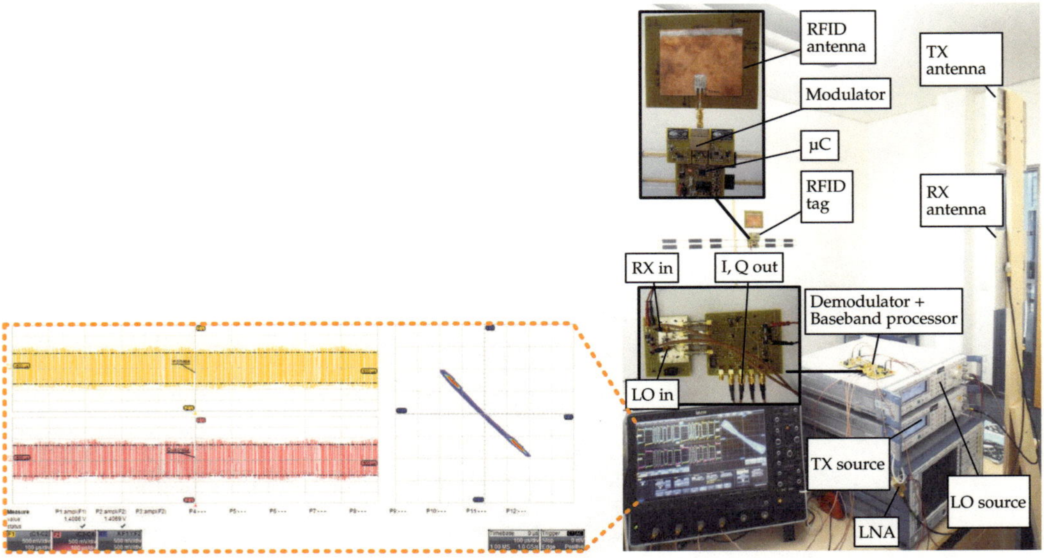

Fig. 6. Setup of CDMA-based UHF-RFID system with a magnification of the oscilloscope's screen

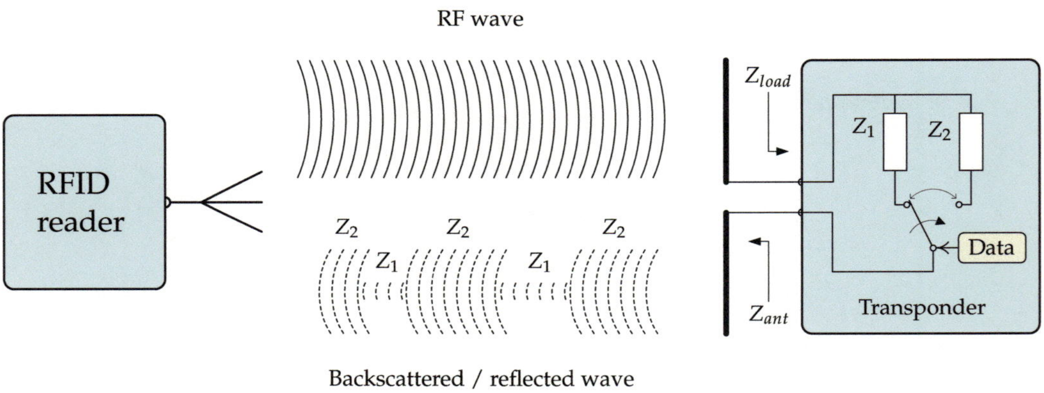

Fig. 7. Basic function of RFID transponder

wants to send. Binary data may be send by altering Z_{load} between Z_1 and Z_2, thus changing the reflection coefficient between Z_{ant} and Z_{load}, which in turn leads to an alteration of the reflection of the RF wave in phase and/or amplitude. Again, this only affects the antenna mode scattering. However, the total resulting backscattered signal is the superposition of the multipath signal, the structural mode scattering and antenna mode scattering effects. Measurements at the end of this paper will show this effects.

Figure 8 shows the basic concept of the CDMA-based semi-passive transponder. A central microcontroller generates the binary output data stream (i.e., the already coded and spreaded user data) to drive the fast RF switch 'S', that alters between two impedance states Z_1 and Z_2; according to the binary state of the output data stream, a logical '1' triggers Z_2, a logical '0' triggers Z_1 to be the corresponding load impedance. Therefore, the data stream directly affects the reflection coefficient. The performance of the uplink (tag to reader radio channel) depends very much on the modulation efficiency η_{mod} of the backscatter modulator (Fuschini et al., 2008; Karthaus & Fischer, 2003; Nikitin & Rao, 2008), which basic calculation is subject of the following paragraph.

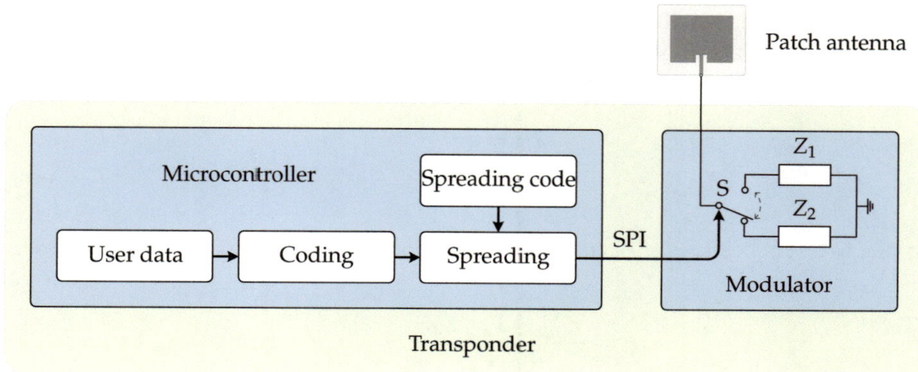

Fig. 8. Concept of CDMA-based semi-passive UHF RFID transponder

4.2.1 Determining load Impedances

Assuming an antenna with complex antenna impedance

$$Z_{ant} = R_a + j\,X_a \tag{4}$$

with $R_a = R_r + R_l$ as the sum of radiation resistance R_r and real antenna losses R_l, and X_a as the imaginary part of the antenna impedance. The complex reflection coefficients $\Gamma_{1,2}$ between the antenna impedance and the load impedances $Z_{1,2}$ can be described as

$$\Gamma_{1,2} = \frac{Z_{1,2} - Z_{ant}^{*}}{Z_{1,2} + Z_{ant}} = \frac{Z_{1,2} - R_a + j\,X_a}{Z_{1,2} + R_a + j\,X_a} \tag{5}$$

According to Rembold (2009) the modulation efficiency η_{mod} can be expressed as

$$\begin{aligned} \eta_{mod} &= \frac{P_{mod}}{P_{max}} = \frac{2}{\pi^2}\,|\Gamma_1 - \Gamma_2|^2 \qquad (6)\\ &= \frac{2}{\pi^2}\left|\frac{Z_1 - R_a + j\,X_a}{Z_1 + R_a + j\,X_a} - \frac{Z_2 - R_a + j\,X_a}{Z_2 + R_a + j\,X_a}\right|^2 \\ &= \frac{8R_a^2}{\pi^2}\left|\frac{Z_1 - Z_2}{(Z_1 + R_a + j\,X_a)\,(Z_2 + R_a + j\,X_a)}\right|^2 \end{aligned}$$

, whereby P_{max} (the maximum receivable power of the antenna) and P_{mod} (the entire power with the information carrying signals) are defined as

$$P_{max} = \frac{1}{8}\,\frac{|U_0^2|}{R_a} = \frac{1}{2}\,|a|^2 \tag{7}$$

$$P_{mod} = \frac{|a|^2}{\pi^2}\,|\Gamma_1 - \Gamma_2|^2 \tag{8}$$

with U_0 as the antenna's open circuit voltage and a being the wave from the antenna impedance to the load impedance (see Rembold (2009) for details).
Maximum modulation efficiency η_{mod} is achieved when the difference of the complex reflections coefficients Γ_1 and Γ_2 is maximum. Supposing two vectors (Γ_1 and Γ_2) in a complex coordinate system, the maximum difference between both vectors is achieved at the point when the phase φ_Γ differs with π under the assumption that the maximum absolute value of any Γ is limited to 1. That determines the complex reflection coefficients $\Gamma_{1,2}$ to

$$\Gamma_1 = e^{j\varphi_{\Gamma,1}} \tag{9}$$

$$\Gamma_2 = e^{j\varphi_{\Gamma,1}+j\pi} \tag{10}$$

Setting $\varphi_{\Gamma,1}$ to 0 sets $\Gamma_{1,2}$ to ± 1. According to Equation (5) this will define the load impedances to

$$Z_{1,2} = \frac{Z^*_{ant} + \Gamma_{1,2}\, Z_{ant}}{1 - \Gamma_{1,2}} \tag{11}$$

$$\rightarrow Z_1 = \frac{Z^*_{ant} + Z_{ant}}{0} = \pm\infty \tag{12}$$

$$\rightarrow Z_2 = \frac{Z^*_{ant} - Z_{ant}}{2} = \frac{-2jX_a}{2} = -jX_a \tag{13}$$

The antenna designed for the RFID transponders is a 50 Ω patch antenna (Figure 10). Therefore the imaginary part (within the specified frequency range) $X_a \approx 0$. This determines $Z_2 = -jX_a \approx 0$. A load impedance of $Z_1 = \infty$ corresponds to an open circuit wheres $Z_2 = 0$ corresponds to a short circuit. Choosing open and short circuit states as desired load impedances, the maximum achievable modulation efficiency is, according to Equation (7), determined to be

$$\eta_{mod} = \frac{2}{\pi^2}\,|+1+1|^2 = \frac{8}{\pi^2} \approx 81\% \tag{14}$$

In order to have the maximum modulation efficiency for the CDMA-based RFID system, the load impedances of the realized semi-passive transponders are set to open and short circuit. By choosing these values as load impedances, one has to keep in mind, that this is only advisable for semi-passive UHF RFID transponders. If passive transponders are designed, one has to consider the power consumption into its calculations. Therefore, open and short circuit values are not suitable, as the backscattered power is, in fact, too high, as the transponder needs a large portion of the incoming power for supplying itself (Dobkin, 2008).

4.2.2 Transponder basics

Figure 9 illustrates the concept of the transponder and Figure 10 shows one of the realized transponders to achieve the previously mentioned tasks. The microcontroller (µC) is powered by a power supply and may be user-controlled using USB or pushbuttons. The µC generates the unique spreading code and subsequently, the spreaded outgoing data. The data are forwarded to the SPI interface to drive the modulator of the transponder with different input voltages to adjust different load impedances respectively reflection coefficients of the modulator. The block diagram of the modulator (Figure 8) shows the principle of the proposed simple backscatter modulator. An example on how to design load modulators can be found in Pardo et al. (2007). However, the incoming spreaded data stream is low-pass filtered to limit

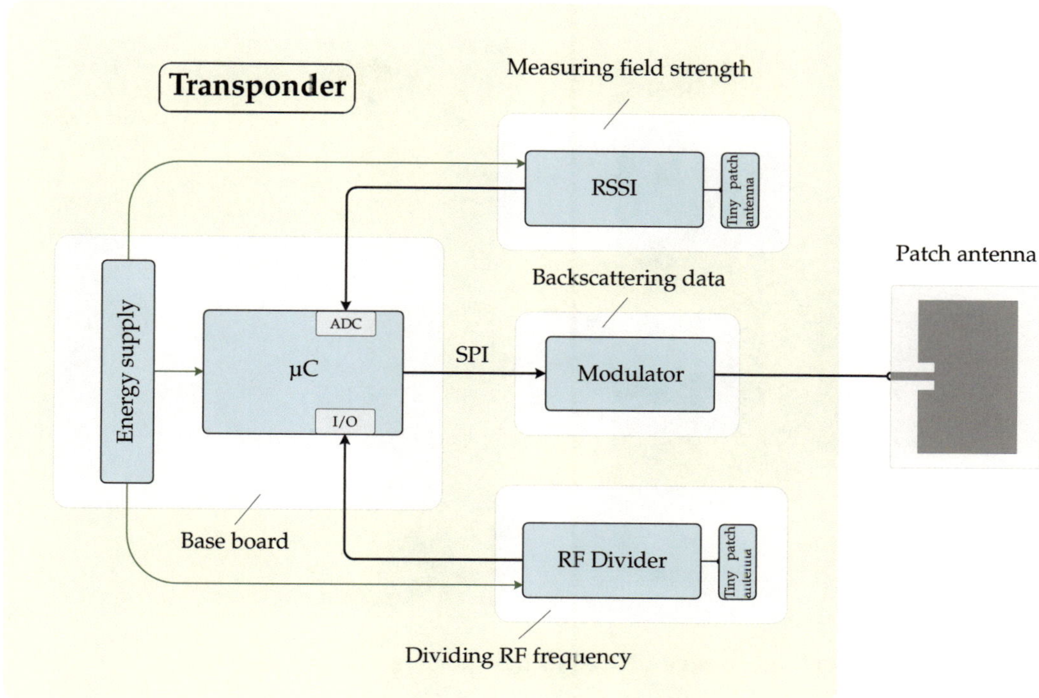

Fig. 9. Block diagram of semi-passive UHF RFID transponder with limited downlink capabilities

the outgoing bandwidth. As the modulator should be as simple as possible, an RF switch 'S' forms the interface between the logic data and the backscattered HF wave. The inputs of the switch are driven by the spreaded data stream with two voltage levels (0 and 2.75 V) given by a buffer driver. One connection of the switch is linked to the patch antenna's microstrip line (50 Ω); the ground connection is linked to the patch antennas ground plane. By triggering the switch's input with the spreaded data to be sent, either Z_1 or Z_2 is connected to the antenna. This modification changes in turn the reflection of an incident electromagnetic wave. The difference of phase and amplitude of the reflection is a direct indicator for the efficiency of a backscattering modulator. As mentioned above the modulators load impedances are set to open and short circuit to achieve maximum modulation efficiency. An exemplary spectral extract of the backscattered output of the transponder, measured at the receiving antenna, is given in Figure 17. On closer inspection, one can see the spreaded data (chip rate is 1.5 Mcps) around the carrier frequency (866.5 MHz). As these data signal levels ($P \approx -90$ dBm $\pm$ 10 dB) are not very high, an accurate implementation of the receiving system becomes necessary.
For a limited downlink (reader to tag) capability the transponders are equipped with a module for measuring the field strength (RSSI) and a module for measuring the frequency (RF Divider) of the incident RF wave emitted by the reader. The *RF Divider* is currently used to indicate the transponder to send its data as soon as a carrier between 865 MHz and 868 MHz is detected. The RSSI module is used for statistical measurements. Anyway, both modules are not part of this work.

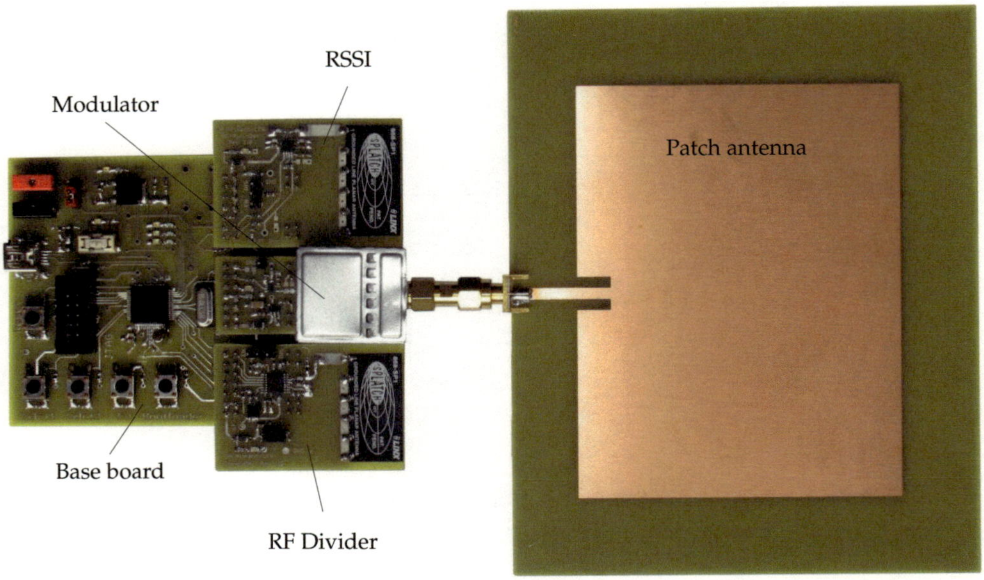

Fig. 10. Prototype of semi-passive UHF RFID transponder

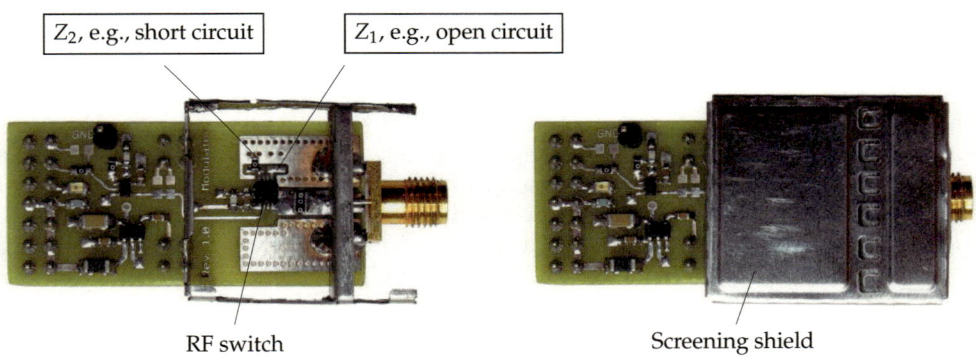

Fig. 11. Backscatter modulator

4.2.3 Modulator

The transponder's modulator is one of the key components of the system. Usually, it effects the energy supply (only for passive working transponders) and the modulation efficiency (for passive and semi-passive working transponders) of transponders. Therefore, it has a direct effect for the maximum achievable range of such a system. The principle of the modulator has been already discussed above, so that this paragraph focuses primarily on the realization. Figure 11 shows the modulator, with and without RF shielding. The left part of the modulator is connected to the transponder's base board, the right SMA plug to the patch antenna as shown in Figure 10. The part within the RF shielding is responsible for the backscattering effects. A part of the incident RF wave is fed into the modulator. The part depends on the antenna (structural and antenna mode) and the reflection coefficient between antenna

impedance Z_{ant} and the load impedance Z_{load} of the connected modulator. This part is fed into the RF switch and the load impedance (either Z_1 or Z_2), which corresponds to the current state of the switch. The state of the switch is defined by a buffered microcontroller output. which itself shows the current voltage of the binary data stream to be sent. In the case of Z_1 (open circuit state), the incident wave is entirely reflected with no phase shift. State Z_2 (short circuit) also corresponds to a total reflection, but with a phase shift of 180°.
Measuring the load impedances of the modulator show a very good accordance with the theoretical results. Figure 12 shows the reflection coefficients within a Smith chart. As one can see the phase difference is not exactly π. Z_2 (short circuit) has nearly short circuit properties; Z_2 (open circuit) has nearly open circuit properties. The frequency range of the measurement was between 852 MHz and 882 MHz.

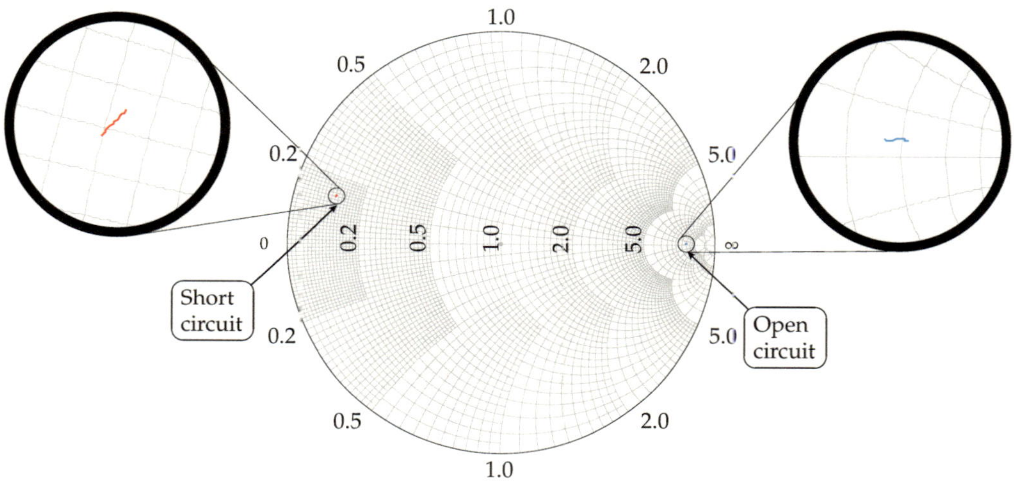

Fig. 12. Smith chart of modulator

4.2.4 Gold codes

The choice of an appropriate set of spreading codes is a key issue when designing CDMA systems. Gold codes seem to be one of the best codes to be used in UHF RFID systems. Mutti & Floerkemeier (2008), for instance, state that Gold codes outperform Kasami codes. Moreover, one Gold code family contains a large number of unique codes, which provides a high probability of finding a well-suited set of codes for a system to be designed.
Gold codes, first introduced by Robert Gold (Gold, 1967b), are commonly used in spread spectrum systems, such as WLAN and UMTS as well as in GPS (C/A code). The generation of Gold codes is quite simple as only two linear feedback shift registers (LFSR) are necessary to create one set of codes. Other advantages of Gold code are:

- Good balance between auto- and cross-correlation
- Flexibility in code length
- No user synchronization necessary, i.e. the transponders need not to be synchronized among each other

Because of above mentioned advantages, the proposed CDMA-based system uses Gold sequences.

However, Gold codes have a length of $2^m - 1$ with m being the order of each linear feedback shift register. For reasons of flexibility a Gold code generator has been implemented on the transponder's 32 bit µC. The choice fell upon a Gold code length of 127 ($m = 7$). The characteristic polynomial is 137_{dec} for the first LFSR and 143_{dec} for the second one. The initial value for the first LFSR is 85_{dec}. By choosing two Gold codes (Code 1 and Code 2) the second LFSR is initialized with 127_{dec} for the first and with 111_{dec} for the second code. Then, a small adjustment was made to the generated Gold codes to be more compatible to the µC. A succeeding binary '0' is added to each code to move it to a length of 128 bit. To show the effect of this '0', the auto-correlation function (ACF) and cross-correlation function (CCF) have been evaluated for both Codes. Figure 13 shows the ACF Φ_{cc} of the original 127 bit Gold codes. Figure 14 illustrates the ACF $\Phi_{cc}(\tau)$ of the adjusted (127+1 bit) Gold codes. The results are slightly higher values beyond the peak value at $\tau = 0$. As not only the auto-correlation counts, the corresponding cross-correlation $\Phi_{1_2}(\tau)$ between the two codes are presented in Figure 15. As expected the values of the adjusted codes are slightly higher compared to the original ones, but without loosing the typical noise-like character. This means, that the effect of the added '0' is negligible for further considerations. However, final system implementations have to consider that fact.

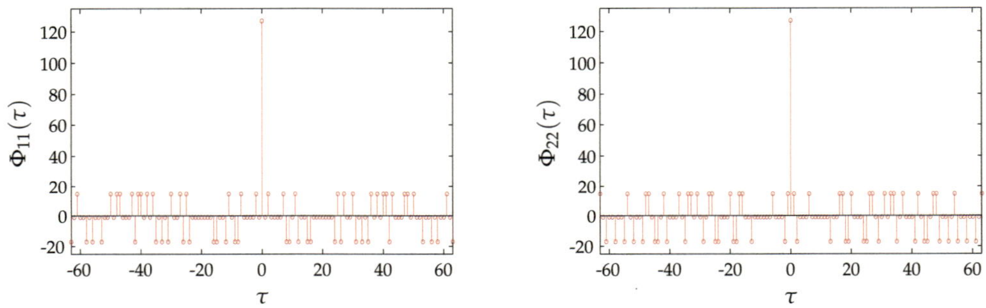

(a) ACF of Code 1 (b) ACF of Code 2

Fig. 13. ACF of original Gold codes

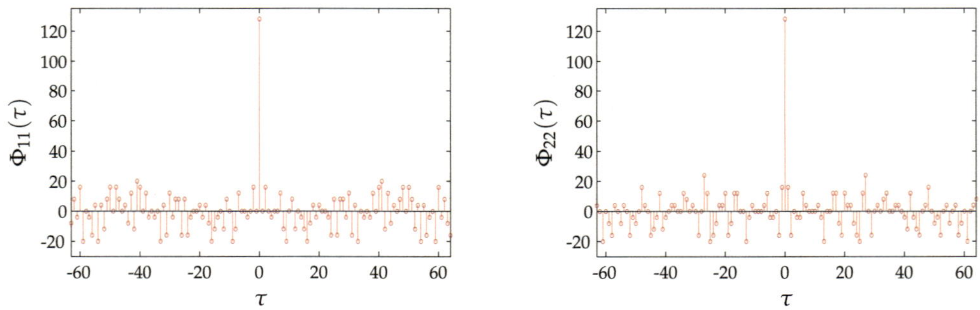

(a) ACF of Code 1 (b) ACF of Code 2

Fig. 14. ACF of adjusted Gold codes

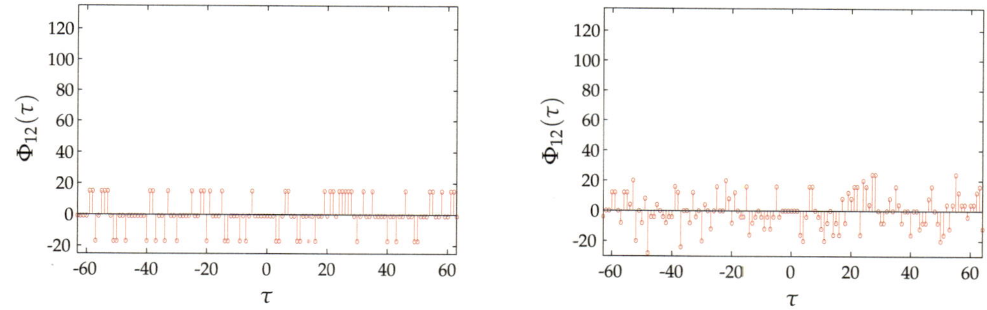

(a) CCF of original Code 1 and Code 2

(b) CCF of adjusted Code 1 and Code 2

Fig. 15. CCF of both, original and adjusted Gold codes

4.3 RX system path

The major tasks of the *Receiving system* are:

- Receive incoming signals from several transponders, i.e., downmixing, analog baseband processing and A/D conversion
- Find separate data streams (transponders) by despreading, demodulating and decoding the signals

The *Receiving system* mainly consists of a hardware part that is needed to mix down the backscattered RF signal, centered at $f_c = 866.5$ MHz, into baseband, despread, demodulate, and decode the baseband signal in order to determine the transponders' data. Figure 16 presents the structure of this receiving part of the RFID reader. The incoming RF signal is caught by a receiving antenna (RX) and amplified by a following low noise amplifier (LNA). A subsequent Zero-IF IQ-Demodulator mixes down the RF signal directly to baseband. The output of the demodulator consists of differential I- and Q-signals, which are band-pass filtered, twice amplified and active low-pass filtered. It has to mentioned that the IQ signals are completely handled differentially throughout the amplifier and filter stages to keep the signal-to-noise ratio (SNR) at a high level. The succeeding Analog-to-Digital conversion (ADC) module samples both, the I- and Q-signal, simultaneously. The A/D converted signals are fed into a digital signal processor (DSP) block with a data rate of 450 Mbps (Sampling of 2 channels with each channel having a resolution of 15 bit (14 data + 1 status bit) including a sampling rate of 15 Msps). The DSP module despreads, demodulates and decodes this data stream. The results are the user data of each recognized transponder.
The following paragraphs focus on the details of the receiving system.

4.3.1 Demodulator

The incoming low-noise amplified signal is fed into the demodulator. The demodulator uses the second RF synthesizer signal (the first is used as RF signal source for the transmit path, see above) as local oscillator (LO) source, to mix down the RF signal directly into baseband (Zero-IF). The demodulator is based on the LT5575 chip (Linear Technology, 2010a) and is 50 Ω-matched between 865 MHz and 868 MHz. The output of the demodulator is differential with 2 I- and 2 Q-signals, respectively.

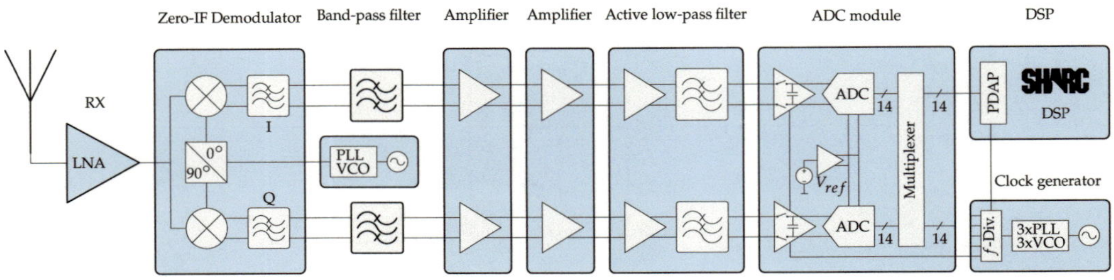

Fig. 16. Architecture of receiving system

4.3.2 Band-pass filter

The differential working band-pass filter, which succeeds the demodulator, is used to suppress the DC-part of the baseband signal, i.e. mainly the non-information carrying down-mixed carrier signal, and high-frequency disturbing signals (from the internal mixer of the demodulator). Therefore the passband is set between 16 kHz and 20 MHz.

4.3.3 Amplifier stage

The following amplifier stage is build upon two differential amplifiers (LTC6421-20 (Linear Technology, 2010d) and LTC6420-20 (Linear Technology, 2010c)), each with a differential voltage gain of 10 V/V.

4.3.4 Active anti-aliasing filter

The last analog signal processing stage is an active anti-aliasing filter for the succeeding ADC module. The cut-off frequency of the 4th order low-pass filter (Chebyshev characteristic) is currently set to 2.5 MHz. This stage is based on an LT6604-2.5 (Linear Technology, 2010b).

4.3.5 A/D conversion

One very important part of the receiving system is a well-designed A/D conversion stage for the baseband signal. The subjective of the ADC module is a time synchron sampling of the differential I- and Q-signals. The module is based on a dual A/D converter of type AD9248 from Analog Devices (2010a). Two channels may be sampled synchronously with a resolution of 14 bit per channel. Maximum sampling rate is 40 Msps. As the fast parallel input of the succeeding DSP module has only 20 bit the internal multiplexer of the A/D converter is used to transmit the I- and Q-data after each other. Therefore one status bit is used to indicate the current transmitted channel data. Here, the A/D converter is driven with 15 Msps per channel, which corresponds to an overall sampling clock rate of 30 MHz. The 14 bit per channel plus the status bit and the sampling rate, generate in total a data rate of 450 Mbps to be handled by the subsequent DSP module.

4.3.6 DSP module

The purpose of the DSP is the handling of all calculations, necessary to evaluate the transponders' user data. Therefore, the following stages are necessary:

- Data acquisition (from ADC module)
- Despreading of baseband signals
- Demodulation of despreaded signals
- Decoding of demodulated data

The following paragraphs give a short introduction to these topics. The data acquisition phase has to be accomplished only once, against what the following stages have to be passed through by every transponder respectively spreading code available.

4.3.6.1 Data acquisition

As the amount of data to handle is quit large (450 Mbps) the data streams are not handled in real time. However, through the usage of this DSP (ADSP-21469 from Analog Devices (2010b)) the processing speed is quite high. The A/D converted data signals are acquired through the DSP's PDAP (Parallel Data Aquisition Port) interface. From there, they are transfered to an internal 8x32 bit buffer. Finally, the data are passed via DMA access to an internal memory. As of limited memory capabilities the data is transferred block-wise to the external memory. As the sampled values are stored as 32 bit values (DWORD), the amount of data for one shot (duration is $T_{shot} \approx 188$ µs) is 90112 samples per channel, so in total 720896 bytes or 704 kbytes.

4.3.6.2 Despreading

The process of despreading is the most calculation intensive operation the DSP has to handle. As this phase needs more time than the data acquisition process the system is, up-to-date not able to work real-time. Parallel processing would be a good solution. The DSP itself has a clock rate of 450 MHz.

Despreading data from the baseband signal has to be done for I- and Q-channel separately. The despreading operation is realized using the cross-correlation between I and Q signals and the origin codes used by every transponder in the field. If $s[k]$ is the I or Q signal and $c[k]$ one of the corresponding codes of one of the transponders, the cross-correlation $\Phi_{s,c}(\tau)$ between these signals is done by multiplying every time instance signal s with code c. Equation (15) shows the corresponding relationship between $c[k]$ and $s[k]$, whereas ★ matches the convolution function:

$$[s \bigstar c][\tau] = \Phi_{s,c}(\tau) = \sum_{t=-\infty}^{+\infty} s^*[t] \cdot c[\tau + t] \qquad (15)$$

A code length of 128 chips corresponds to 1280 samples (R_{chip} = 1.5 Msps and R_{sample} = 15 Msps) and 90112 samples per channel for I and Q. This results into 230,686,720 multiplications and 180,224 additions.

One goal was to reduce this high amount of operations. This is realized through estimation of the time moments the chips appear within the IQ signals. This estimation method works as follows. The IQ baseband signal is sampled and correlated among the first $2 \cdot 1280 = 2560$ samples. This results in 6,553,600 multiplications and 5120 additions. The first maximum, corresponding to the first peak indicates the initial index i_0 to start the despreading process. The following peaks are estimated by jumping from i_0, 1280 samples ahead. As certain incertitudes (oscillators, etc.) will lead to synchronization errors, the correlation is not only made at sample index $i_0 + n \cdot 1280$, but at 5 samples before and after the estimated time index. That means, the second peak is determined by executing the cross-correlation $\Phi_{i,1}(\tau)$ as given in Equation (16).

$$\Phi_{i,1}(\tau) = \sum_{t=i_0+1280-5}^{i_0+1280+5} s^*[t] \cdot c[\tau + t] \qquad (16)$$

The result is 11 correlations per peak and a new synchronization index, as the new peak indicates the next starting point for the succeeding peak estimation. With 70 data peaks within one shot and 1 within the initial guess, the total number of correlations per channel is $2560 + 69 \cdot 11 = 3319$. This leads to 8,496,640 multiplications and 6,638 additions in total for both channels. This is only 3.6% of the full correlation.

4.3.6.3 Demodulation

The process of demodulation inherits the merge of the I and Q signals. According to their signal quality, estimated through the maximum correlation values, the signals are weighted and superimposed. This process of demodulation is beyond this paper's scope and not further described.

4.3.6.4 Decoding user data

The demodulated signal stream is Manchester coded (Loeffler et al., 2010) and needs to be decoded accordingly. The resulting data stream corresponds to the transponder's respectively the user data.

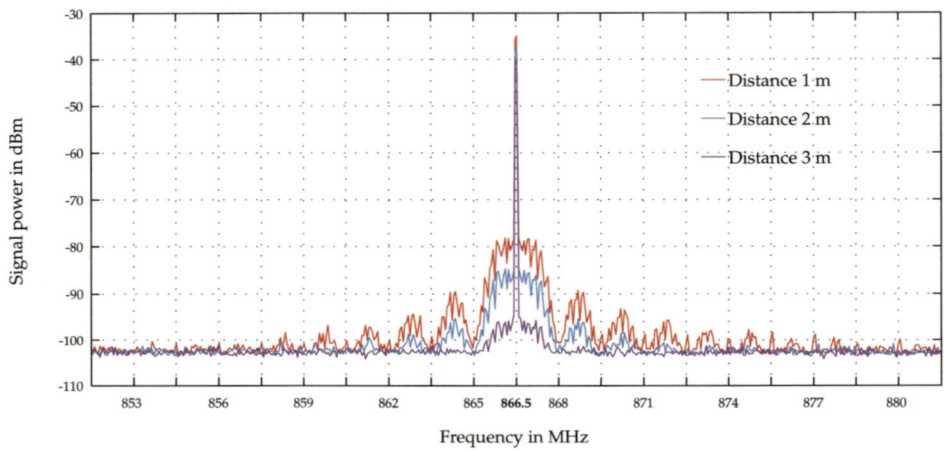

Fig. 17. Spectrum of backscattered signal from transponder

5. Measurements

This section presents measurements of various parts of the system, including transponder, analog baseband processing and DSP.

5.1 Transponder measurements

Figure 17 shows the spectrum of the backscattered transponder signals. For this measurement an RF signal ($P_{TX} = 10$ dBm, $f_{\text{carrier}} = 866.5$ MHz) is fed into the linear polarized transmit antenna. One transponder is placed at a distance of 1, 2 and 3 m. The resulting reflected signal spectrum after the receiving antenna is shown in Figure 17. As expected, the backscattered signal parts drop with increasing distance from the reader's antennas.
The IQ constellation diagrams of the received RF signal are shown throughout Figure 18(a) to Figure 18(c). It can be shown that the backscattered signals show a mixture between ASK and PSK modulation. For instance, as in Figure 18(a), the mean of the data points (from the two states of the one transponder) is not the origin (0,0). This discrepancy is the effect of multipath and structural antenna mode scattering. Same applies for Figure 18(b) with 2

transponders, generating $2^2 = 4$ constellation points, and Figure 18(c) with 3 transponders, generating $2^3 = 8$ constellation points. The number of constellation points for n transponders is 2^n because all n transponders have 2 states sharing the same coherent RF signal from the reader.
However, as expected the transponders show a near exact BPSK modulation (as configured in Subsubsection 4.2.3), if the ASK part is neglected.

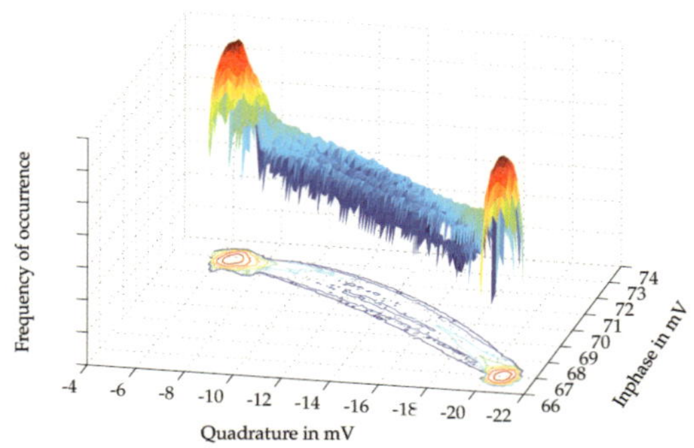

(a) IQ constellation diagram for 1 transponder

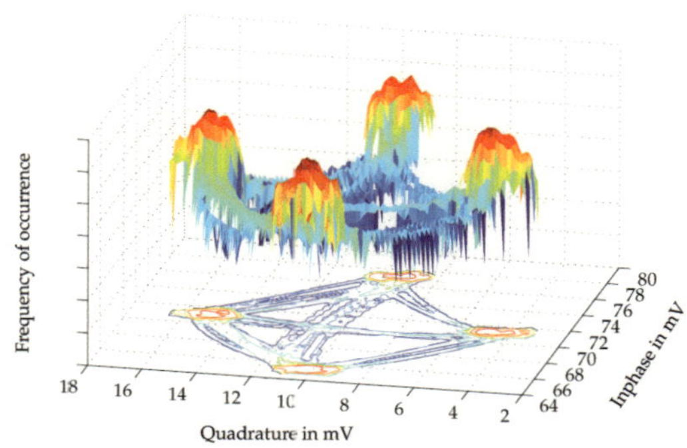

(b) IQ constellation diagram for 2 transponders

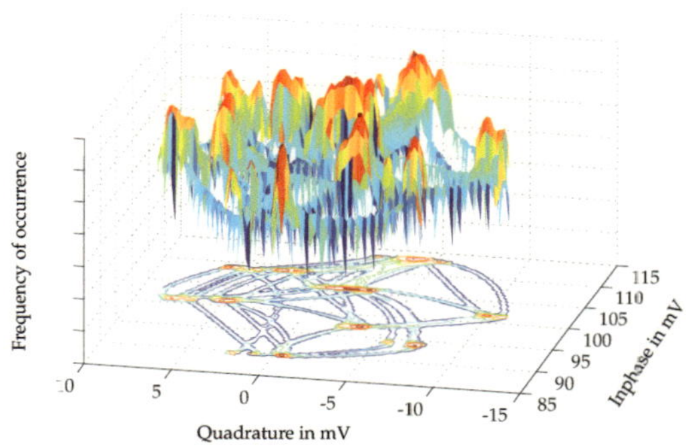

(c) IQ constellation diagram for 3 transponders

Fig. 18. Various IQ constellation diagrams for 1, 2 and 3 transponders in the field of the reader

5.2 RX measurements

Two measurements have been carried out to show the basic working principle of the analog baseband processing module. The goal of this module is the signal conditioning for the succeeding ADC module. Figure 19(a) shows the output of the demodulator, i.e. the I- and Q-signals. As mentioned above these signals are handled differentially (I_+, I_-, Q_+ and Q_-). To simplify matters the differential signals have been put together ($I = I_+ - I_-$ and $Q = Q_+ - Q_-$) . The signals are amplified and filtered with a resulting signal as shown in Figure 19(b). The signals were recorded with 2 transponders in the field. As in the IQ measurements before, 2 transponders generate $2^2 = 4$ different signal levels (evaluated from Figure 19(b)) leading to a quasi QPSK-like signal with an elliptic distribution of the absolute values:

$$\begin{aligned} 0.1\text{ V} + j0.2\text{ V} &\equiv 0.23\ e^{+j49.4^\circ} \equiv 0.23\ e^{j0^\circ} \\ 0.3\text{ V} - j0.4\text{ V} &\equiv 0.55\ e^{-j50.5^\circ} \equiv 0.55\ e^{j260.1^\circ} \\ -0.2\text{ V} - j0.2\text{ V} &\equiv 0.27\ e^{-j123.7^\circ} \equiv 0.27\ e^{j186.9^\circ} \\ -0.4\text{ V} + j0.5\text{ V} &\equiv 0.59\ e^{-j233.6^\circ} \equiv 0.59\ e^{j77.0^\circ} \end{aligned} \tag{17}$$

Although the phase relations between the different states is about 90° in this measurement, usually the phase is randomly distributed, being dependent on the geometric formation between transponder and reader antennas. This snapshot was taken because of easy visibility.

5.3 DSP measurements

The DSP module comes with some debugging functionalities. One of these functionalities is able to provide the DSP values, from its internal or external memories, via USB to a host PC. Figure 20 shows the results of a full cross-correlation. For simplicity the CCFs have been normalized to one. The values show the maximum number of samples (90112) and the peaks, with each peak describing a bit. The value of the bit may be positive (+1) or negative (−1). The difference between the peaks and the noise floor is an indicator for the quality of the communication link.

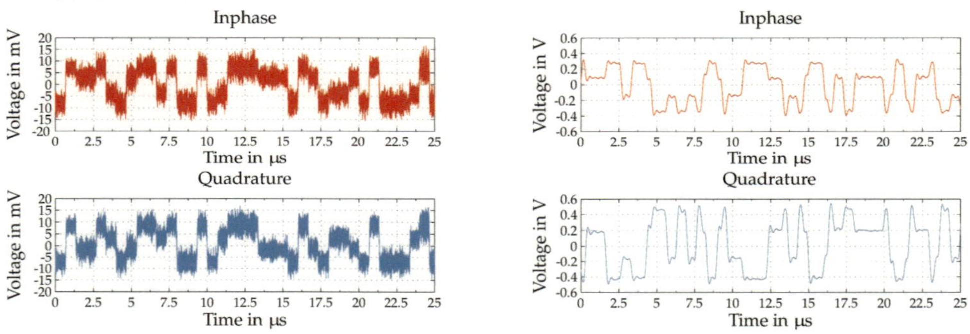

(a) IQ signal after demodulator / 2 Transponders

(b) IQ signal after baseband processing / 2 Transponders

Fig. 19. IQ signals after demodulator (right) and after baseband processing (left)

6. Results

According to the measurements the proposed system worked as expected. It was proved that the UHF RFID system for broadcasting information data using a CDMA method worked out very good. During the experiments there was a maximum distance to the antennas being around 15 m. The transmitted RF-power at 866.5 MHz was 20 dBm. The introduced transponders are semi-passive, which means that the communication link is still passive, whereas the data generation (on the transponder's side) is active, driven by 3.3 V power supplies.

Smaller problems arose, when various transponder had a different path length to the antennas. In that case one transponder (the nearest) dominated the second transponder (more far away) which often occurred to a non-detection of transponder two. This problem is known

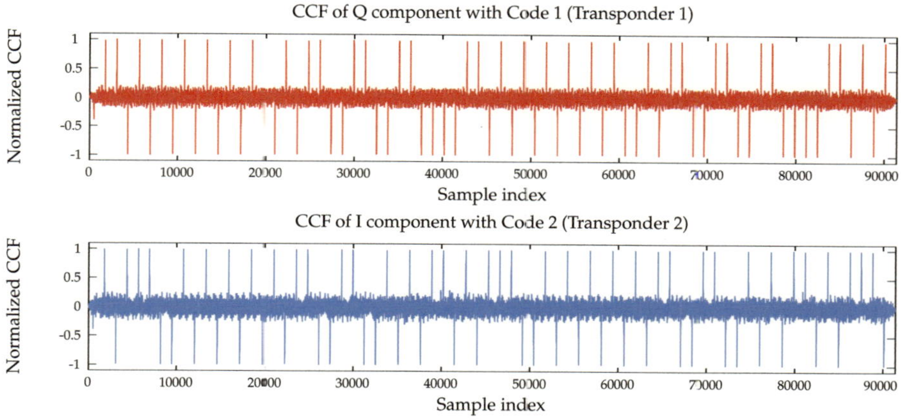

Fig. 20. Cross-correlation of signals with origin spreading codes - Process of despreading / 2 Transponders

in CDMA systems and is referred to as near-far problem (Andrews, 2005). One possibility to reduce the near-far effect is the usage of Huffman sequences (Liu & Guo, 2008). But this approach asks for more than 2 states of the load impedance of the transponder's modulator. Nevertheless, carried out indoor experiments showed that the near-far effect of the proposed system is, in fact, very low.
Also, theoretical work, which states an advantage (this statement is only valid for certain cases) of CDMA-based RFID systems compared to state-of-the-art RFID systems based on TDMA methods, complies with the measured results of the proposed CDMA-based UHF RFID system.

7. Conclusion

This article presented an implementation of a CDMA-based RFID system working in the UHF region. At the beginning the article gave a short introduction to anti-collision methods used in RFID technology. Subsequently, a performance comparison was made to show the effect of using CDMA in RFID. It could be stated, that CDMA does outperform traditional TDMA methods, but only in particular fields of applications. The implemented RFID system itself is build upon a *Transmitting system* providing a continuous electromagnetic wave. This emitted RF carrier is backscattered through one or more designed UHF tags. Each of these semi-passive operating transponders generate a unique spreading sequence. The proposed spreading sequences are Gold codes providing a good orthogonality. A simple modulator on the transponder generates the desired backscatter signal. The *Receiving system* captures this signal by down mixing the RF signal to baseband. Further analog signal processing and subsequent A/D conversion gives the DSP the chance to despread, demodulate and decode the desired transponder signals.
The significant advantage of such a structure compared to present systems lies in the ability to avoid particular TDMA-based anti-collision schemes. Certainly, this will lead to less time needed for *inventorizing* RFID tags, as this can be achieved within one time slot. However, the number of tags to be read this way, is somewhat limited (due to the usage of CDMA), whereas TDMA methods may recognize a huge amount of transponders, indeed, at the expense of time to identify. Finally, one can say, that the deployment of CDMA is useful in cases where the number of transponders has an upper limit or is fixed. For such cases the time for detection

may be minimized using appropriate spreading codes. Fields of application mainly include closed systems, e.g., found in industrial facilities.

8. Acknowledgment

I would like to thank Fabian Schuh, and in particular Ingo Altmann, without whom this publication would not have been possible. His ideas, work, and research on this topic made a big contribution to this chapter. Also, I would like to thank my colleagues for their very productive ideas and valuable discussions.
To my wife Sonja, my daughters Jenny and Jolina, and my son Tom, for having the patience with me, despite my long periods in the office which decrease the amount of time I can spend with them.

9. References

Abramson, N. (1970). THE ALOHA SYSTEM: another alternative for computer communications, *Proceedings of the November 17-19, 1970, fall joint computer conference*, AFIPS '70 (Fall), ACM, New York, NY, USA, pp. 281–285. URL: *http://doi.acm.org/10.1145/1478462.1478502*

Aein, Joseph M. (1964). Multiple Access to a Hard-Limiting Communication-Satellite Repeater, *Space Electronics and Telemetry, IEEE Transactions on* 10(4): 159–167. URL: *10.1109/TSET.1964.4337583*

Analog Devices (2010a). AD9248: Dual 14-Bit, 20/40/65 MSPS, 3 V Analog-to-Digital Converter. URL: *http://www.analog.com/en/analog-to-digital-converters/ad-converters/ad9248/products/product.html*

Analog Devices (2010b). ADSP-21469: High Performance Fourth Generation DSP. URL: *http://www.analog.com/en/embedded-processing-dsp/sharc/adsp-21469/processors/product.html*

Andrews, J. (2005). Interference cancellation for cellular systems: A contemporary overview, *IEEE Wireless Communications* 12(2): 19–29.

Bang, O., Kim, S. & Lee, H. (2009). Identification of RFID tags in dynamic framed slotted Aloha, *Advanced Communication Technology, 2009. ICACT 2009. 11th International Conference on*, Vol. 01, pp. 354 –357.

Bertsekas, D. & Gallager, R. (1992). *Data networks (2nd ed.)*, Prentice-Hall, Inc., Upper Saddle River, NJ, USA.

Choi, J. H., Lee, D. & Lee, H. (2007). Query tree-based reservation for efficient RFID tag anti-collision, *Communications Letters, IEEE* 11(1): 85 –87.

Cui, Y. & Zhao, Y. (2009). A modified Q-parameter anti-collision scheme for RFID systems, *Ultra Modern Telecommunications Workshops, 2009. ICUMT '09. International Conference on*, pp. 1 –4.

Dobkin, D. (2008). *The RF in RFID: passive UHF RFID in practice*, Newnes.

EPCglobal Inc. (2008). Class 1 Generation 2 UHF Air Interface Protocol Standard "Gen 2" v. 1.2.0.

Finkenzeller, K. (2003). *RFID handbook*, Wiley West Sussex, England.

Fuschini, F., Piersanti, C., Paolazzi, F. & Falciasecca, G. (2008). On the Efficiency of Load Modulation in RFID Systems Operating in Real Environment, *Antennas and Wireless Propagation Letters, IEEE* 7: 243 –246.

Gold, R. (1967a). Optimal binary sequences for spread spectrum multiplexing (corresp.), *Information Theory, IEEE Transactions on* 13(4): 619 – 621.

Gold, R. (1967b). Optimal binary sequences for spread spectrum multiplexing (Corresp.), *Information Theory, IEEE Transactions on* 13(4): 619 – 621.

Gopalan, S., Karystinos, G. & Pados, D. (2005). Capacity, throughput, and delay of slotted ALOHA DS-CDMA links with adaptive space-time auxiliary-vector receivers, *Wireless Communications, IEEE Transactions on* 4(1): 79 – 92.

Hansen, R. (1989). Relationships between antennas as scatterers and as radiators, *Proceedings of the IEEE* 77(5): 659 –662.

IPICO (2009). IPICO's IP-X RFID Air-interface Protocol.
URL: *http://www.ipico.com/*

Karthaus, U. & Fischer, M. (2003). Fully integrated passive uhf rfid transponder ic with 16, $7\mu w$ minimum rf input power, *IEEE* 38(10): 1602–1608.

Kleinrock, L. & Tobagi, F. (1975). Packet Switching in Radio Channels: Part I–Carrier Sense Multiple-Access Modes and Their Throughput-Delay Characteristics, *Communications, IEEE Transactions on* 23(12): 1400 – 1416.

Lee, D., Bang, O., Im, S. & Lee, H. (2008). Efficient dual bias Q-Algorithm and optimum weights for EPC Class 1 Generation 2 Protocol, *Wireless Conference, 2008. EW 2008. 14th European*, pp. 1 –5.

Linear Technology (2010a). LT5575 - 800MHz to 2.7GHz High Linearity Direct Conversion Quadrature Demodulator.
URL: *http://www.linear.com/pc/productDetail.jsp?navId=H0,C1,C1011,C1725,P36240*

Linear Technology (2010b). LT6604-2.5 - Dual Very Low Noise, Differential Ampli?er and 2.5MHz Lowpass Filter.
URL: *http://www.linear.com/pc/productDetail.jsp?navId=H0,C1,C1154,C1008,C1148, P85251*

Linear Technology (2010c). LTC6420-20 - Dual Matched 1.8GHz Differential Amplifiers/ADC Drivers.
URL: *http://www.linear.com/pc/productDetail.jsp?navId=H0,C1,C1154,C1009,C1126, P80614*

Linear Technology (2010d). LTC6421-20 - Dual Matched 1.3GHz Differential Amplifiers/ADC Drivers.
URL: *http://www.linear.com/pc/productDetail.jsp?navId=H0,C1,C1154,C1009,C1126,P80589*

Linnartz, J.-P. M. & Vvedenskaya, N. D. (2009). DS-CDMA Packet Network with Random Access.
URL: *http://www.wirelesscommunication.nl/reference/chaptr06/stack/cdmastck.htm*

Liu, D., Wang, Z., Tan, J., Min, H. & Wang, J. (2009). ALOHA algorithm considering the slot duration difference in RFID system, *RFID, 2009 IEEE International Conference on*, pp. 56 –63.

Liu, H. & Guo, X. (2008). A passive UHF RFID system with Huffman sequence spreading backscatter signals, *Proceedings of the 1st international conference on The internet of things*, Springer-Verlag, pp. 184–195.

Liu, Q., Yang, E.-H. & Zhang, Z. (2001). Throughput analysis of CDMA systems using multiuser receivers, *Communications, IEEE Transactions on* 49(7): 1192 –1202.

Liu, Z. & El Zarki, M. (1994). Performance analysis of DS-CDMA with slotted ALOHA random access for packet PCNs, *Personal, Indoor and Mobile Radio Communications, 1994. Wireless Networks - Catching the Mobile Future., 5th IEEE International Symposium on*, Vol. 4, pp. 1034 –1039 vol.4.

Lo, F. L., Ng, T. S. & Yuk, T. (1996). Performance analysis of a fully-connected, full-duplex CDMA ALOHA network with channel sensing and collision detection, *Selected Areas in Communications, IEEE Journal on* 14(9): 1708 –1716.

Loeffler, A., Schuh, F. & Gerhaeuser, H. (2010). Realization of a CDMA-based RFID System Using a Semi-active UHF Transponder, *Wireless and Mobile Communications (ICWMC), 2010 6th International Conference on*, pp. 5 –10.

Maguire, Y. & Pappu, R. (2009). An Optimal Q-Algorithm for the ISO 18000-6C RFID Protocol, *Automation Science and Engineering, IEEE Transactions on* 6(1): 16 –24.

Makwimanloy, S., Kovintavewat, P., Ketprom, U. & Tantibundhit, C. (2009). A novel anti-collision algorithm for high-density RFID tags, *Electrical Engineering/Electronics, Computer, Telecommunications and Information Technology, 2009. ECTI-CON 2009. 6th International Conference on*, Vol. 02, pp. 848 –851.

Mutti, C. & Floerkemeier, C. (2008). CDMA-based RFID Systems in Dense Scenarios: Concepts and Challenges, *RFID, 2008 IEEE International Conference on*, pp. 215–222.

Nikitin, P. & Rao, K. (2008). Antennas and propagation in uhf rfid systems, *RFID, 2008 IEEE International Conference on*, pp. 277 –288.

Pardo, D., Vaz, A., Gil, S., Gomez, J., Ubarretxena, A., Puente, D., Morales-Ramos, R., Garcia-Alonso, A. & Berenguer, R. (2007). Design criteria for full passive long range uhf rfid sensor for human body temperature monitoring, *RFID, 2007. IEEE International Conference on*, pp. 141–148.

Penttila, K., Keskilammi, M., Sydanheimo, L. & Kivikoski, M. (2006). Radar cross-section analysis for passive RFID systems, *Microwaves, Antennas and Propagation, IEE Proceedings -* 153(1): 103 – 109.

Pupunwiwat, P. & Stantic, B. (2010). A RFID Explicit Tag Estimation Scheme for Dynamic Framed-Slot ALOHA Anti-Collision, *Wireless Communications Networking and Mobile Computing (WiCOM), 2010 6th International Conference on*, pp. 1 –4.

Rembold, B. (2009). Optimum modulation efficiency and sideband backscatter power response of RFID-tags, *Frequenz - Journal of RF-Engineering and Telecommunications* 63(1 -2): 9 –13.

Roberts, L. G. (1975). ALOHA packet system with and without slots and capture, *SIGCOMM Comput. Commun. Rev.* 5: 28–42.
URL: *http://doi.acm.org/10.1145/1024916.1024920*

Sakata, A., Yamazato, T., Okada, H. & KATAYAMAt, M. (2007). Throughput Comparison of CSMA and CDMA slotted ALOHA in Inter-Vehicle Communication, *Telecommunications, 2007. ITST '07. 7th International Conference on ITS*, pp. 1 –6.

Sastry, A. (1984). Effect of Acknowledgment Traffic on the Performance of Slotted ALOHA-Code Division Multiple Access Systems, *Communications, IEEE Transactions on* 32(11): 1219 – 1222.

van Nee, R., van Wolfswinkel, R. & Prasad, R. (1995). Slotted ALOHA and code division multiple access techniques for land-mobile satellite personal communications, *Selected Areas in Communications, IEEE Journal on* 13(2): 382 –388.

Wang, L.-C. & Liu, H.-C. (2006). A Novel Anti-Collision Algorithm for EPC Gen2 RFID Systems, *Wireless Communication Systems, 2006. ISWCS '06. 3rd International Symposium on*, pp. 761 –765.

Zhang, Z., Lu, Z., Pang, Z., Yan, X., Chen, Q. & Zheng, L.-R. (2010). A Low Delay Multiple Reader Passive RFID System Using Orthogonal TH-PPM IR-UWB, *Computer Communications and Networks (ICCCN), 2010 Proceedings of 19th International Conference on*, pp. 1 –6.

16

An Unconditionally Secure Lightweight RFID Authentication Protocol with Untraceability

Hung-Yu Chien[1], Jia-Zhen Yen[2] and Tzong-Chen Wu[2,3]
[1]Department of Information Management, National Chi-Nan University
[2]Department of Information Management, National Taiwan University of Science and Technology
[3]Taiwan Information Security Center (TWISC) at National Taiwan University of Science and Technology
Taiwan

1. Introduction

Radio frequency identification (RFID) is a wireless technology that uses radio signals to identify objects automatically and remotely. The most popular tags are passive devices owing to their low cost. Nowadays, RFID devices are widely deployed in many applications, such as supply chain management, inventory control, contactless credit card and so on, due to the low-cost and convenience in identifying objects with non-line-of sight reading, However, there are many potential security threats around the tiny RFID tags attached to users. The carrying items or privacy information contained in these tags might be compromised. Furthermore, low-cost makes these tags very resource-limited, which makes it very challenging to design secure protocols for these tags.

From the point of end user's side, a secure RFID system should provide the capability of location/content privacy protection, anonymity, untraceability and availability [2]. Several RFID lightweight authentication protocols like [4-10] have been developed, but not all of them satisfy all the security requirements. All the previously proposed protocols are designed to be computationally secure, i.e., the security depends on the hardness of solving mathematical problem. Recently, Alomair *et al.* [1] proposed an unconditionally secure lightweight RFID (UCS-RFID for short) protocol, and claimed that their protocol achieved unconditional secrecy and unconditionally integrity. The security of the UCS-RFID protocol depends on the freshness of the keys. However, the UCS-RFID protocol does not achieve backward untraceability, even though it does achieve forward untractability.

Forward and backward untraceability are important privacy properties for RFID authentication protocol [4]. Forward untraceability requires that even if the adversary reveals the internal state of a tag at time τ, the adversary still cannot know whether a transaction after time $\tau + \delta$ (for some $\delta > 0$) involves the same tag or not, provided that the adversary does not eavesdrop on the tag continuously after time τ. Backward untraceability requires that even if the adversary reveals the internal state of a tag at time τ, the adversary is not able to tell whether a transaction before time τ involves the same tag or not [3]. These two properties are important for the RFID systems that the equipped tags are low-cost and potentially prone to being captured and compromised.

Notation	Description
R	RFID reader
T_i	i-th RFID tag
S	Back-end database
p	A 2N-bit prime integer, where N is …..
Z_p	The finite integer ring with usual addition and multiplication modulo p
Z_p^*	The multiplicative group modulo p, Z_p^* contains all non-zero elements of Z_p; that is, $Z_p^* = Z_p \setminus \{0\}$
$n^{(m)}$	n denotes a 2N-bit random number which is drawn uniformly from the Z_p^*, m denotes that it is used in the m-th session
$n_l^{(m)}$	The left N most significant bits of $n^{(m)}$
$n_r^{(m)}$	The right N least significant bits of $n^{(m)}$
$K_i^{(m)}$	The secret keys of the RFID tag T_i . They consist of five subkeys, i.e., $K_i^{(m)} = (k_{a\ i}^{(m)}, k_{b\ i}^{(m)}, k_{c\ i}^{(m)}, k_{d\ i}^{(m)}, k_{e\ i}^{(m)})$. The superscript m denotes the m-th run, and the subscript i denote the i–th tag T_i.
$k_{a\ i}^{(0)}$	A subkey which is initially drawn independently and uniformly from Z_{2^N}
$k_{b\ i}^{(0)}$	A subkey which is initially drawn uniformly from Z_p
$k_{c\ i}^{(0)}$	A subkey which is initially drawn independently and uniformly from Z_p^*
$k_{d\ i}^{(0)}$	A subkey which is initially drawn independently and uniformly from Z_{2^N}
$k_{e\ i}^{(0)}$	A subkey which is initially drawn independently and uniformly from Z_p^* that will be used for updating the secret keys to maintain certain properties

Table 1. Notations or Symbols

In this book chapter, we first examine the USC-RFID protocol, and show that the USC-RFID protocol does not achieve backward untraceability. After that, we will extend the USC-RFID protocol to an enforced one with untraceability.

2. The UCS-RFID protocol

The UCS-RFID procotol [1] is a lightweight RFID authentication protocol and is the first RFID protocol providing unconditional security for low-cost tags. The UCS-RFID protocol has the merits that it does not require tags to support random number generation and it requires only one simple multiplication on tags. The security of this protocol mainly relies on the RFID reader's capability to deliver random numbers to RFID tags in an authenticated and secure way.

The UCS-RFID protocol consists of four phases: the tag identification phase, the reader authentication phase, the tag authentication phase, and the key updating phase (see Fig. 1 for more details). For the convenience of describing the UCS-RFID protocol, we first introduce the notations or symbols shown in Table 1. Initially, each tag T_i has a secret key set $K_i^{(0)}$ shared with the back-end database. In the following, we describe the m-th run of the protocol.

Tag identification phase

i. The reader R sends a *Hello* message to the tag T_i.

ii. T_i sends its message $A^{(m)}$ to R, and R forwards this message $A_i^{(m)}$ to the back-end database S.

iii. S looks up the database for the secret key $K_i^{(m)}$ corresponding to the message $A_i^{(m)}$. If the $A_i^{(m)}$ could be identified as a valid identifier, then S sends back the tag's secret key $K_i^{(m)}$ to R. Otherwise, the tag T_i is rejected.

Reader Authentication Phase

i. R generates a random number $n^{(m)}$, computes $B^{(m)} \equiv n^{(m)} + k_{b\ i}^{(m)} \bmod p$ and $C^{(m)} \equiv n^{(m)} \times k_{c\ i}^{(m)} \bmod p$, and then sends these two messages ($B^{(m)}$, $C^{(m)}$) to T_i.

ii. After receiving $B^{(m)}$ and $C^{(m)}$, T_i extracts $n^{(m)} \equiv (B^{(m)} - k_{b\ i}^{(m)}) \bmod p$, and then verifies its integrity via checking whether the equation $(B^{(m)} - k_{b\ i}^{(m)}) \times k_{c\ i}^{(m)} \equiv C^{(m)} \bmod p$ holds. If so, R is authenticated; otherwise, the tag aborts the protocol.

Tag Authentication Phase

i. T_i computes $D^{(m)} = n_l^{(m)} \oplus k_{d\ i}^{(m)}$ and returns this value.

ii. After receiving the value, R verifies whether the equation $D^{(m)} \overset{?}{=} n_l^{(m)} \oplus k_{d\ i}^{(m)}$ holds. If so, the tag is authenticated; Otherwise, the tag is rejected.

Key Updating Phase: After a successful mutual authentication between the tag and the reader, the secret key and the tag identifier are updated at the back-end database and the tag respectively as specified in Fig. 1. Fig. 1 depicts the protocol for the m-th run.

The above protocol cannot deter possible denial-of-service attacks (DOS attacks), and Alomair et al. had extended the above protocol to prevent DOS attacks and possible key exposure problem. Since these extensions are not relevant to our improvements, we will not discuss these parts for easy presentation, and interested readers are referred to [1] for details.

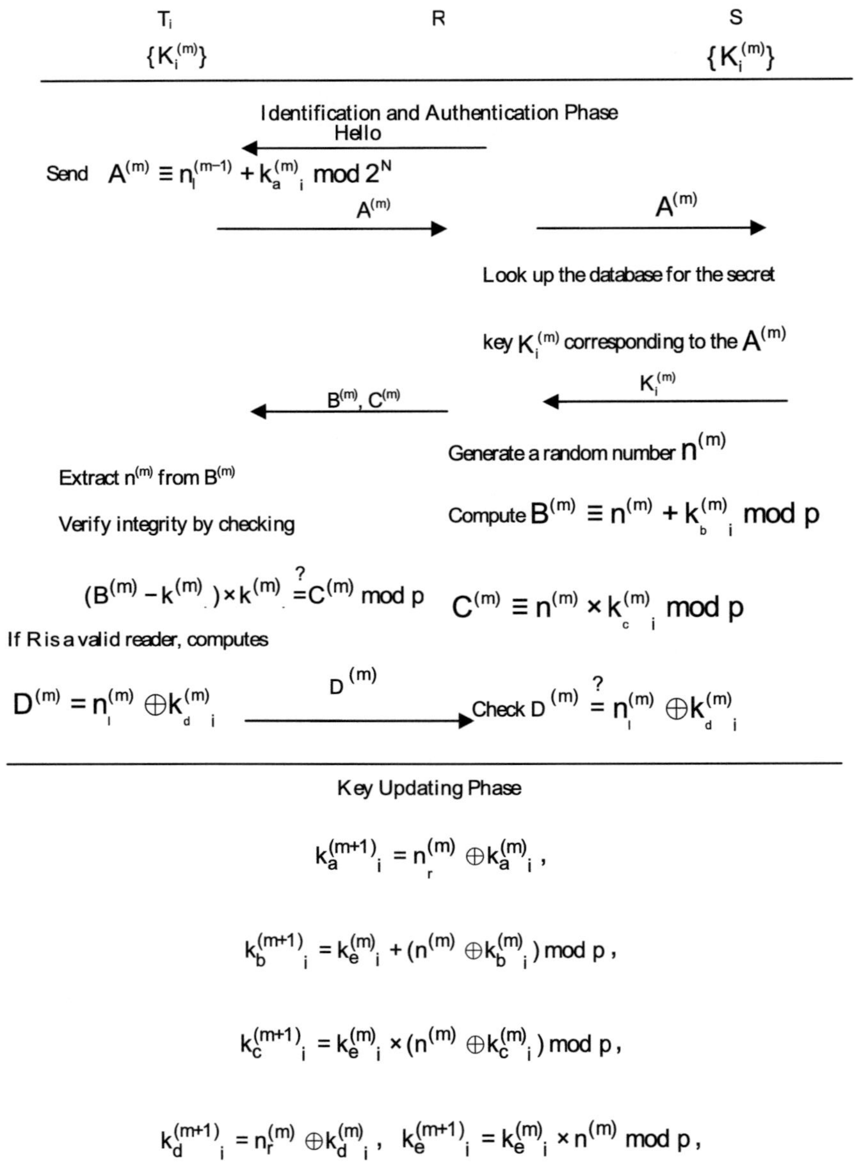

Fig. 1. The UCS-RFID protocol.

3. Extending the USC-RFID to untraceability

In Section 3.1, we examine the untraceability of the USC-RFID protocol, and then provide an improved scheme to enhance its untraceability.

3.1 Untraceability of the UCS-RFID protocol

Here we show that the UCS-RFID protocol does not provide backward untraceability as follows.

Suppose the tag T_i has been compromised and the internal secrets $A^{(m)} \equiv n_l^{(m-1)} + k_{a\ i}^{(m)} \bmod 2^N$ and $K_i^{(m)}$=($k_{a\ i}^{(m)}$, $k_{b\ i}^{(m)}$, $k_{c\ i}^{(m)}$, $k_{d\ i}^{(m)}$, $k_{e\ i}^{(m)}$) are revealed at time τ. Let (*A*, *B*, *C*, *D*) be one eavesdropped message. Then we can tell whether the message (*A*, *B*, *C*, *D*) comes from the same tag or not as follows.

1. Derive $n_l^{(m-1)} = A^{(m)} - k_{a\ i}^{(m)} \bmod 2^N$.
2. Derive $k_{d\ i}^{(m-1)} = D \oplus n_l^{(m-1)}$, $n_r^{(m-1)} = k_{d\ i}^{(m)} \oplus k_{d\ i}^{(m-1)}$ and $n^{(m-1)} = n_l^{(m-1)} \,||\, n_r^{(m-1)}$.
3. Now we can derive the previous internal state $k_{a\ i}^{(m-1)} = n_r^{(m-1)} \oplus k_{a\ i}^{(m)}$, $k_{e\ i}^{(m-1)} = k_{e\ i}^{(m)} \times (n^{(m-1)})^{-1} \bmod p$, $k_{b\ i}^{(m-1)} = (k_{b\ i}^{(m)} - k_{e\ i}^{(m-1)} \bmod p) \oplus n^{(m-1)}$, $k_{c\ i}^{(m-1)} = (k_{c\ i}^{(m)} \times (k_{e\ i}^{(m-1)})^{-1} \bmod p) \oplus n^{(m-1)}$ and $k_{d\ i}^{(m-1)} = n_r^{(m-1)} \oplus k_{d\ i}^{(m)}$.
4. Now we check whether the two equations $B \stackrel{?}{=} n^{(m-1)} + k_{b\ i}^{(m-1)} \bmod p$ and $C \stackrel{?}{=} n^{(m-1)} \times k_{c\ i}^{(m-1)} \bmod p$ hold. It is obvious that if the two equations hold, then the message (*A*, *B*, *C*, *D*) is the $(A^{(m-1)}, B^{(m-1)}, C^{(m-1)}, D^{(m-1)})$ from the compromised tag.

   We can recursively apply the above steps to trace the messages from the same tag for *i*-th run, where $i \le m-1$. That is, the USC-RFID protocol cannot provide backward untraceability.

Even though the USC-RFID protocol does not satisfy backward untraceability, it does provide forward untraceability. This is because, in forward untraceability, if the adversary reveals the internal state of a tag at time τ, it is required that the adversary does not eavesdrop on the tag *continuously* after time τ. It is this break of eavesdropping that makes the USC-RFID satisfy forward untraceability.

3.2 Enhancing the untraceability

The key to find the link in our backward traceability is that the equation $A^{(m)} = n_l^{(m-1)} + k_{a\ i}^{(m)} \bmod 2^N$ contains only one unknown value $n_l^{(m-1)}$ when the adversary learn the internal state $A^{(m)}$ and $K_i^{(m)}$=($k_{a\ i}^{(m)}$, $k_{b\ i}^{(m)}$, $k_{c\ i}^{(m)}$, $k_{d\ i}^{(m)}$, $k_{e\ i}^{(m)}$); therefore, the adversary can derive $n_l^{(m-1)} = A^{(m)} - k_{a\ i}^{(m)} \bmod 2^N$ and the other values accordingly. We also notice that each of the other key updating equations in the key updating phase contains at least two unknown values. Therefore, we can amend the protocol by simply modifying this equation $A^{(m)} = n_l^{(m-1)} + k_{a\ i}^{(m)} \bmod 2^N$ to contain two unknowns. One simple suggestion is that $A^{(m)} = n_l^{(m-1)} + k_{a\ i}^{(m-1)} \bmod 2^N$. With this modification, the adversary should solve two unknowns in each equation to derive the secret even assume he has learned the current state ($A^{(m)}$, $k_{a\ i}^{(m)}$, $k_{b\ i}^{(m)}$, $k_{c\ i}^{(m)}$, $k_{d\ i}^{(m)}$, $k_{e\ i}^{(m)}$). It, therefore, cannot provide adversaries a unique and deterministic link to trace the tag.

4. Conclusion

In this book chapter, we have shown that the UCS-RFID protocol which is the first unconditionally secure mutual authentication protocol for RFID systems cannot satisfy backward untraceability, and we have proposed a simple amendment to enhance its backward untraceability. The unconditional secure RFID protocol is very promising approach for RFID security. In this book chapter, we have enhanced the first unconditional secure RFID protocol to satisfy untraceability. Our future work is to further analyze and improve the security of unconditional secure RFID protocols.

5. References

[1] B. Alomair, A. Clark, J. Cuellar, and R. Poovendran, Securing Low-Cost RFID Systems: an Unconditionally Secure Approach, 2010 Workshop on RFID Security - RFIDsec'10 Asia, 2010.

[2] H. -Y. Chien and C. -S. Laih, ECC-Based Lightweight Authentication Protocol with Untraceability for Low-Cost RFID, Journal of Parallel and Distributed Computing. 69 (10) (2009) 848-853.

[3] R. C. -W. Phan, J. Wu and K. Ouafi, Privacy Analysis of Forward and Backward Untraceable RFID Authentication Schemes, 2008. Available from : <http://www.cacr.math.uwaterloo.ca/~dstinson/papers/bfrfid-2.pdf/>.

[4] A. D. Henrici, and P. MÄuller, "Hash-based Enhancement of Location Privacy for Radio-Frequency Identification Devices using Varying Identifiers," In the Proceedings of PerSec'04 at IEEE PerCom, 2004, pp.149-153.

[5] S. Karthikeyan, M. Nesterenko, "RFID security without extensive cryptography," Proceedings of the 3rd ACM workshop on Security of ad hoc and sensor networks, 2005, pp. 63-67.

[6] D. Molnar and D. Wagner, "Privacy and security in library RFID: Issues, practices, and architectures," Conference on Computer and Communications Security - CCS'04, 2004, pp. 210–219.

[7] M. Ohkubo, K. Suzki and S. Kinoshita, "Cryptographic Approach to 'Privacy-Friendly' Tags," In RFID Privacy Workshop, 2003.

[8] S. A. Weis, "Security and Privacy in Radio-Frequency Identification Devices," Masters Thesis MIT, 2003.

[9] G. Avoine, E. Dysli, and P. Oechslin, "Reducing time complexity in RFID systems," The 12th Annual Workshop on Selected Areas in Cryptography(SAC), 2005.

[10] H. Y. Chien, "SASI: A New Ultra-Lightweight RFID Authentication Protocol Providing Strong Authentication and Strong Integrity", IEEE Transactions on Dependable and Secure Computing 4(4), pp. 337-340, October, 2007.

17

Application of Monte Carlo Method for Determining the Interrogation Zone in Anticollision Radio Frequency Identification Systems

Piotr Jankowski-Mihułowicz and Włodzimierz Kalita
Department of Electronic and Communications Systems,
Rzeszów University of Technology
Poland

1. Introduction

Current problems that occur in the field of anticollision Radio Frequency IDentification (RFID) prototype systems are solved in experimental way (De Blasi et al., 2010; Lehto et al., 2009; Polivka et al., 2009; Brown, 2007; Clarke et al., 2006; Penttilä et al., 2006; Jones & Chung, 2007). The low efficiency coefficient of identification for the multiple objects localized in the space Ω_{ID} doesn't allow to realize practical projects, such as, the identification of Fast Moving Consumer Goods (FMCG) – Fig. 1. In the light of nascent and modified legal communications standards, like for example, Electronic Product Code (EPC) in the area of UHF and HF ISO 18000-6, ISO 15693, ISO 18000-3 normalizations, there is a necessity to continue complex theoretical research and experimental investigations in the range of simultaneous analysis of EM field, communication protocols, and electric aspects of operating conditions of efficiency identification in anticollision RFID systems.

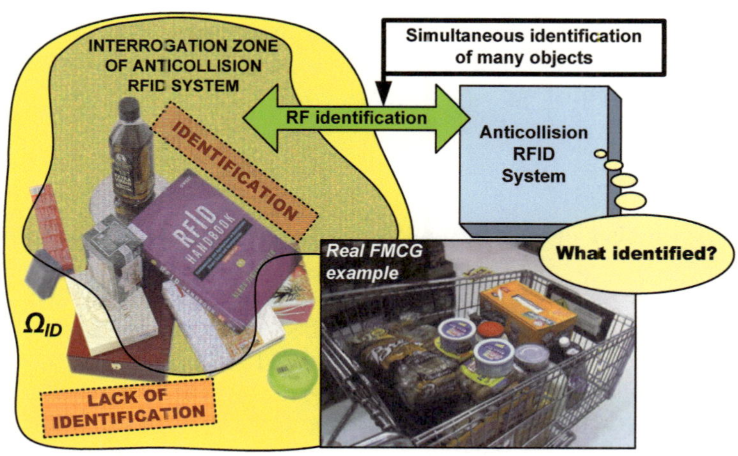

Fig. 1. Illustration of RFID automatic identification process

To generalise, the operation of passive anticollision inductive- (LF, HF), and also propagation (UHF) coupling RFID system is characterized by the interrogation zone (IZ) which is estimated in any direction of 3D space for a group of electronic tags. The elements of algorithm of identification of interrogation zone for anticollision RFID system with the consideration of the energetic (i.e. field and electrical) and communication aspects of operation conditions have been presented in the following chapter. For calculations of the interrogation zone the algorithm based on Monte Carlo (MC) method and a computer program with the use of Mathcad 14 (called *JankoRFIDmc'IZ*) has been utilized.

2. Determining the interrogation zone using MC method

Unequivocal estimation of the interrogation zone for anticollision RFID system depends on automatic identification process. In accordance with the conditions of the correct operation of any RFID system, different locations of many tags strongly change the functioning of an antenna unit array: read/write device (RWD) and individual tags. The problem of determining the interrogation zone is related to two cases. In the first of them, an assumption is made that the location of the n-tags group is determined, whereas in the second case, all possible locations of the group of n-tags in a space around the RWD antenna are going to be analyzed. The problem connected with the first case is realizable by the assumption that the process of determining the interrogation zone in RFID system will be carried out in a few feedback cycles which allow to find the proper location of tags. The statement "few feedback cycles" is related to the time which is accepted for determining the interrogation zone for all n-tags. The mentioned feedback cycles in the carried out simulation include a modification of the tags location which don't fulfil conditions of the correct RFID system operation. The problem from the second case is almost impossible to solve because the prolonged process of calculations would be ineffective. Seemingly, in that case, a method of "trial and error" during the search of the interrogation zone of RFID system might be easier to apply, however, the presented MC method is a well-founded alternative.

The presented premises lean towards the necessity of application of the techniques which make use of random numbers (Kalos & Whitlock, 2008). The result of this is the solution of the problem of the n-tags group location, and testing the functional efficiency of the antenna unit array: read/write device-tags, that is an estimation of anticollision RFID system interrogation zone for given efficiency of identification η_{ID}. The percentage of identification efficiency is given by the equation:

$$\eta_{ID} = \frac{l_{IDOK}}{n} \cdot 100\% \tag{1}$$

where l_{IDOK} is the number of tags for which the desired read/write operations have been properly done.

The problem contained in the MC method has a probabilistic nature, and it's solution is obtained by simulation of the given object (Rubinstein & Kroese, 2007). The simulation object is represented by the antenna unit array: RWD-tags with the consideration of a synthesis of this antenna unit array and according to all equations which are going to be determined during the synthesis of its electric model in an anticollision RFID system.

For a laboratory process of automatic objects identification the solution of the problem consists in finding the interrogation zone of given RFID system, with its shape, location and

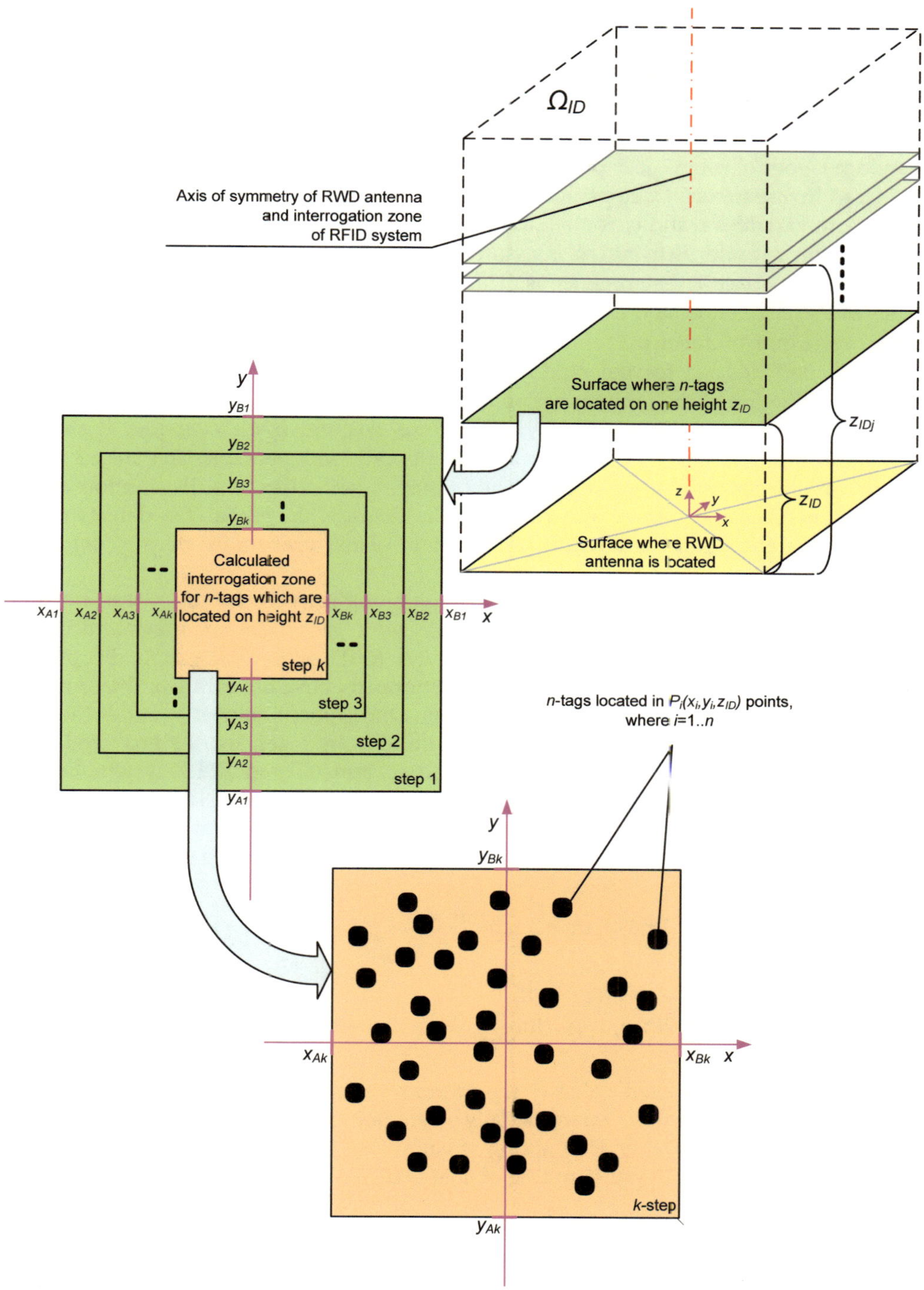

Fig. 2. Graphic representation of the process of determining the interrogation zone in anticollision RFID system using MC method

orientation in 3D space assumed. In the conducted research it was assumed that the demanded area should be square shaped and situated at the z_{ID} height, whereas it's location should be axially - symmetrical and parallel to RWD antenna (Fig. 2). Such an assumption results from the orientation of tags that are parallel to symmetrical RWD antenna which has, for example, circular or square shape in inductive coupling RFID systems.

A random layout of n-tags at P_i points of Cartesian space at (x_i,y_i,z_{ID}) has been assumed in - considered in a sequence - k steps during the search of the RFID system interrogation zone. The random variables x_i and y_i, for i=1..n obtain various values which can't be predicted, but for which the definite distribution is assumed. The electromagnetic field in any point of communication space is heterogeneous. This effect becomes the clearer, the nearer to the surface of RWD antenna, and the farther from its centre a point is situated. This knowledge allows to make a uniform (rectangular) distribution in intervals: $\langle x_{Ak},x_{Bk}\rangle$ for the random variable x_i, and $\langle y_{Ak},y_{Bk}\rangle$ for the random variable y_i, in k-step for the analyzed area. For uniform distribution of the random variables x_i and y_i, and for the definite values of x and y, the distribution functions are given by: $(x-x_{Ak})/(x_{Bk}-x_{Ak})$ and $(y-y_{Ak})/(y_{Bk}-y_{Ak})$. It should be noticed that the random variables x_i and y_i are mutually independent. This means that the random variables x_i and y_i are stochastically independent, since the distribution of the x_i does not depend on the value y_i and vice versa. In this case, the probability density of a pair of random variables (x_i,y_i) is equal to the product of the probability density (x_i) and (y_i) independently.

In order to determine that the RFID system is functioning correctly for given tags locations it is not enough to achieve the efficiency of identification η_{ID}=100% for n-tags and fulfill all conditions for a correct operation of anticollision RFID system. It cannot be predicted whether for k area in which all the conditions mentioned above are fulfilled, the coordinates sampling of tags locations on the surface of their arrangement, allows to fulfill the border case of a correct operation of the whole RFID system. In k-step for the analyzed area in which all the conditions of a correct operation of anticollision RFID system for given efficiency of identification are fulfilled, the practical use of the law of large numbers (Kalos & Whitlock, 2008) is the solution to this problem. For the random variables x_i and y_i independently, the strong law of large numbers for the analyzed case is given by:

$$PP\left(\lim_{m\to\infty} S_m(x_i) = \lim_{m\to\infty} \sum_{i=1}^{n\cdot m} \frac{x_i}{n\cdot m} = p = \frac{x_{Ak}+x_{Bk}}{2}\right) = 1 \tag{2}$$

$$PP\left(\lim_{m\to\infty} S_m(y_i) = \lim_{m\to\infty} \sum_{i=1}^{n\cdot m} \frac{y_i}{n\cdot m} = p = \frac{y_{Ak}+y_{Bk}}{2}\right) = 1 \tag{3}$$

where p denotes the expected value of the random variables x_i and y_i (which are equal to zero because the interrogation zone is axially - symmetrical and parallel to RWD antenna), and PP denotes the probability of sampling of variables for m approaching to infinity, but m denotes the number of multiple sampling of tags location (i.e. random variables x_i and y_i) for k analyzed area.

What follows from the equations (2) and (3) is that the sequences of random variables $S_m(x_i)$ and $S_m(y_i)$ converge with probability "1" to the expected value p=0 of the random variables x_i and y_i. It can be found that the m-tuple increase of the number of the random variables x_i and y_i sampling in k-step for the analyzed area lengthens the calculation process during the simulation of an antenna unit array. In accordance with the law of large numbers, the

probability of a correct estimation of the interrogation zone for RFID system increases. First of all, this is connected with the examination of a larger number of localized n-tags cases. If the conditions of the correct operation of anticollision RFID system are not fulfilled in any of m multiple sampling of tags location for k analyzed area, then the next process of multiple sampling should be stopped, and it becomes necessary to examine the next $(k+1)$ - smaller area of tags location in the x-y plane. The MC solution for the analyzed object completes a procedure which confirms the fulfillment of all conditions for the correct operation of anticollision RFID system. The procedure is correct for the given efficiency of identification, and for the area in which all the m multiple sampling of tags location lead to a positive calculation result of the antenna unit array: read/write device-tags.
Correct selection of the m number, which will be satisfactory under the experiment, as well as adequate to calculation time and probability of possible tags locations, is a problem. From equations (2) i (3) which describe the strong law of large numbers, the dependence of probability PP for the random variables x_i and y_i (which are stochastically independent, and which have a uniform distribution) in function of the $m\cdot n$ has been presented in Fig. 3.

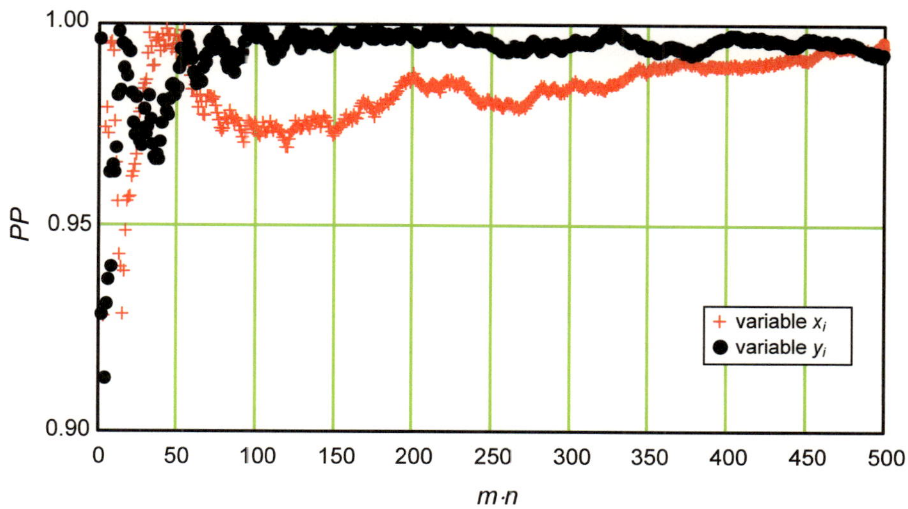

Fig. 3. Example result of probability PP of sampling of independent random variables x_i and y_i in function of numbers product: multiple sampling of m and n-tags location

Assuming that the probability PP exceeds the value 0.95 independently for the random variables x_i and y_i, the value $m\cdot n$=250 was determined during the calculation of interrogation zone in automatic identification process. These parameters were determined for 10^6 sampling of 250 independent random variables x_i and y_i which have a uniform distribution. For every sampling, the minimum value of probability PP has been searched. The determined value of $m\cdot n$=250 is compatible with a central limit theorem which states that the sum of a sufficiently large number of identically distributed independent random variables, each with finite mean and variance, is going to be approximately normally distributed (Rice, 2006). Uniform distribution of the random variables x_i and y_i is in fact different from a normal distribution, but - for this determined value of $m\cdot n$=250 - random variables x_i and y_i are convergent to a normal distribution. In this case, the obtained compatibility with a central limit theorem confirms the correctness of product $m\cdot n$=250.

The presented idea of n-tags analysis at a specifically determined z_{ID} height, results from a practical demand for realization of automatic identification process with the anticollision RFID systems. The identification of single products which are located inside a container on a pallet can be the practical example of this process. Identification of single objects separately is impossible in this situation, but their location on a pallet is mostly scheduled - because logistic system has to work satisfactorily (Mo & Lorchirachoonkul, 2010; Shaoping Lu et al., 2007; March, 2005; Jones & Chung, 2007). The development of the presented MC solution on an area Ω_{ID} in the x-y-z space requires investigation of every j-surface independently where tags will be located on the whole area Ω_{ID} at points $P_{ij}(x_i, y_i, z_{IDj})$ - (Fig. 2). If all n-tags are in a disordered state in the space Ω_{ID}, then the stochastic independence of all the coordinates x_i, y_i, z_i should be assumed. However, this idea is very complicated because it is not enough to assume that all the pairs of random variables are independent. Taking into consideration the practical requirements of different automatic identification processes, the example presented above is marginal, yet very interesting from a scientific point of view.

3. Conditions of correct operation of anticollision RFID system with inductive coupling

Passive RFID systems with inductive coupling are widespread (ID World, 2009; Wolfram et al., 2008; Jones & Chung, 2007; Paret, 2005). These systems can operate in individual and anticollision regime (Finkenzeller, 2003), and the need to design such systems appears more often nowadays. Functioning of RFID systems with inductive coupling is based on the use of energy which is stored in a magnetic field (Chen & Thomas, 2001; Rautio, 2003; Troyke & Edgington, 2000). The kind of executed operation in individual phases of the exchange of data between units of a system is essential during communication in anticollision identification process (Jankowski-Mihułowicz et al., 2008). The basic condition of effective operation of the system is the proper supply of each tag in a heterogeneous magnetic field created by RWD antenna loop. A minimal value of energy (necessary for proper read/write operation of the tag) is determined by minimal value of magnetic induction B_{min} in each point $P(x,y,z)$ of its location (Fig. 4).

Analysis of the general RFID system schema allows for the determination of the complete impedance of RWD antenna Z_R, taking into account an influence of all coupled tags. Maximal change of the impedance Z_R under the influence of the tags is expressed by the maximum value of difference in impedance arguments $\Delta\varphi_{Rmax}$ without the tags and with them, respectively. The value $\Delta\varphi_{Rmax}$ is limited to assure the correct operation of the system.

A determination of communication conditions for proper operation of RWD-tags antenna set is also possible on the basis of the schema analysis, taking into consideration the properties of data transmission process (transmitted frequency band, data flowability and required time relations in selected communication protocol).

The quality factor Q is a measure of tag antenna unit functioning efficiency in areas of energy transfer and communication conditions in RFID systems with inductive coupling (Newman et al., 1975; Redinger et al., 2003). These conditions are expressed by maximal values of quality factors of RWD and tag antennas: Q_{Rmax} and Q_{Tmax}, respectively. In this case, it is necessary to observe that selection of value Q_{Tmax} is compromised by energy (utilization of magnetic field energy) and communication requirements (Jankowski-Mihułowicz & Kalita, 2009).

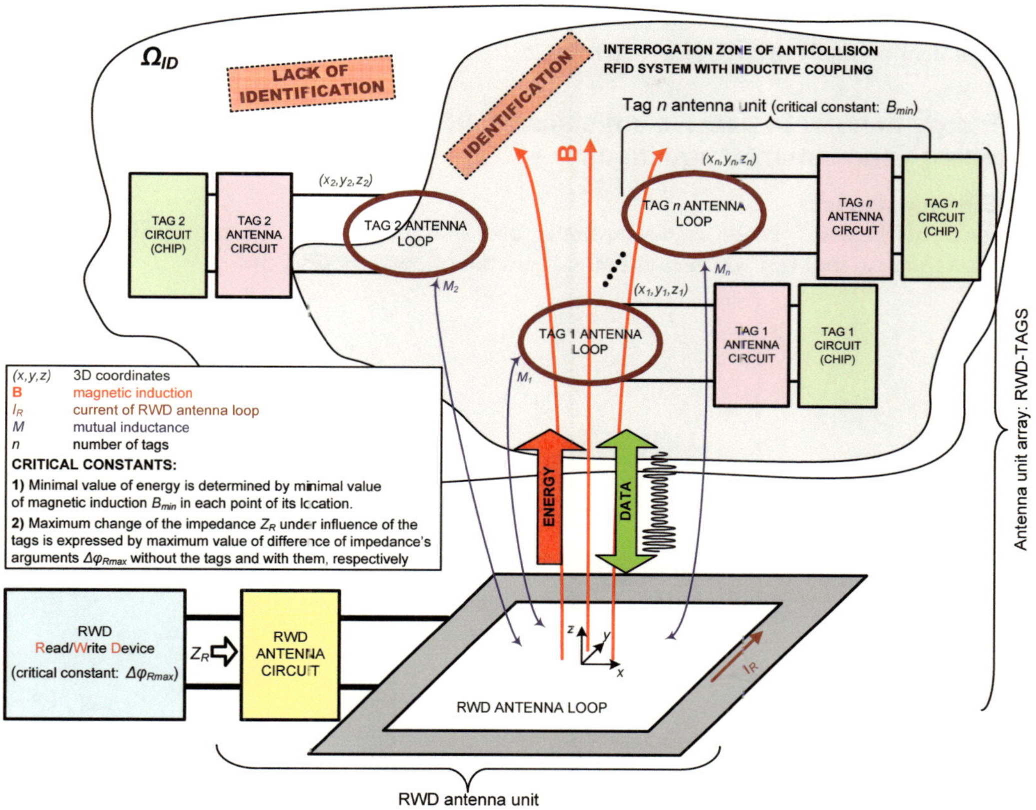

Fig. 4. Block diagram of anticollision RFID system with inductive coupling

During the synthesis of tag antenna, the changes of this parameter can be made by inductance change (indirectly - by changing the effective resistance of antenna loop), adjusted to requirements of the shape and geometrical sizes of an electronic tag. The proper synthesis of interrogation zone of RFID system is closely connected with three aspects which concern the maximum value of Q factor for the operating tag. The first of these aspects is related to the correct operating of the tag supply system, that is, the possibility of radio communication appearing. The second aspect concerns the necessity to obtain the required data transmission bit-rate (and also the bandwidth) in direction: tag-read/write device. The third aspect concerns the impulse and step response of tag circuit in case of reverse data transmission, to provide a correct identification of commands sent from the RWD. The last two aspects should result directly or indirectly from the electronic tag chip specification for which the antenna is going to be projected. The first aspect should be considered at the stage of antenna synthesis, and the value of Q factor should contain all of the mentioned limitations of operating passive tag. It is essential to ensure the homogeneous proper interrogation zone of RFID system that is a mutual overlay of zones that result from conditions of tag supply (by absorbing the energy of magnetic field) and the radio communication carried in the system (realized with the suitable value of signal-to-noise).
Paying attention to the maximum work distance between elements of the RFID system, in particular for systems working in the RFID far field, it is necessary to estimate the simulated

and built antenna set RWD-tags in relation to the obligatory normalizations of communication and EMC (ETSI EN 300 330, 2010; Jankowski-Mihułowicz, 2010).

4. Energy transfer in passive anticollision RFID system with inductive coupling – fundamental equations

4.1 EM field aspects

Analysis of the read/write device antenna unit allows to make an assumption that the antenna loop current (I_R) is constant along the whole flow way (Fig. 5).

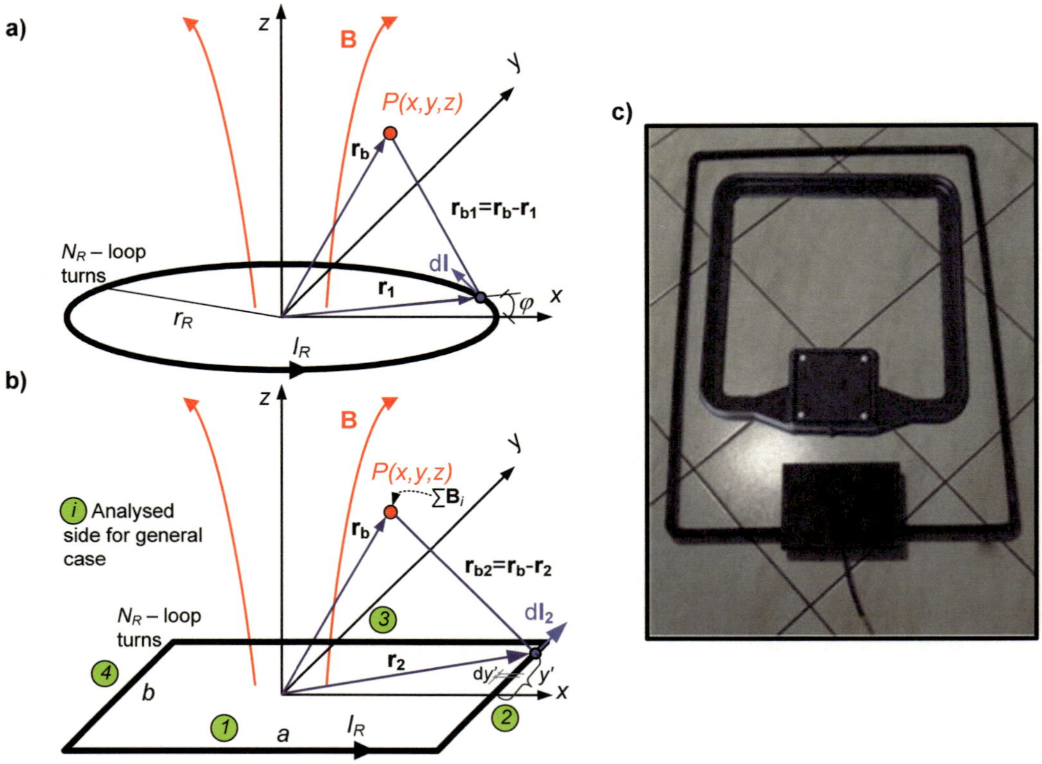

Fig. 5. Analyzed cases of RWD antenna loop: a) circular loop, b) loop of polygon shape, c) some realizations of tested RWD antennas

Change of the electric charge density in time equals zero, so in that case the electric current density divergence equals zero as well. Making these assumptions permits to apply the magnetostatic laws to magnetic field analysis for any RWD shape. In accordance with vector Biot-Savart law, the magnetic induction value of **B** in any space point $P(x,y,z)$ is given by the equation:

$$\mathbf{B} = \frac{\mu_0 I_R N_R}{4\pi} \oint \frac{\mathrm{d}\mathbf{l} \times \mathbf{r}_{b1}}{\left|\mathbf{r}_{b1}\right|^3} \tag{4}$$

where: $\mu_0 = 4 \cdot \pi \cdot 10^{-7}$ H/m.

Application of the Biot-Savart law for the RWD's antenna loops is possible by the additional assumptions that the wire diameter of RWD's antenna is negligible in relation to the geometrical loop sizes, and also that there is full inductive coupling between the individual loop turns (N_R).

For the circle-shaped loop (Fig. 5-a), the axial symmetry permits convenient change from Cartesian to cylindrical coordinates. The vector describing the dl location at P point, in which the value of magnetic induction is calculated, is given by the formula:

$$\mathbf{r}_{b1} = \mathbf{r}_b - \mathbf{r}_1 \tag{5}$$

where the vector describing dl element location that changes in φ angle function, and the vector describing location of point P, are given as follows:

$$\mathbf{r}_1 = \begin{pmatrix} r_R \cdot \cos(\varphi) \\ r_R \cdot \sin(\varphi) \\ 0 \end{pmatrix} \tag{6}$$

$$\mathbf{r}_b = \begin{pmatrix} x \\ y \\ z \end{pmatrix} \tag{7}$$

The unit vector connected with $\mathbf{dl} = r_R \cdot \mathrm{d}\varphi$ is given by the formula:

$$\mathbf{u}_\varphi = \begin{pmatrix} -\sin(\varphi) \\ \cos(\varphi) \\ 0 \end{pmatrix} \tag{8}$$

Changing the coordinate system leads to final equation, which describes the magnetic vector at any space location with (x,y,z) coordinates for circle-shaped RWD loop:

$$\mathbf{B} = \frac{\mu_0 I_R N_R}{4\pi} \int_0^{2\pi} \frac{r_R \cdot \mathbf{u}_\varphi \times \mathbf{r}_{b1}}{\left|\mathbf{r}_{b1}\right|^3} \mathrm{d}\varphi \tag{9}$$

In the case of RWD antenna loop, constructed as polygon (Fig. 5-b, c), Biot-Savart law with principle of superposition permits to add at location P vectors, that descend from individual antenna parts. In this case, the total magnetic induction is calculated from the equation:

$$\mathbf{B} = \sum_i \mathbf{B}_i \tag{10}$$

where i denotes analysed side for RWD antenna loop, constructed as polygon.

The obtained vector equations permit numerical calculating of the value of magnetic induction separately for individual components in directions x, y and z (B_x, B_y, B_z). The components of magnetic induction B in any space point $P(x,y,z)$ are given for a circular loop shape (Fig. 5-a) by the following equations:

$$B_x = \frac{\mu_0 I_R N_R}{4\pi} \int_0^{2\pi} \frac{z \cdot r_R \cdot \cos(\varphi)}{\left[\left(x - r_R \cdot \cos(\varphi)\right)^2 + \left(y - r_R \cdot \sin(\varphi)\right)^2 + z^2\right]^{3/2}} \mathrm{d}\varphi \tag{11}$$

$$B_y = \frac{\mu_0 I_R N_R}{4\pi} \int_0^{2\pi} \frac{z \cdot r_R \cdot \sin(\varphi)}{\left[\left(x - r_R \cdot \cos(\varphi)\right)^2 + \left(y - r_R \cdot \sin(\varphi)\right)^2 + z^2\right]^{3/2}} \mathrm{d}\varphi \tag{12}$$

$$B_z = \frac{\mu_0 I_R N_R}{4\pi} \int_0^{2\pi} \frac{-r_R \cdot \sin(\varphi) \cdot \left(y - r_R \cdot \sin(\varphi)\right) - r_R \cdot \cos(\varphi) \cdot \left(x - r_R \cdot \cos(\varphi)\right)}{\left[\left(x - r_R \cdot \cos(\varphi)\right)^2 + \left(y - r_R \cdot \sin(\varphi)\right)^2 + z^2\right]^{3/2}} \mathrm{d}\varphi \tag{13}$$

and for any polygon (e.g. rectangular; Fig. 5-b) shape:

$$B_x = \frac{\mu_0 I_R N_R}{4\pi} \left[\int_{\frac{-b}{2}}^{\frac{b}{2}} \frac{z}{\left[\left(x - \frac{1}{2}a\right)^2 + \left(y - y'\right)^2 + z^2\right]^{3/2}} dy' + \int_{\frac{b}{2}}^{\frac{-b}{2}} \frac{z}{\left[\left(x + \frac{1}{2}a\right)^2 + \left(y - y'\right)^2 + z^2\right]^{3/2}} dy' \right] \tag{14}$$

$$B_y = \frac{\mu_0 I_R N_R}{4\pi} \left[\int_{\frac{-a}{2}}^{\frac{a}{2}} \frac{-z}{\left[\left(x - x'\right)^2 + \left(y + \frac{1}{2}b\right)^2 + z^2\right]^{3/2}} dx' + \int_{\frac{a}{2}}^{\frac{-a}{2}} \frac{-z}{\left[\left(x - x'\right)^2 + \left(y - \frac{1}{2}b\right)^2 + z^2\right]^{3/2}} dx' \right] \tag{15}$$

$$\begin{aligned} B_z = \frac{\mu_0 I_R N_R}{4\pi} \cdot \Bigg[& \int_{\frac{-a}{2}}^{\frac{a}{2}} \frac{y + \frac{1}{2}b}{\left[\left(x - x'\right)^2 + \left(y + \frac{1}{2}b\right)^2 + z^2\right]^{3/2}} dx' + \int_{\frac{-b}{2}}^{\frac{b}{2}} \frac{-x + \frac{1}{2}a}{\left[\left(x - \frac{1}{2}a\right)^2 + \left(y - y'\right)^2 + z^2\right]^{3/2}} dy' + \\ & + \int_{\frac{a}{2}}^{\frac{-a}{2}} \frac{y - \frac{1}{2}b}{\left[\left(x - x'\right)^2 + \left(y - \frac{1}{2}b\right)^2 + z^2\right]^{3/2}} dx' + \int_{\frac{b}{2}}^{\frac{-b}{2}} \frac{-x - \frac{1}{2}a}{\left[\left(x + \frac{1}{2}a\right)^2 + \left(y - y'\right)^2 + z^2\right]^{3/2}} dy' \Bigg] \end{aligned} \tag{16}$$

Many practical solutions for identification are characterized by the parallel location of tag antenna and RWD loop (Fig. 6).

This location of individual tags allows the magnetic induction B_{min} to reach its minimum value only in relation to z-magnetic induction components (vectors $\mathbf{B_{z1}} \div \mathbf{B_{zn}}$ for 1 to n tags). This case applies to places in which tags working in anticollision process has been located ($P_1(x_1, y_1, z_1) \div P_n(x_n, y_n, z_n)$).

The presented approach creates numerous limitations connected with the decrease of the interrogation zone in RFID system. This results from too low value of the perpendicular magnetic induction component in relation to tag antenna loop plane. The efficient use of communication space in which anticollision process is going to take place, and also specification of the object marked by passive RFID tag, requires consideration of any tag orientation with regard to the individual components of magnetic induction vector.

The issue of any tag orientation in three dimensions x-y-z comes down to the tag deviation of α and β angles from parallel location of RWD-tag antenna loops (Fig. 7-a). In accordance with presented model (Fig. 7-b and Fig. 7-c), deviation of α angle occurs in z-x plane, however deviation of β angle occurs in α-y plane. Calculating the value of perpendicular magnetic induction component for tag, which is deviated of α and β angles ($B_{\alpha\beta}$) has been divided in two parts. By application of the superposition theorem in the first part, after tag deviation of α angle, the perpendicular magnetic induction component is given by:

$$B_{xz\alpha} = B_{x\alpha} + B_{z\alpha} \tag{17}$$

where the values of vector components are given by:

$$B_{x\alpha} = B_x \cdot \sin(\alpha) \tag{18}$$

$$B_{z\alpha} = B_z \cdot \cos(\alpha) \tag{19}$$

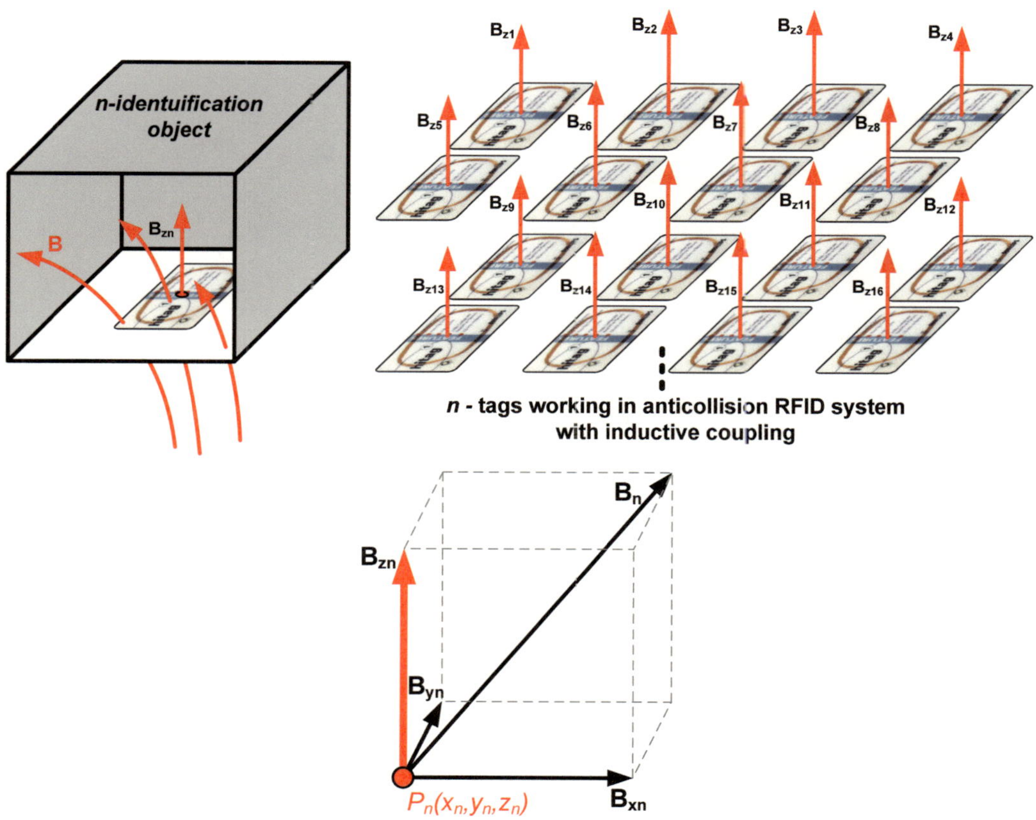

Fig. 6. Typical orientation of tags working in anticollision RFID system in relation to components of magnetic induction vector

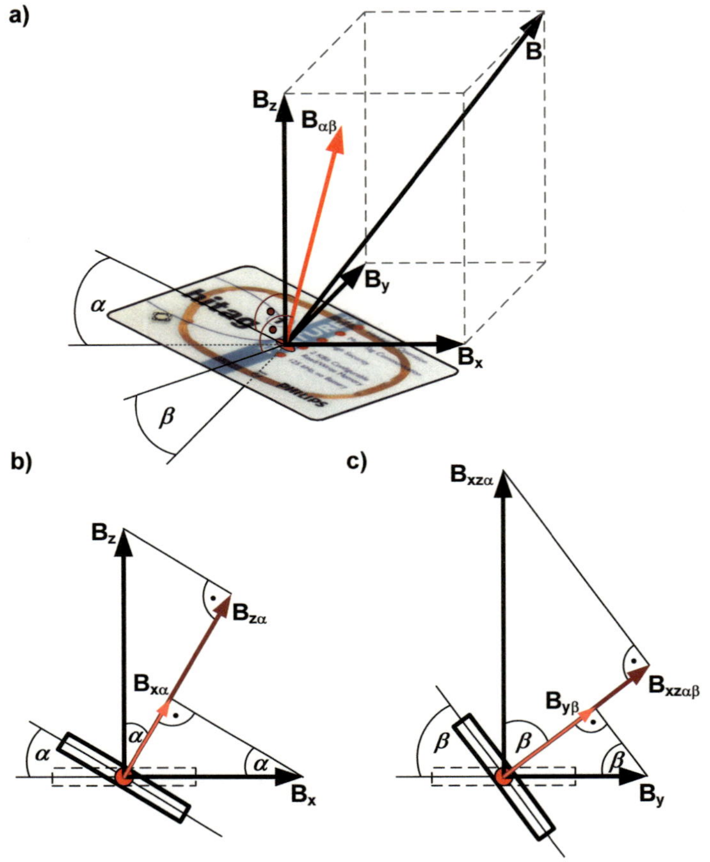

Fig. 7. Orientation of a tag, which is deviated of α and β angles from components of magnetic induction vector: a) deviation in 3D coordinate, b) deviation of α angle on z-x plane, c) deviation of β angle on α-y plane

By application of the superposition theorem in the second part again, after tag deviation of β angle, the perpendicular magnetic induction component is given as follows:

$$B_{\alpha\beta} = B_{y\beta} + B_{xz\alpha\beta} \tag{20}$$

where the values of vector components are given by:

$$B_{y\beta} = B_y \cdot \sin(\beta) \tag{21}$$

$$B_{z\alpha} = B_z \cdot \cos(\alpha) \tag{22}$$

From the equations (17)÷(22) comes, that the perpendicular magnetic induction component for passive tag, which is deviated of α and β angles, is given by:

$$B_{\alpha\beta} = B_z \cdot \cos(\alpha) \cdot \cos(\beta) + B_x \cdot \sin(\alpha) \cdot \cos(\beta) + B_y \cdot \sin(\beta) \tag{23}$$

Knowing the magnetic induction separately for individual components in directions x, y and z (B_x, B_y, B_z), the obtained equation (23) permits calculation of the perpendicular magnetic induction component. The aforementioned necessity of changing the tag orientation should be carried out for assurance of correct tag work in the individual space point $P(x,y,z)$. In this way it is possible to calculate the system interrogation zone, which is forced by the specification of identified object, that is the necessity of individual tag location on marked object.

4.2 Electric aspects

The second essential stage of energy transfer from RWD to tags includes an elaboration of electrical model of the whole anticollision RFID system with inductive coupling for the full frequency range (LF - typically 125 kHz, 138 kHz and HF - typically 13.56 MHz). The basic part of the system is circuitry of RWD - tag antenna set in which an adequate operation states should be taken into consideration. The idea of electrical model of analyzed RFID system is based on the circuitry of RWD - tag antenna units (Fig. 8).

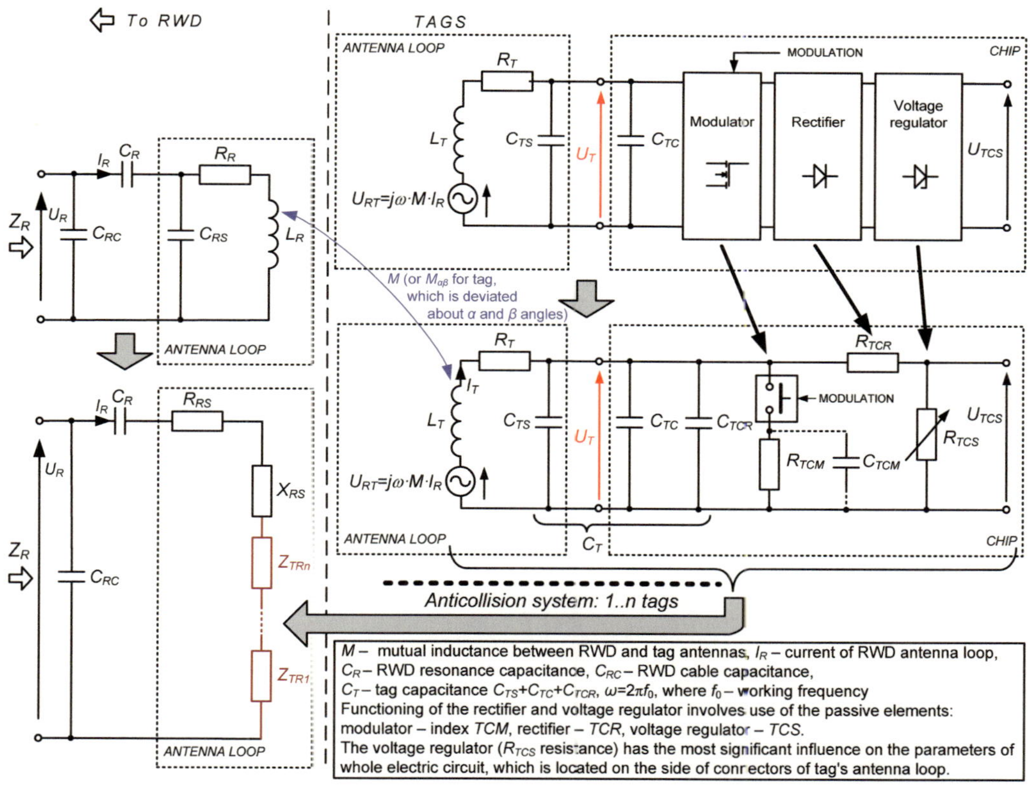

Fig. 8. Equivalent schematic diagram of anticollision RFID system with inductive coupling

The correct operation of tag internal integrated circuit in communication process phases is conditioned by the voltage U_T which is inducted on the trimming tag antenna. Functioning of the memory and microprocessor of passive tag circuit is connected with the occurrence of the stabilization voltage U_{TCS} inside its chip (Villard, 2002; Friedman et al., 1997).

Taking into consideration the fact that the stabilizer is on input to the part of the tag chip and it influences tag's antenna unit, it is necessary to make the synthesis of a module which was divided into two parts: rectifier and voltage regulator. Rectification of the voltage induced in the tag antenna loop, takes place in half-wave and full-wave rectifier (Jamali, 2006), however, it is controlled by a voltage regulator (Friedman et al., 1997). Functioning of the rectifier and voltage regulator involves the use of other passive elements to the tag's schematic diagram (modulator - index *TCM*, rectifier - *TCR*, voltage regulator - *TCS*). The voltage regulator (represented by the R_{TCS} resistance) has the most significant influence on the parameters of the whole electric circuit which is located on the side of connectors of the tag's antenna loop.
The inductive coupling between the antenna loops is expressed by mutual inductance M determined in relation to field conditions at the given system efficiency for arbitrary location and orientation of the tags:

$$M = \frac{N_T}{I_R} \int_0^{r_T} \int_0^{2\pi} B_z\left(r\cos(\varphi) + x, r\sin(\varphi) + y, z\right) r \mathrm{d}\varphi \mathrm{d}r \tag{24}$$

and:

$$M_{\alpha\beta} = \frac{N_T}{I_R} \int_0^{r_T} \int_0^{2\pi} B_{\alpha\beta}\left(r\cos(\varphi) + x, r\sin(\varphi) + y, z\right) r \mathrm{d}\varphi \mathrm{d}r \tag{25}$$

where N_T is number of tag antenna turns of wire.
In tag antenna loop is induced the voltage U_T (as effect of the coupling):

$$U_T = \frac{j\omega \cdot M \cdot I_R}{1 + \left(\frac{1}{(R_{TCR} + R_{TCS}) \| R_{TCM}} + j\omega C_T \right) \cdot (j\omega L_T + R_T)} \tag{26}$$

where the symbol || denotes the parallel connection of resistances representing the generalized circuit of rectifier and voltage regulator (R_{TCR}, R_{TCS}) and the modulator (R_{TCM}) inside the chip structure.
Assuming that the module of U_{TCS} is constant, the variable resistance of voltage regulator is given by (27). This equation is true when $U_T > U_{TCS}$:

$$R_{TCS} = \left| \frac{U_{TCS}\left(\omega^2 L_T C_T R_{TCR} - R_{TCR} - j\omega L_T - j\omega C_T R_T R_{TCR} - R_T\right)}{U_{TCS}\left(1 + j\omega C_T R_T - \omega^2 L_T C_T\right) - j\omega M I_R} \right| \tag{27}$$

The minimum value of U_T (U_{Tmin}) is the base for determining the interrogation zone for the anticollision system. The value of voltage U_{Tmin} clearly impacts the analytical relation, which allows for expression of the minimum value of magnetic induction B_{min} – the elementary parameter defining the interrogation zone:

$$B_{min} = \left|U_{Tmin}\right| \cdot \left| \frac{\left[1 + \left(\frac{1}{R_{TCR} + R_{TCS}} + j\omega C_T \right) \cdot (j\omega L_T + R_T) \right]}{j\omega \cdot N_T \cdot S_T} \right| \tag{28}$$

The induction B_{min} is differentiated on the basis of value U_{Tmin}; for the direction of data transmission and the kind of operations in internal tag memory. For experimental verification of calculated value B_{min} the special laboratory stand has been made. The stand allows to measure the maximum distance between the RWD and tag antennas, for which the correct operation of the RFID system is ensured.

Operating of the anticollision RFID system with inductive coupling means that working tags, which are located in the system interrogation zone, are going to have an influence on the parameters of RWD antenna unit. The synthesis of the electrical model of an anticollision RFID system includes the replacement of all tag electric circuits by the impedance Z_{TR} (Fig. 8). Those circuits are coupled by the mutual inductance M. Location of a specific number of n-tags in the interrogation zone of RFID system means that the working tags' quantitative participation in the influence on the RWD is going to be represented by the sum of all impedances Z_{TR} (from Z_{TR1} to Z_{TRn}). Such an influence of impedances Z_{TR} on the read/write device leads to the adverse effect of detuning the RWD antenna. Additionally, the value of antenna loop current I_R decreases in a read/write devices without output current stabilization. In a read/write device with current stabilization, the antenna voltage U_R increases to the limit, and after an overflow, the value of antenna loop current I_R also decreases as in the previous case. The consequence of the influence of operating tags on the RWD in each case is the reduction of the system interrogation zone which is caused by the energetic conditions of devices' operation. Changes are the stronger the larger is the value of the mutual inductance between tags antenna loop and RWD.

This problem can be illustrated by the total impedance of RWD antenna Z_R, taking into account the influence of all coupled tags. For an equivalent schematic diagram from Fig. 8, this impedance is given by the following equation:

$$Z_R = \left[\frac{(j\omega L_R + R_R) \cdot \dfrac{1}{j\omega C_{RS}}}{j\omega L_R + R_R + \dfrac{1}{j\omega C_{RS}}} + \frac{1}{j\omega C_R} \right] \Bigg\| \frac{1}{j\omega C_{RC}} + \sum_n Z_{TRn} \tag{29}$$

where the equation describes the impedance Z_{TR}, and also quantitative influence of working tags on the RWD, that is given by:

$$Z_{TR} = \frac{\omega^2 M^2}{R_T + j\omega L_T + \dfrac{(R_{TCR} + R_{TCS}) \| R_{TCM}}{1 + j\omega C_T \left[(R_{TCR} + R_{TCS}) \| R_{TCM} \right]}} \tag{30}$$

An influence of working tags causes the RWD antenna unit to detune of the value Δf_R from the frequency f_0. This detuning is revealed by the change of impedance's argument $\Delta\varphi_R$ for the working frequency of RFID system. The border values of these parameters determine the last directive to the efficiency of antenna unit array research for given read/write device. There is a necessity to maintain $\Delta\varphi_R$ within the limits in order to provide correct functioning of the RWD and of the whole anticollision RFID system. Such an assumption is practically used in external antenna tuning devices dedicated to tuning the long-range antennas working in LF and HF inductive coupling RFID systems (Feig, 2005, 2006; Philips, 1996; Texas Instruments, 2002).

5. Results

The experimental research have been carried out for different RFID elements using the laboratory system which allows to determine single and anticollision identification process for all frequencies in RFID systems with inductive coupling (Fig. 9).

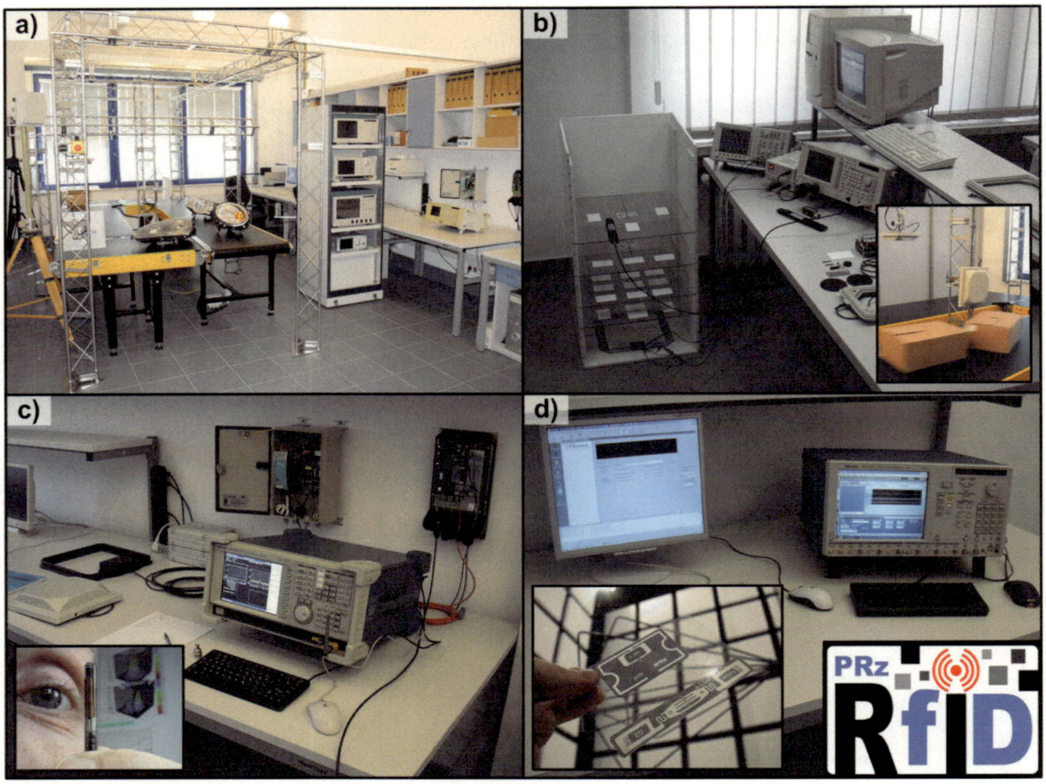

Fig. 9. RFID laboratory in the Rzeszów University of Technology: a) dynamic test stand, b) static test stand, c) example of long range read/write devices, d) example of measuring equipment

During the search of the interrogation zone of RFID system for a given efficiency of identification η_{ID} (1), the appearance of condition $\Delta\varphi_R > \Delta\varphi_{Rmax}$ makes a correct identification impossible. In MC calculation the parameter $\Delta\varphi_R$ is calculated on the basis of the total impedance Z_R of RWD antennas arrangement (29), taking into consideration influence of functioning tags on this antenna - $Z_{TR1...n}$ calculated from the equation (30). For example, the limit phase value was $\Delta\varphi_{Rmax}=\pm 15^{\circ}$ for the Philips HITAG RM 800 read/write device, working at frequency f_0=125 kHz. The technical documentation available on the basis of an agreement reached between the Department and the Philips Semiconductors has been used in the investigations.

For the correct energy transfer in the anticollision RFID system with inductive coupling, assuming the possibility of using identical tags in automatic identification process, the specified value of minimum magnetic induction B_{min} will be the parameter that limits tags' correct operation area. This parameter should be calculated from the equation (28) for the individual single tag construction and the kind of operations executed in its internal memory (read/write of tags memory). If the value of perpendicular component of magnetic induction vector at point of the location of tag is smaller than his parameter B_{min}, then the correct functioning of this tag in anticollision system is impossible. This denotes that the tag is in the area where communication with RWD is impossible, and efficiency of identification η_{ID} is lowered.

Process of determining the interrogation zone using MC method has been preceded by measurement and calculation of B_{min} conducted during the process of reading information from the internal memory of tag. The results of these measurements and calculations were presented in the table 1.

In the simulation and measuring part of the experiment respectively, the calculated and measured values B_{min} were the minimum limit of the correct operation of a single tag located in the area of field conditions of functioning of the whole RFID system. In both parts of the experiment locations (in points P_i of cartesian space at (x_i,y_i,z_{ID}) coordinates) of ten tags of a chosen type were selected randomly 25 times (from chapter 2: $n \cdot m$=250).

Tag	Measuded z_{max} 1)	Measured B_{min} 2)	Calculated B_{min} 3)
-	m	µT	µT
HITAG 1 ISO CARD	0.52	0.74	0.74
HITAG 1 WORLD TAG 50	0.44	1.16	1.16
HITAG 1 WORLD TAG 30	0.27	3.72	3.73
HITAG 1 WORLD TAG 20	0.22	5.63	5.62

1) The measurement of the maximum working distance z_{max} from the center on axis of symmetry of RWD antenna loop for square read/write device antenna (where a=0.3 m, N_R=32, I_R=0.213 A) – this is the result of the positive identification of the tag serial number.
2) The measurement by means of analyser Advantest R3132 and Rohde & Schwarz HZ-14 near field probe (Rohde & Schwarz, 2003).
3) The values calculated in the *JankoRFIDmc'IZ v. 4.08* application (Jankowski-Mihułowicz, 2007) - on basis of electrical model - equation (28).

Table 1. Measured and calculated values B_{min} for tags selected to investigations

The example results of the calculated and measured interrogation zone (Fig. 10), were placed on the plane at (x, y, z_{ID}) coordinates. The measured interrogation zone is the result of the positive identification of all n=10 tags serial numbers, during conducted experiment, all m=25 multiple sampling of their location. For every multiple sampling of the location of tags in measuring chamber, spatial measurements of z component of magnetic induction **B** vector were made. On the basis of (Rohde & Schwarz, 2003), the measurement of the component of the vector **B** perpendicular to the area of the antenna loops of tags was conducted in the 625 points (the resolution of 2 cm on 0.5 m x 0.5 m x-y surface – the movable platform in the measuring chamber – Fig. 9-b).

All of the calculations and measurements were performed for square antenna of the RWD unit which was tuned in the measuring chamber without the influence of tags, and the achieved value was $\Delta\varphi_R$=2.5°. In all studied cases, the border value of $\Delta\varphi_{Rmax}$, wasn't crossed. Thanks to this, the efficiency of identification for the height z_{ID} was 100 % in the area of fulfillment of the condition of the magnetic induction minimum value. Difference between the calculated and measured interrogation zone (in the worst case, for the smallest heights z_{ID}, on the level ±1.5 cm), is caused mainly by applying an approximate geometrical model of the antenna loop of the RWD. These differences are caused by the fact that the RWD antenna loop was build as loose turns of wire, and that was assumed during synthesis of the geometrical model of the RWD antenna loop.

The measurements in the RWD - tags antennas arrangement required applying many direct and indirect measuring methods. The obtained results always contained certain dispersion of the values, which can always be - in a justified way - ascribed to measured sizes. The multiple results were obtained from many measuring sets.

Generally, the problem of the uncertainty of determining the interrogation zone of the anticollision RFID system with the inductive coupling, has two aspects: simulations and measures. In the process of evaluation of the uncertainty of determining the interrogation zone in the measuring part of the experiment, essential factors are uncertainties of the magnetic induction components $u(B)$ measurements:

$$u(B)=\sqrt{\left(\left|\frac{\partial B}{\partial H}\right|\cdot u(H)\right)^2+\left(\left|\frac{\partial B}{\partial \mu}\right|\cdot u(\mu)\right)^2} \tag{31}$$

where:

$$u(H)=\sqrt{\left(\left|\frac{\partial H}{\partial V_0}\right|\cdot u(V_0)\right)^2+\left(\left|\frac{\partial H}{\partial AF}\right|\cdot u(AF)\right)^2} \tag{32}$$

where $u(V_0)$ - standard uncertainty of voltage measured by means of Advantest R3132 spectrum analyzer and the R&S HZ-14 near magnetic field probe.

This uncertainty includes the systematic influences which cannot be removed during the conducted experiment. They are represented by the set of coefficients read from prepared tables and graphs in the Advantest R3132 spectrum analyzer user manual. $u(AF)$ denotes the uncertainty of antenna coefficient read for measuring frequency (f_0). For the spatial, multipoint measurements which were made in the measuring chamber of the investigative set, the standard relative uncertainty for the magnetic induction $u_{\%}(B)$ was on the level 1-2 %.

In the process of evaluation of uncertainty of the interrogation zone estimation in the simulating part of the experiment, the component factors of the complex uncertainty of the entrance data measurements and output data calculations were considered. They were taken into account in the process of estimating the efficiency of the system antennas arrangement with the MC method, which is made by the *JankoRFIDmc'IZ* application (Jankowski-Mihułowicz, 2007).

Explaining this problem, function f which represents the interrogation zone exhibits significant nonlinearity. Therefore, regarding the error propagation, the higher terms in the Taylor's expansion should be taken into account. Their form is as follows:

$$\sum_{i=1}^{n}\sum_{j=1}^{n}\left[\frac{1}{2}\cdot\frac{\partial f}{\partial x_i}\cdot\frac{\partial^2 f}{\partial x_i \partial x_j}+\frac{\partial f}{\partial x_i}\cdot\frac{\partial^3 f}{\partial x_i \partial^2 x_j}\right]\cdot u^2(x_i)\cdot u^2(x_j) \tag{33}$$

where: i,j=1..n.

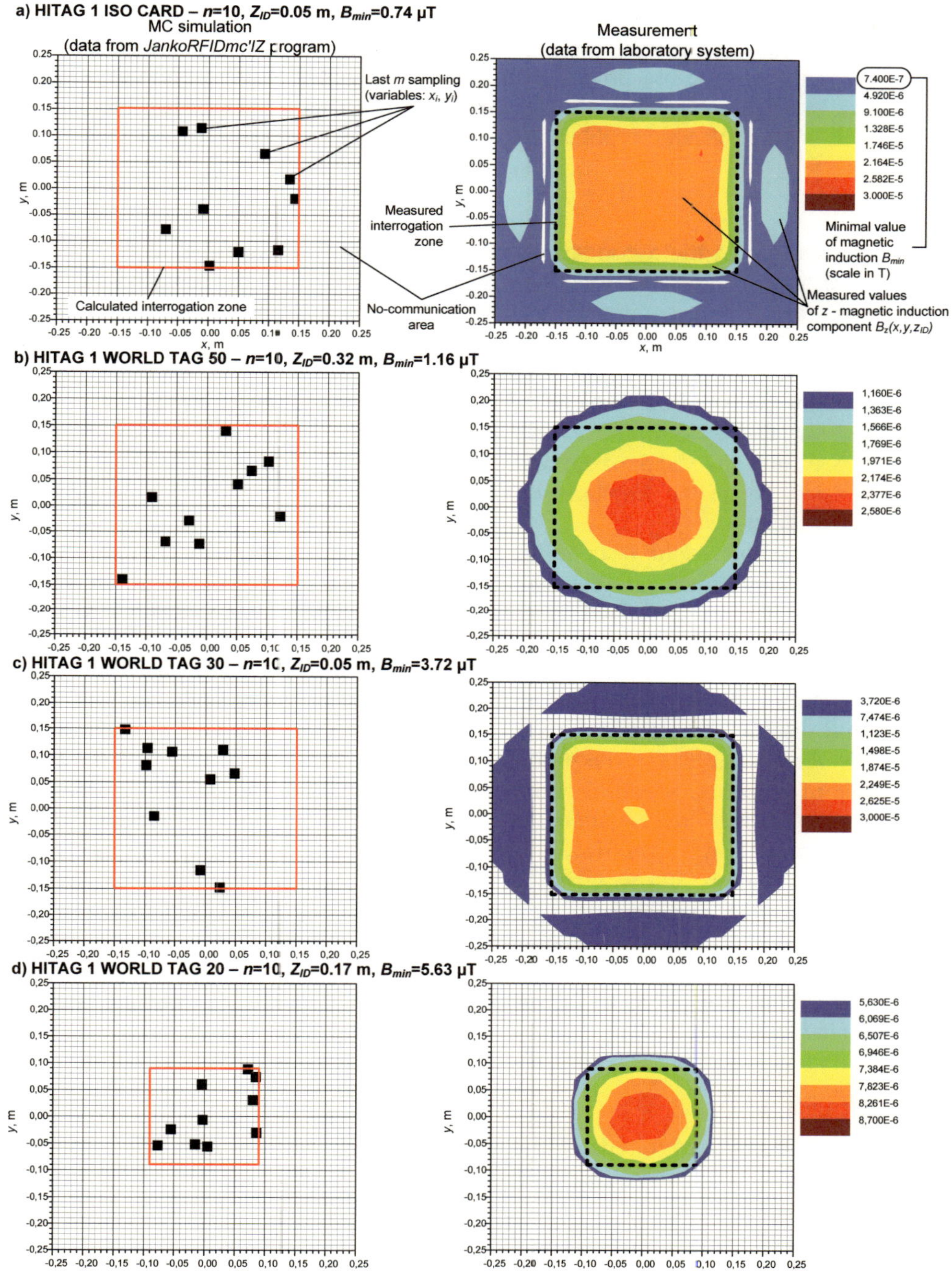

Fig. 10. Description of example elements of calculated and measured characteristics of interrogation zone for HITAG 1: a) ISO CARD (Z_{ID}=0.05 m, B_{min}=0.74 μT), b) WORLD TAG 20 (Z_{ID}=0.32 m, B_{min}=1.16 μT), c) WORLD TAG 30 (Z_{ID}=0.05 m, B_{min}=3.72 μT) and d) WORLD TAG 30 (Z_{ID}=0.17 m, B_{min}=5.63 μT)

In indirect measurements every size, calculated or measured directly, brings the different contribution to the uncertainty *u(f)*. The determination of suitable weighting factors resulting from the uncertainty propagation law for the considerably nonlinear function *f*, according to the higher terms in the Taylor's expansion, is a complicated mathematical question. This is a complicated problem at the present stage of works.

6. Conclusion

The efficient leading of the automatic identification processes, such as: forwarding mail, materials, articles (in industry); identification of valuable minerals, samples for analysis (in science and medicine), requires the use of a modern radio methods of the simultaneous identification of many objects. The mentioned processes generally belong to the automatic identification group, in which RFID electronic tags are replacing, for example, barcodes. This is caused by the well-known technical limitations of the objects identification methods used nowadays. The accessibility of electronic tags, the continuous reduction of their production costs and the standardization of work conditions of RFID technology, allows to make a decision about the implementation of quite a new method in the process of automatic identification.

The laboratory research and tests fully confirm the correctness and usefulness of the elaborated (in Department of Electronic and Communication Systems at Rzeszów University of Technology), method of synthesis of anticollision RFID system, where the essential component, based on Monte Carlo method, is the determination of interrogation zone for the system with suitably located tags. It should be noted that the synthesis procedure includes the simultaneous analysis of electromagnetic field, communication protocols and electric aspects of operation conditions in the process of system efficiency identification. Presented part of the problem of interrogation zone synthesis is the base for practical use of projected identification systems, required for specific anticollision RFID applications. The future investigations will be focused on the analysis of efficiency and interrogation zone of the anticollision RFID systems operated in dynamic conditions (speed changes of orientation of suitably located tags). Additionally, the extension of *JankoRFIDmc'IZ* program on a propagation coupling RFID system is planned. The elements of algorithm of interrogation zone identification for anticollision RFID system taking into consideration the energetic (i.e. field and electrical) and communicational aspects of operation conditions are going to be supplemented by elements of antennas and wave propagation in UHF.

7. Acknowledgment

This work was partly supported by the Project "Developing research infrastructure of Rzeszów University of Technology" within the Operational Program Development of Eastern Poland 2007-2013 of the Priority Axis I Modern Economics of Activity I.3 Supporting Innovation, Contract No. POPW.01.03.00-18-012/09-00.

8. References

De Blasi, M.; Mighali, V.; Patrono, L. & Stefanizzi, M. L. (2010). Performance Evaluation of UHF RFID Tags in the Pharmaceutical Supply Chain, In: *The Internet of Things. 20th*

Tyrrhenian Workshop Digital Communications, Giusto, D.; Iera, A.; Morabito, G. & Atzori, L. (Ed.), pp. 283-292, Springer, ISBN 978-1441916730

Brown, D. (2007). *RFID Implementation*, McGraw-Hill, ISBN 978-0072263244

Chen, S. C. Q. & Thomas, V. (2001). Optimization of inductive RFID technology, *Proceedings of the IEEE International Symposium Electronics and the Environment*, pp. 82-87, ISBN 978-0780366558, Denver, CO , USA, May 7-9, 2001

Clarke, R. H.; Twede, D.; Tazelaar, J. R. & Boyer, K. K. (2006). Radio frequency identification (RFID) performance: the effect of tag orientation and package contents, *Packaging Technology & Science*, Vol. 19, No. 1, Jan./Feb. 2006, pp. 45-54, ISSN 0894-3214

ETSI EN 300 330 (2010). *Short Range Devices (SRD); Technical characteristics and test methods for radio equipment in the frequency range 9 kHz to 25 MHz and inductive loop systems in the frequency range 9 kHz to 30 MHz*

Feig Electronic GmbH (2005). Dynamic Antenna Tuning Board ID ISC.DAT, *Technical Documentation*, 7.02.2011, Available from:

http://www.feig.de/files/FEIG_DOCS/OBID_DOCS/Products/i-scan/i-scan%20HF/DE/id_isc_dat_e.pdf

Feig Electronic GmbH (2006). Static Antenna Tuning Controller for long-range antennas ID ISC.SAT.C-A, *Technical Documentation*, 7.02.2011, Available from:

http://www.ti.com/rfid/docs/manuals/pdfSpecs/RR-IDISC-SAT-C-A.pdf

Finkenzeller, K. (2003). *RFID Handbook: Fundamentals and Applications in Contactless Smart Card and Identification*, SE, Wiley, ISBN 978-0470844021

Friedman, D.; Heinrich, H. & Duan, D. W. (1997). A Low-Power CMOS Integrated Circuit for Field-Powered Radio Frequency Identification Tags, *Proceedings of the IEEE ISSCC'97*, pp. 294-295, ISBN 978-0780337213, San Francisco, CA , USA, Feb. 6-8, 1997

ID World (2009). The Players of the Auto ID Industry, Section RFID, *ID World*, Vol. Dec. 2009, pp. 42-63

Jamali, B.; Ranasinghe, D. & Cole, P. (2006). Design and optimisation of power rectifiers for passive RFID systems in monolithic CMOS circuit, In: *Microelectronics: Design, Technology, and Packaging II*, Hariz, A. J. (Ed.), Vol. 6035, SPIE, ISBN 978-0819460660

Jankowski-Mihułowicz, P. (2010). Field Conditions of Interrogation Zone in Anticollision Radio Frequency Identification Systems with Inductive Coupling, In: *Radio Frequency Identification Fundamentals and Applications, Bringing Research to Practice*, Turcu, C. (Ed.) pp. 1-26, InTech, ISBN 978-9537619732, Vienna, Austria

Jankowski-Mihułowicz, P. & Kalita, W. (2009). Efficiency of Tag Antenna Unit in Anticollision Radio Frequency Identification Systems with Inductive Coupling, *Acta Electrotechnica et Informatica*, Vol. 9, No. 2. pp. 3-7, ISSN 1335-8243

Jankowski-Mihułowicz, P.; Kalita, W. & Pawłowicz, B. (2008). Problem of dynamic change of tags location in anticollision RFID systems, *Microelectronic Reliability*, Vol. 48, Issue 6, pp. 911-918, ISSN 0026-2714

Jankowski-Mihułowicz, P. (2007). *Creation conditions of antenna array efficiency of anticollision Radio Frequency Identification systems with inductive coupling*, Dissertation, AGH University of Science and Technology, Kraków, Poland

Jones, E. C. & Chung, C. A. (2007). *RFID in Logistics - A Practical Introduction*, CRC Press, ISBN 978-0849385261

Kalos, M. H. & Whitlock P. A. (2008). *Monte Carlo Methods*, 2-rd Ed., Wiley-VCH, ISBN 978-3527407606

Lehto, A.; Nummela, J.; Ukkonen, L.; Sydänheimo, L. & Kivikoski, M. (2009). Passive UHF RFID in Paper Industry: Challenges, Benefits and the Application Environment. *IEEE*

Transactions on Automation Science and Engineering, Vol. 6, Issue 1, pp. 66-79, ISSN 1545-5955

March, V. (2005). Built-in RFID for electronic products, *Global Identification Magazine*, Vol. Oct, pp. 38-39

Mo, J. P. T. & Lorchirachoonkul W. (2010). RFID Infrastructure for Large Scale Supply Chains Involving Small and Medium Enterprises. In: *Sustainable Radio Frequency Identification Solutions*, Turcu, C. (Ed.) pp. 1-22, InTech, ISBN 978-9537619749, Vienna, Austria

Newman, E.; Bohley, P. & Walter, C. (1975) Two Methods for the Measurement of Antenna Efficiency, IEEE Transactions on Antennas and Propagation, Vol. 23, Issue 4, pp. 457-461, ISSN 0018-926X

Paret, D. (2005). *RFID and Contactless Smart Card Applications*, Wiley, ISBN 978-0470011959

Penttilä, K.; Keskilammi, M.; Sydänheimo, L. & Kivikoski, M. (2006). Radio frequency technology for automated manufacturing and logistics control. Part 2: RFID antenna utilisation in industrial applications, *The International Journal of Advanced Manufacturing Technology*, Vol. 31, No. 1-2, pp. 116-124, ISSN 0268-3768

Philips Semiconductors (1996). HT OT840 – HITAG Antenna Tuning Device, *Technical Documentation*, 7.02.2011, Available from: http://www.nxp.com/acrobat_download/various/HTOT840_N_1.pdf

Polivka, M.; Svanda, M.; Hudec P. & Zvanovec, S. (2009). UHF RF Identification of People in Indoor and Open Areas, *IEEE Transactions on Microwave Theory and Techniques*, Vol. 57, Issue 5, pp. 1341-1347, ISSN: 0018-9480

Rohde & Schwarz (2003). Probe Set for E and H Near-Field Measurements 9 kHz to 1 GHz HZ-14, *Operating Manual 1026.7744.03*

Rautio, J. B. (2003). Electromagnetic Analysis Speeds RFID Design, *Microwaves & RF*, Vol. 42, No. 2, February 2003, pp. 55-62, ISSN 0745-2993

Redinger, D.; Farshchi, R. & Subramanian, V. (2003). An all-printed passive component technology for low-cost RFID, *Proceedings of the 33th IEEE DIGEST'03*, pp. 187- 188, ISBN 0780377273, June 23-25, 2003

Rice, J. (2006). *Mathematical Statistics and Data Analysis*, 3-rd edn. Duxbury Press, ISBN 978-0534399429, Belmont, CA, USA

Rubinstein, R. Y. & Kroese, D. P. (2007). *Simulation and the Monte Carlo Method*, 2-rd Ed.,Wiley-Interscience, ISBN 978-0470177945

Shaoping Lu; Yaohua Wu & Yongtao Fu (2007). Research and design on pallet-throughout system based on RFID. *Proceedings of the IEEE International Conference on Automation and Logistics*, pp. 2592-2595, ISBN 978-1424415311, Jinan, China, Aug. 18-21, 2007

Texas Instruments (2002) TI RFID Series 2000 Antenna Tuning Indicator RI-ACC-ATI2, *Technical Documentation*, 7.02.2011, Available from: http://www.ti.com/rfid/shtml/prod-accessories-RI-ACC-ATI2.shtml

Troyke, P. R. & Edgington, M. (2000). Inductive Links and Drivers for Remotely-Powered Telemetry Systems, *Proceedings of the Antennas and Propagation Symposium*, pp. 60-62, ISBN: 0780363698, Salt Lake City, UT , USA, July 16-21, 2000

Villard, P.; Bour, C.; Dallard, E.; Lattard, D.; de Pontcharra, J.; Robert, G. & Roux, S. (2002). A low-voltage mixed-mode CMOS/SOI integrated circuit for 13.56 MHz RFID applications, *Proceedings of the IEEE SOI'02*, pp 163-164, ISBN: 0780374398

Wolfram, G.; Gampl, B. & Gabriel, P. (2008). The RFID Roadmap: The Next Steps for Europe, Springer, ISBN 978-3540710189

18

Iterative Delay Compensation Algorithm to Mitigate NLOS Influence for Positioning

Koji Enda and Ryuji Kohno
Yokohama National University
Japan

1. Introduction

Wireless sensor networks are attracting considerable attention in recent years as constituent elements of next-generation wireless networks. Determining the position information of sensor tags is extremely important, and hence, position estimation using RFIDs for sensor networks is a widely studied topic. In order to estimate these RFID tag's position, TDOA positioning algorithm is focused on because each sensor tag is desirable of plain hardware configuration. Tag's position is estimated to measure arrival time from tag to some reception nodes. In case of executing positioning process, Sensor tag is not necessary to synchronize with node, it is necessary to synchronize in time domain with only each node. Therefore, these features of TDOA positioning algorithm fulfill that sensor tag should be simple, independent and low power consumption. We use the NEWTON method because of its fast conversion property and its ability to yield the minimum square difference with few computations. The non-line-of-sight (NLOS) problem must be taken into consideration when employing positioning methods that involve the use of time-domain data. The problem is characterized by the fact that in addition to direct waves, reflected or diffracted waves are also incident on the target, resulting in the geometrical stretching of the obtained paths along the normal direction and a positive bias in the travel time. The resulting effect is a difference in the arrival time which, in turn, causes deterioration in the positioning accuracy. In this paper, in order to mitigate the influence of the NLOS propagation, we propose the iterative delay compensation algorithm based on NEWTON algorithm which improves the accuracy of positioning using the DCF and shift vector compensation (SVC) algorithm. In the proposed method, hypothetical coordinates are estimated by using the conventional NEWTON method. Then, the node positions and distances are derived from the estimated coordinate information. DCF is used to compensate for the difference between the calculated reception time and the actual measured time. The propagation delay included in the measured value is reduced step-by-step by repeatedly applying the compensation function. This helps in minimizing the effect on the line-of-sight (LOS) node, resulting in improved positioning accuracy. Next, the estimation accuracy is improved by compensating the influence vector caused by NLOS delays in the temporarily estimated positions by using the node distributions and geometrical relations among the estimated positions. The iterative algorithm using DCF and SVC fulfills high accuracy of positioning even in an NLOS environment. Furthermore, we make an experiment of TDOA tracking system using tag and node. The experiments show that tracking accuracy is improved and abnormal tracking position estimation is reduced.

2. Positioning system model and positioning algorithm

In this section, we state positioning system model of TDOA and a principle of the TDOA positioning algorithm.

2.1 System Model

Let us assume that positioning is to be carried out in a two-dimensional field. The component elements comprise mobile devices defined as tags, which are the targets for positioning, and fixed devices with known positions, defined as nodes. The tags send only signal and nodes receive only messages from the tags. The nodes need be synchronized among themselves. Time distance of arrival information is extracted by means of reception signal from the tags to nodes. In concrete terms, tag sends a signal to each node (x_i, y_i) [i = $1...M$], and calculates the distance difference on the basis of the time required for the signal to receive. M denotes the number of nodes. This information is transmitted to the master node where the position is estimated by using signal processing. Signal losses that occur during the signal transmission are not considered. The distance between the nodes and tags is obtained as follows: Let us assume that T_{start} is the transmission time, T_r is the reception time, and c is the speed of light. The reception time T_i from the tag to the node is as shown below:

$$T_i = T_r - T_{start} \tag{1}$$

Then, the propagation distance D_i is

$$D_i = c\, T_i \tag{2}$$

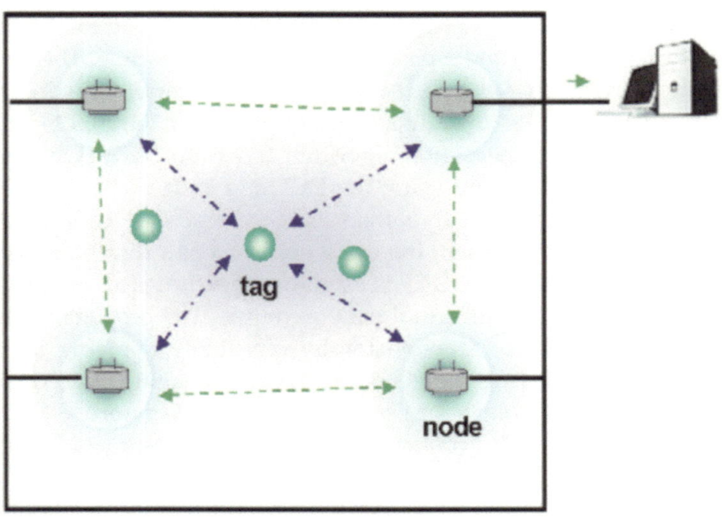

Fig. 1. Positioning situation

Since we are assuming that the distance between a node and a tag is calculated on the basis of signal transmission between the two, we can assume that the preamble portion of the packet is used. Distance calculation is based on the synchronization of the preamble. In the case of multiple incoming signals, the time of arrival of the signal taking the longest path is considered to be the reception time.

2.1 Principle of TDOA positioning algorithm

In this section, we show TDOA positioning method.
The positioning system configuration considered in this paper is shown in Fig.2.

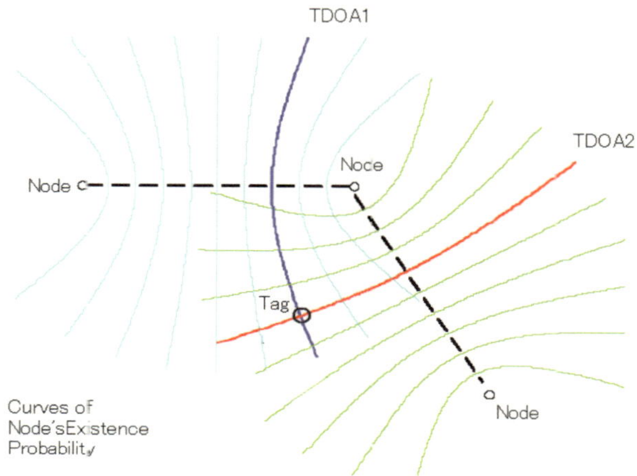

Fig. 2. TDOA principle

The number of node is M and these nodes have the information of position of x_i, y_i. Each node receives the signal of tag and tag's position is estimated using TDOA of each node. In TDOA system, the synchronization between tag and each node is not necessary. Therefore, the distance information is derived by comparing the received time of nodes and multiplying speed of light.

2.2 NEWTON algorithm

NEWTON algorithm is a linear search and an iterative algorithm. It is the algorithm of converging into true position by deriving the relative shift value from the gradient information. In other words, it starts with an initial guess, and improves the estimate at each step using least-squares. At first, distance difference of arrival (DDOA) R_{ij} computed by each combination of nodes is given by

$$R_{ij} = R_i - R_j = c(t_i - t_j) \quad i(1, M-1),\ j(2, M),\ i < j \tag{3}$$

where R_i is distance between tag and each node, t_i is arrival time of each node. First, likelihood computation at arbitrary position $P(m_0, n_0)$ is performed.

$$D_{ij}(m_0, n_0) = \sqrt{(x_i - m_0)^2 + (y_i - n_0)^2} - \sqrt{(x_j - m_0)^2 + (y_j - n_0)^2} \tag{4}$$

$$\Delta R_{ij} = R_{ij} - D_{ij}\ (m_0, n_0) \tag{5}$$

The gradient of R_{ij} evaluated in a initial position is expressed as

$$\frac{\partial R_{ij}}{\partial x}\Big|_{x=x0}\ \hat{x}\ + \frac{\partial R_{ij}}{\partial y}\Big|_{y=y0}\ \hat{y} \tag{6}$$

where

$$\frac{\partial R_{ij}}{\partial x} = -\frac{x_i - x_0}{\sqrt{(x_i - m_0)^2 + (y_i - n_0)^2}} + \frac{x_j - x_0}{\sqrt{(x_j - m_0)^2 + (y_j - n_0)^2}} \tag{7}$$

And similarly for $\frac{\partial R_{ij}}{\partial y}$

Let G be the gradient matrix given by

$$G = \begin{bmatrix} \frac{\partial R_{12}}{\partial x} & \frac{\partial R_{12}}{\partial y} \\ \frac{\partial R_{13}}{\partial x} & \frac{\partial R_{13}}{\partial y} \\ \vdots & \vdots \\ \frac{\partial R_{(M-1)M}}{\partial x} & \frac{\partial R_{(M-1)M}}{\partial y} \end{bmatrix} \tag{8}$$

Let $\Delta(x,y)$ be the adjustment matrix defines as

$$\Delta(x,y) = (G^T G)^{-1} G^T \Delta R \tag{9}$$

such that $x_0 \leftarrow \Delta x + x_0, y_0 \leftarrow \Delta y + y_0$. The process is repeated iteratively till $\Delta(x,y) \cong (0,0)$.

2.2 NLOS problem and delay modeling

If there are no differences among the arrival times of the signal from different nodes, the tag positions can be estimated very accurately. However, in general, this is not the case because node clock differences, the time resolution of the devices, and the NLOS problem. NLOS is the geometrical enlargement of the propagation path that occurs because of the presence of obstacles between the transmission and reception points. The fact that only reflected or diffracted waves arrive instead of the direct waves is responsible for the geometrical enlargement of the propagation time and the positive bias in the arrival time. This is illustrated in Fig.3.

This effect causes an error in the measurement of the arrival time, and results in the deterioration of positioning performance. In addition, a reception time error exists at the nodes and is expressed as a Gaussian error (Additive White Gaussian Noise: AWGN). The error is caused by factors such as time resolution limitation, jitter, and internal clock offset; this error also results in the deterioration of positioning accuracy. Therefore, the arrival time t_i can be written as

$$t_i = T_0 + T_A + T_N \tag{10}$$

Here, T_O is the true arrival time, T_A is the AWGN error, and T_N is the error caused by NLOS. The multiplication of these parameters with $\boldsymbol{c}$ yields distances, and the distance of arrival R_i is expressed as

$$R_i = R_0 + R_A - R_N \tag{11}$$

Here, R_O is the arrival distance, R_A is the error in the arrival distance, and R_N is the error caused by NLOS delay. Hereafter, for the sake of uniformity of units, our analysis will be carried out after converting all time parameters into distances. As previously mentioned, the errors in the distance calculated on the basis of TOA are expressed as AWGN, and their probability density function (PDF) is expressed as a Gaussian function of the form shown below:

$$P(R_A \mid x) = \frac{1}{\sigma}\exp(-\frac{x^2}{2\sigma^2}) \tag{12}$$

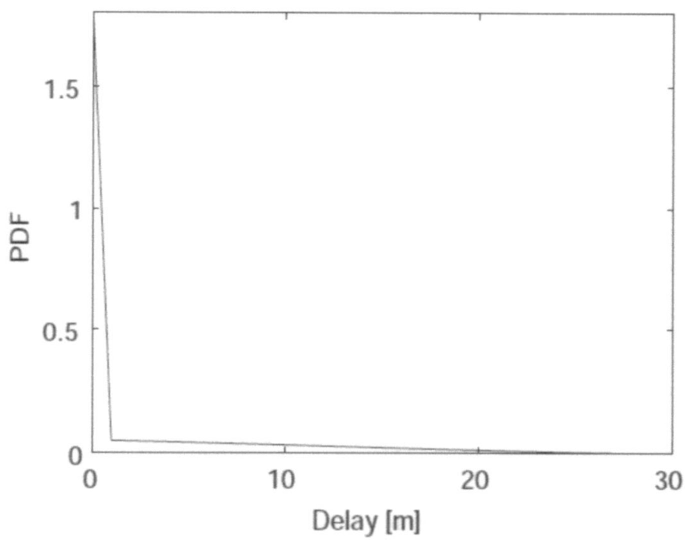

Fig. 4. IEEE.802.15.4a propagation PDF CM1(LOS)

Here, σ denotes the variance. The NLOS delay measurements are supposed to be carried out in an indoor environment using UWB (more specifically Home CM1/CM2). The probability distribution function used for modelling delays is the one based on IEEE.802.15.4a for UWB analysis. The actual reception time is the time taken for receiving the waves that travel along the path associated with the largest peak of the received signal. For R_N, the sum of the Generalized Extreme Value (GEV) distribution and Lognormal Distribution function (shown in Fig.4) is used for obtaining the LOS (CM1), while the PDF expressed as a Weibull distribution function, shown in Fig.5, is used for NLOS (CM2). Whether the value of R_N to be added to the received time in each node is LOS or NLOS depends upon a parameter called NLOS Rate. This parameter is based on the probability that the node is in an NLOS environment

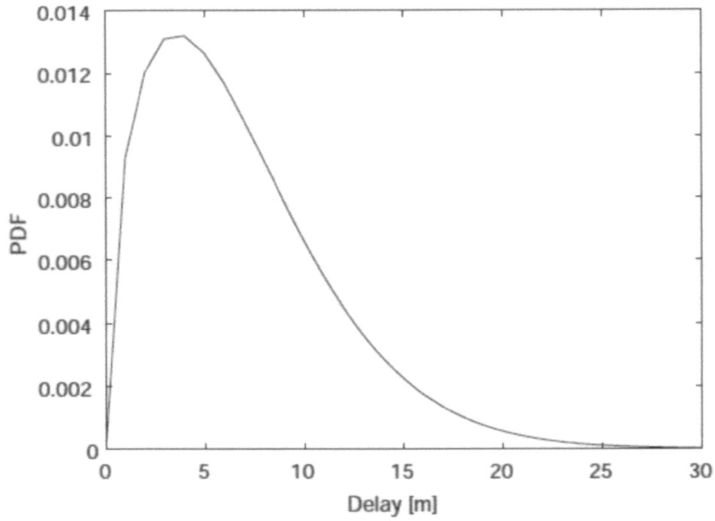

Fig. 5. IEEE.802.15.4a propagation PDF CM2(NLOS)

3. Iterative NLOS delay compensation algorithm and shift vector compensation algorithm

In order to compensate for the time delay caused by NLOS propagation, we consider a procedure in which the delays added to NLOS nodes are compensated in a step-by-step manner; we try to avoid modifying the parameters associated with LOS nodes. First, on the basis of the information obtained from all the nodes, the NEWTON method is used to obtain a preliminary estimate of the coordinates. Then, the transmission time is obtained by reverse calculation by using these coordinates. We assume that the greater the difference between this time value and the measured value, the larger is the effect of NLOS propagation on the measured value. An appropriate function derived using the above results is used to compensate for the time delay. In the absence of an error in the preliminary estimated coordinates, the derived NLOS delay is also correct. However, in practice, there is an error in the preliminary estimated coordinates, and therefore, there is no guarantee that a correct NLOS delay can be estimated if the correction is performed by considering the above-mentioned assumption. As previously mentioned, a naive correction may affect not only NLOS nodes but also LOS nodes, and therefore, the positioning accuracy cannot be improved to a satisfactory level. In order to resolve this problem, positioning estimation is carried out for minimizing the effect of delays on LOS nodes by performing compensation in a step wise manner, starting from large NLOS delays.

3.1 Delay compensation function

In this section, we discuss the modeling of a function that can be used for correcting the NLOS delay. Basically, the propagation time is estimated by calculating the distance from the nodes to the preliminary coordinates of a tag. In addition, the preliminary NLOS delay is obtained by subtracting the distance between the preliminary estimated position and the position of the node from the distance calculated by multiplying the measured time multiplied by the speed of light. Here, the closer the preliminary estimated position is to the true value, the more accurate is the estimated NLOS delay. Therefore, it is not desirable to correct all NLOS delays simultaneously. By correcting large NLOS delays first and then proceeding gradually to smaller NLOS delays, the errors in preliminary estimated positions also decrease in a gradual manner, enabling a more appropriate compensation of the NLOS delay. Below is a more detailed description.

The preliminary NLOS delay (distance) is expressed by the following equations:

$$D^{NLOS}{}_{i} = R_i - D_i \tag{13}$$

$$D_i = \sqrt{(x_i - X_{tem})^2 + (y_i - Y_{tem})^2} \tag{14}$$

As explained in section 2.2 a node under the influence of NLOS has a positive value added to its true arrival time. Therefore, there is a higher probability that the coordinates of preliminary estimated positions shift in a direction opposite to the direction of influence of NLOS. As a consequence, for example, an error in this preliminary estimated position may produce the following influence.

- $D^{NLOS}{}_i$ of a LOS node on the opposite side of the NLOS node outputs a positive value.
- $D^{NLOS}{}_i$ of a LOS node on the same side as the NLOS node outputs a negative value.

Therefore, it is difficult to perform adequate compensation simply by correcting the preliminary NLOS delay that is the output here.

Proper compensation requires the setting of a reference value that can be used in the positioning NLOS delay equation. The reference value is expressed as D_{basis}, and is changed according to the rules given below:

First, the largest of all $D^{NLOS}{}_i$ values is selected;

$$D^{NLOS}{}_{\max} = \max\left[R_i - D_i\right]_{i=1\ldots M} \tag{15}$$

Similarly, smallest of all $D^{NLOS}{}_i$ values is selected;

$$D^{NLOS}{}_{\min} = \min\left[R_i - D_i\right]_{i=1\ldots M} \tag{16}$$

As the maximum value approaches the reference value, only $D^{NLOS}{}_i$ of the nodes affected by large NLOS delays tend to assume positive values; this has negligible influence on the LOS nodes. On the other hand, the nodes that are affected by other NLOS delays are not adequately compensated. In contrast, if the reference is close to the minimum value, $D^{NLOS}{}_i$ of almost all the nodes become positive; thus making it possible to compensate for the NLOS delays for all the nodes. However, if the error in the preliminary estimated positions is large, its influence on the LOS nodes tends to be significant. Therefore, by setting the reference value closer to the maximum value $D^{NLOS}{}_{\max}$ in the first iteration, compensating the NLOS delay, and dynamically setting the reference to values closer to zero, it is possible to correct only the NLOS error and lessen the influence on the LOS nodes since the error in the preliminary estimated positions decreases at later stages. This is shown in equation 17.

$$D_{basis} = D^{NLOS}{}_{\max} - (D^{NLOS}{}_{\max} - D^{NLOS}{}_{\min})\frac{IN_k}{IN_{\max}} \tag{17}$$

Here, INmax is the total number of iterations, and INk denotes the k-th iteration. Then, using Dbasis, the equation for the compensation value DCi is reconstructed as

$$D^{C}_{i} = D^{NLOS}_{i} - D_{basis} \tag{18}$$

The delay compensation function (DCF) for each node DCF(DNLOSi) corrects only the positive component, and is expressed as follows:

$$DCF(D^{NLOSi}) = \begin{cases} 0 & (D^{c}_{i} < 0) \\ D^{c}_{i} & (D^{c}_{i} > 0) \end{cases} \tag{19}$$

This value is arranged such as

$$R_i^{'} \leftarrow R_i - DCF(D^{NLOS}_{i}) \tag{20}$$

and next positioning process is performed using this R'i. Finally, these processes are repeated (INk=1~INmax).

3.2 Compensating the positioning shift resulting from the relative position between node and tag

3.2.1 Basic principle

As previously mentioned, the existence of NLOS delay causes a positive bias to be added to the actual distance, deteriorating the positioning accuracy. However, a problem arises: the bias tends to push the estimated position farther away from the node under consideration and with respect to the actual position. In other words, the effect is similar to that of a vector whose reference is the line joining each node to the tag position. In the present paper, the above vector is referred to as "shift vector". It is possible to partially alleviate this effect by means of the NLOS delay-compensation process on the basis of a DCF. However, if the error associated with the initial estimation that is based on raw information from all nodes is too large, it becomes difficult to alleviate the effect of NLOS in a satisfactory manner. Therefore, it is necessary to alleviate the effect through an analysis of geometrical relations. In addition, enhancing the synergy effect that exists with respect to DCF by improving the precision of the initial positioning estimation, appears to be possible. Hereinafter, the present algorithm is referred to as Shift Vector Compensation (SVC).

3.2.2 Mathematical expression

First, as in the case of the previously mentioned algorithm, TDOA positioning is carried out using raw data obtained from all the nodes. The determined position is X_{tem}, Y_{tem}. The shift vector that results from a delay originating from an arbitrary node i and j is expressed by the following equation:

$$V_i = \begin{bmatrix} X_{tem} - x_i \\ Y_{tem} - y_i \end{bmatrix} \tag{21}$$

$$V_j = \begin{bmatrix} X_{tem} - x_j \\ Y_{tem} - y_j \end{bmatrix} \tag{22}$$

This unit vector is expressed as follows:

$$\frac{V_i}{|V_i|} = \begin{bmatrix} \dfrac{X_{tem} - x_i}{\sqrt{(X_{tem} - x_i)^2 + (Y_{tem} - y_i)^2}} \\ \dfrac{Y_{tem} - y_i}{\sqrt{(X_{tem} - x_i)^2 + (Y_{tem} - y_i)^2}} \end{bmatrix} \tag{23}$$

Similarly, $\frac{V_j}{|V_j|}$ is computed, too.

When distance difference of arrival between node A and node B in Fig.6 is computed, if NLOS delay is added into node B, each vector influenced by NLOS is represented as vector *a* and *b*. Vector *b* is the unite vector from position B to TEP and vector *a* is the unite inverse vector from position A to TEP. Additionally, the sum vector of each unit shift vector is represented as vector *c*.

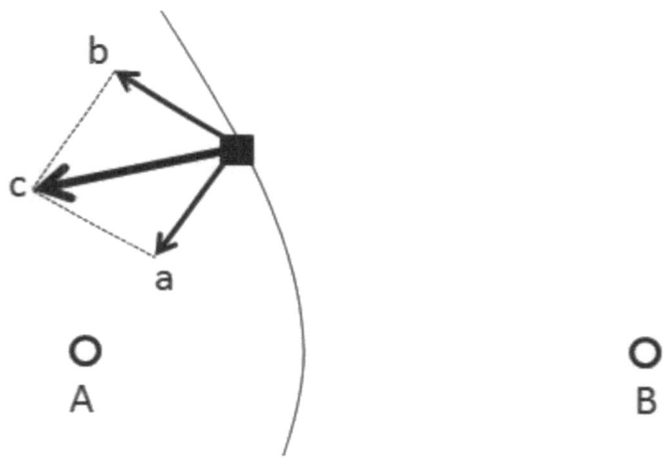

Fig. 6. Shift vector principle

Therefore, the resultant shift vector is obtained by a summation of the unit shift vector divided by the number of nodes multiplied by the preliminary NLOS delay. Number of combination of nodes is ${}_MC_2$=M(M-1)/2. Then, the sum of absolute value of all shift vectors is represented as

$$Z = \frac{1}{{}_MC_2} \sum_{i=1}^{M-1} \sum_{j=i+1}^{M} \left| \frac{V_i}{|V_i|} - \frac{V_j}{|V_j|} \right| \tag{24}$$

where this value Z is sum of scalar value. Next, the iteration process of shift vector is performed same as the process using DCF.

$$D^T{}_i = D^{NLOS}{}_i - (D^{NLOS}{}_{\max} - D^{NLOS}{}_{\min})(1 - \frac{INV_L}{INV_{\max}}) \\ if(D_i^T < 0) D_i^T = 0 \tag{25}$$

If $D^T{}_i$ is smaller than 0, $D^T{}_i$ becomes 0. Therefore, the $D^T{}_i$ is the extracted value over basis value. INV_L and $INV_{\max}$ denotes the L-th iteration and total iteration number respectively in SVC process.
Therefore, shift vector is represented as

$$\begin{bmatrix} X_s \\ Y_s \end{bmatrix} = \frac{1}{{}_M C_2} \frac{1}{Z} \frac{INV_L}{INV_{\max}} \sum_{i=1}^{M-1} \sum_{j=i+1}^{M} (D^T{}_i - D^T{}_j)\left(\frac{V_i}{|V_i|} - \frac{V_j}{|V_j|}\right)$$
$$\text{if}\left(D^{NLOS}{}_i < 0\right) D^{NLOS}{}_i = 0,\ \text{if}\left(D^{NLOS}{}_j < 0\right) D^{NLOS}{}_j = 0 \tag{26}$$

Furthermore, it is possible to further adjust the estimated position to a value closer to the true position by subtracting the shift vector from the estimated position. Next, X_S and Y_S is compensated from X_{tem} and Y_{tem} respectively.

$$\begin{aligned} X_{tem} &\leftarrow X_{tem} - X_s \\ Y_{tem} &\leftarrow Y_{tem} - Y_s \end{aligned} \tag{27}$$

The equation (26) and (27) are repeated (INV_L=1~$INV_{\max}$). Finally, compensated position is output.

3.2.2 Embedding into the iterative process

We now describe a method to embed SVC into the iterative process described in the previous section. Basically, the portion used in the SVC was taken after excluding the portion compensated by the delay compensation algorithm. Using compensated arrival distance R'_i of equation (20), SVC algorithm is performed. In other words, this equation shows that as the iterative process advances, the portion to be compensated by using SVC decreases with an increase in the compensation by using DCF. In addition, because of the synergy effect of the iterative DCF and SVC, delay compensation is more effective than that achieved by these methods separately, possibly resulting in better estimation accuracy.

4. Simulation

A simulation is carried out to compare the different approaches presented so far. Shown below is a description of the simulation method used. A comparison is carried out among the iterative algorithm using the DCF function (referred to as Iteration), the SVC algorithm, and a combination of the iterative algorithm and SVC (referred to as Iterative SVC)
The structure of the room may be a conventional cube or cuboid; in either case, from a geometrical perspective, the effect is more pronounced if the nodes taken as references are uniformly distributed with sufficiently large distances between them. On the contrary, if the reference nodes are distributed only along a straight line, the obtained accuracy may not be as expected.

4.1 Evaluating changes introduced in AWGN parameters

In this evaluation, the iterative algorithm is evaluated in terms of the appropriate number of iterations. If the number of iterations is small, the compensation is carried out while the preliminary estimated position is under the influence of NLOS delay, and this is used as a reference for further compensations. For that reason, the highly reliable LOS nodes are also

Field	30x30 [m]
Tag position distribution	At random within the field
Number of trials	10000
Node position error	0 [m]
AWGN parameter	0.3 [m]
Number of nodes	9
Number of iterations	5
NLOS rate	0.5

Table 1. Simulation parameters

The nodes are arranged in the system of coordinates illustrated in Figure 7.

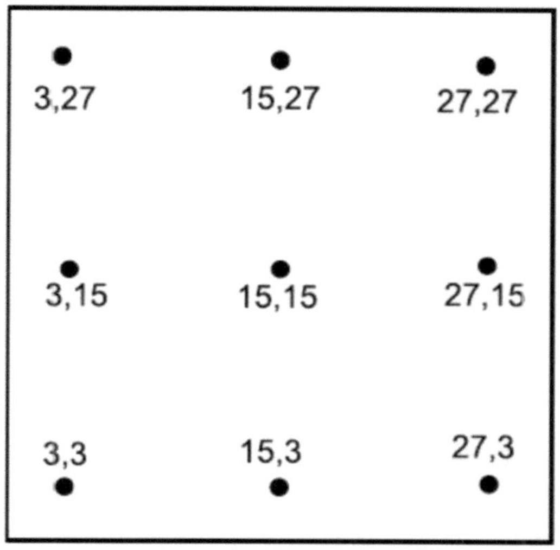

Fig. 7. Node distribution

affected, resulting in an error in the estimation of the final positioning. On the other hand, if the number of iterations is large, the influence on LOS nodes is reduced however the computational cost may increase. Fig.8 illustrates the effect of the number of iteration on the actual estimation.

As shown in Fig.8, the results indicate that for the Iteration and Iterative SVC algorithms the accuracy improves as the number of iterations increases. A possible reason for this increase is that by increasing the number of estimation correction steps, the amount of compensation required per step decreases, alleviating the effect on NLOS nodes. However, it is worth noting that the accuracy does not increase significantly when the number of iterations exceeds 5~7 especially in Iterative SVC algorithm.

The reason is that NLOS errors can be compensated enough in a few steps by the synergy effect with Iterative algorithm and SVC algorithm.

4.2 Evaluating changes introduced in AWGN parameters

Fig.9 shows the results of changes introduced in the AWGN parameters of each node. The first conclusion that the characteristics of Iteration, SVC, and Iterative SVC algorithms improve than NEWTON algorithm. As a general rule, the Iterative SVC case exhibits the best characteristics in small AWGN error, however as the AWGN error increases, the difference between the characteristics of different methods tends to decrease. When the AWGN error is large, the Iteration algorithm outperforms the Iterative SVC algorithm. A possible explanation for this is that the large AWGN error causes a general drop in the reliability, which in turn deteriorates the reliability of the shift vector itself.

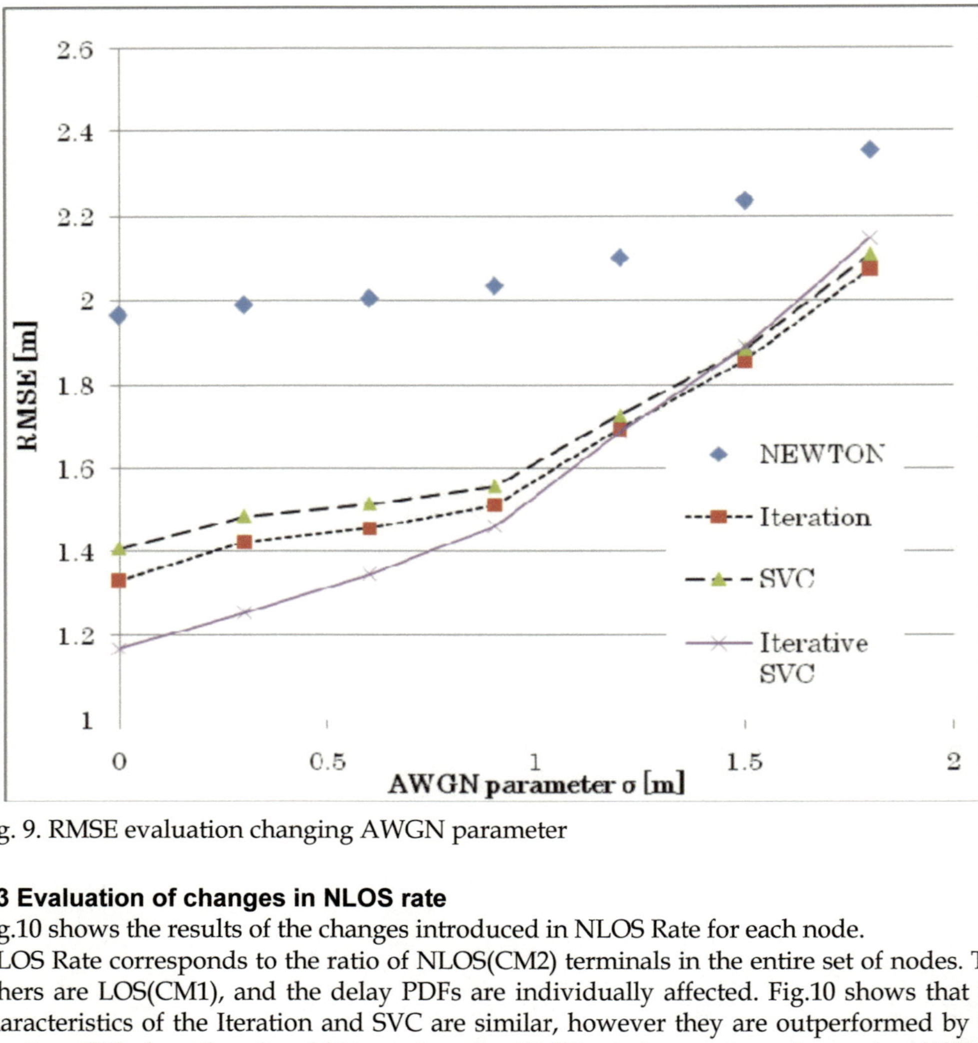

Fig. 9. RMSE evaluation changing AWGN parameter

4.3 Evaluation of changes in NLOS rate

Fig.10 shows the results of the changes introduced in NLOS Rate for each node.

NLOS Rate corresponds to the ratio of NLOS(CM2) terminals in the entire set of nodes. The others are LOS(CM1), and the delay PDFs are individually affected. Fig.10 shows that the characteristics of the Iteration and SVC are similar, however they are outperformed by the Iterative SVC algorithm. In addition, when the NLOS rate is zero, i.e., when only AWGN is present at each node, this result changes slightly. This happens because the algorithm performs some correction for any NLOS delay that may exist, however its performance decreases in other environments. In concrete terms, in the Iteration algorithm, a part of AWGN is interpreted as NLOS even though there is no NLOS delay. In the Iterative SVC algorithm, even when no geometrical shifts exist, compensation is performed in another direction. However, these problems do not cause a significant deterioration.

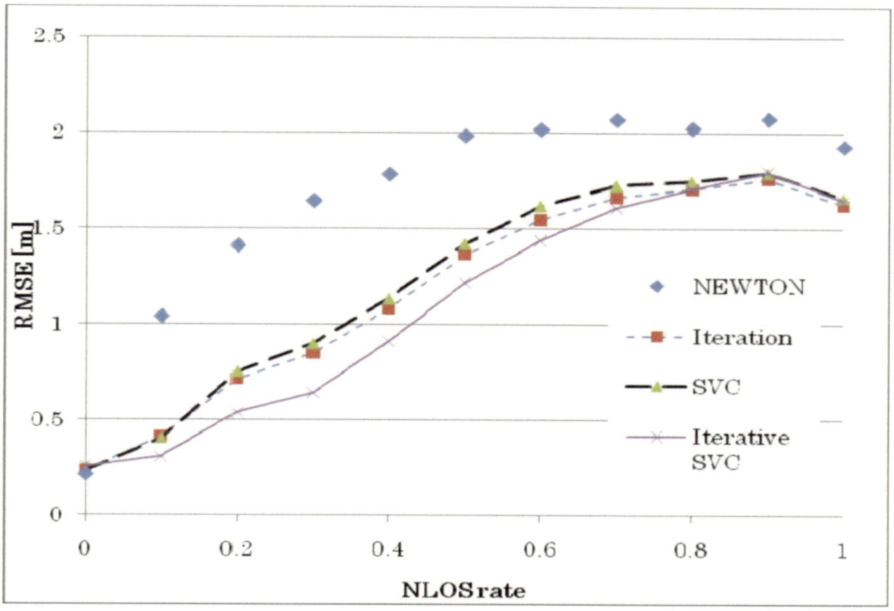

Fig. 10. RMSE evaluation changing NLOS rate

5. Tracking experiment by transmission tag and reception node

In this section, we perform the tracking experiment by using prototype device in NLOS environment. These devices are made in Fujitsu Co., Ltd. and Fujitsu Component Co,. Ltd. This appearance of tag is shown in Fig.11 and the appearance of node is shown in Fig.12.

Fig. 11. Tag appearance

Fig. 12. Node appearance

Reception nodes are distributed fixedly and the position of each node is listed in table \ref{tb:experiment}.

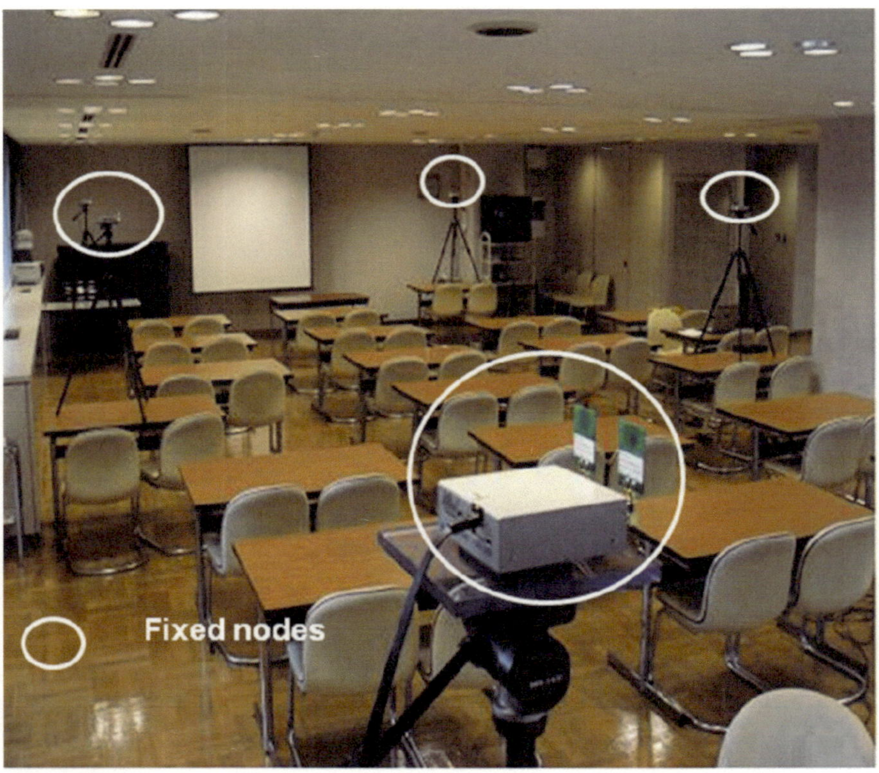

Fig. 13. Tracking situation

Node 1	0 [m]	0 [m]
Node 2	4.228 [m]	0 [m]
Node 3	8.456 [m]	0 [m]
Node 4	12.684 [m]	0 [m]
Node 5	12.684 [m]	5.436 [m]
Node 6	8.456 [m]	5.436 [m]
Node 7	4.228 [m]	5.436 [m]
Node 8	0 [m]	5.436 [m]

Table 2. Each node position for tracking expriment

Height of each node is 2 [m].

And, tag's position is estimated in the case of low-grade NLOS which tag is held up over human head, and in the case of serious NLOS which tag is held on breast side of human body. Tag is moved along the sides of the drawn rectangle. If signal from tag to node is vanished, available node is reduced, and if the number of available node is less or equal three, system outputs impossibility to estimate position. If estimated position is greatly exceeds the range covered by all nodes, previous once estimated result is output. Under this condition, NEWTON method which is conventional method and Iterative SVC algorithm which is proposed method is shown in Fig.13 and 14.

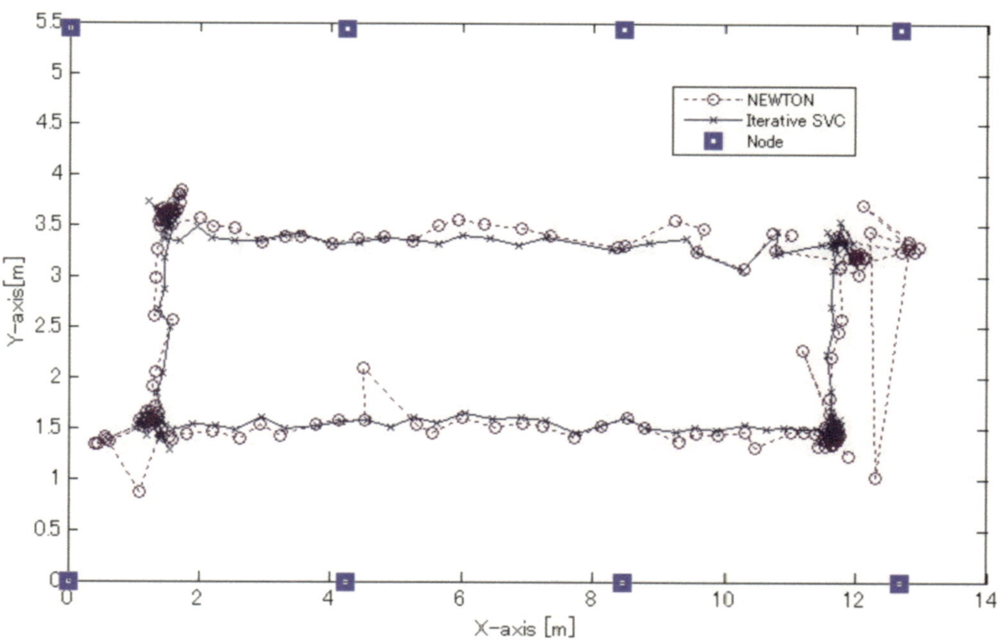

Fig. 14. Tracking result in low-grade NLOS environment

Fig.14 is low-grade NLOS situation, and Fig.15 is serious NLOS situation. These results show that positioning error occurs at several position in conventional method, however proposed method can decrease the error, and can improve the accuracy of positioning and tracking.

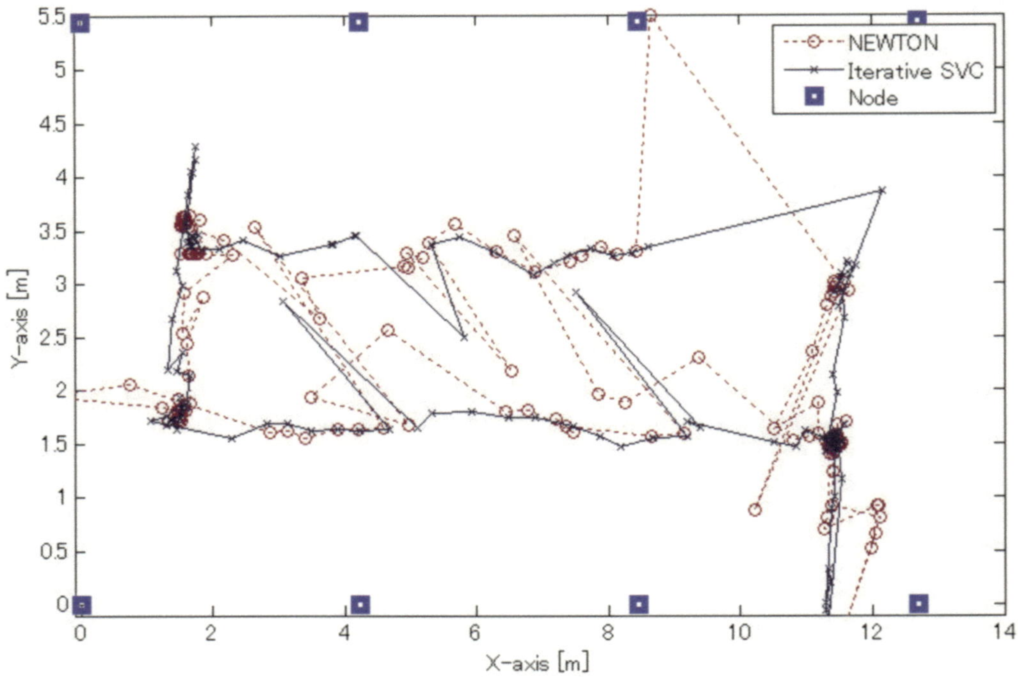

Fig. 15. Tracking result in serious NLOS environment

6. Conclusions

In the present paper, we proposed novel TDOA algorithm for reducing the error in the positioning estimation by using a positioning system in an UWB environment. In this process, a provisional position is first estimated using the NEWTON method. We then considered an NLOS delay compensation and a compensating function to alleviate the effect on LOS propagation. Then, we developed an adaptive system in which the function is adaptively renewed according to the number of iterations. This way, the vector expressing the relative positions of nodes and tags as well as the vector that corrects the vector resulting from NLOS delay were corrected. By repeating the above process for a certain number of times, an algorithm to gradually improve the positioning accuracy was developed. As previously shown in the explanation of the system model, the basic application scope of the present algorithm is an indoor environment using UWB with small signal loss. For each node, we assume the existence of the influence of a delay distribution on the basis of AWGN components and IEEE.802.15.4a, however the present algorithm may be extended to other types of delay distribution environments as well. Finally, we perform the experiment of tag tracking using TDOA system. This result shows that proposed method can mitigate abnormal tracking and improve accuracy.

The topics to be dealt with in the future are tag displacement, and extension to three dimensions. Extending to three dimensions would require the calculation of the height coordinate, which would influence the amount of computation and accuracy. It seems possible to use the existing positioning algorithm for three-dimensional analyses.

7. Acknowledgment

Part of the present research received support from the Global COE Program "Creating innovation by the integration of Medicine and Engineering using information and communication" of Yokohama National University. We express our deepest gratitude to all the people.

8. References

Kentaro TANIGUCHI, Ryuji KOHNO, "Positioning Algorithm Based on TDOA Measurements Using Layered Particle Filter in Sensor Network", IEICE transaction A Vol. J89-A No.12 pp.1068-1078

Denis B., Keignart J., Daniele N., "Impact of NLOS propagation upon ranging precision in UWB systems",Ultra Wideband Systems and Technologies, 2003 IEEE Conference,pp.379-383

Maali, A. Mimoun, H. Baudoin, G. Ouldali, A. EMP, Commun. Syst. Lab., Algiers, "A new low complexity NLOS identification approach based on UWB energy detection",RWS'09.IEEE,pp.675 - 678

A.F. Molisch, K. Balakrishnan, and C.C. Chong, IEEE 802.15.4a channelmodel-finalreport. \\ http://www.ieee802.org/15/pub/TG4a.

Kozo SAKAWA, ?Non Line-of-Sight Microwave Propagation Characterization for Personal Communications with High-Tier Base Station Antenna?, IEICE TRANSACTIONS Vol.E85-A No.7 pp.1569-1577

Wei-Kai CHAO Kuen-Tsair LAY , "Mobile Positioning and Tracking Based on TOA/TSOA/TDOA/AOA with NLOS-Reduced Distance Measurements", IEICE TRANSACTIONS on Communications Vol.E90-B No.12 pp.3643-3653

S.Al-Jazzar, J.Caffery, "ML and Bayesian TOA Location Estimators for NLOS Environments", VTC2002-Fall pp.1178-1181

Wuk KIM Jang-Gyu LEE Gyu-In JEE, "Estimation of NLOS Propagation-Delay Error Improves Hybrid Mobile Station Location",

IEICE TRANSACTIONS on Fundamentals of Electronics, Communications and Computer Sciences Vol.E85-A No.12 pp.2877-2880

Yangseok JEONG , Heungryeol YOU , Dae-Hee YOUN , Chungyong LEE, "A New Method for Calibration of NLOS Error in Positioning Systems", IEICE TRANSACTIONS on Communications Vol.E85-B No.5 pp.1056-1058

K. Yu and Y. J. Guo, "Improved positioning algorithms for nonline-of-sight environments", IEEE Trans. Vehicular Technology, vol. 57, no. 4, 2342-2353,July 2008.

H. Miao, K. Yu, and M. Juntti, ?Positioning for NLOS propagation: algorithm derivations and Cramer-Rao bounds,? IEEE Trans. Vehicular Technology, vol. 56, pp. 2568?2580, Sept. 2007.

W. H. Foy, "Position-Location Solutions by Taylor-Series Estimation," IEEE Trans. Aerospace and Electronic Systems, Vol. AES-12, NO. 2, pp.187-193.

Zhu XIAO, Ke-Chu YI, Bin TIAN and Yong-Chao WANG,"UWB Localization for NLOS under Indoor Multipath Channel: Scheme and TOA Estimation", IEICE Transactions on Communications 2008 E91-B(10):3391-3394; doi:10.1093/ietcom/e91-b.10.3391

19

Efficient Range Query Using Multiple Hilbert Curves

Ying Jin[1], Jing Dai[2] and Chang-Tien Lu[3]
[1]*Cold Spring Harbor Lab*
[2]*IBM T. J. Watson Research Center*
[3]*Virginia Polytechnic Institute and State University*
USA

1. Introduction

Indoor location tracking based on RFID has been widely discussed and applied. RFID reading process is efficient and reliable, therefore it is suitable for discovering locations inside buildings where GPS signals are usually unreachable. In general, there are two approaches for location sensing using RFID. 1) Deploying RFID tags at fixed locations and RFID readers attached to moving objects (Willis, 2004). Each tag represents a reference point in the space, and a reader determines its location by the set of tags being detected. 2) Deploying RFID readers (and tags) at fixed locations and RDIF tags attached to moving objects (Hightower, 2001; Ni, 2004). The readers report to the system when a tag is detected, and the system identifies the location of this tag by the set of readers that have reported and their corresponding signal strength.

These location management systems require multi-dimensional access methods to allow efficient handling of spatial queries. Because there is no total ordering of locations that preserves the spatial locality between objects, it is difficult to design multi-dimensional access method in the way as traditional one-dimensional access methods. However, mapping multi-dimensional data into a single dimension makes it possible to utilize the extensively exploited B/B+-tree as the index and its associated concurrency control and recovery mechanisms.

Space-filling curves (SFCs) (Simmons, 1963) have been widely used to map the multi-dimensional data points into a linear order. It was first introduced by Peano (Peano, 1890) to map from a unit interval to a unit square. SFC can link all cells with passing through each of them only once, so it provides a way of generating a total linear ordering of all grids in a multi-dimensional space. Many SFCs have been proposed in the literature, such as Peano curve (or Z-order) (Orenstein & Merrett, 1984; Peano, 1890), Hilbert curve (Hilbert, 1891), Gray curve (Gray, 1953), Sweep, and Scan. The multi-dimensional data are transformed to a set of one-dimensional integer values using SFC mapping schemes. The transformed data can be stored in a traditional one-dimensional database based on the linear orders, and indexed by B-trees or B+-trees. Then the spatial queries, such as range query, kNN query, and spatial join, can be processed. Using SFCs to enable processing spatial queries based on traditional one-dimensional indices is proposed in (Faloutsos, 1988; Faloutsos & Rong, 1991; Faloutsos & Roseman, 1989; Jagadish, 1990).

Spatial range query identifies spatial objects located within a given area. For example, "find all hospitals in city A" is a common range query for a GIS application. In data mining applications, range queries are used to discover the characteristics of a specific region. For instance, a data set contains different environmental variables of the areas which are the habitat of some kinds of bird. A user may submit a range query like "find how many lakes or rivers within that area", in order to identify associations between water and bird. In addition, many spatial query operations, like k-nearest neighbor query, and spatial join query, rely on range queries. Developing an efficient range query processing scheme will contribute to improving overall spatial query operations.

Several works have been conducted on utilizing SFCs to solve range queries. Gray and Hilbert SFCs are used to handle range queries in (Faloutsos, 1988), and (Jagadish, 1990), respectively. Given a range query, the number of continuous runs on Gray, Z-order, and Hilbert are evaluated in (Jagadish, 1990). Hilbert curve is used as a spatial access method in (Faloutsos & Rong, 1991; Faloutsos & Roseman, 1989), where the data is stored in a one-dimensional disk based on Hilbert values. Hilbert curve is also used as multi-dimensional indexing method (Lawder & King, 2000) and spatio-temporal indexing methods (Jensen et al, 2004). Recently, concurrent spatial indexing methods based on Hilbert curve are proposed in (Dai & Lu, 2007, 2009). More importantly, those experiments show that space-filling curve approach is more preferable in high dimensional space than the R-tree family. Previous works demonstrated that among these SFCs, Hilbert curve has the optimal clustering property (Abel & Mark, 1990; Jagadish, 1990; Mokbel et al, 2003) over a variety of computing conditions. In other words, Hilbert curve provides the best linear mapping to preserve the locality between multi-dimensional objects in one-dimensional space. Given that the data objects are physically arranged on disk according to their SFC values, Hilbert curve is more likely to organize the spatially adjacent data into the same disk page or consecutive disk pages, and hence reduces the required disk accesses for a spatial query. Nevertheless, even using Hilbert curves, a range query may cover several discrete clusters (set of cells with consecutive SFC values), which leads to multiple index tree traversals. The reason is that linear mapping loses spatial relationship between spatial objects. Although objects near to each other in a linear ordering must also be close in the spatial space, the opposite is not always true. In some cases, two neighbors in a spatial space may be far away from each other in the one-dimensional ordering. Recently, Partitioned Hilbert curve has been used to reduce the search range for spatial queries including range and kNN queries in (Zheng et al, 2004). However, the number of clusters covered by a query window is not reduced, and some cells outside the query window may still be in the search range. In order to improve the query response time by reducing the number of clusters (or continuous runs), approximate NN query based on multiple shifted Hilbert curves is proposed by (Liao et al, 2001).

In this chapter, an efficient spatial range query method is designed for compensating the lost of spatial relationship by the linear mapping mechanisms. Hilbert space-filling curve is chosen to map spatial space into the one-dimensional domain, because of its best clustering property. Different from previous work, the proposed method uses multiple copies of Hilbert curves, and each has different rotations or shift. When a range query comes, the Hilbert curve that generates the minimum number of clusters will be selected. Theoretical proofs are provided to show that the rotation of Hilbert curve can reduce the number of clusters significantly. The experiments conducted on real data sets demonstrate that the proposed approach is efficient and scalable, and the combination of rotation and shift outperforms applying any of them independently.

2. Motivation

A range query searches all objects that overlap with a given region (also called query window). When Hilbert curve is used as indices, an intuitive approach to answer a range query is to search on the Hilbert values of the cells that overlap with the query window. This procedure consists of three steps, mapping, filtering, and refinement. First, the query window is mapped to a set of cell numbers according to the Hilbert curve traversal. The cells connected consecutively by the curve forms a cluster that indicates a single continuous query. Second, for each cluster, the ranges of cell numbers, i.e., Hilbert values are used to query the corresponding B+-tree. All leaf entries of the B+-tree with their key values exactly the same as the cell numbers within the query window are identified. These entries point to the disk pages that store the objects that potentially overlap with the query window. Third, each object retrieved from those disk pages will be validated to make sure that the object actually overlaps with the query window.

To map multi-dimensional points into one-dimensional values using Hilbert space-filling curve, there is an existing algorithm with the time complexity of $O(kn)$ (Lawder et al., 2001), where k is the order of the curve and n is the dimensionality of the data space. By contrast with the light CPU computation in the mapping step, I/O cost for index traversals in the filtering step is a dominating factor of the time cost of a spatial range query. The I/O cost of the filtering step depends on the number of clusters covered by the range query. *Fig. 1*(a) gives an example of a range query in a two-dimensional space. The data space is divided by 64 uniform grid cells ordered according to Hilbert values. The shaded area represents a range query *A*. *A* overlaps with 8 cells (4,5,6,7,56,57,58,59) that consist of 2 clusters (4-7), (56-59). Using traditional methods, it is necessary to traverse the B+-tree at least 2 times to find all candidate cells that overlap with A. The underlying reason is that some spatial relationships between spatial objects are lost by applying linear mapping. In this example, the two clusters are adjacent in the two dimensional space, but far away from each other in the Hilbert ordering. To reduce the tree traversal times, one intuitive solution is to expand two clusters to form a big cluster, i.e., (4-59). However, a problem of this solution is that it will cause many unnecessary data accesses, i.e., data pages between 8 and 55. Another solution could be enlarging the cell size, but the overhead of accessing additional data space and filtering will be increased accordingly. The following sections focus on reducing the number of clusters of each range query, and meanwhile remaining the refinement overhead. It is observed that the number of clusters covered by a query window varies under Hilbert curves with different orientations and shift. The following two subsections describe the observations in detail.

2.1 Rotation

A range query covers different number of clusters when using Hilbert curves with different orientations. Hilbert curve is a recursive space-filling curve. Specifically, in two-dimensional space, the ith-order curve is derived by replacing each quadrant with the $(i-1)^{th}$-order curve, and two of them are rotated 90 degree clockwise and anticlockwise respectively. A Hilbert curve in a two-dimensional space has four orientations. The same query window on differently oriented Hilbert curves may derive different number of clusters. Fig. 1(a) and (b) show an example. Hilbert curves with two orientations are applied on the same data space. For the same range query *A*, the two curves result in different numbers of clusters: two in (a) and only one in (b). Note that this change on the number of clusters within a given query window may vary with different query positions.

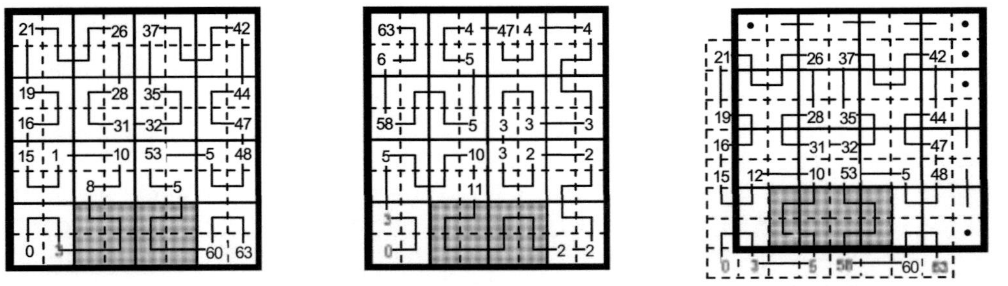

(a) Original (b) Rotated (c) Shifted

Fig. 1. Range Query A.

To study the relations between the orientations of a Hilbert curve and the number of clusters covered by a query window, we start from the simplest case which is described in Fig. 2. The same notations in (Moon et al., 2001) for the orientation of Hilbert curves are used here. In Fig. 2, a 16-cell grid space is shown in the leftmost. Assuming the size of the query window is 2 * 2, the total number of different query positions is 9, indicated by the numbers shown on the curve in the figure. Fig. 2 (a), (b), (c), and (d) use a 3 * 3 matrix to present the number of clusters contained in the query window, on 2+-orientation, 1+-orientation, 2--orientation, and 1--orientation, respectively. In this matrix, each entry corresponds to a query position, such as the top left entry stands for the position 1. As can be seen, when the query window is located in the top row of the space, the orientation represented in 2(a) gives the fewest clusters. However, when the query window is located in the bottom row of the space, the orientation represented in 2(c) gives the fewest clusters. Based on the matrices of the four orientations, the minimum number of clusters for each position can be calculated, as illustrated in the final matrix 2(e). In 2(e), all positions except the center position have only one cluster. Based on this observation, a fact can be found is that using Hilbert curves with different orientations may reduce the number of clusters for a query window which are not located in the center position.

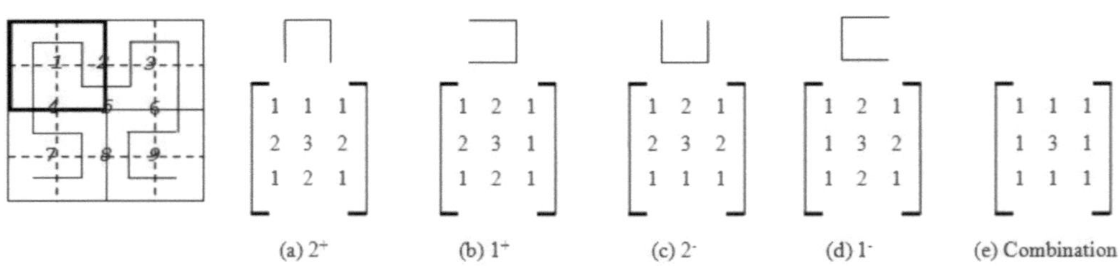

Fig. 2.Clusters with different orders.

2.2 Shift

The number of clusters may be decreased when the Hilbert curve is shifted diagonally. The example presented in Fig. 2 shows that when the query window is located at the center of the grid space, the number of clusters can not be reduced with multiple orientations. A similar example can be found in Fig. 3 (a) and (b), where the number of clusters in query window B remains unchanged after the rotation. The reason is that in a two-dimensional case, whenever the order of a Hilbert curve increases, it splits the whole space quarterly, replaces each sub-space using the original or rotated Hilbert curves, and uses three connection edges to link the four quadrants. Among the three connection edges, only one is contributed to the center of the entire space, while the other two are for boundaries. However, when the Hilbert curve is diagonally shifted, as shown in Fig. 3 (c), the same query window will cover only one cluster.

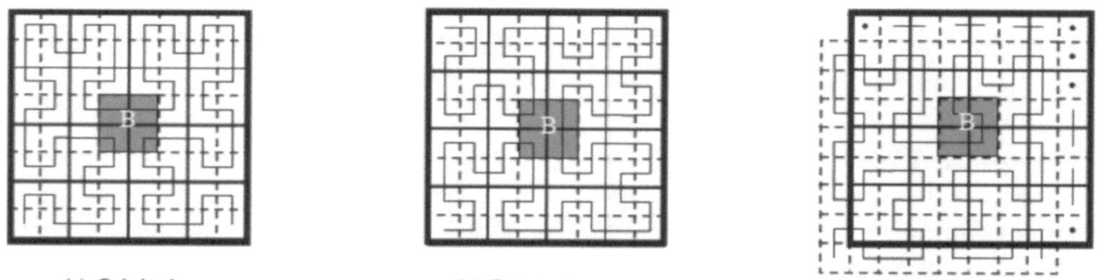

(a) Original (b) Rotated (c) Shifted

Fig. 3. Range query B.

2.3 Hybrid

Symbols	Definition
H_{k+n}	2-dimensional Hilbert curve with size $2^{k+n} * 2^{k+n}$
H_k	2-dimensional Hilbert curve with size $2^k * 2^k$
t_n	Number of connection edges within the top boundary of a 2^+-*oriented* H_{k+n}.
b_n	Number of connection edges within the bottom boundary of a 2^+-*oriented* H_{k+n}.
s_n	Number of connection edges within one side boundary of a 2^+-*oriented* H_{k+n}.
$c_{t,n}$	Number of connection edges between H_k sub regions in top boundary and in other areas of a 2^+-*oriented* H_{k+n}.
$c_{b,n}$	Number of connection edges between H_k sub regions in bottom boundary and in other areas of a 2^+-*oriented* H_{k+n}.
$B_{i+/-,n}$	Number of $i^{+/-}$-*oriented* H_k in the bottom boundary of a 2^+-*oriented* H_{k+n}.
$T_{i+/-,n}$	Number of $i^{+/-}$-*oriented* H_k in the top boundary of a 2^+-*oriented* H_{k+n}.
h_k	Number of horizontal edges in a 2-*oriented* H_k.
v_k	Number of vertical edges in a 2-*oriented* H_k.
$N_{i+/-,t}$	Average number of clusters within $2^k * 2^k$ query region in the top boundary of $2^{k+n} * 2^{k+n}$ $i^{+/-}$-*oriented* H_{k+n}.
$N_{i+/-,b}$	Average number of clusters within $2^k * 2^k$ query region in the bottom boundary of $2^{k+n} * 2^{k+n}$ $i^{+/-}$-*oriented* H_{k+n}.

Table 1. Notations.

From the above observations, a range query can have different numbers of clusters under Hilbert curves with different orientations or shifts. Thereby, using multiple copies of rotated or shifted Hilbert curves should reduce the number of clusters for range queries in general. However, if only rotation is used, there is a "center position" problem. As described previously, shift can reduce the number of clusters when the query window is located in center position. However, it is not always better to shift for other range queries. For example, in Fig. 1 (c), the shift result of (a), still has the same number of clusters for range query A. It can be concluded from these observations that, in some cases, rotation is better than shift, but some cases may be opposite. Therefore, combining the rotation and shift can be more effective than applying any single one of them independently.

3. Spatial range query algorithms

In this section, we provide a theoretical analysis on the first observation, and then derive a formula to measure the improvement of applying multiple copies of Hilbert curves with different orientations. We also introduce a new spatial range query algorithm designed based on the combination of rotations and shift.

3.1 Theoretical proofs

In this section, formulas will be derived to calculate the average number of clusters for a given query region in the top and bottom boundary of a 2+-oriented Hilbert curve. And then we prove that the average number of clusters within given query region on 2--oriented Hilbert curve is smaller than the average number of clusters on 2+-oriented Hilbert curve, when the queries are located on the bottom boundary of the space. This proof can be extended to queries located in other areas and Hilbert curves with other orientations. Specifically, we assume that the query window is a region with size $2^k * 2^k$, and the size of the grid space is $2^{k+n} * 2^{k+n}$. The notations used in the proof are listed in Table 1. We define connection edge in a $2^{k+n} * 2^{k+n}$ Hilbert curve as the edge that connects two sub curves, each with size $2^k * 2^k$.

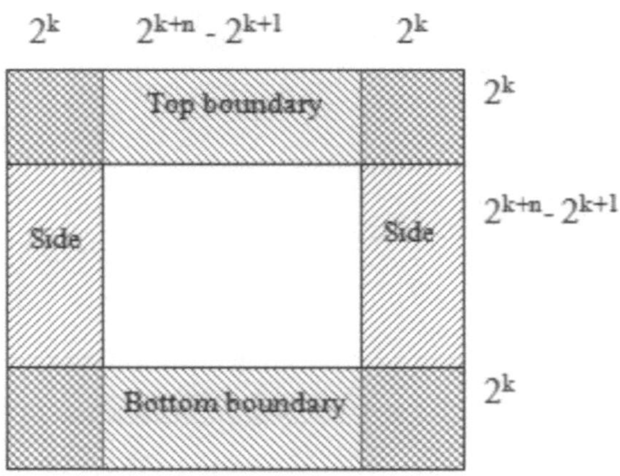

Fig. 4. H_{k+n} divided into 9 subregions.

The grid space of H_{k+n} is divided into nine sub regions, as shown in *Fig. 4.* The smaller side length of each sub region on the boundary is 2k. Then, the $2^{k+n} * 2^{k+n}$ grid region H_{k+n} can be considered as a collection of 2^{2n} H_k, each of which connects to one or two neighbors by connection edges. The following proves are deducted from parts of the conclusions in (Moon et al., 2001).

By definition of Hilbert curve, 2+-oriented Hilbert curve and 2--oriented Hilbert curve are symmetrical, when given the curve-space, order, so for given query region, the average number of clusters in the top boundary of a 2+-oriented Hilbert curve is equal to the average number of clusters in the bottom boundary of a 2--oriented.

Remark 1: *The difference of the average number of clusters between 2+-oriented Hilbert curve and 2--oriented Hilbert curve when queries are located in the bottom boundary of the curve-space is equal to the difference between those of the bottom boundary and the top boundary of 2+-oriented Hilbert curve, for the same query region.*

From (Moon et al., 2001), we have 1) which gives formula to calculate the number of connection edges in the top boundary, and the relationship between the number of connection edges in the bottom boundary and those in the side boundary; 2) which states there is only 2+-oriented H_k on the top boundary of 2+-oriented H_{k+n}, and no 2--oriented H_k on the bottom boundary of 2+-oriented H_{k+n}; 3) which presents the relationship between the numbers of differently oriented H_k in the bottom boundary of 2+-oriented H_{k+n}. Based on this, the formulas to calculate the exact number of connection edges in bottom and side boundary, and the number of H_k in the bottom boundary are derived in the following Lemma 1 and Lemma 2, respectively.

Lemma 1: For any positive integer n,

$$b_n = (2^{n+1} + (-1)^n)/3 - 1, s_n = (2^{n+2} - 3 - (-1)^n)/6.$$

Proof.

$$\begin{aligned} & s_n = s_{n-1} + 2s_{n-2} + 1 \\ & \Rightarrow s_n + s_{n-1} = 2(s_{n-1} + s_{n-2}) + 1 \Rightarrow s_n + s_{n-1} = 2^n - 1 \\ & \Rightarrow s_1 = (2^{n+2} - 3 - (-1)^n)/6. \end{aligned}$$

Lemma 2:

$$B_{2^+,n} = \frac{2^n + (-1)^n 2}{3}, B_{1^+,n} = B_{1^-,n} = \frac{2^n - (-1)^n}{3}.$$

These can be proved in the similar way as Lemma 1.

So far, the number of connection edges and the number of H_k inside the top or bottom boundary are derived. Next, the number of connection edges connecting the top or bottom boundary to the other areas need to be obtained.

Lemma 3:

$$c_{t,n} = 2^n, \quad c_{b,n} = (2^{n+1} - 2(-1)^n)/3.$$

Proof.

There are only *2+-oriented* H_k in the top boundary of a *2+-oriented* H_{k+n}. Each of them has two end points (one incoming point and one outgoing point). One end point connects to the adjacent *2+-oriented* H_k in the top boundary and another connects to a sub curve inside boundaries or center area. Accordingly, $c_{t,n}$ is equal to the number of *2+-oriented* H_k in the top boundary, i.e., 2^n.

Similarly, $c_{b,n}$ is equal to the sum of the numbers of *1+-oriented* and *1--oriented* H_k in the bottom boundary of a *2+-oriented* H_{k+n}, because the *2+-oriented* H_k does not contribute to connections to the other areas and there is no *2--oriented* H_k in the bottom boundary.

It is known that the number of clusters within a query region is equal to half the number of edges cut by the boundary of the region. Each connection edge in the top and bottom boundary is horizontal and cut twice by the left and right sides of query windows; each horizontal edge in a H_k of the top or bottom boundary is also cut twice by the left and right sides of query windows; each edge connecting the top and bottom boundary to the center area is vertical and is cut 2^k times by the top or bottom sides of query windows, except those edges in the two side boundary, which is cut once only.

As defined in *Table 1*, h_k and v_k denote the number of horizontal and vertical edges in a *2-oriented* H_k, so they indicate the vertical and horizontal edges in a *1-oriented* H_k, respectively.

In the top boundary of the H_{k+n}, the total number of the possible positions of the query window $2^k * 2^k$ is $2^{k+n}-2^k+1$. Therefore, we derive the formula for calculating the average number of clusters of the query window located on the top/bottom boundary of *2⁺-oriented* Hilbert curve as follows.

Theorem 1: *The average number of clusters of a $2^k * 2^k$ query window located in the top boundary and bottom boundary of a $2^{k+n} * 2^{k+n}$ grid space which is 2^+-oriented H_{k+n} are equal to*

$$N_t = \frac{2(T_{2+,n} * h_k + t_n + 1) + 2^k(c_{t,n} - 2)}{2(2^{k+n} - 2^k + 1)} = \frac{2^{k+n} + 2^{n+1} * h_k + 2^n + 2 - 2^{k+1}}{2(2^{k+n} - 2^k + 1)}$$

$$N_b = \frac{2(B_{2+,n} * h_k + (B_{1+,n} + B_{1-,n}) * v_k + b_n) + 2^k(c_{b,n} - 2) + 2}{2(2^{k+n} - 2^k + 1)}$$

$$= (\frac{2^n + 2(-1)^n}{3} h_k + \frac{2^{n+1} - 2(-1)^n}{3} v_k + \frac{2^{n+1} + (-1)^n}{3} + \frac{2^{k+n} - 2^k(-1)^n}{3} - 2^{k+1}) / (2^{k+n} - 2^k + 1).$$

Note. For a $2^{+/-}$-oriented H_k, the number of vertical edges is one more than the number of horizontal edges by definition.

Corollary 1: *The difference between the average number of clusters on top boundary and bottom boundary for a 2^+-oriented H_{k+n} can be derived:*

$$N_b - N_t = \begin{cases} \dfrac{2^{n+1} + 2^{n-1} - 2 + h_k + 2^{k+n-1} - 2^{k+1}}{3(2^{k+n} - 2^k + 1)}, & n \text{ is even,} \\ \dfrac{2^{n+1} + 2^{n-1} - 3 + h_k + 2^{k+n-1} + 2^{k+1}}{3(2^{k+n} - 2^k + 1)}, & n \text{ is odd.} \end{cases}$$

The number of clusters for the side boundary can be derived with the similar idea. Although the above formula expresses the calculation on *2⁺-oriented* Hilbert curve, it is still applicable to all 2-dimensional Hilbert curves with other orientations. From the above theorem, we note that the top boundary of the *2⁺-orientation*, the bottom boundary of the *2⁻-orientation*, the right side boundary of the *1⁺-orientation*, and the left side boundary of the *1⁻-orientation* contains the fewest clusters for a given query window size comparing with the any other orientations at the same position.

3.2 Algorithms

3.2.1 Index construction

According to the first observation, we create four B+-trees for the same data set based on the four Hilbert curves with different orientations. These curves have identical curve-space, order, and the cell size (granularity). The B+-trees and corresponding Hilbert curves are

Algorithm for spatial range query
Procedure RANGEQUERY (*qw*)
Input: *qw*: query window
Output: *RS*: set of objects which overlap with *qw*.
1. *ClusterList[]* ← empty

//Mapping query window into clusters based on every Hilbert curve, (H_0: *Origin*, H_1: *Right*, H_2: *Left*, H_3: *Down and* H_4: *Shift*).
2. **for** $i = 0$ to 4
3. *Cells[]* ← Hilbert codes of cells of H_i overlap with *qw*
4. *Clusters[][]* ← group cells into clusters
5. *ClusterList[i]* ← <H_i, *Clusters*>
6. **end for**

//Select the Hilbert curve with the smallest numbers of clusters, and load the corresponding B+-tree.
7. Find *k*, such that *ClusterList[k]* = Min(*ClusterList[0]* ... *ClusterList[n]*)
8. *T* ← load B+-tree corresponding to H_k

//Traverse B+-tree, find out candidate data objects.
9. **for** each cluster C_j in *ClusterList[k].Clusters*
10. Traverse tree *T*, Get data page *p*,
11. *Objs[]* ← every objects stored in *p*
12. **end for**

//Refinement
13. **for** every object in *Objs[]*
14. **if** *Objs[i]* overlaps with *qw* then
15. *RS* ← *Objs[i]*
16. **end if**
17. **end for**
18. **return** *RS*

Fig. 5. Range query algorithm.

named in terms of the orientation of the Hilbert curves. Specifically, the *2^+-oriented* Hilbert curve and the corresponding B+-tree are named as "Origin", the *2^--oriented* Hilbert curve, and the corresponding tree are named as "Down", similarly, the *1^+-oriented* as "Right" and the *1^--oriented* as "Left". According to the second observation, another B+-tree, "Shift", is also created for the same data set. For instance, if the original data space is of the range [0, 1] on each dimension, the shifted range will be [s, 1+s] on all dimensions respectively, where *s* is the side length of a cell. To calculate the cells located in the area $[1, 1+s]^d$, (*d* represents dimension), the Hilbert curve space needs to be enlarged to [0, 2] on each dimension, meanwhile the order will be increased by 1. Therefore, "Shift" is generated using the same cell size as the original Hilbert curve, and doubled curve space. In "Shift", each data point is shifted up-right by one cell. For example, a point in original data space is *p(x, y)*, it will be changed to *p'(x+s, y+s)* before calculating the Hilbert value, and then be inserted into "Shift" with the new Hilbert value as the key. Although multiple indices are created for one data set, the data objects are stored in disk based on their "Origin" Hilbert curve values. Reasonably, we assume that there is a page buffer to reduce additional data page seek time by sorting the addresses of data pages before accessing them physically.

3.2.2 Mapping and filtering

The detailed algorithm for processing range query based on multiple copies of Hilbert curves is presented in Fig. 5. The example shown in Fig. 1 can be used to illustrate this algorithm. In this example, the whole data space is [0, 8] * [0, 8]; the cell size is 1; the order of the curve is 3; and the query window A is < (2, 0), (6, 2)>. The clusters covered by *A* on the five Hilbert curves are calculated at first. To compute the clusters under shifted Hilbert curve, the region of the query window needs to be recalculated, since the whole data space has been shifted. For example, the query window *A*< (2, 0), (6, 2)> is transformed to *A'*< (3, 1), (7, 3)>. A data structure *ClusterList* is used to store the cluster information. Each entry of the list represents clusters for one Hilbert curve, in the form of <Curve name, [cluster1]… [clusterN]>. In this example, the *ClusterList* contains three entries, <"Origin", ([4-7][56-59])>, <"Right", ([12-19])>, and <"Shift", ([6-9][54-57])>. The index corresponding to the entry with fewest clusters is selected, e.g., the index "Right" is used for answering range query *A*. In case that more than one Hilbert curves produce the fewest clusters, the one that has smaller sum of gaps between clusters will be selected. Because when the gap between two clusters is small, the corresponding leaf nodes of the second cluster can be located quickly from the first cluster, by just following links between leaf nodes.

3.2.3 Refinement

After the data objects are obtained from the filtering step, further validation is needed to check the overlaps between query window and these retrieved objects. If an object overlaps with the query window, it will be put into the result set. Otherwise, the object will be removed. This step is similar to the refinement of the traditional spatial range query processing approach.

4. Experiment

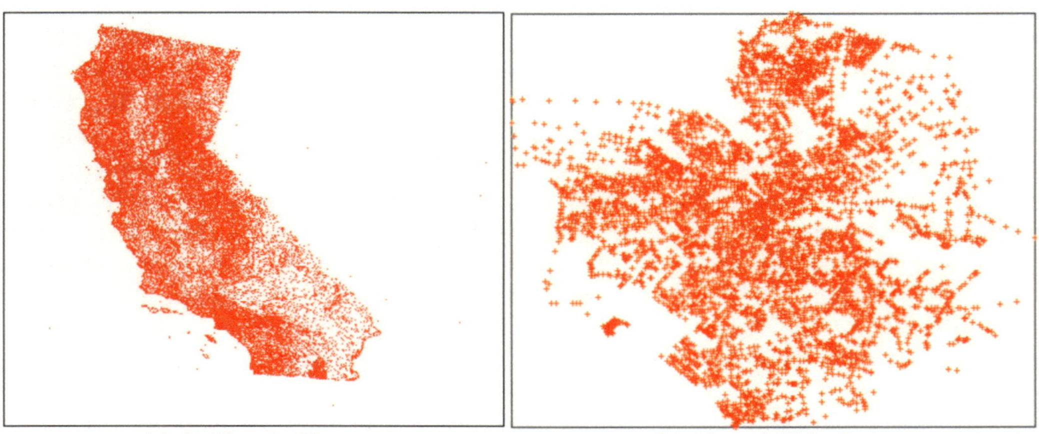

(a) California Places (b) City of Oldenburg

Fig. 6. Datasets.

To demonstrate the efficiency of the proposed algorithm and the correctness of the analysis, we conducted experiments to evaluate the performance of range queries by comparing with access method using only one Hilbert curve. The I/O costs of range queries with various sizes and positions are examined on the proposed method with different combinations of rotations and shift. The objective of our experiments is to assess the efficiency of different combinations of rotations and shift.

4.1 Experiment design

The experiment is performed on point data sets downloaded from (Sequoia 2000) and collection of real road network (R-tree portal; Li et al., 2005). The two data sets are shown in Fig. 6. The one from Sequoia 2000 is composed of more than 62 thousand 2-dimensional points, which represents places in California; another, from a collection of road network, presents about 6,000 road network's nodes in the city of Oldenburg. The experiments are conducted as illustrated in Fig. 7. The average number of page access for several range queries with difference size are compared based on the different copies of Hilbert curves. The size of the query window ranges from 1% to 15% of the whole data space. To obtain exact measurements of the average number of clusters, all possible positions for different range query sizes are examined over the whole grid space. Multiple B+-trees are constructed based on Hilbert codes of data points computed from Hilbert curves with variant orientation and shift. We compared the performance achieved by multiple Hilbert curves to that of the original approach, which uses only one Hilbert curve, as well as the performance of different combinations of rotation and shift. The performance is measured by the average number of page accesses in the B+-tree for a range query.

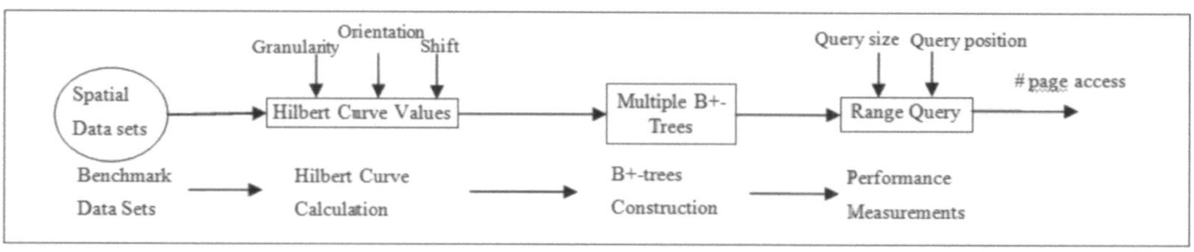

Fig. 7. Experimental design.

4.2 Experiment results

4.2.2 Effect of different number of rotations

Fig. 8 shows the comparison of range queries on different numbers of rotations. The query window size varies from 1% to 15% over the whole data space for both data sets. As shown in both Fig. 8 (a) and (b), the average number of page accesses increases with the growth of the query size. Consistent to theoretical analysis, when multiple Hilbert curves with variant orientations are used, the average number of page accesses is less than that of only one Hilbert curve. Moreover, the I/O cost saved by applying multiple Hilbert curves is enhanced with the increase of the query size. Observed from the results, using four orientations definitely reduces more I/O cost than two orientations. However, the performance gained by using two orientations from one orientation is more remarkable than the performance improved by using four orientations from two orientations. Based on this conclusion, there is a tradeoff between the performances improvement by using multiple Hilbert curves and the storage space required to store additional copies of indices. It depends on different applications to determine how many orientations are most appropriate. For space sensitive applications, two orientations may be deployed rather than using all four orientations, considering the additional space requirement. However, for the applications in which query efficiency is most crucial, applying all four orientations may be a better choice.

4.2.2 Effect of shift

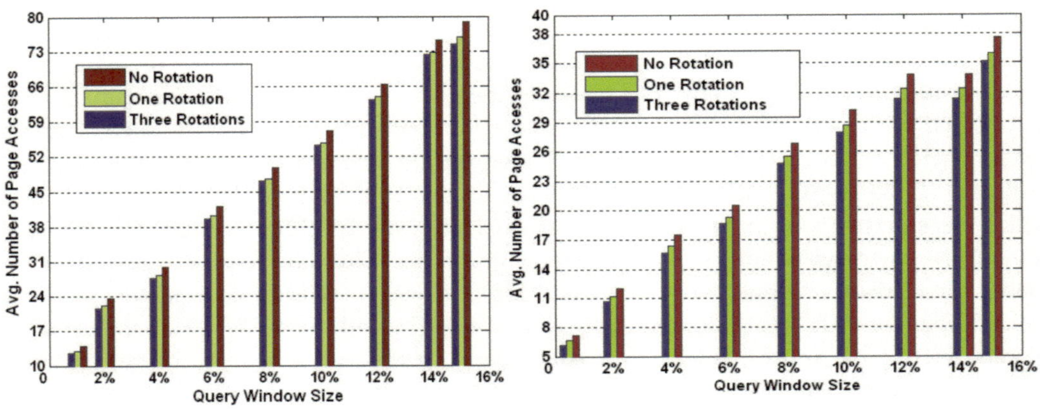

(a) California Places (b) City of Oldenburg

Fig. 8. Comparison of different rotations.

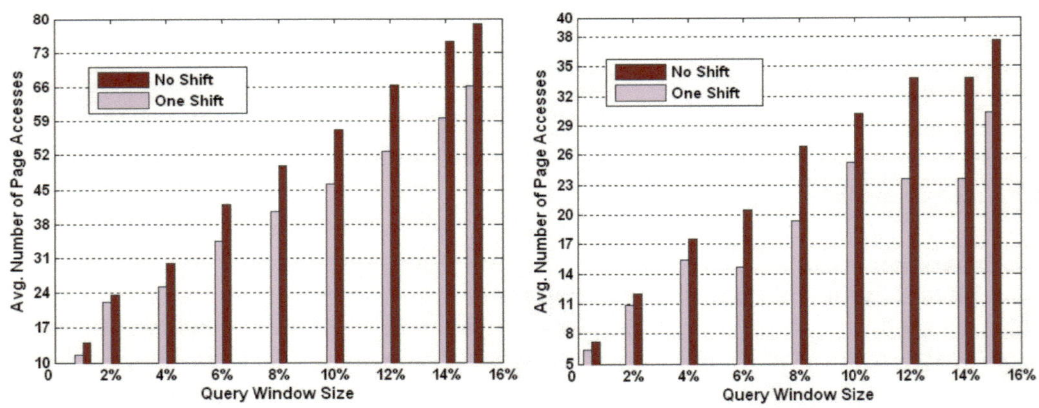

(a) California Places (b) City of Oldenburg

Fig. 9. Comparison on shift.

Fig. 9 describes the effect of using one additional copy of the Hilbert curve with shift. We set the same parameter values such as query size, position, order of Hilbert curve, as the first experiment. The average number of page accesses is significantly reduced by using shift comparing to using only one Hilbert curve. For instance, when the query size is over 12%, the average number of page accesses is reduced up to 30%. As a similar trend observed here as the effect of rotations, with the query size increasing, the number of reduced page accesses by shift also increases. However, the average number of page accesses of different query size presents a zigzag form in the case of using shift technique on the second data set. It is observed that the average number of page accesses decrease when the size of a query window happens to consist of integral number of 2*2 cell-blocks. For example, in Figure 9(b), when the size of query window is 6% of the whole space, the side length of the query window is 8, so that it contains 16 2*2 cell-blocks. The reason is that when the query range size meets the above condition, cells covered by the query window tend to be grouped in the same cluster along the Hilbert curve, by choosing shifted or original space. The shift technique can increase the probability that a range query contains only one cluster, even if it has several clusters on the original Hilbert curve. While in case of multiple rotations, if the range query contains multiple clusters on the original Hilbert curve, it can not consist only one cluster with any rotations. Fig. 3 is an example. The size of the query B is 2*2, and it contains 3 clusters with any rotations, but only one cluster on the shift.

4.2.2 Effect of hybrid

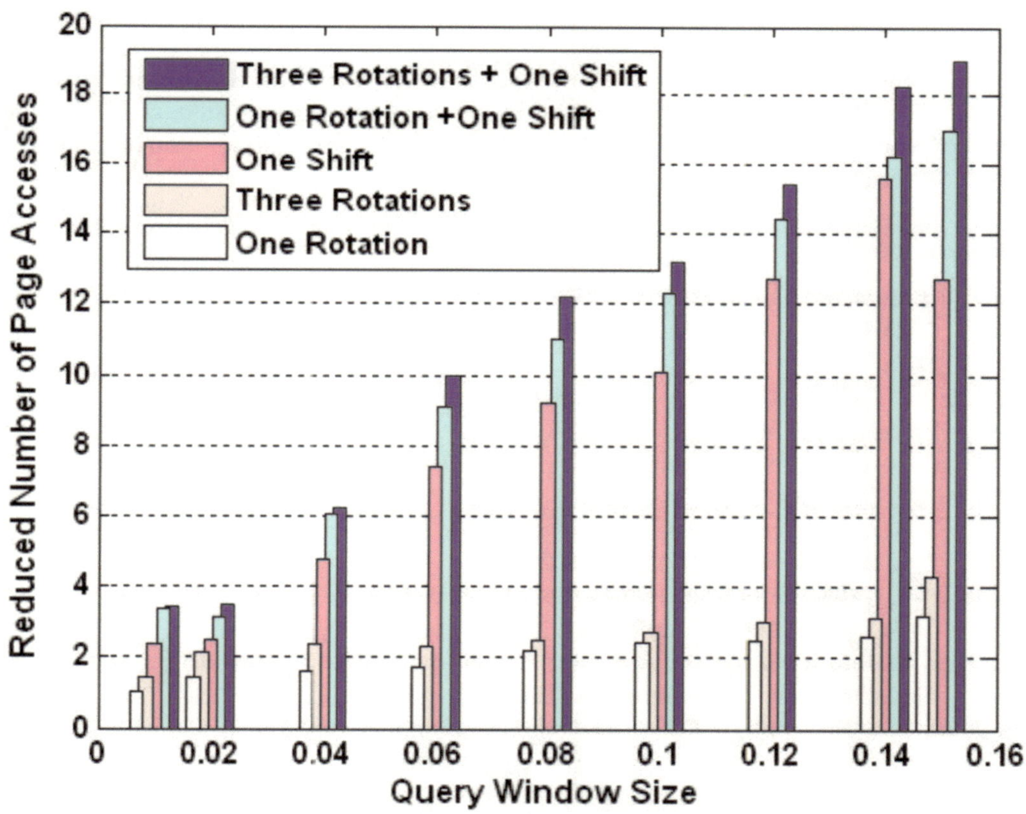

Fig. 10. Efficiency comparisons.

Fig. 10 illustrates the comparisons between different combinations of rotations and shift, rotations only and shift only. Comparisons are based on the number of page accesses reduced comparing to the original approach on California Places. As can be observed from the figure, the different combinations can be ordered by the number of page accesses as follows: One Rotation < Three Rotations < Shift < One Rotation + Shift < Three Rotations + Shift. Along this ordered sequence of the combinations, the gap between Shift and Three Rotations are the largest. This indicates that rotations do not reduce I/O cost as significantly as shift does. However, combining all rotations and shift performs better than applying any one of them independently, consistent to what we deduced in Section 2.3.

5. Conclusions

This chapter proposes an efficient spatial range query processing method based on rotation and shift techniques. Facts are observed that the same query on Hilbert curve with different orientations and shift obtains different numbers of clusters. Theoretical analysis is also provided to prove that multiple copies of Hilbert curves with different orientations can reduce the number of clusters of a range query. The experiments on two real data sets demonstrate that the proposed method reduces I/O costs of range queries. The results show that the combinations of rotation and shift in general provide the better performance than applying any one of them independently.

Future directions from this work include: investigation on jumps between clusters to further improve query performance, theoretical analysis on the effectiveness of shift, and designing spatial operations such as KNN, spatial join, and moving object queries utilizing multiple Hilbert curves.

6. References

Abel, D. J. & Mark, D. M. (1990). A Comparative Analysis of Some Two-Dimensional Orderings, International J. Geographical Information Systems, Vol. 4, No. 1, pp. 21-31.

Dai, J. & Lu, C.-T. (2007). CLAM: Concurrent Location Management for Moving Objects, Proceedings of the 15th ACM International Symposium on Advances in Geographic Information Systems (ACMGIS), pp. 292-299.

Dai, J. & Lu, C.-T. (2009). A Concurrency Control Protocol for Continuously Monitoring Moving Objects, Proceedings of the 10th International Conference on Mobile Data Management (MDM), pp. 132-141.

Faloutsos, C. (1988). Gray Codes for Partial Match and Range Queries, IEEE Transactions on Software Engineering, Vol. 14, No. 10, pp. 1381-1393.

Faloutsos, C. & Rong, Y. (1991). A Spatial Access Method Using Fractals, Proceedings of the International Conference on Data Engineering (ICDE), pp. 152-159.

Faloutsos, C. & Roseman, S. (1989). Fractals for Secondary Key Retrieval, Proceedings of the 8th ACM SIGACT-SIGMOD-SIGART Symposium on Principles of Database Systems (PODS), ACM Press, New York, NY, pp. 247-252.

Gray, F. (1953). Pulse Code Communications, US Patent 2632058, 1953. Hightower, J., Vakili, C., Borriello, C., & Want, R. (2001). Design and calibration of the SpotON

ad-hoc location sensing system, University of Washington Technical Report CSE 00-02-02.
http://www.cs.washington.edu/homes/jeffro/pubs/hightower2001design/hightower2001design.pdf

Hilbert, D. (1891). Ueber stetige abbildung einer linie auf ein flashenstuck, Mathematishe annalen, pp. 459-460.

Jagadish, H. V. (1990). Linear Clustering of Objects with Multiple Attributes, Proceedings of the ACM SIGMOD Conference on Management of Data, ACM Press, New York, NY, pp. 332 - 342.

Jensen, C. S., Lin, D., Ooi B.C. (2004). Query and Update Efficient B+-Tree Based Indexing of Moving Objects, Proceedings of the 30th International Conference on Very Large Data Bases (VLDB), pp. 768-779.

Lawder, J. K. & King, P. J. H. (2000). Using Sapce-filling Curves for Multi-dimensional Indexing, Proceedings of the 17th British National Conference on Databases (BNOD), pp. 20-35.

Lawder, J. K. & King, P. J. H. (2001). Using State Diagrams for Hilbert Curve Mappings, International Journal of Computer Mathematics, Vol. 78, No. 3, pp. 327-342.

Li, F., Cheng, D., Hadjieleftheriou, M., Kollios, G., and Teng, S.-H. (2005). On Trip Planning Queries in Spatial Databases, Proceedings of the 9th International Symposium on Spatial and Temporal Databases (SSTD).

Liao, S., Lopez, M. A., & Leutenegger, S. (2001) High Dimensional Similarity Search with Space-Filling Curves, Proceedings of the International Conference on Data Engineering (ICDE), pp. 615-622.

Mokbel, M. F., Aref, W. G., & Kamel, I. (2003) Analysis of Multi-dimensional Space-Filling Curves, GeoInformatica, Vol. 7, No. 3, pp. 179-209.

Moon, B., Jagadish, H. V., Faloutsos, C., & Saltz, J. H. (2001). Analysis of the Clustering Properties of the Hilbert Space-Filling Curve, IEEE Transactions on Knowledge and Data Engineering, Vol. 13, No. 1, pp. 124-141.

Ni, L. M., Liu, Y., Lau, Y. C., & Patil, A. P. (2004). LANDMARC: Indoor Location Sensing Using Active RFID, Wireless Networks, Vol. 10, pp. 701–710.

Orenstein, J. A. & Merrett, T. H. (1984). A Class of Data Structures for Associative Searching, Proceedings of the 3rd ACM SIGACT-SIGMOD-SIGART Sympsium on Principles of Database System (PDOS), ACM Press, New York, NY, pp. 326-336.

Peano, G. (1890). sur une courbe qui remplit toute une air plaine, Mathematishe Annalen, Vol. 36, pp. 157-160.

Simmons, G. F. (1963). Introduction to Topology and Modern Analysis, New York, McGraw-Hill Book Company, 1963.

Willis, S. & Helal, S. (2004). A Passive RFID Information Grid for Location and Proximity Sensing for the Blind User, University of Florida Technical Report, TR04-009.
http://www.cise.ufl.edu/tech_reports/tr04/tr04-009.pdf

Zheng, B., Lee, W.-C., & Lee, D. L. (2004). Spatial Queries in Wireless Broadcast Systems, Wireless Networks, Vol. 10, No. 6, pp. 723-736.
R-tree portal: http://www.rtreeportal.org/.
Sequoia 2000: http://s2k-ftp.cs.berkeley.edu:8000/sequoia/.

Part 5

Case Studies/Applications

20

The Study on Secure RFID Authentication and Access Control

Yu-Yi Chen[1] and Meng-Lin Tsai[2]
[1]Department of Management Information System
National Chung Hsing University
[2]Department of Computer Science and Engineering
National Chung Hsing University
Taiwan

1. Introduction

In recent years, Radio Frequency Identification (RFID) technology is rapid progress and has been widely used in daily life. RFID systems consist of three components: radio frequency (RF) tags, RF readers and a back-end database server. A passive RFID tag is a microchip capable of transmitting a static identifier or serial number for a short distance. Readers query tags for their contents by broadcasting an RF signal. Tags respond with resident data, such as a unique serial number. Tag data may be read automatically without line of sight. RFID systems have many applications in supply chain managements, inventory control, anti-counterfeiting, ticketing systems, healthcare and smart home developments.

However, it may bring up some privacy threats. Anyone can easily access tagged items and collect data without line of sight that personal privacy under threat. The most concerned issues are the tracking and the location privacy. Based on the characteristic of outstanding traceability, the history of the tag's location might be identified as a tag's information is intercepted and collected by the attacker in different location. For instance, the unique tag's EPC data can be used to trace a person or an object carrying a tag in time and space. The collected information can be merged and linked in order to generate a person's profile. It will be a serious problem as RFID tags are widely used.

Without privacy protection, a person with carried RFID tags can be tracked and profiled by unauthorized people. The unique information of the items may be indicated that a customer carrying those tags is subject to track from unauthorized readers.

Ideal RFID systems used in product lifecycle should satisfy high confidentiality, anonymity, integrity and high availability (Gao et al., 2004; Pisarsky, 2004). The product life cycle is a procedure that the product from manufacture to be recycled. This procedure from the perspective of commerce can be divided into five stages(Figure 1): (1)&(2) are the stage of "production to retail store" (business-to-business) , (3) is the stage of "retail store to customer" (business-to-customer), (4) is the stage of "individual sales" (customer-to-customer), (5)&(6) are the stage of "after-sales service", and (7) is the stage of "recycling" (reverse logistics). Since a tag is embedded in the product, security risks such as privacy threats may be occurred in each stage of the product life cycle.

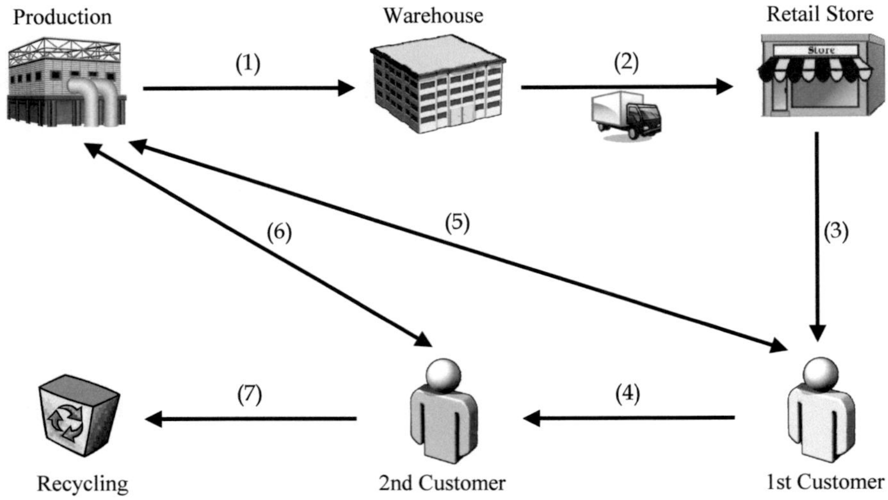

Fig. 1. The product life cycle.

To our desirable point, researchers need to pay more effort to develop object identication throughout the life cycle with guaranteeing the corporate and personal privacy, illegal tracking, unauthorized profiling, impersonating, cloning, and illegal reading/writing. This article is not purpose of an exhaustive literature survey but summarizes some aspects of RFID authentication and access control in the proposed studies.

2. Basic RFID tags

In most RFID systems, tags automatically emit their unique serial numbers upon reader interrogation without alerting their users. The challenge in providing security for RFID tags is such kinds of low-cost device unable to perform basic cryptographic operations. Basic RFID tags just have a little rewritable memory, even have no programmable-supported computing capability. At best, such RFID tags may include security functions supporting keyed reads and keyed writes which essentially just like PIN-controlled data accesses. In this section, we show how privacy and authentication may be considerably improved in low-cost RFID tags with only a small enhancement of their capabilities.

2.1 Killing and sleeping

The "kill command" method is a straightforward approach to make a tag no longer functional. This approach proposed by the AutoID Center is indeed for tags to be killed upon purchase of the tagged product. A tag can be killed by sending it a special "kill command" with a short PIN (Sarma et al., 2002; Weis et al., 2003). As the tag receives the "kill" command, its state changes into the inoperative state. Kill the tag technique is to restrict the use of a tag by removing its identity. As shown in Fig. 2, the killed tag has no way to change back to the inventoried state. It cannot be identified for more detailed information again. For example, purchased goods would be killed at checkout clerks such that no one would contain active RFID tags for protecting the consumer privacy. This solution is simple and effective but the tag can not be reused. Clearly, the tag's lifecycle is end and it cannot be applied for after-sale purposes.

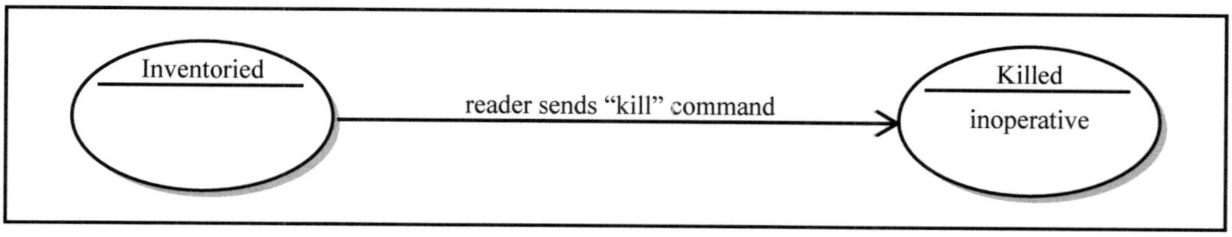

Fig. 2. The state changing of the tag in killing approach

Another kind of solution is using the "sleeping" mechanism. As the reader sends a "sleep" command to the tag, the tag will temporarily inactive. The sleeping tag can be waked as the tag receives PIN from the reader. The state changing of the tag is shown in Fig. 3. The tag's state can be switched between inventoried and sleep. For controlling the tag's access, the tag's owner has to manage the PINs of all tags on purchased good. Unfortunately, passwords may be overheard or collected by spoofing a tag. This approach also pose other problems: a set of tags use a single generic PIN which can be easily defeated, but each tag use a unique PIN which could be uniquely identified by the adversary.

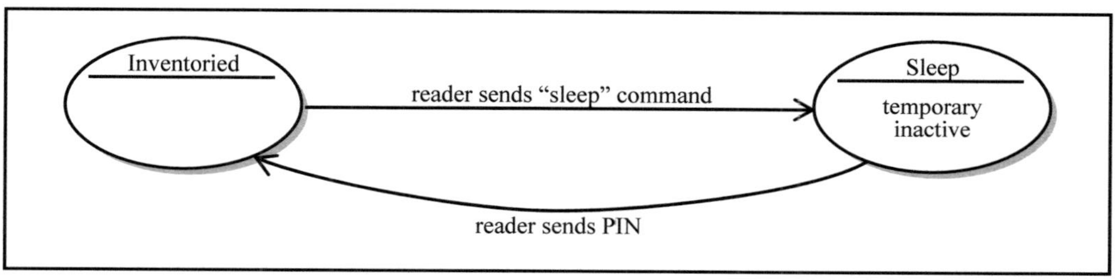

Fig. 3. The state changing of the tag in sleeping approach

2.2 Renaming approach

The solutions of relabeling or re-encrypting the tag's serial number were proposed for minimal security requirements. This approach takes into account the natural computational limitations of RFID tags, it involves no computational operations but only relatively little storage. The relabelled or re-encrypted serial number is overwritten to the tag at checkout for protecting the consumer's privacy. This is possible for current generation tags and would prevent the unauthorized compilation of bibliographic directories. However, even if the relabelled or re-encrypted identifier emitted by an RFID tag has no intrinsic meaning, it can still be tracked since the relabelled or re-encrypted identifier is just a static meta-identifier. Therefore, point-to-point tracking is possible if the meta-identifier is not changed over time. For this reason, this approach does not solve the problem of privacy.

2.2.1 Relabeling

Sarma et al. (2003) proposed an idea to protect the tracking problem (Sarma et al., 2003). As a customer purchases goods, the reader sends a "delete" command at the point of sale such that the tags' unique serial number is erased. Only the product code information of the tag is retained for later use. The state changing of the tag is shown in Fig. 4. However, the tracing problem is still existed to distinguish individual by a fixed group RFID-tagged products. For example, someone is a fan of a particular brand will always take the brand's shoes, watch and bag such that tracking is still possible by associating these kinds of particular tag types with holder identities.

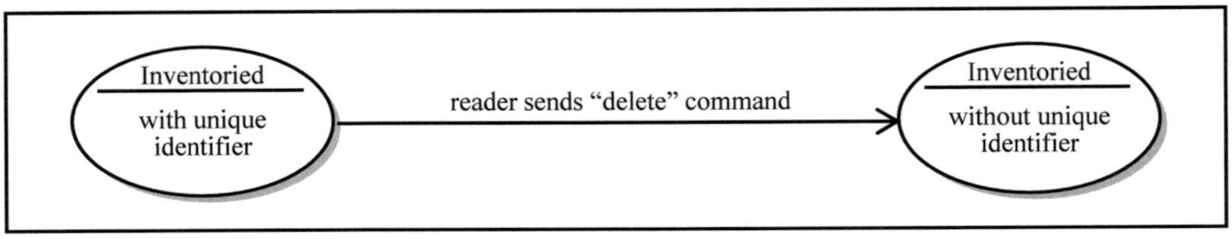

Fig. 4. Sarma's idea for erasing the tag's unique identifier

Inoue & Yasuura (2003) proposed another relabling approach to offer users the identifier's controllability for protecting privacy (Inoue & Yasuura, 2003; Inoue et al., 2002). Each tag has a read-only memory (ROM) and an electrically-erasable programmable read-only memory (EEPROM). These two memories are used exclusively. The state changing of the tag is shown in Fig. 5. A unique and permanent identity is stored in the tag's ROM by the producer. As the tag remains on ROM mode, the permanent identity can be read. The tag can provide unlimited identification with ROM mode for total management at its production, distribution, and sale stage. For purchased goods, the owner can set a private and temporary identity in EEPROM. As switching to EEPROM mode, the tag cannot operate the permanent object identification. Even the temporary identity can be read by anyone, no one can recognize the tag since the information about the object in the network is distributed accompanying the permanent identity on the ROM as a key. Therefore, the adversary has nothing to do with the temporary identity. The object can be identified only by the owner. Moreover, the tag can be switched to ROM mode again by certificating the owner or restricting the change only via contacted communication. This approach remains the permanent identity for life cycle of the object. As the object is discarded, the scrap merchant can make the tag to be switched to ROM mode to operate the permanent object identification and utilize it for recycling. However, the temporary identity is unique and cannot avoid the point-to-point tracing problem since it could be uniquely identified by the adversary.

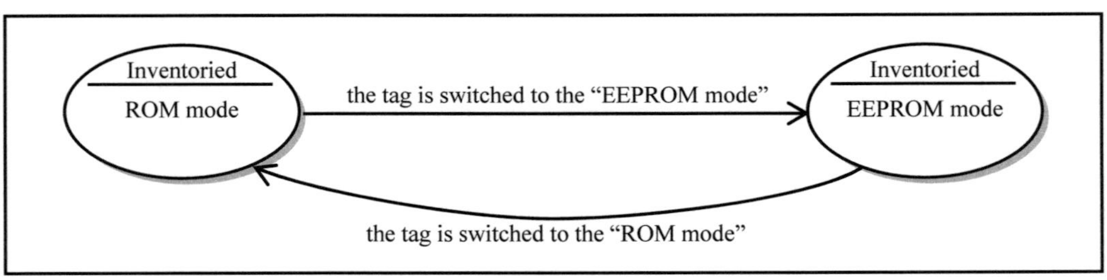

Fig. 5. Inoue's double mode tag

Kinosita et al. (2003) proposed another approach to rewrite the tag (Kinosita et al., 2003). As a customer purchases the product on checkout, the reader rewrites a new random number to the tag. Fig. 6 shows the state changing of the tag. However, the random identifier is unique and cannot avoid the point-to-point tracing problem since it could be uniquely identified by the adversary.

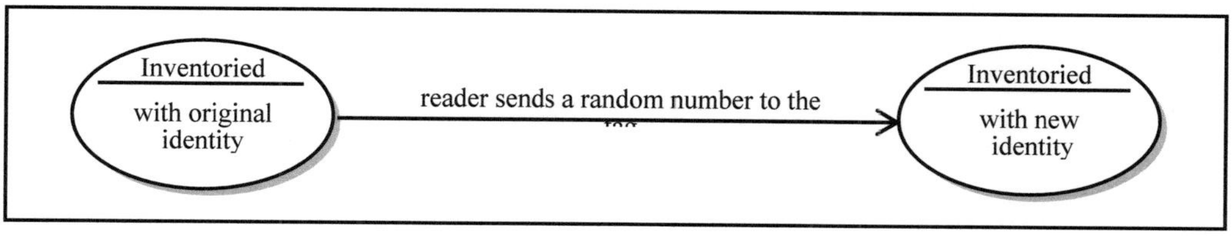

Fig. 6. Kinosita's approach to rewrite the tag

2.2.2 Re-encryption

Juels & Pappu's (2003) proposed an approach based on re-encryption concept (Juels & Pappu, 2003). The public key cryptosystem is used in this scheme. The data of a banknote is arranged into optical and radio frequency areas. A unique serial number and a signature are printed on the banknote. The banknote serial number and signature are encrypted by the law-enforcement's public key. The resulting ciphertexts are stored in the banknote's tag. Clearly, the tag can be authenticated as the ciphertexts are decrypted by the law-enforcement for verifying the signature of serial number. For rendering multiple appearances of the tag unlinkable, these ciphertexts are re-encrypted with a new encryption factor by the law-enforcement's public key after each access session. The encryption-operation requires high computational loading which is performed by the reader not the tag. The change in each appearance is designed for preventing the tracing problem. Fig. 7 shows the state changing of the tag. However, the ciphertexts keep constant (Ohkubo et. al, 2003) such that the tag still can be traced between twice re-encryptions. It means the tag must be rewritten often. This makes re-encryption approach unsuitable in practical. Basing on the re-encryption concept, a similar scheme proposed by Golle et al. (Golle P et al., 2004) known as universal re-encryption mechanism. It is essentially a special extension of the ElGamal cryptosystem (Elgamal T., 1985) in which re-encryption is possible without knowledge of public keys. However, this universal re-encryption mechanism has a practical drawback of requiring the role of agent to perform re-encryption.

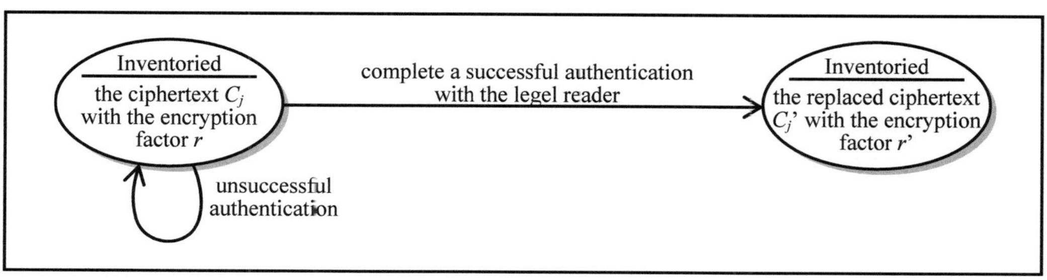

Fig. 7. Juels & Pappu's re-encryption approach

2.3 Distance measurement

Fishkin et al. proposed an approach to measure the distance between the reader and the tag (Fishkin et al., 2004). An adversary usually interrogates the tag in the far distance. Fishkin et al. observes and analyzes the energy of the received signal by the tag. The distance between the reader and the tag can be estimated by the signal-to-noise ratio. This distance information is used as a variable in a tiered authentication scheme, where the tag releases general or specific information to the reader according to the distance variable. Fig. 8 shows the state changing of the tag.

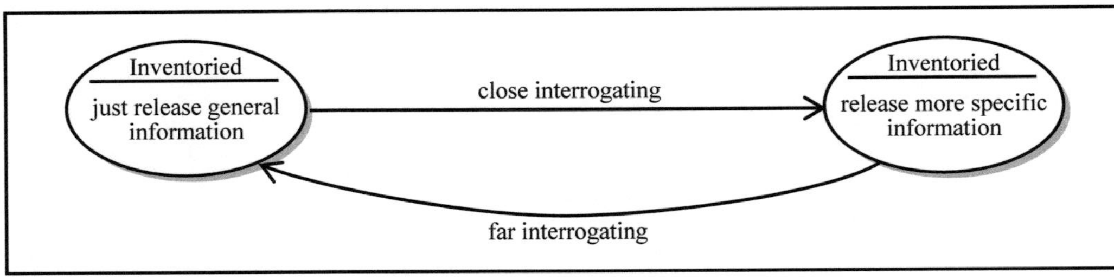

Fig. 8. Fishkin's approach

2.4 Blocking & soft blocking

Juels et al's (2003) proposed a mechanism to interfere with the readers' interrogation by a blocker tag (Juels et al., 2003). The blocker tag simulates all possible RFID tags to prevent the malicious identification of the target tag. This privacy protection scheme depends on adding a privacy bit to the tag. While inside a store, the tag's privacy bit usually is set to 0, indicating public access to the tag's identification. While during checkout, this privacy bit is changed to 1, denoting the tag is entering restricted access. Then the tag must interact with another tag known as the "blocker tag" (Juels et al., 2003). The blocker tag broadcasts radio signals to block/disrupt nearby RFID readers could work. It is accomplished through non-standard interaction with the anti-collision protocols employed in tag-reading session (Auto-ID Center, 2003; Sarma, 2001). The blocker tag will manipulate the query result of a normal tag by scrambling the bits of certain tags determined by their privacy bit (Juels & Brainard, 2004). The state changing of the tag is shown in Fig. 9. As the privacy bit is set to 0, the tag can be unrestricted scanned and the blocker tag doesn't interrupt the reading of tag. As the privacy bit is set to 1, the tag is private with restricted access under the cover of blocker tag. Juels and Brainard proposed an enhancement mechanism called soft blocking (Juels & Brainard, 2004). The soft blocker tag transmits a policy statement to enforces and monitors the reader not violate the security policies. However, blocker tag is expensive (Cavoukian, 2004) and suffers from the heterogeneity of current RFID systems using different frequencies, air protocols, etc. The blocker tag and its variants have limited applicability.

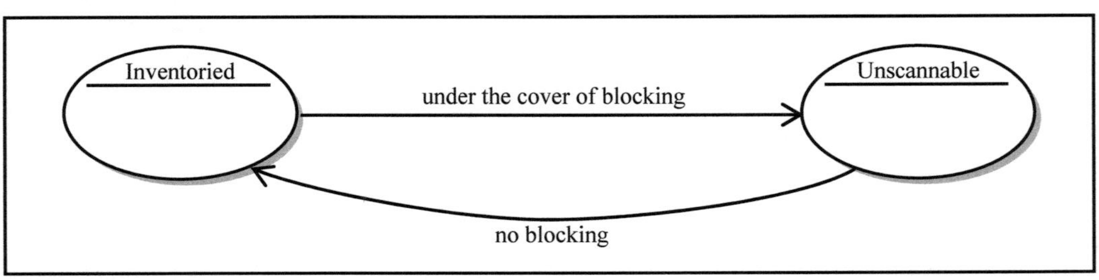

Fig. 9. Blocking approach

3. Symmetric-key tags

Symmetric-key tags are considered as the type of security obtainable with a small amount of rewritable memory, but very limited computing capability. Such RFID tags may be expected to perform some basic computational operations, but not conventional cryptographic ones. Many approaches have been proposed to achieve private authentication in such RFID systems. The proposals usually include hash function, silent tree-walking, or other light cryptography-based approaches to prevent the unauthorized reading of RFID tags. Most researchers devoted to show that standard cryptographic functionality is not needed to achieve stronger security in RFID tags. Since the communication between the reader and the tag is using RF signals, which make an RFID system vulnerable to various attacks such as eavesdropping, traffic analysis, spoofing and denial of service. Within the scanning range, a malicious reader can perform bogus authentication with detected tags to retrieve sensitive information. The sensitive information may be disclosed and hence infringe on the user's privacy. Traceability is another type of privacy violation, the relation between the user and the tag can be found will cause the tracing of the tag makes the tracing of the user possible (Avoine & Oechslin, 2005). The proliferation of RFID applications (Ni et al., 2003) raises an emerging requirement - protecting user privacy (Robinson & Beigl, 2003) in RFID authentications.

As the relationship is illustrated (Fig. 10) in Weis's paper (Weis et al, 2003), the forward channel (reader-to-tag) is assumed to be easily monitored by an adversary since the signal broadcasted by the reader is strong enough, the backward channel (tag-to-reader) is relatively much weaker and may only be monitor by an adversary within the tag's shorter operating range. The reader-to-tag (forward) channel and the tag-to-reader (backward) channel are assumed not secure, but eavesdroppers may only monitor the forward channel without detection.

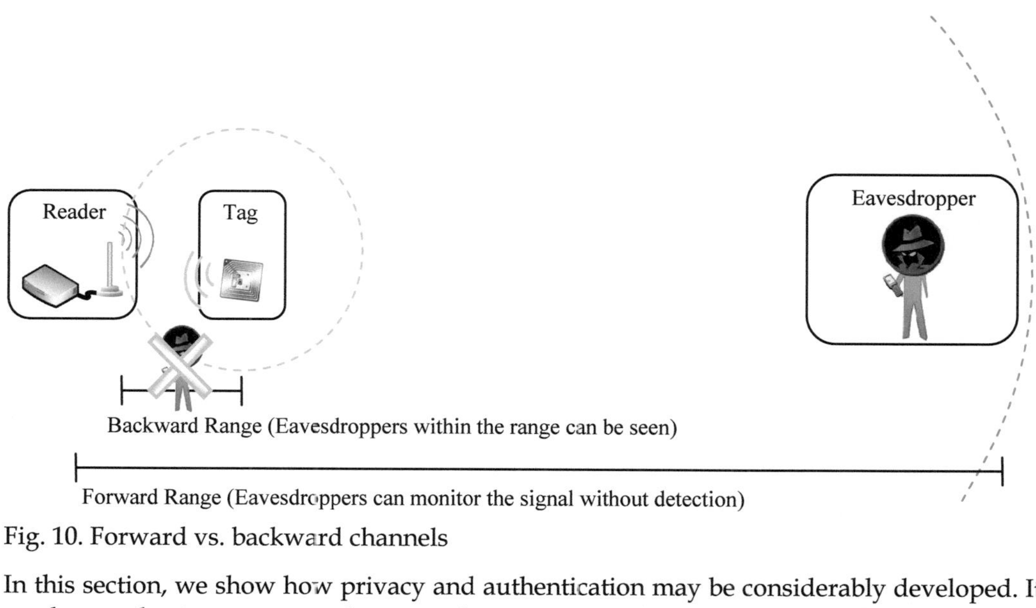

Fig. 10. Forward vs. backward channels

In this section, we show how privacy and authentication may be considerably developed. It needs to take into account the natural computational limitations and the likely attack scenarios. The challenge in providing security for low-cost RFID tags is that they are computationally weak devices, unable to perform even basic symmetric-key cryptographic operations.

3.1 Non-indexed key-search approach

The general approach of key search for RFID-tag identification was proposed by Weis et al. (2003). Upon receiving a query from the reader, the tag first sends the hash value of its key with a random nonce. Without any index, the reader must compute for all keys until it identify the tag. As the tag responds with different values every time, the reader must exhaustively search until it finds the matched one. The scheme is not scalable for a huge number of tags since many computations must be performed at the back-end.(Rhee et al., 2005; Weis et al., 2003)

Weis et al. (2003) proposed two simple hash-based access control protocols, the hash-lock scheme and the randomized hash-lock scheme (Weis et al., 2003). Fig. 11 shows the randomized hash-lock scheme. Each tag has its initial ID_i is issued by the back-end database server. As the reader tries to access the tag, the tag's response is a hash value $\alpha = h(ID_i \| R)$ generated by hashing the tag's ID_i concatenated with a random number R. If the reader is legal, it can ask the back-end database server to provide all tags' identities. Then the reader performs a brute-force searching comparison between α and $h(ID_k \| R)$ to find the corresponding record. This scheme is not scalable since the reader's computational loading is $O(n)$.

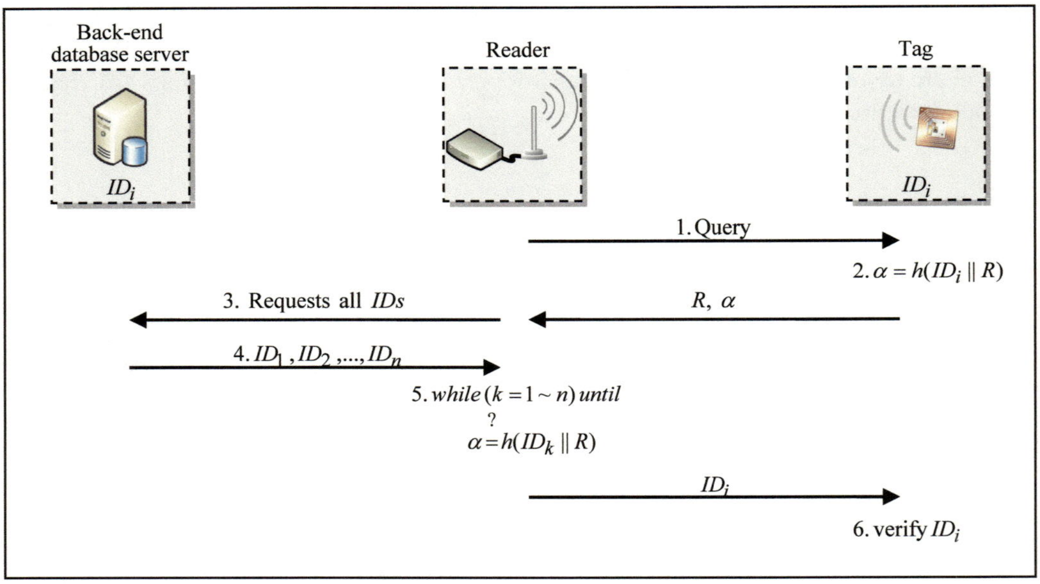

Fig. 11. Weis's randomized access control scheme

The motivation of this scheme is to make the tag's response message not predictable to prevent the tracing of individual. To randomizes tag responses instead of a invariable tag response in order to protect location privacy. However, the tag still can be traced as shown in the following use-case diagram (Fig. 12). An adversary can eavesdrop on the legal reader's broadcasts ID_i for collecting to its own database. As the target tag's identity is collected, the adversary immediately realizes the tag had appeared on the location. In addition, the adversary may interrogate a tag to get its response message (R, α) for making a brute-force searching comparison between α and $h(ID_i || R)$ to figure out which collected identity ID_i is matched. Therefore, any collected identity can be traced.

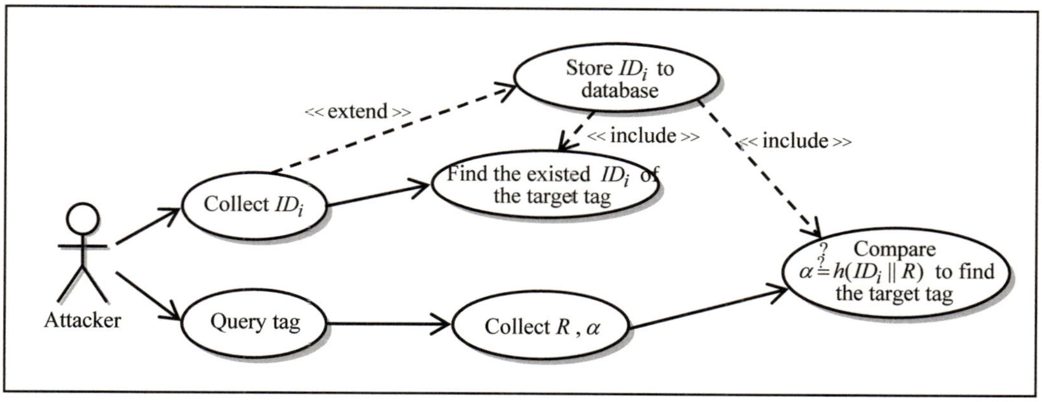

Fig. 12. The attack on Weis's randomized access control scheme

3.2 Indexed key-search approach

The major sticking point with the non-indexed key-search approach is that the reader's computational loading is $O(n)$. Under the practical consideration, it is not scalable since the process of key search can be prohibitively costly if the set of tags is large. For reducing the cost of key search, the tag's first reply message must be the index for key-searching. As the reader has sent the right response being the "key", then the tag reveals its identity. Unfortunately, the invariable index value will cause the tag traceable. (Chien, 2006; Huang, 2009; Weis et al., 2003)

3.2.1 Weis's hash-based access control scheme

Weis et al. (2003) proposed the hash-lock scheme (Weis et al., 2003), shown in Fig.13. Each tag has a hash value $metaID_i$ of its Key_i as it is issued by the back-end database server. The

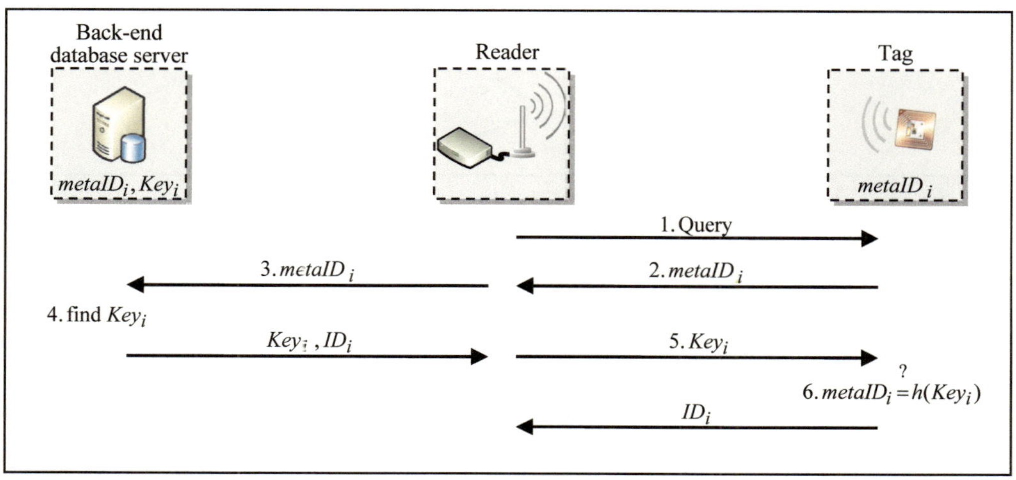

Fig. 13. Weis's hash-based access control scheme

reader can only get this hash value $metaID_i$ as it tries to access the tag. If the reader is legal, it can ask the back-end database server to retrieve the corresponding Key_i. After the tag receives the correct Key_i from the reader, the tag's information can be accessed by the reader. Unfortunately, the scheme not offers location privacy since the tag can be uniquely identified by its hash value. Another drawback is that the plain key is sent over the forward channel which can be eavesdropped in the RF-signal range.

In this scheme,the tag can be traced as shown in the following use-case diagram (Fig. 14). The adversary can eavesdrop on the legal reader's broadcasts Key_i for collecting to its own database. As the target tag's key is collected, the adversary realizes the tag had appeared on the location. Moreover, the adversary may interrogate a tag to get its response message $metaID_i$ for making a comparison between $metaID_i$ and $h(Key_i)$ to figure out which collected Key_i is matched. Since a tag's response message is an invariable $metaID_i$, it can be treated as an identifier, for the adversary to trace individuals. This scheme supports data privacy but can not protect location privacy of the tag since the invariable hash value is used in each time.

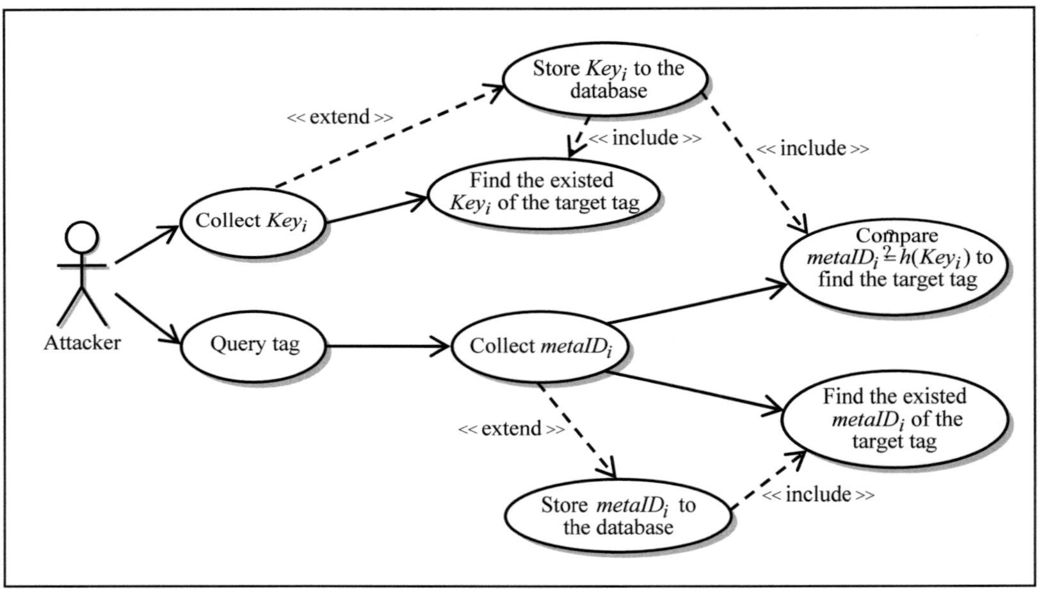

Fig. 14. The attack on Weis's hash-based access control scheme

3.2.2 Chien's hash-based access control scheme

Chien (2006) proposed another hash-based access control scheme (Chien, 2006), shown in Fig. 15. The back-end database server's master secret key is K_{svr}, and each tag's unique key is $Key_i = h(K_{svr} \,||\, ID_i)$ Each tag has a hash value $metaID_i$ of its Key_i as it is issued by the back-end database server. As the reader tries to access the tag, it can get this hash value $metaID_i$ and the current $date$. If the reader is legal, it can ask the back-end database server to retrieve the corresponding ID_i for generating the right Key_i. Then the reader generates a hash value $h(Key_i \oplus date)$ by the tag's Key_i and the received current $date$. After the tag receives the correct hash value $h(Key_i \oplus date)$ from the reader, the tag's information can be accessed by the reader.

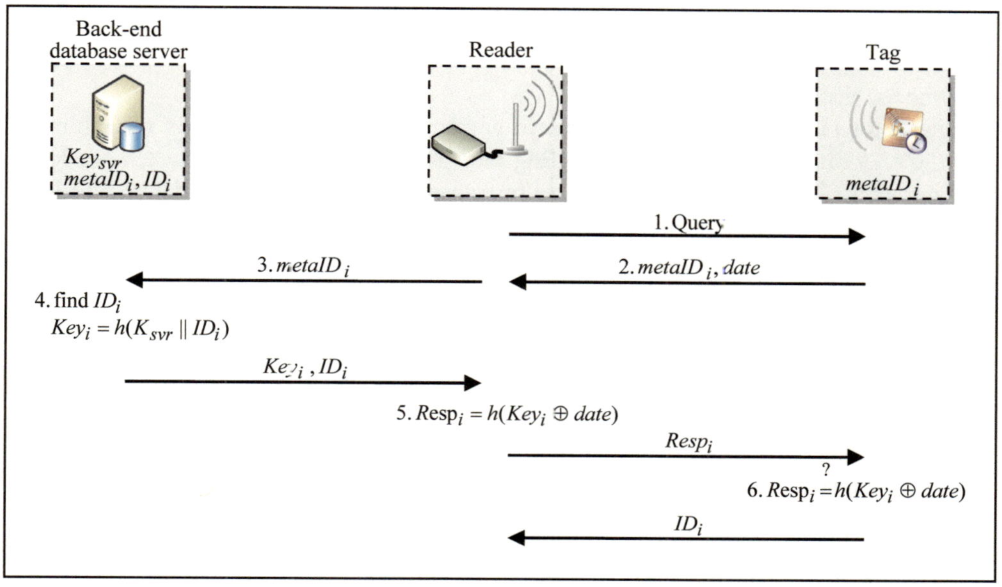

Fig. 15. Chien's hash-based access control scheme

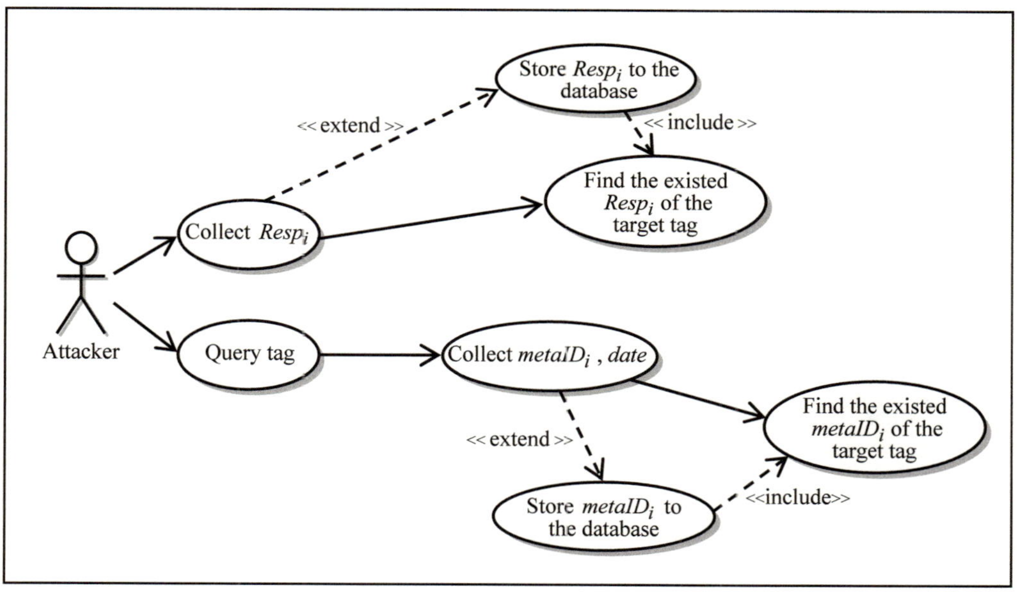

Fig. 16. The attack on Chien's hash-based access control scheme

Fig. 16 shows the use-case diagram of this scheme's weaknesses. The adversary can eavesdrop on the legal reader's broadcasts $Resp_i$ for collecting to its own database. As the target tag's response is collected on the same day, the adversary realizes the tag had appeared on the location. Moreover, the adversary may interrogate a tag to get its response message ($metaID_i$, $date$). Since a tag's response message has an invariable $metaID_i$, it can be treated as an identifier for the adversary to trace individuals.

3.3 Synchronization approach

The general idea is to change the tag's identifier after each access session. By refreshing both of the tag's identifier and the corresponding back-end database record in each session, the identifier cannot be employed for tracking purposes. The adversary can only eavesdrop or intercept a single, unreliable message exchange, it seems to provide the tag with location privacy. The literature explores several variants of this principle. Ohkubo, Suzuki, and Kinoshita (OSK) propose the conceptually simplest approach. Henrici and Müller propose to resolve the synchronization problem. Dimitriou proposes a scheme that eliminates the issue of desynchronization entirely. (Avoine & Oechslin, 2005; Dimitriou, 2005; Henrici & Muller, 2004; Joaquin et al., 2011; Juels, 2004; Lee et al., 2005, 2006; Ohkubo et al., 2003; Osaka et al., 2006)

3.3.1 Henrici & Muller's hash-based ID variation scheme

Henrici & Muller (2004) proposed a hash-based ID variation scheme (Henrici & Muller, 2004), shown in Fig. 17. A tag with initial ID_i , transaction number TID_i and last successful

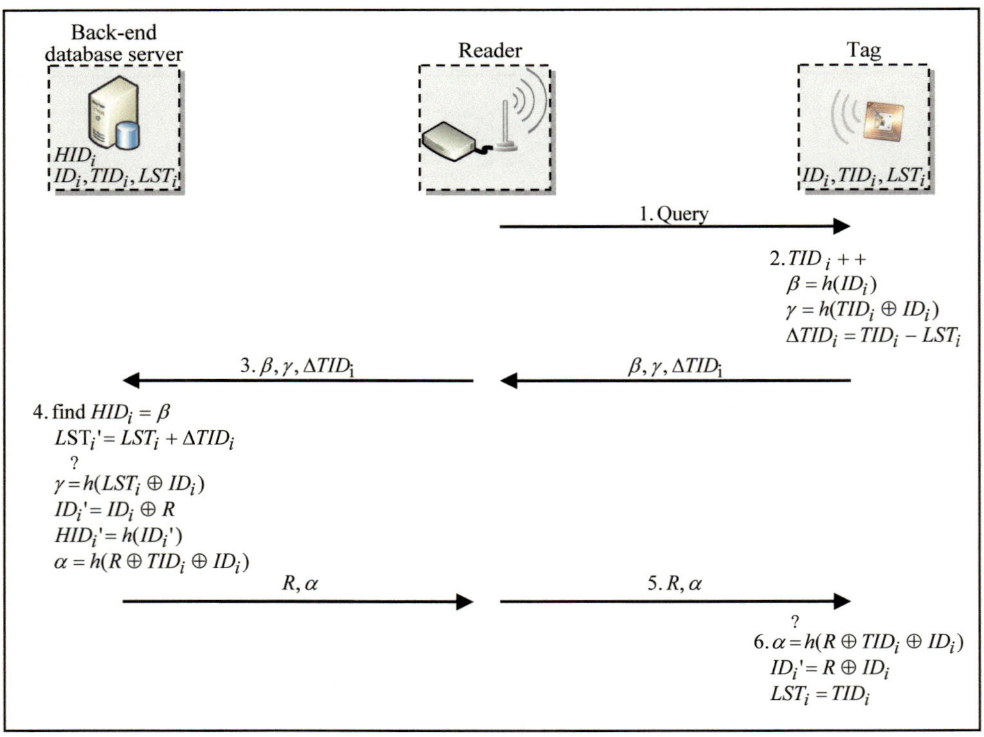

Fig. 17. Henrici & Muller's hash-based ID variation scheme

transaction number LST_i are issued by the back-end database server. As the tag is queried, the tag's transaction number TID_i is increased progressively and the message ($\beta = h(ID_i)$, $\gamma = h(TID_i \oplus ID_i)$, $\Delta TID_i = TID_i - LST_i$) is responded to the reader. If the reader is legal, it can ask the back-end database server to use $\beta = h(ID_i)$ identifying the tag. Then the back-end database server's response is the hash value $\alpha = h(R \ominus TID_i \oplus ID_i)$ generated by the transaction number TID_i, tag's identity ID_i, and a random number R. After the tag receives the correct hash value α from the reader, the tag's information can be accessed by the reader.

In this scheme, the tag updates its ID_i after each successful access. It seems to make the tag's response message $\beta = h(ID_i)$ not predictable to prevent the tracing of individual. However, the design of identity variation not really guarantees the location privacy. Fig. 18 shows the use-case diagram of this scheme's weaknesses. The adversary may interrogate a tag to get its response message $\beta = h(ID_i)$ and $\Delta TID_i = TID_i - LST_i$ for collecting to its own database. If $\Delta TID_i \geq 2$, it means the last transaction is not successful and the tag's identity ID_i is not updated. As the target tag's hash value $\beta = h(ID_i)$ once again collected, the adversary immediately realizes the target tag appeared.

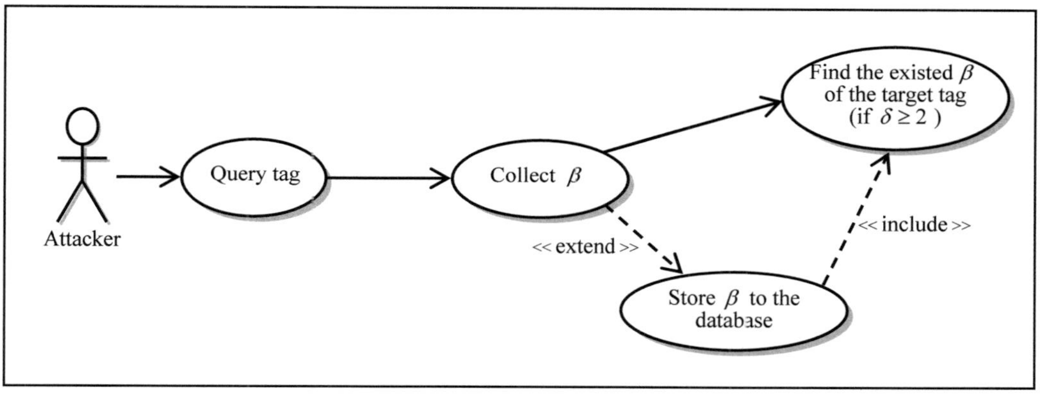

Fig. 18. The attack on Henrici & Muller's hash-based ID variation scheme

3.3.2 LCAP scheme

Lee et al. (2005) proposed a low-cost RFID authentication protocol (LCAP) (Lee et al., 2005), shown in Fig.19. A tag with initial ID_i is issued by the back-end database server. The back-end database server always maintains a previous-session record and a current-session record for a tag. Each record has the fields (*HaID*$_i$, *ID*$_i$, *TD*$_i$). *HaID*$_i$ value is the hash value of *ID*$_i$ used for identifying or addressing the tag. *TD*$_i$ is used to link the previous-session record and the current-session record each other in order to synchronize the tag and the database in case of incompletion of the current session. As the reader tries to query the tag with a random number R, the tag emits a hash value $\beta = h(ID_i)$ and the left-half hash value $\alpha_L = f_L(h(ID_i \mid\mid R))$. If the reader is legal, it can ask the back-end database server to use $\beta = h(ID_i)$ identifying the tag. Then the back-end database server's response is the right-half hash value $\alpha_R = f_R(h(ID_i \mid\mid R))$. After the tag receives the correct right-half hash value α_R from the reader,the tag's information can be accessed by the reader.

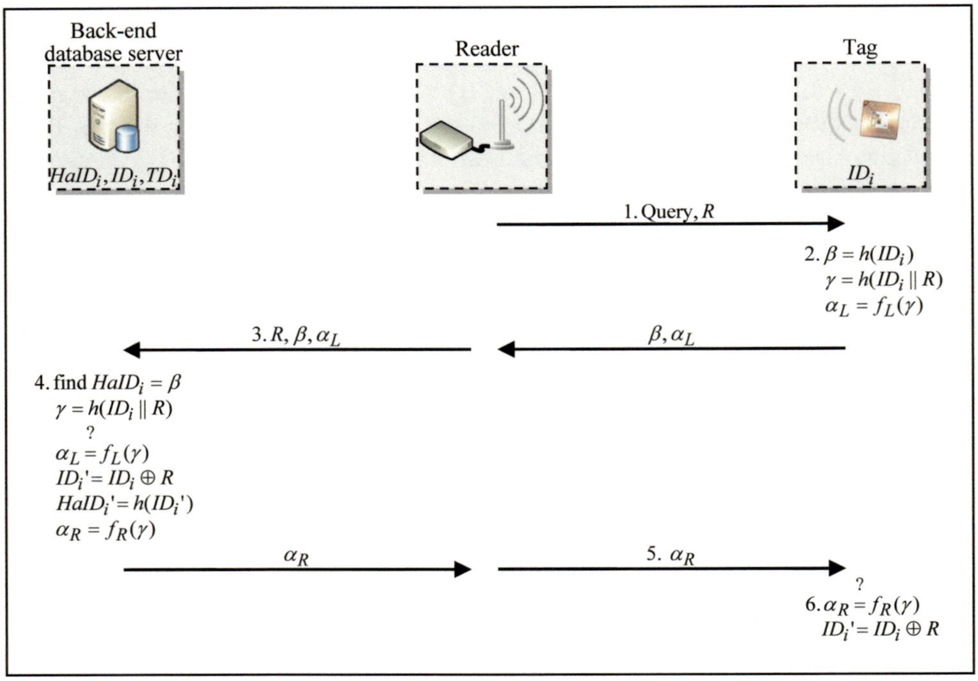

Fig. 19. LCAP scheme

In this scheme, the tag's identity ID_i is refreshed simultaneously by the tag and the back-end database server after each successful access. It seems to make the tag's response message $\beta = h(ID_i)$ not predictable to prevent the tracing of individual. However, the design of "dynamic" identity not really guarantees the location privacy. Fig. 20 shows the use-case diagram of this scheme's weaknesses. Gildas Avoine and Philippe Oechslim had described an attack based on refreshment avoidance (Lee et al., 2005). In the attack, an adversary always makes a tag unable to refresh its identity and hence can trace the tag. For example, the adversary interrogates all tags with the same number R to get the response message $\beta = h(ID_i)$ and the left-half hash value $\alpha_L = f_L(h(ID_i \mid\mid R))$ for collecting to its own database. As the target tag's hash value $\beta = h(ID_i)$ once again collected, the adversary immediately realizes the target tag appeared.

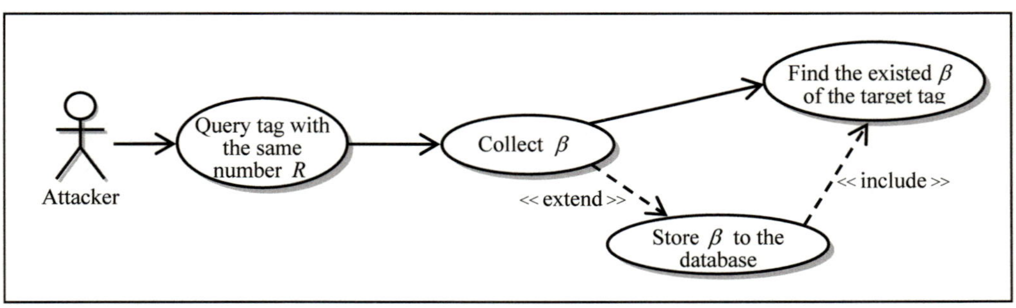

Fig. 20. The attack on LCAP

3.4 Tree-based approach

In the tree-based approach, each tag is not just assigned with a single key but associated with a unique leaf node. In fact, a sequence of keys from the root to the leaf node are defined for the associated tag. The tag's authentication response is performed by the sequence of keys such that it can be identified by the reader using a breadth-first search in the key tree. Based on the logarithmic complexity of tree-based key search, the tree-based identification is efficient to support a large scale system. (Bringer et al., 2008; Dimitriou, 2006; Lu et al., 2007; Molnar & Wagner , 2004; Molnar et al., 2005; Wang et al., 2007; Yeh et al., 2008)

3.4.1 Dimitriou's tree-based tag identification scheme

Dimitriou (2006) proposed a tree-based tag identification scheme (Dimitriou, 2006). Each edge is defined with a secret value Key_i . Key_i in the path from the root to the leaf node are hereby distributed to this tag. If the tree depth is d , each tag contains d keys. Fig. 21 shows a binary key tree with eight tags. For example, T_4 has keys k_4^1, k_4^2 and k_4^3.

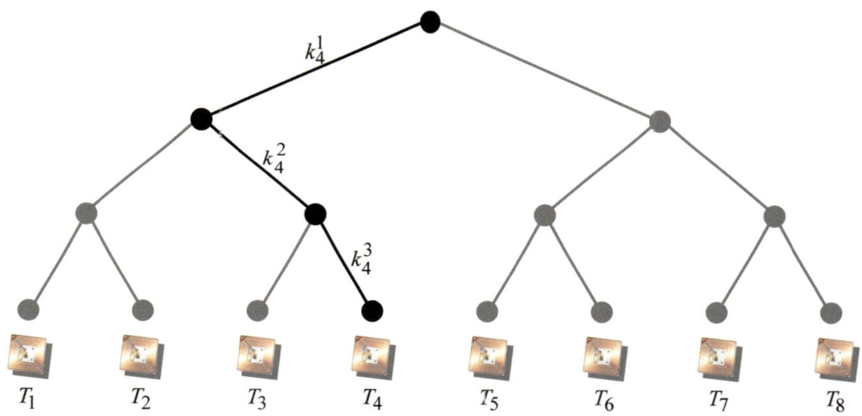

Fig. 21. A binary key tree with eight tags

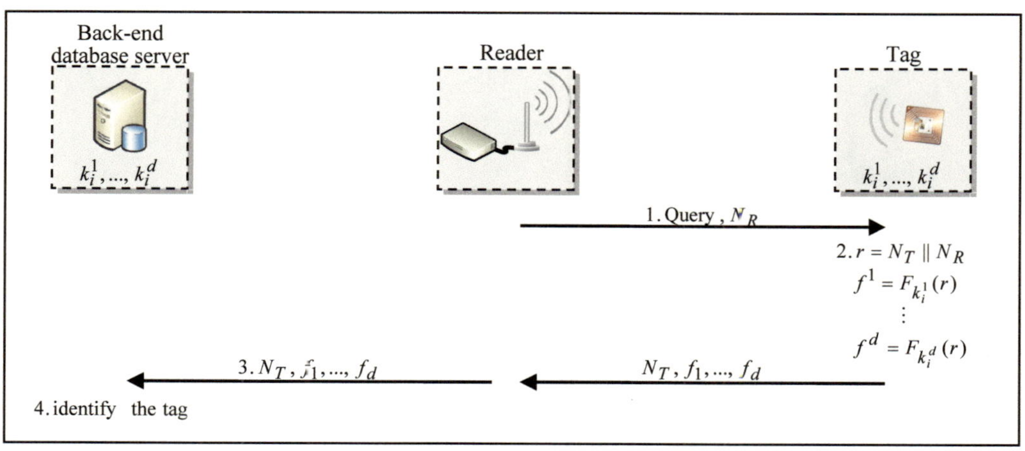

Fig. 22. Dimitriou's tree-based tag identification scheme

The procedure of Dimitriou's tree-based tag identification scheme is shown in Fig. 22. As the reader tries to query the tag with a random number N_R, the tag generates a random number N_T and computes the message $(f^1, f^2, \ldots, f^d)$ by all its keys. The back-end database server has to find out the keys in the trees from the root to the leaf node for identifying the tag. If the path exists, the back-end database server regards the tag as a valid tag.
In this scheme, tag searching using the idea of the tree walking algorithm is efficient for the reader. However, it may not be afforded for low-cost tags without enough computing capability to generate the responses by the sequence of keys in a transaction.

3.4.2 Wang et al.'s tree-based authentication scheme

Wang et al.'s (2007) proposed a Storage-Aware Private Authentication protocol (SAPA) (Wang et al, 2007). This scheme uses a sparse tree structure to organize keys of all tags. In the tree, only the root and the leaf node store a key. Each tag T_i is arranged to a leaf node and has a key triple (k_h, k_m, k_r). k_h, is the key assigned to the root. k_r is the key assigned to the leaf node. k_m represents the path from the root to the leaf node which is expressed in 0 and 1. For example fig.23 shows a sparse binary tree of three levels, the key triple (k_h, k_m^2 and k_r^2) is assigned to T_2.

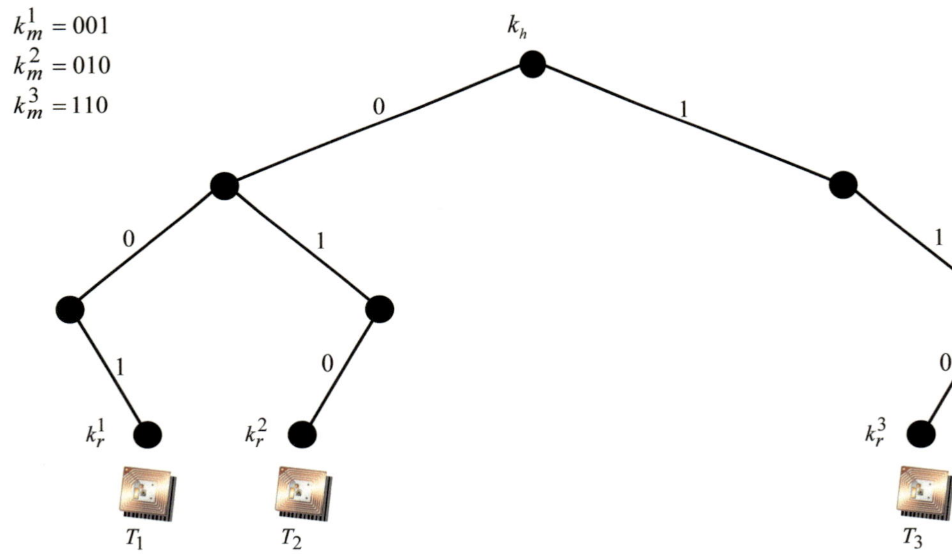

Fig. 23. A sparse binary tree of three levels

The procedure of Wang et al.'s tree-based authentication scheme is shown in Fig. 24. As the reader tries to query the tag with a random number r_1, the tag generates a random number r_2 and computes a sequence of hash chains $(M_0, M_1, \ldots, M_l, M_i)$. Then the back-end database server first verifies the message M_0 to authenticate the tag. After, the back-end database server performs a recursive algorithm to identify the tag through the path.

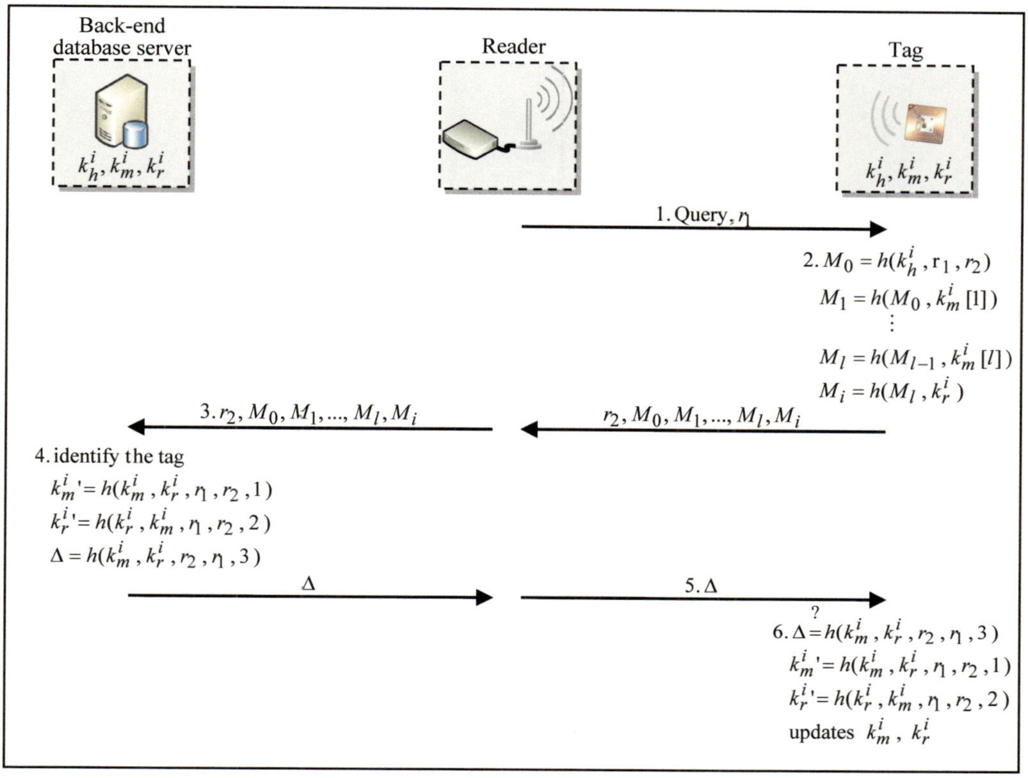

Fig. 24. Wang et al.'s tree-based authentication scheme

In this scheme, both the tag and the back-end database server update key triple (k_h, k_m', k_r') for each successful authentication. It may cause a collision for the new path k_m' assigned to the tag. The asynchronous attack may happen as the adversary blocks Δ sent back to the tag. These drawbacks cause the scheme impractical.

3.5 Chen et al.'s indefinite-indexed approach

Chen et al. (2011) proposed the indefinite-indexed access control scheme (Chen et al., 2011), shown in Fig.25. As a tag is issued by the back-end database server, the $index_i$, Key_i and a square matrix ω are stored in the tag. The tag's serial number $index_i$ is specified as a pair of values (x_i, y_i) which can also be regarded as a coordinate. For the purpose of keeping the tag's location private, the serial number cannot be emitted directly. Infinite possibilities exist to select two un-parallel lines crossed on the coordinate. If the tag is allowed to freely determine the two un-parallel lines, it means $index_i$ can be represented randomly. The first line can be determined by the tag's serial number (x_i, y_i) and any point (x_1, y_1). Then the second point (x_2, y_2) can be randomly selected on this line. Later, the other two points (x_3, y_3) and (x_4, y_4) can also be determined similarly. The values of these four coordinates will be re-arranged into a matrix π and performs the matrix product $\pi \cdot \omega$ as the response for the reader. Therefore, only the back-end database server holds the inverse matrix of ω can obtain the matrix π and figure out the tag's serial number.

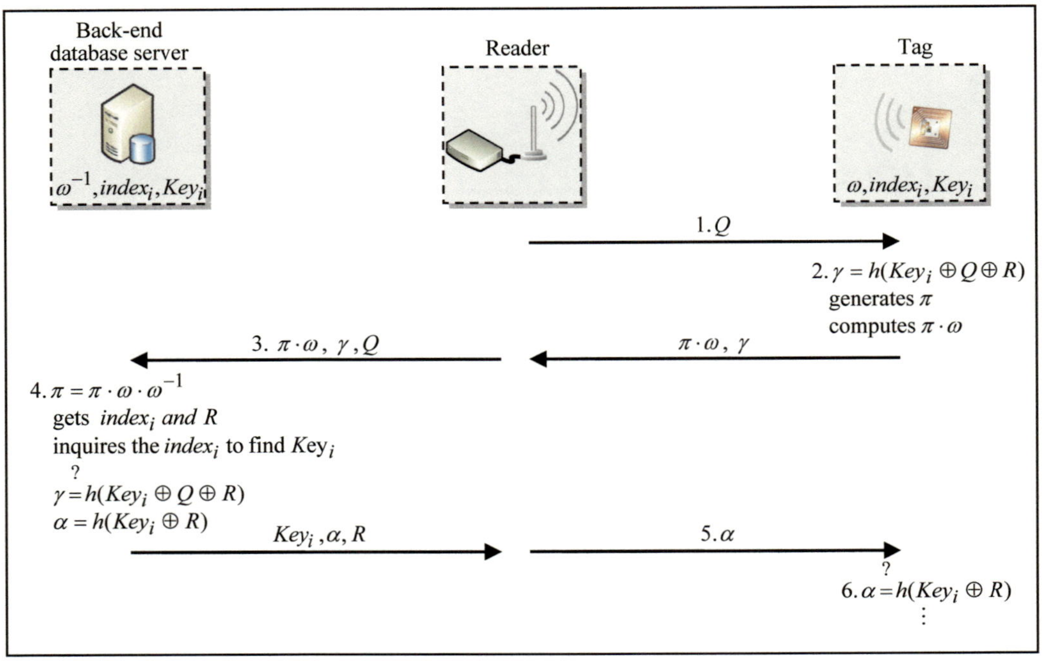

Fig. 25. Chen's indefinite-indexed access control scheme

The motivation of this scheme is to make the tag's response message not predictable to prevent the tracing of individual. In other word, the tag's response message in each access cannot be recognized it is emitted by the same tag. In this scheme, the tag's serial number is regarded as a coordinate. Infinite possibilities exist to select two un-parallel lines crossed on the coordinate. Therefore, the tag's serial number can be represented differently in each access and not useful to identify the tag. Moreover, the other messages emitted between the tag and the reader are also randomized and not useful to trace the tag. Therefore, the tag's location privacy can be guaranteed. In addition, this scheme also guarantees mutual authentication and resists the man-in-the-middle attack, the spoofed reader attack, and the spoofed tag attack.

4. Conclusions

Modern RFID systems are creating a new era of ubiquitous information society. It allows almost everything to be uniquely numbered by embedding a RFID tag. Then the process automation efficiency and usability could be improved (Chang, 2005; Garfinkel et al., 2005). It allows objects to be scanned and identified without the need for visual or physical contact. However, due to the powerful tracking capability of RFID tag, it poses a potentially widespread threat to consumer privacy (McCullagh, 2003). In the world of RFID tags widespread deployment, anyone with an RFID reader can potentially discover individuals' informational preferences without their permission.

Without access control, anyone can read the information stored on current generation RFID tags. The static unique identifiers stored on tags can be traced for linking the tagged items to the individuals who carry the item. Therefore, security and privacy in RFID systems are an important aspect that needs particular attention. Current researches in RFID technology not just concentrate on the identification scheme. Secure and efficient authentication and access control mechanisms have received much attention in the proposed researches. This article examines the main privacy concerns: information leakage of a tag, traceability of the person and impersonation of a tag. The impersonation problem is always the first one to be analyzed and solved in each scheme. Otherwise, the adversary can collect the information sent by the tag and the adversary can try a spoofing or replay attack to impersonate a target tag. For further consideration, the disclosure of information arising during a transmission of data possibly reveals various personal details without awareness of the holder. Most of the proposed schemes were well designed to prevent the problem of tag's information leakage. However, most of the proposed schemes can not really avoid the problem of traceability. The adversary may try to distinguish whether the response is transmitted by the target tag or not. Once a link is established between the response and the target tag, the adversary can monitor the person's location. For those schemes analyzed in this article, state diagram and use-case diagram are used to figure out the schemes' weaknesses. Through this way, the security requirements in RFID applications can be clearly understood to know which mechanism actually brings which feature. We expect it is more beneficial those researchers as just devoting to the RFID security studies.

5. References

Auto-ID Center (2003). 13.56 MHz ISM Band Class 1 Radio Frequency Identification Tag Interference Specification: Recommended Standard, Version 1.0.0, Technical Report, Auto-ID Center.

Avoine G. (2004). Privacy Issues in RFID Banknote Protection Schemes, *in Proc. 6th Conference on Smart Card Research Advanced Application*, pp. 33–48.

Avoine G. & Oechslin P. (2005). A Scalable and Provably Secure Hash Based RFID Protocol, *2nd IEEE International Workshop on Pervasive Computing and Communication Security*, pp. 110-114.

Avoine G. & Oechslin P. (2005). *RFID Traceability: A Multilayer Problem, Financial Cryptography*.

Bringer J., Chabanne H. & Icart T. (2008). Improved Privacy of the Tree-Based Hash Protocols Using Physically Unclonable Function, *Proc. of the 6th International Conference on Security and Cryptography for Networks – SCN 2008*, pp. 77-91.

Cavoukian A. (2004). Tag, You're It: Privacy Implications of Radio Frequency Identification (RFID) Technology, Information and Privacy Commissioner/Ontario.

Chang G.C. (2005). A Feasible Security Mechanism for Low Cost RFID Tags, *International Conference on Mobile Business*, pp. 675–677.

Chen Y.Y., Tsai M.L. & Jan J.K. (2011). The Design of RFID Access Control Protocol using the Strategy of Indefinite-Index and Challenge-Response, *Computer Communications*, Vol. 34, No. 3, pp. 250-256.

Chien H.Y. (2006). Secure Access Control Schemes for RFID Systems with Anonymity, *Proceedings of the 7th International Conference on Mobile Data Management (MDM 2006)*.

Dimitriou T. (2005). A Lightweight RFID Protocol to Protect Against Traceability and Cloning Attacks, *Proceedings of the First International Conference on Security and Privacy for Emerging Areas in Communications Networks*, pp. 59-66.

Dimitriou T. (2006). A Secure and Efficient RFID Protocol that could make Big Brother (partially) Obsolete, *Proceedings of the Fourth Annual IEEE International Conference on Pervasive Computing and Communications (PERCOM'06)*, Mar. 13-17.

Elgamal T. (1985). A Public Key Cryptosystem and a Signature Scheme Based on Discrete Logarithms, IEEE Transactions on Information Theory, Vol. 31, pp. 469–472.

Fishin K., Roy S. & Jiang B. (2004). Some Methods for Privacy in RFID Communication, *in Proc. 1st Eur. Workshop on Security in Ad-hoc and Sensor Networks.*

Gao X., Xiang Z., Wang H., Shen J., Huang J. & Song S. (2004). An Approach to Security and Privacy of RFID System for Supply Chain. *Proceedings of the IEEE International Conference on E-Commerce Technology for Dynamic E-Business.*

Garfinkel S.L., Juels A. & Pappu R. (2005). RFID Privacy: An Overview of Problems and Proposed Solutions, *IEEE Security & Privacy*, pp. 34–43.

Golle P., Jakobsson M., Juels A. & Syverson P. (2004). Universal Re-encryption for Mixnets, *in Proc. RSA Conference - Cryptographers' Track (CTRSA)*, pp. 163–178.

Good N., Han J., Miles E., Molnar D., Mulligan D. & Quilter L. (2004). Radio Frequency Id and Privacy with Information Goods, *in Proc. Workshop on Privacy in the Electronic Society*, pp. 41-42.

Henrici D. & Muller P. (2004). Hash-based Enhancement of Location Privacy for Radio-Frequency Identification Devices using Varying Identifiers, *Proceedings of the Second IEEE Annual Conference on Pervasive Computing and Communications Workshops*, pp. 149-153, Mar.

Huang Y.C. (2009). Secure Access Control Scheme of RFID System Application, *Proceedings of the 2009 Fifth International Conference on Information Assurance and Security*, pp. 525-528.

Inoue S. & Yasuura H. (2003). RFID Privacy using User-Controllable Uniqueness, *in Proc. RFID Privacy Workshop*, Nov.

Inoue S., Konomi S. & Yasuura H. (2002). Privacy in the Digitally Named World with RFID Tags, *Workshop on Socially-informed Design of Privacy-enhancing Solutions in Ubiquitous Computing.*

Joaquin G.A., Guillermo N.A., Ana C. & Jean L. (2011). Secure and Scalable RFID Authentication Protocol, 5th International Workshop on Data Privacy Management and Autonomous Spontaneous Security, pp. 231-243.

Juels A. (2004). Minimalist Cryptography for Low-Cost RFID Tags, *Security in Communication Networks*, pp. 149-164.

Juels A. & Brainard J. (2004). Soft Blocking: Flexible Blocker Tags on The Cheap, *in Proc. Workshop on Privacy in the Electronic Society*, pp. 1–7.

Juels A. & Pappu R. (2003). Squealing Euros: Privacy Protection in RFID-Enabled Banknotes, *in Proc. Financial Cryptography, Lecture Notes in Computer Science*, Vol. 2742, pp. 103-121.

Juels A., Rivest R.L. & Szydlo M. (2003). The Blocker Tag: Selective Blocking of RFID Tags for Consumer Privacy, *in Proc. 8th ACM International Conference on Computer Communication Security*, pp. 103–111.

Kinosita S., Hoshino F., Komuro T., Fujimura A. & Ohkubo M. (2003). Nonidentifiable Anonymous-ID Scheme for RFID Privacy Protection, *to appear in CSS 2003 in Japanese.*

Lee S.H., Asano T.Y. & Kim K.G. (2006). RFID Mutual Authentication Scheme Based on Synchronized Secret Information, *Symposium on Cryptography and Information Security*, January.

Lee S.M., Hwang Y.J., Lee D.H. & Lim J. I. (2005). Efficient Authentication for Low-Cost RFID Systems, *International Conference on Computational Science and its Applications - ICCSA 2005*, pp. 619-627.

Lu L., Han J., Hu L., Liu Y. & Ni L.M. (2007). Dynamic Key-Updating: Privacy-Preserving Authentication for RFID Systems, *Fifth Annual IEEE International Conference on Pervasive Computing and Communications*, pp. 13-22, Mar. 19-23.

McCullagh D. (2003). RFID Tags: Big Brother in Small Packages, *CNET News*, http://news.com.com/2010-1069-980325.html.

Molnar D. & Wagner D. (2004). Privacy and Security in Library RFID: Issues, Practices, and Architectures, *Conference on Computer and Communications Security – CCS 2004*, pp. 210–219.

Molnar D., Soppera A. & Wagner D. (2005). A Scalable, Delegatable Pseudonym Protocol Enabling Ownership Transfer of RFID Tags, *Selected Areas in Cryptography – SAC*, pp. 276-290, Aug..

Ni L.M., Liu Y., Lau Y.C., & Patil A. (2003). LANDMARC: Indoor Location Sensing Using Active RFID, *in Proceedings of IEEE PerCom.*

Ohkubo M., Suzuki K. & Kinoshita S. (2003). Cryptographic Approach to Privacy-Friendly Tag, *RFID Privacy Workshop*, MIT, MA, USA, November.

Osaka M., Takagi T., Yamazaki K. & Takahashi O. (2006). An Efficient and Secure RFID Security Method with Ownership Transfer, *2006 International Conference on Computational Intelligence and Security*, pp. 1090-1095, Nov. 3-6.

Pisarsky G.M. (2004). RFID Technology: An Analysis of Privacy and Security Issues, *20th Computer Science Seminar*, pp. 1–5.

Rhee K., Kwak J., Kim S. & Won D. (2005). Challenge-Response Based RFID Authentication Protocol for Distributed Database Environment, *International Conference on Security in Pervasive Computing - SPC 2005*, pp. 70-84.

Robinson P. & Beigl M. (2003). Trust Context Spaces: An Infrastructure for Pervasive Security in Context-Aware Environments, *in Proceedings of SPC.*

Sabaragamu Koralalage K.H.S., Mohammed Reza S., Miura J., Goto Y., & Cheng J. (2007). POP Method: An Approach to Enhance the Security and Privacy of RFID Systems Used in Product Lifecycle with an Anonymous Ownership Transferring Mechanism, *Proceedings of the 2007 ACM Symposium on Applied Computing*, pp. 270-275, Mar. 11-15.

Sarma S.E.(2001). Towards The Five-Cent Tag, Technical Report, Auto-ID Center.

Sarma S.E., Weis S.A. & Engels D.W. (2002). Radio-Frequency Identification Systems, Workshop on Cryptographic Hardware and Embedded Systems - CHES' 02, LNCS, Vol. 2523, pp. 454–469.

Sarma S.E., Weis S.A. & Engels D.W. (2003). RFID Systems and Security and Privacy Implications, *In Workshop on Cryptographic Hardware and Embedded Systems*, pp. 454-469.

Wang W., Li Y., Hu L. & Lu L. (2007). Storage-Awareness: RFID Private Authentication based on Sparse Tree, *Third International Workshop on Security, Privacy and Trust in Pervasive and Ubiquitous Computing (SecPerU 2007)*, July 19.

Weis S., Sarma S., Rivest R. & Engels D. (2003). Security and Privacy Aspects of Low-Cost Radio Frequency Identification Systems, *in 1st Intern. Conference on Security in Pervasive Computing (SPC)*, pp. 50-59, March.

Yeh K.H., Lo N.W. & Winata E. (2008). An Efficient Tree-Based Tag Identification Protocol for RFID Systems, *22nd International Conference on Advanced Information Networking and Applications - Workshops*, pp. 996-970, Mar. 25-28.

Attacks on the HF Physical Layer of Contactless and RFID Systems

Pierre-Henri Thevenon[1], Olivier Savry[1],
Smail Tedjini[2] and Ricardo Malherbi-Martins[1]
[1]*Leti, MINATEC, CEA Grenoble*
[2]*LCIS Lab, Grenoble-INP Valence*
France

1. Introduction

During the past few years, RFID technology has strongly penetrated in our lives. Nowadays public transportation ticketing, passports, ID cards, driving licenses and credit cards are using the electromagnetic waves to improve the quickness of the exchanged data. RFID devices can be divided in two main classes: the contactless cards which are smartcards with a wireless inductive interface compliant to the ISO14443 or ISO15693 standards, and the RFID tags which can have an HF or UHF interface compliant to the ISO18000 standard which now includes the EPCGlobal contribution. RFID tags are mainly dedicated to identification of objects. These exhibit a large reading distance but provide poor computational and processing resources. RFID devices and smartcards have a common characteristic; their contactless interface adds threats in term of security and privacy. This chapter will deal with this specificity by moving apart the well-known physical threats on smartcards like side channel attacks. Indeed, it is worth pointing out that the RF channel opens new potential vulnerabilities which could jeopardize security and as a consequence they should be listed and studied:

- Bidirectional data communication over the air:
  The transactions can be easily eavesdropped by a spying probe within a distance of several meters. Due to the low resources feature of such a device, encryption remains difficult to implement.
- Unidirectional power transfer over the air:
  The device is not the master of its energy which should be provided by the reader or by the attacker opening a backdoor for denial of service.
- Clock transfer over the air (especially for HF interface):
  The sequencer of the card can be monitored by the reader or the attacker. Pauses or accelerations of the processor can be achieved.
- Passive devices and no ON/OFF switch:
  The owner of the card or the tag is not able to switch off his device involving a main threat for its privacy.
- Load based retro-modulation:

The communication from the tag to the reader is really weak and performed in a passive way without emission of electromagnetic field but with modulation of the load at the terminals of the tag antenna. It can be easily blurred or modified.

- Singulation or Anti-collision protocol:
  The reader should have to deal with numerous tags or cards in its field. It requires a kind of identification which could endanger privacy.

This chapter proposes an overview of all these physical layer attacks.

2. Security & privacy

The vulnerabilities introduced by the contactless standards should be seen as vectors for attacks and as causes for risks on the security of the system and on privacy of people. Those two latter issues could be considered as antagonists. On the one hand, companies which deploy or use RFID systems naturally target profits and as a consequence try to nullified fraud which could be a severe competitor, to protect their business. On the other hand, for privacy, the point of view has changed: the security is no longer seen from the eyes of the provider but with the eyes of the user. More and more, users will live in a digital world with one or several digital doubles. So, the issue becomes individual freedom and more specifically in this case the protection of personal data and the insurance of not being spied or traced. Tracking a person by scanning tags or cards on him, using the access card without the agreement of its owner to enter in a secure building, all these attacks can be currently done by using information in contactless cards or RFID tags memory. For these reasons, contactless technology is often associated with privacy invasions, population under surveillance. The interests of the user and of the provider could be shared if the latter realizes that privacy is framed by regulatory matters sometimes dedicated to RFID like European recommendations and that it is a condition of a large scale deployment of RFID. Risk analysis should be performed on these two main topics: prevention of economical fraud and preservation of privacy. The targeted assets and the motivation of attackers differ even if countermeasures could help both. Vulnerabilities and attacks to security or privacy lead mainly to four risks:

- Eavesdropping on the communication:
  In the field of privacy, identifier of tags could be listened enabling tracking or impersonation of tags. For the security of the system, secret data like session keys could leak.
- Remote activation without the consent of the owner:
  This is the main threat to privacy since silent physical tracking and inventorying of people possessions could be carried out. This risk is also the basis of the relay attack which is able to circumvent any cryptographic protocol.
- Denial of service: the system becomes inoperative:
  Due to the weak signal answered by the tags, it is easy to blur it. The simple destruction of the tag is also possible by applying an over estimated field. Many solutions exist that could lead to an out of order system.
- Unique identifier which is a pointer on a database:
  The fact that each items bought in a supermarket will be tagged with an unique identifier will enable to trace it and to fill all the properties of the object and of the owner in a database.

All those risks require to study in detail the attacks which are at their origin.

3. Eavesdropping

Eavesdropping is a passive attack, which consists in secretly listening a private communication between a reader and a card (Figure 1). This attack, particularly simple to realize, is a true threat because the attacker can analyze transmitted data between the reader and the card to recover confidential information.

Fig. 1. Eavesdropping attack

3.1 State of art

First experiments on eavesdropping attacks were published by the NIST (National Institute of Standard and Technology) in 2004. Researchers have succeeded in recovering e-passport private data situated at 9 metres from their spy. Despite the lack of details in the description of the measurement protocol, it seems that only the forward communication (communication from the reader to the card) has been eavesdropped (Hoshida, 2004). Furthermore, it seems that ISO14443-B standard is more sensitive to eavesdropping attacks than devices using ISO14443-A (ISO/IEC14443-2, 2001).

In 2004, Finke and Kelter of BSI (German federal office for information security) have presented results demonstrating that a communication between an NXP contactless reader and a card can be intercepted at 2 metres (Finke & Kelter, 2004). The main feature of their attack is the use of a specific position of the spy antenna called second Gauss position (see part 3.2).

A report from the FOIS (Federal Office for Information Security) has described all threats specific to the contactless link. No experience is described in this paper, but the main features of the attack are given. Anti-collision protocols amplify the risk factor because confidential data are repeated during theses protocols. Based on theoretical studies, it seems that an attacker may listen the uplink communication up to few dozens of metres and only 50 cm for the downlink communication (FOIS, 2004).

In 2006, researchers of the NIST have realized experiments using an NXP reader, compliant to the ISO14443-A standard (Guerrieri & Novotny, 2006). Their work shows the influence of the spy antenna positions; two position, called Gauss, positions are described. They succeed in spying a communication up to 6.5 metres in the first position and up to 15 metres in the second position. The characteristics of these positions will be explained in the section 3.2.

Hancke has presented experiments on the main attacks that occurred on the physical layer. His paper gives a lot of information on the measurement protocol, particularly on the used equipment (Hancke, 2006). The results show that the entire communication (forward and backward) can be eavesdropped at a distance of 4 metres. The author has completed this first article with a new paper by adding new results and conclusions in 2008. The measurement protocol is well detailed and all HF standards are studied. During these experiments, the results are sampled then processed on a computer in order to enlarge the spying distance. It shows that forward communication is easier to recover than the downlink communication (Hancke, 2008a, 2008b).

3.2 Theoretical study on Gauss positions

The position of the attacker antenna with respect to the reader antenna has an important influence on the amplitude of the signal recovered by the spy antenna. Two positions are particularly important; they are called Gauss positions and are used in few attacks described in the state of art (previous section). To enlarge the eavesdropping distance of an attacker, a theoretical study will be made on the Gauss position. A loop antenna can be considered as a magnetic dipole antenna when the diameter of the emission antenna is much smaller than the distance between the antenna and the observation point (Figure 2a).

Equations 1, 2 and 3 give magnetic and electric fields seen as a distance r of the emission loop antenna.

$$H_\theta = \frac{I.S.\sin\theta}{4\pi r^3}(1 + j\frac{2\pi r}{\lambda} - \frac{4\pi^2 r^2}{\lambda^2})e^{j(wt2\pi r/\lambda)} \quad (1)$$

$$H_r = \frac{I.S.\cos\theta}{2\pi r^3}(1 + j\frac{2\pi r}{\lambda})e^{j(wt2\pi r/\lambda)} \quad (2)$$

$$E_\phi = j\pi\frac{I.S.\sin\theta}{w\varepsilon_0\lambda^2 r^2}(1 + j\frac{2\pi r}{\lambda})e^{j(wt2\pi r/\lambda)} \quad (3)$$

Equations are used to predict the magnetic field in the case of the two Gauss positions, i.e. with θ = 0° for the first position and θ = 90° for the second position. The results on Figure 2b show that the first gauss position is more interesting when the attacker is situated at a distance smaller than 8 metres. When distance is larger than 8 metres, the second Gauss position will allow an attacker to obtain the highest RF field amplitude on the spying antenna.

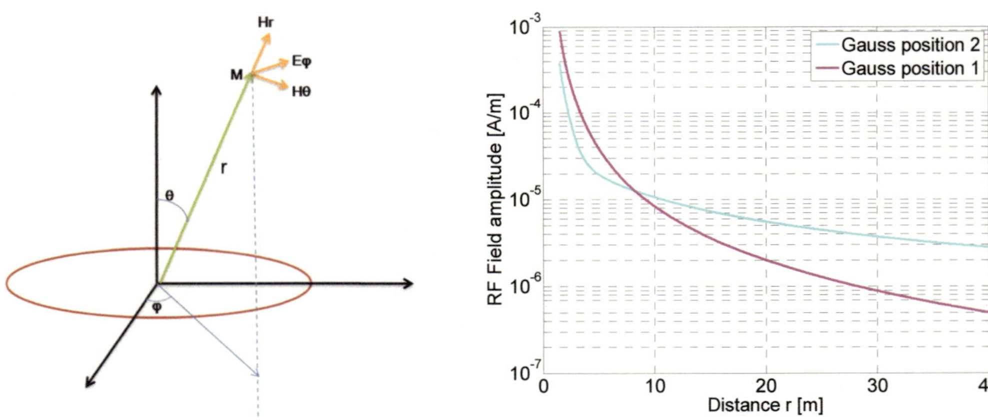

Fig. 2. a: Magnetic and electrical field seen as a distance r of the antenna; b: Results of the theoretical approach

Figure 3 gives the positions of the antennas in the case of the two positions de Gauss and conclude on their use in function of the eavesdropping distance. In the first Gauss position, an axis perpendicular to the reader antenna passes through the centre of the reader antenna and the spying antenna. In the second Gauss position, an axis parallel to the reader antenna passes through the centre of the reader antenna and the spying antenna.

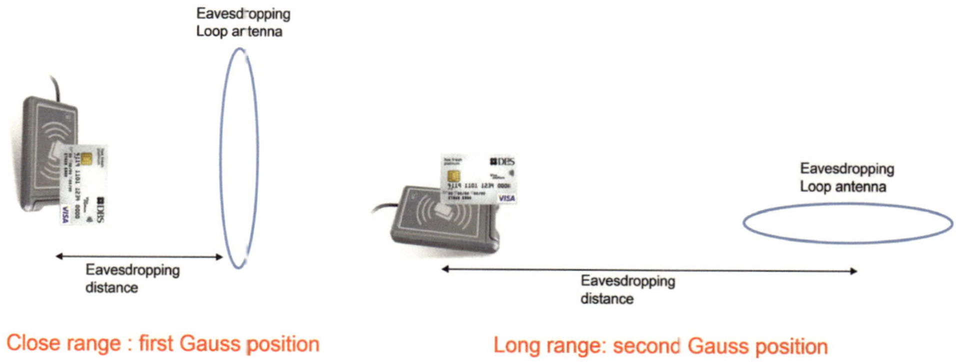

Fig. 3. Antennas positions for the two Gauss positions

3.3 Measurement protocol

Eavesdropping on a communication is a simple attack; a tuned antenna and an oscilloscope are sufficient to analyse signals transmitted between two contactless devices. Improvements on the signal processing can increase the eavesdropping distance. In these experiments, the reader and the card are products compliant to contactless ISO14443 standards; the reader is connected to a loop antenna compliant with the standard ISO10373-6 (ISO/IEC10373-6, 2011). The measured field at the centre of the reader antenna is 3.1A/m, i.e. the average range of the standard. Attacker antennas are one-turn inductive loop with a diameter of 30 and 50 cm tuned at 13.56MHz or 14.4 MHz (centred on the subcarrier frequency of the retro-modulated signal) and made with coaxial cable. The antenna signal is 60dB amplified, filtered with a band-pass filter at 13.56MHz and finally recorded on a scope (Figure 4).

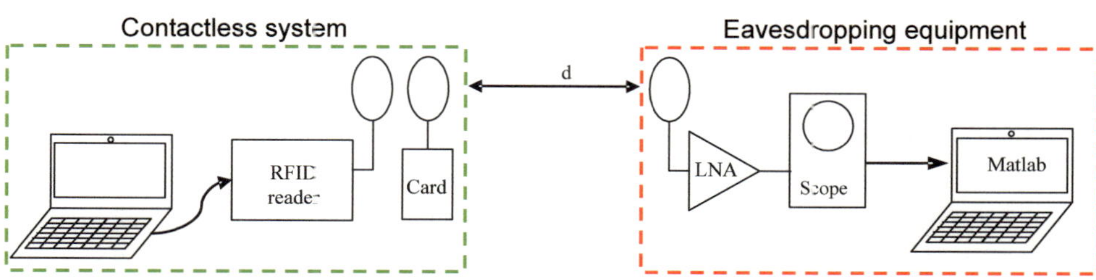

Fig. 4. Eavesdropping bench test

The recorded signals are then processed under Matlab: pass-band filtering, synchronous demodulation and detection (Figure 5).

3.4 Experiments

The forward and backward communication signals must be processed to recover a maximum of information. The contactless cards are passive, and the way to transmit data from the card to the reader via the backward link is by the retro modulation of the reader signal. This implies that the distance to listen to the card is definitely smaller than for the forward link. The figure 6 gives an outline of analyzed signal. The forward communication can definitely be eavesdropped further because of the modulation used type.

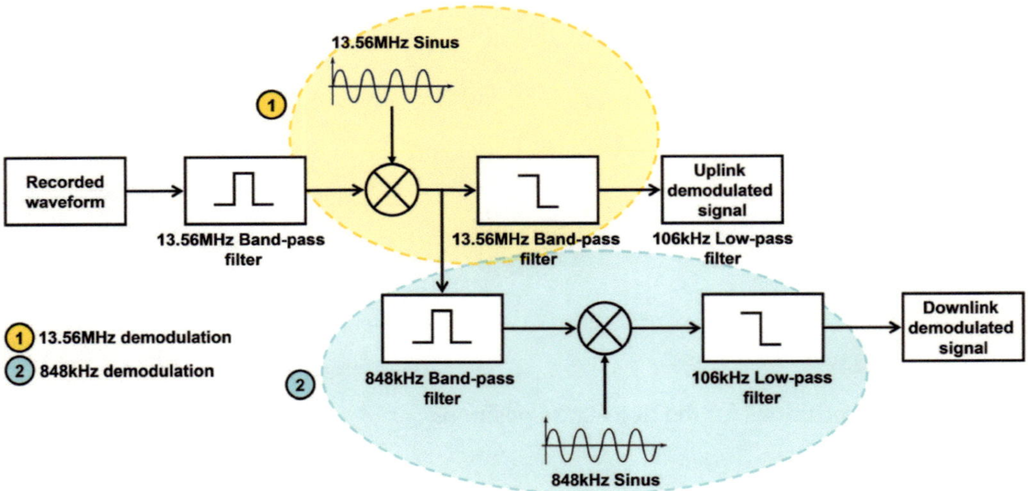

Fig. 5. Signal processing with Matlab

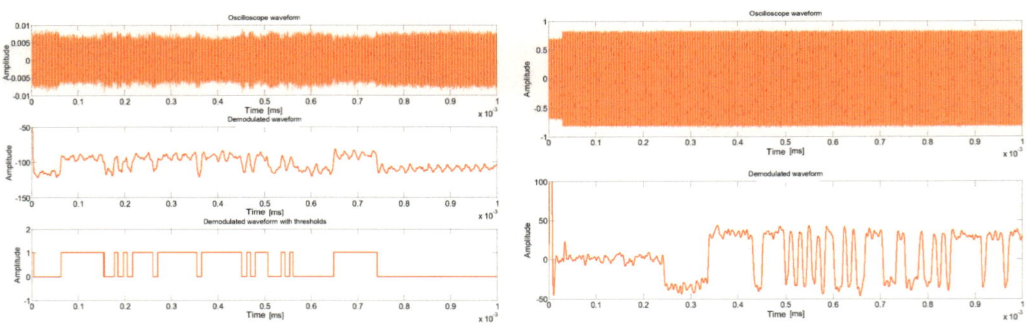

Fig. 6. Contactless forward communication link at d= 22 m and backward communication link at 3.5 m

A magnetic antenna will be used in the most of the realized experiments but the capability of using an electrical antenna to eavesdrop the HF signal gives information on the equipment that an attacker could use. A very simple antenna, an electric dipole has been used for this experiment. The results on the figure 7 show that the eavesdropping is noisier with an electrical antenna and that only the forward communication could be recovered at 4 metres.

First experiments on eavesdropping were realized in outdoor to avoid disturbances due to the environment. However, the attacker can not have a clean environment and it is important to understand the way in which an indoor environment can help an attacker to recover data. Two experiments were realized in indoor to answer to this question. During the first experiment, an antenna, used in EAS (Electronic Article Surveillance) system, generates a rotating magnetic field. Then the RF field amplitude has been listed in few locations next to this antenna. It was demonstrated that signal voltage at the level of the antenna can be larger when the eavesdropping antenna is located further. In the same way, the second experiment was the analysis of the RF field of a badge antenna used in access control fixed on a laboratory door. It was possible to listen and record data several floors under in the lower part of the building with more than 8 m of vertical distance.

After the analysis of these experiments, it was concluded that wirings, wall materials as reinforced concrete or metal framings of the doors appeared as very effective antennas relays.

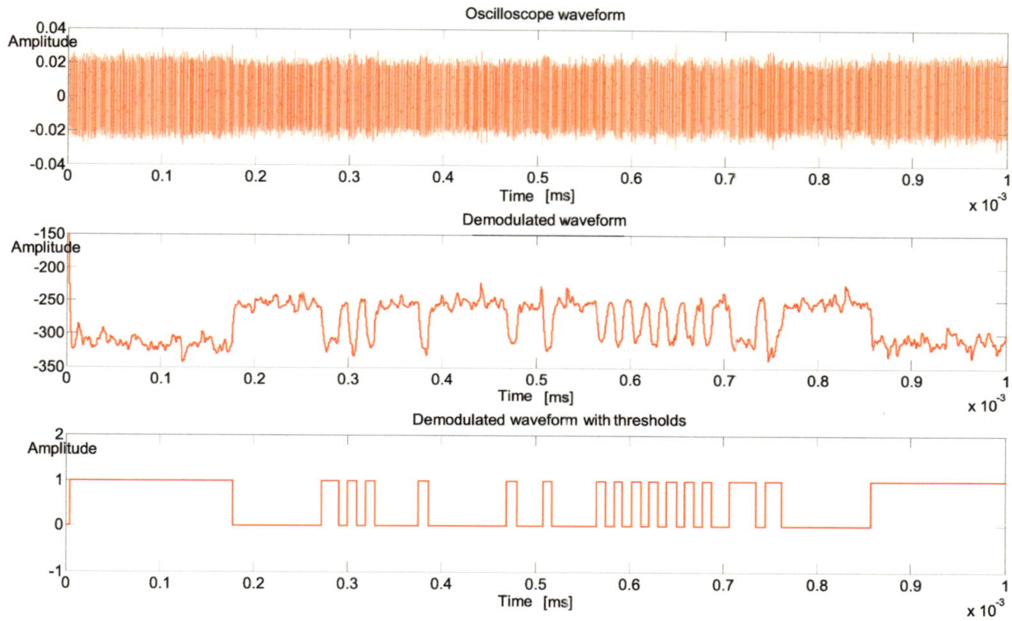

Fig. 7. Measured signals with an electric dipole at 4 metres from the emission

4. Skimming

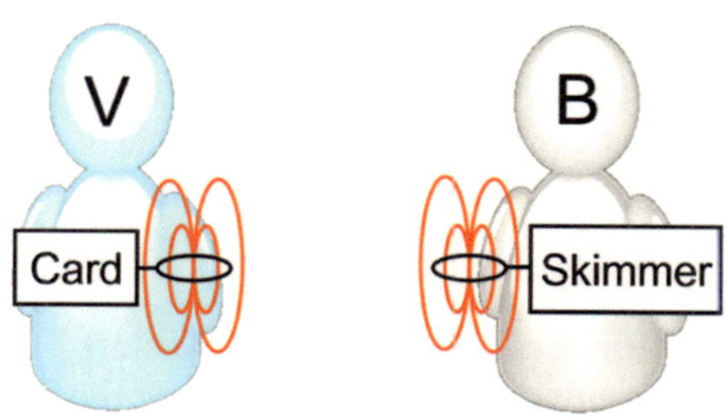

Fig. 8. Skimming attack

The skimming attack is to activate a card without its owner's agreement. In this active attack, the hacker needs to power the card, well modulates the field in the forward channel, and be capable to well process the load modulation of the backward channel in order to communicate with the card.

4.1 State of art

Many publications describe the features of the skimming attack. However only few of them describe practical scenarios or details of the experiments. Hancke has shown some interesting and detailed results on the skimming attack (Hancke, 2010). He has considered two different distances, the activation distance and the distance to retrieve the backward channel. Using different antenna sizes and different power levels, he has analysed different ways to activate the card and eavesdrop the communication. An important contribution of his paper is that the activation range do not increase in the same way as the distance of he could retrieve the token response. On the one hand, the best result of the retrieve distance was 2 m but with a skimming range of 15 cm. On the other hand he achieved a skimming range of 27 cm, however with less than 2 m of retrieval distance. In 2006, Kirschenbaum and Wool (Kirschenbaum & Wool, 2006), have already demonstrated almost the same skimming range. Using a cooper tube loop antenna and a power amplifier, they have demonstrated a theoretical and experimental setup to activate a card within a distance of 25 cm. Moreover, NXP (Tobergte & Bienert, 2007) has published that the skimming distance of ISO14443 systems is limited to approximately 30 cm. In addition, Kfir and Wool have demonstrated that beyond 50 cm the attack is hardly feasible, because the power requirements become increasingly important (Kfir & Wool, 2005). To conclude, lot of information is available about HF antennas. Application notes such as Texas antenna cook book (HF Antenna Cookbook) and Microchip antenna circuit design (Youbok, 1999), combined with some knowledge of ISO14443 systems, are enough information to know how to build a low cost skimmer device.

4.2 Theoretical study

4.2.1 Theoretical activation distance

Based on the Biot-Savart law, Equation 4 describes the link between the current I in a circular antenna and the magnetic field H function of the distance d between the reader and the transponder, r the radius of the circle and N the spires number of the antenna.

$$H(d) = \frac{r^2}{(r^2 + d^2)^{3/2}}.N.I \tag{4}$$

To keep the compliance with ISO standards, the field at the level of the transponder must be higher than 1.5 A/m. Figure 9 describes the behaviour of the field in the case of an antenna with one spire and 0.45 m radius parameters for different current in the circular loop.
Theoretical curves show that an attacker can hardly power and then activate a card situated at one metre from the reader.

4.2.2 Identifying the key parameters for the card activation

In order to identify the critical parameters, some aspects of the communication must be run through. The energy transfer can be improved and the attacker power optimized for a given frequency and communication range. Regarding RFID tokens which use high frequencies

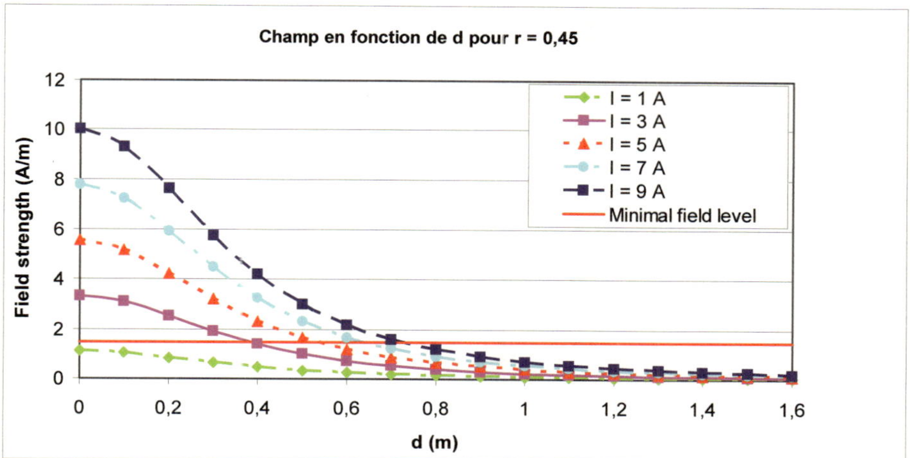

Fig. 9. Field amplitude versus distance between the reader and the transponder

and short range communication, the technique in this case is an inductive coupling. With the aim of activating the card, the hacker's inductive antenna converts an electrical signal into a magnetic signal transmitted over the air. The interaction between the reader and the card is governed by the mutual inductance. The token will harvest all of its power from the energy emitted by the hacker's antenna. Then, it can read, write and retransmit data through this magnetic field. Figure 10 describes the principle of coupling between two circuits with inductive loops.

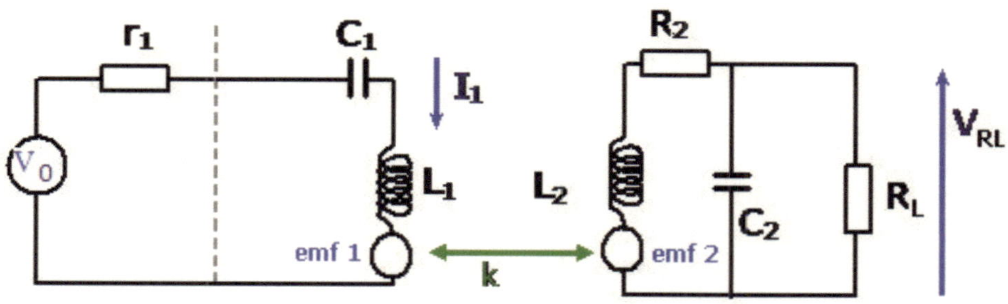

Fig. 10. Power transfer between the reader and the transponder

The mutual inductance between two circuits is defined as the ratio of the partial flux enclosed by the inductive loop of the card on the current I_1 passing through the loop of the reader (Reinhold, 1993)(Equation 5).

$$M_{12} = \frac{\psi_{12}(I_1)}{I_1} = \frac{\mu_0 N_2 H_1(I_1) A_2}{I_1} \tag{5}$$

Thus the mutual inductance between the antennas depends on many aspects of the card such as N_2 the number of turns in the card antenna, μ_0 the magnetic constant, and A_2 the area of the card. It also strongly depends on the hacker's antenna field strength $H_1(I_1)$. To measure the efficiency of the coupling we use the coupling coefficient k, defined in Equation 6.

$$k = \frac{M_{12}}{\sqrt{L_1 L_2}} \tag{6}$$

Theoretical values of coupling coefficient could vary between the two worst cases $0 \le k \le 1$. However, in most of the standard HF RFID communications, the coupling coefficient average is close to 3%. (Finkenzeller, 2003).
As far as the coupling coefficient is greater than 0, an electromotive force emf will be created in the token circuit. This emf will empower the card and is calculated in the equation 7.

$$emf_2 = j\omega M_{12} I_1 \tag{7}$$

And the voltage across the load is given by the equation 8:.

$$V_{RL} = \frac{j\omega M_{12}.I_1}{1 + (j\omega L_2 + R_2).\left(\frac{1}{R_L} + j\omega C_2\right)} \tag{8}$$

Most of the contactless card needs 10 mW to be activated. The Q factor of the circuit is defined in Equation 9 (Malherbi Martins et al., 2010).

$$Q = \frac{1}{\frac{R_2}{\omega L_2} + \frac{\omega L_2}{R_L}} \tag{9}$$

4.2.3 Improving the card activation

The main point for the hacker is to make the attack without being detected by the victim. Analysing the equations of the previous section, with the purpose to increase the operational range the hacker must increase the mutual inductance between his antenna and the victims token. In order to do that, he must increase his antenna power and the field strength emitted by it.
For each distance d, there is an optimal hacker antenna size. For typical forms of antenna, such as rectangular and circular antennas we can easily find the optimal radius r to obtain the maximum field strength H at a distance d (Youbok, 1999) (Equation 10).

$$H'(R) = \frac{d}{dR} H(R) \tag{10}$$

Using the optimal radius, the hacker increases the field strength H and the mutual inductance between his antenna and the victim's card.
To increase this power, the hacker will try to have the strongest current flowing through his antenna. For that, he will design his system with a high Q factor. However, this increase has a limit for two reasons. Firstly, a tuned circuit acts as a band pass filter, so with the purpose of recover the data of the backward channel, he must choose a compromise between power and bandwidth. For the ISO14443-A standard, the subcarrier is at 847 kHz, and the data rate

is 106kBits/s. In this way, the attacker must choose a low Q to maintain a high bandwidth BW (Equation 11).

$$Q \le \frac{f_c}{2 \times BW} \tag{11}$$

Secondly, for different coupling coefficient values k there is an optimal Q factor for the antennas. The relation between the Q factor and k coefficient coupling is defined in the Equation 12:

$$n = k \cdot \sqrt{Q_2 Q_1} \tag{12}$$

For example a well designed antenna to activate a card at a distance d_1 could not activate a card at a distance d_2. In fact, the activation distance depends of the value of n the system could have different behaviours.

For each value of n the system will has a specific coupling regime, as shown in the figure 11:

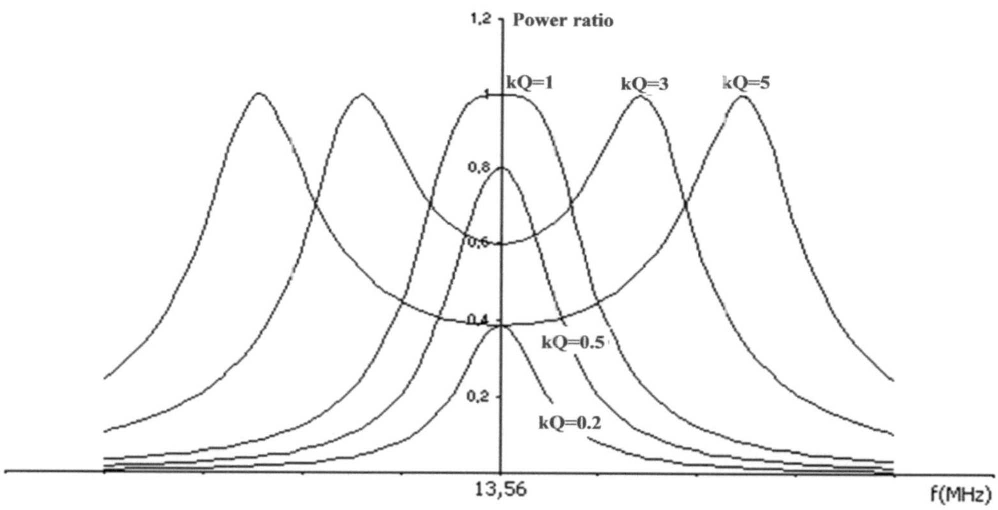

Fig. 11. The transferred power function of the product n

To have the maximum power across the R_L, the parameter "n" must be equal to 1. Using all these information a hacker could try to activate a card at important distances.

5. Relay and man-in-the-middle attack

The main objective of relay attacks consists in the setting of a communication between a true reader and a true contactless card. This attack is based on the Grand Master Chess problem described by Conway in 1976. The later shows how a person, who does not know the rules of this game, could win against one of two grand masters by challenging them in a same play. The relay attack is just an extension of this problem applied to the security field. By relaying information between a reader and a card outside the reader field, an attacker can circumvent the authentication protocol. This attack requires two devices, a mole which

pretends to be the true reader and a proxy which claims to be the true card. These two devices can communicate and thus relay data between the reader and the card. This communication can be wired or wireless.
Scenario:
In Japan or in USA, contactless systems are already used for payment applications. Under this scheme, an attacker can, by using relay attack, charges his purchases by a victim situated in the running area of the relay. To build this attack, the attacker is near the payment terminal and his accomplice near the victim. During the payment, the attacker places the proxy close to the payment terminal and the accomplice places the mole near the pocket of the victim; a relay is created between the terminal and the valid contactless card (Figure 12). The valid reader communicates with the card because it believes it is nearby him, so it debits the victim's account.

Fig. 12. Payment scenario: the relay attack setup

The man-in-the-middle attack is often mistaken for relay attack even if it is more advanced. The main features of the two attacks are the same: to allow the data transition through the relay between two wireless devices which are not in their communicating range. The distinctive feature is that the bits stream between the communicating devices can be modified during the relay by the attacker in the case of a man-in-the-middle attack.

5.1 State of art

5.1.1 Relay attack

To define the features of a relay attack, three different parts should be characterized: the mole, the proxy and the relay between them. The proxy acts as skimmer that has to activate and power the attacked contactless card and to communicate with it. Skimming a card is an attack as itself and is already discussed in this document.
The mole is able to eavesdrop on the communication with a real reader and to talk back to it. The eavesdropping on the communication is also well discussed in this document. Regarding the fake card response, active modulation could be used instead of a load modulation that requires a close coupling and so a short range.
The larger the distance between the different elements is, the more efficient is the relay. The theoretical maximum distance between the reader and the proxy is 50 m and 50 cm between the mole and the card. The distance between the mole and the proxy is not limited; it just depends on the used technology. (Kfir & Wool, 2005).

The wireless relay developed by Hancke can transmit requests and replies between an honest reader and an honest card separated by 50 metres (Hancke, 2005). The delay, introduced by such a relay is more than 15 µs. However, many others communication channels can be used to link the mole and the proxy to increase the relay distance (Lishoy et al., 2010).

In (Oren & Wool, 2009), the authors have presented their work on relay attacks realized on an Israeli e-voting scheme. They have demonstrated that low-cost relays can compromise this system, the privacy and the security of voters. They have inserted a communication link between the voting terminal inside the voting booth and the ballot box that carried votes which were already cast. This attack permits to read, modify or suppress votes while the attacker is in the voting booth. It is carried out that relay attacks are easy to realize, difficult to detect and very dangerous for the privacy of data recorded on smartcards.

At the physical layer, this attack is the most dangerous for many reasons:

- With its capability of attacking from a long distance without the consent of the user and of bypassing the encryption of the contactless transaction, the relay attack appears to be one of the main threats for RFID systems. The card is activated and transmits information when it is powered, without the agreement of the victim. Anyone can be this victim because the attacker has just to be enough close to you to control your card.
- Relay attack is an attack on the physical layer; the relay transmits coded bits without knowledge about the frame significance. The ISO9798 standard describes an authentication protocol to prove that the actors of the communication know the secret key (ISO/IEC9798-3, 1998). For the eavesdropping or skimming attacks, the use of this kind of protocol limits the risks. For the relay attacks, it's not required to know this key. In fact, a relay does not modify information of the frame, and has not to know the frame meaning, it just transmits the data. The encrypted data are transmitted like a plain text.
- Contactless standards such as ISO14443 impose time constraints to synchronize data sent by many cards at the same time, particularly during the anti-collision protocol. However, these constraints are not enforced by the majority of cards. If we consider the ISO14443-A standard, the card shall reply after a precise time after the request. These requirements would complicate the relay attack if they are really applied. Carluccio et al. have realized have realized experiences which show that the reader accepts card reply starting within a time slice of 2.5 us every 9.44 us (Carluccio et al., 2006). Moreover, the token can specify the response time after its selection; this is the delay that a card can take to reply after a reader request. This time has a default value which is 4.8 ms and a theoretical maximum value of 5 s. These values let plenty of time to the attacker to relay the information and to modify the data in the relay too (Halvac & Rosa, 2007). Hancke et al. have presented results about experimental response timeouts; its value could be set to 19.7 s during the communication after the card selection (Hancke et al., 2009).

5.1.2 Man-in-the-middle attack

The man-in-the-middle attack is able to modify transmitted data without demodulating and decoding the signal (Verdult, 2008). However, this solution does not present much interest because it does not allow changing any bit for a specific coding and the attacker does not know the significance of its modifications. Another solution consists in demodulating and decoding the signal, analysing the frame and modifying this frame according to the data that the attacker wants to transmit. By relaying information between a reader and a card without decoding the signal, an attacker can circumvent the authentication protocol; it is the main strength of the relay attack. The man-in-the-middle does not have this strength; the attacker has to know the frame significance to modify data. In the case of a ciphered communication, he must know the cryptographic algorithm and especially the secret key to decode the signal and discover its significance. The attacker knows the bits he can modified thanks to the decoded data. The new frame is then coded and modulated as a standard RFID signal. This attack is a real challenge if the attacker does not want to change arbitrary bits. Compared to relay attacks, the man in the middle attacks takes more time but the 5 seconds timeout defined by the ISO standard is enough to demodulate and compute any cryptographic algorithms.

5.2 Presentation of two new relays

The delay in current relays is mainly due to the use of components such as microcontrollers or RFID chips. This kind of components is used for the reconstruction of the decoded signals. Then, the original signal becomes compatible with other protocols, like Wifi or GSM, used in the wireless communication between the mole and the proxy. All these signal processes lead adding delays in the relay. Another solution is the use of analog components only. Attack scenarios with wired relays must be considered because they can introduce very low delays. Moreover, this kind of relays is simple to realize with few low-cost components. Even if they seem to be unlikely, they can be effective in a queue for example.

Passive wired relay: This attack consists of a coaxial cable of length l and an inductive antenna at each side, both matched and tuned at 13.56MHz (Figure 13a). The simple design allows very low delays. This delay is due to the length of the cable and the establishment time of the signal. This attack is a major threat because the delay may be less than a period of the 13.56 MHz carrier.

Wireless super heterodyne system: This relay, shown on figure 13b, is quite similar to the relay attack developed by Hancke because it is not restricted by a wired link. Unlike Hancke's relay, our wireless relay does not use digital components such as microcontrollers or RFID chips to process the signal. The delay introduced by this relay should be shorter. To do so, the reader signal of frequency fc is mixed with another signal of frequency F, generated by a local oscillator. It results a signal of frequency fc+F, easier to amplify and to send further. A PLL is used as a local oscillator to have the same frequency in the modulation and demodulation circuit.

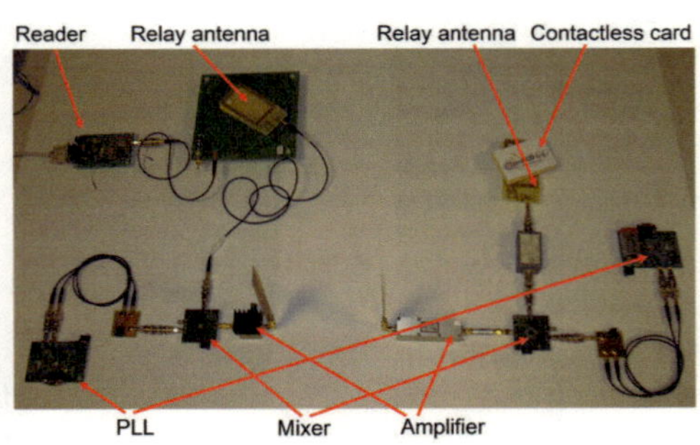

Fig. 13. a: Potential use of a wired relay in a hallway, an attacker creates a relay between an access terminal and a true contactless card; b: Experimental setup of the uplink system of a wireless relay

5.3 Experiments on relays

Two different experiments were realized. The first was to prove that all these relays work and to determinate their characteristics such as complexity, cost or values based on their performances. The delays introduced by these relays are so short that it is not possible to determinate them with unaided eye and it is necessary to use signals with precise properties to correlate them. For that, the second experience has for objective to measure values of these delays with the best accuracy. For all these tests, our reference is the relay developed by Hancke which is the most detailed in the literature (Hancke, 2005).

The bench test is roughly the same for the two experiments; the reader is connected to the computer in order to send commands, the card is placed few meters away of the reader. A relay is positioned between the reader antenna and the card. Two calibration coils are placed up on either side of the relay; they allow visualizing and recording signals on the oscilloscope.

5.3.1 Demonstration of relay efficiency

For these tests, only equipment compliant with the standard ISO14443-A is used because the main objective is to prove that our systems can relay data between a true reader and a true card (Figure 14). The reader is connected to an antenna conform to the ISO standard which has a quality factor of 10.

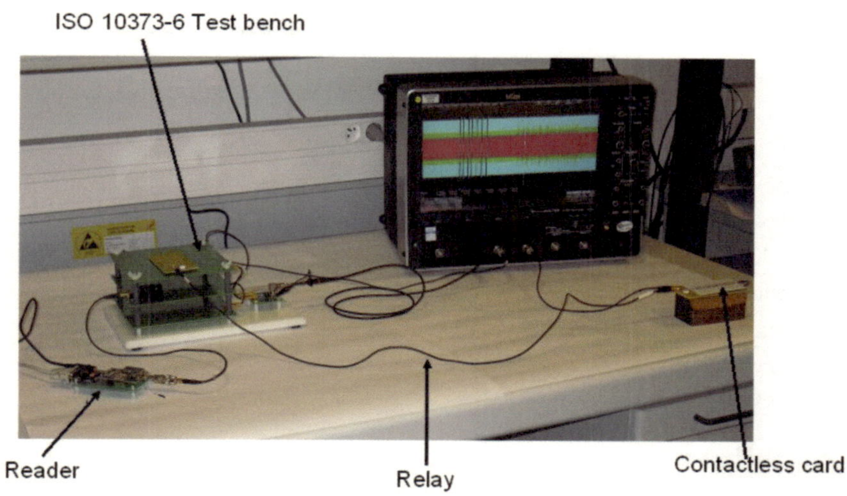

Fig. 14. Test for the wired relay: experimental setup to measure the load modulation amplitude

It is possible to vary different parameters such as the physical length of the relay, the distance between the reader and the proxy or the distance between the card and the mole.

Varying the distance reader-proxy is not really interesting because the attacker could put the proxy as near to the reader antenna as he can. Conversely, the mole-card distance is an important indication on the relay performance as the attacker can not really control the distance between his mole and the card he wants to activate. For each relay attack, we focus on the maximum distance of activation of the card. For a contactless system without relay, the activation distance is close to 10 cm

The cost and the complexity are important to determine the resources an attacker need to put in place a relay attack.
The ISO standard imposes a load modulation amplitude at least $30/H^{1.2}$ (mV peak) where H is the (rms) value of magnetic field strength in A/m (ISO/IEC14443-2, 2001). In our test, the value of H is 2A/m so the load modulation shall be superior to 13 mVp rms. Our two relays have load modulation amplitude higher than the standard level, which proves they are in accordance with this standard and undetectable by a subcarrier amplitude computation method. The table 1 summarizes all the results of the experiments realized on relay attacks.

5.3.2 Delay measurements

The main objective of the second experience is to measure the delays introduced by the two described relays using correlation computation. Measuring delays close to the period of the subcarrier is difficult with a standard contactless system so we have used an open reader that we developed. With this reader, it was possible to send signal with specific modulation properties easier to correlate. The reader sends a fixed signal through a relay; this signal is recorded directly on two calibration coils located close to the two relay antennas.
This experience is reproduced for different coupling between the reader antenna and the first antenna of the relay. To vary the coupling between the antennas, we increase the distance between them; 3 distances are chosen: 1, 3 and 8 cm.

		Contactless system with relay				Contactless system
		Wired relay	Wireless relay	Hancke's relay		
Cost		*	****	***		
Complexity		*	****	****		
Activation distance		4.5 cm	5 cm	No data		10 cm
Load modulation (mVp)		14.17	14	No data		> 13 mVp rms
Relay maximum length		> 10 m	> 50 m	50 m		
Distance reader antenna – relay antenna						
delay	1 cm	295 ns	566 ns	15 µs		
	3 cm	442 ns	454 ns			
	8 cm	442 ns	652 ns			

Table 1. Characteristics for each relay

The correlation of the two recorded signals permits to compute the temporal shift between them. Results in table 1 show that the delay sometimes decreases for a same relay when the distance increases between its antennas. Generally, the delay in a contactless system depends on three parameters: the establishing time in antennas which is function of the bandwidth, the propagation time in the air and the signal processing in case of relay presence. The propagation time in the air is short, i.e. less than 300 ps. The important parameter is the establishing time of the modulation amplitude variations which strongly depends on the coupling between the two antennas and of the bandwidth of the antenna,

i.e. the quality factor. This factor can introduce disparities in the cross-correlation computation.
Figure 15 gives a global view of the computed delays. Each type of relay is characterized by a time distribution. Wireless relays and wired relays have roughly the same delays because the mix of the signals is very fast. The computed delay for the relay developed by Hancke is more important than the others due to the time duration of each step of the signal processing.

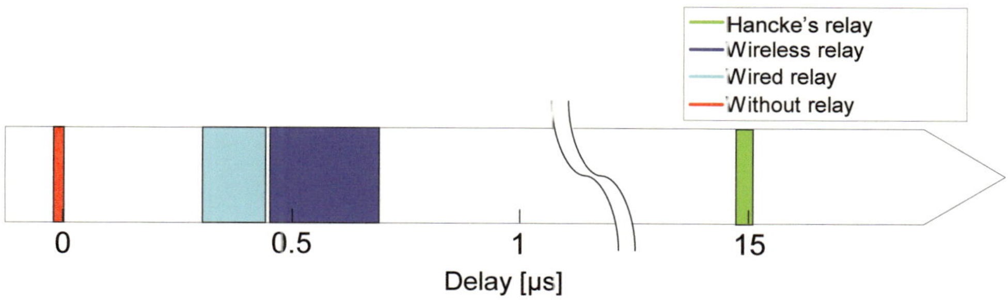

Fig. 15. Computed delays with correlation technique versus delay introduced by hancke's relay

6. Denial of service

The goal of denial of service attacks is to somehow deny a given service (e.g. identification) to valid users. Denial of service attacks are easy to accomplish and difficult to guard against. They can be divided into the four categories:

- Interferences in the anti-collision protocols
- Reader- and card jamming
- Faraday cage
- Destruction or deactivation

6.1 Interferences in the anti-collision protocols

Anti-collision protocol is an algorithm which avoids the communication of different cards in the same time. A reader is not able to decode data from multiple sources. To manage the collision of transponders, there exist two different protocols: a probabilistic protocol and a deterministic one. These protocols can be used as denial of service device by forbidding the access at one or many transponders.

6.1.1 Deterministic protocol: the Tree Walking algorithm (ISO/IEC14443-3, 2001)

This protocol is able to calculate the exact time required by the reader to know the UID (Unique IDentifier) of every tag depending on the number of tags in the reader field. All possible UID can be viewed as the leaves of a binary tree of depth equal at the UID bit length. The reader initiates the singulation protocol at the root of the binary tree and requests the UID of all tags in its field. A collision occurs when two tags simultaneously send a different bit. In this case, the tree-walking interrupts on the child collision node. Then, the reader requests all tags with an UID starting by the bits recorded before the collision to start at the previous node and selects the next branch '0' or '1'. Only tags with UID starting with this bit path reply to the reader. If another collision occurs, the reader repeats the last sequence until it obtained a complete UID (Figure 16).

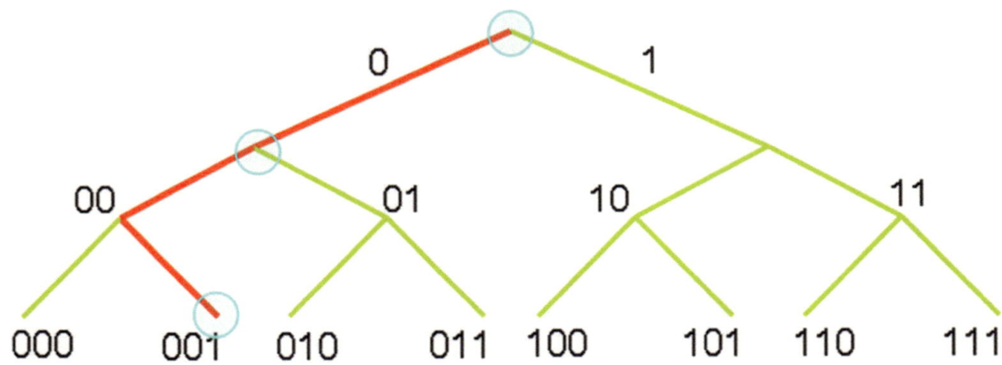

Fig. 16. Tree Walking for a 3-bit UID to obtain the "001" UID

The blocker tag is a device developed by Juels et al. to block the tree walking protocol and then prevent your card access by an intrusive reader (Juels et al., 2003). To jam the protocol, the blocker tag simulates a real tag and emits both '0' and '1' (which requires two antennas) at each reader anti-collision request. This creates a collision and the reader can not complete the algorithm and find out the tags UID. In most cases, such a device is used by the consumer to protect against unwanted scanning. However, the blocker tag can be used by an attacker to generate a denial of service in a legitimate system. We can even assume that a blocker tag is always malicious since it cannot be selective and forbids the reading of one tag whereas it authorizes the reading of others.

The attack is area limited since the blocker tag acts as a passive tag. It needs to be in the reader field to operate. Then it must be located close to the reader to have the best efficiency. However, an active tag can be realized to improve the blocking area of the system.

6.1.2 Probabilistic protocol: the Aloha method (ISO/IEC14443-3, 2001)

The second protocol assesses the probability to obtain the UIDs of a certain number of tags in a given time without giving any guarantee concerning its maximum value. This algorithm is known as time slots or Aloha method. The principle is simple: each tag can reply to the reader in fixed time slots during one or many rounds. A time slot is a period of time during one or more tags can reply. At the beginning of the first round of the anti-collision protocol, the reader sends the number of time slots used during the round. A tag can only reply during an single random time slot in a same round. The reader analyses each time slot:

- When a tag is alone to reply during a time slot, the reader retrieves and records its UID.
- When at least two tags reply during the same time slot, there is a collision.

If there is at least one collision during a round, the reader must throw again a new round without identified tags. Then, during the new round, the probability that two tags reply in the same time slot is lower. The reader throwes again rounds until there is no collision during a round and the reader has the UID of all tags in its field.

An attacker can use the weaknesses of this anti-collision protocol to jam it and block the communication between the reader and the card. The principle is close to the blocking protocol of the Tree-walking algorithm. As the reader creates a new round if a collision occurs during one, an attacker just has to create a collision during each round. In each time slot of a round, the attacker simulates a tag and sends a random UID. The reader is blocked in its anti-collision protocol and it can not collect UID cards (Figure 17).

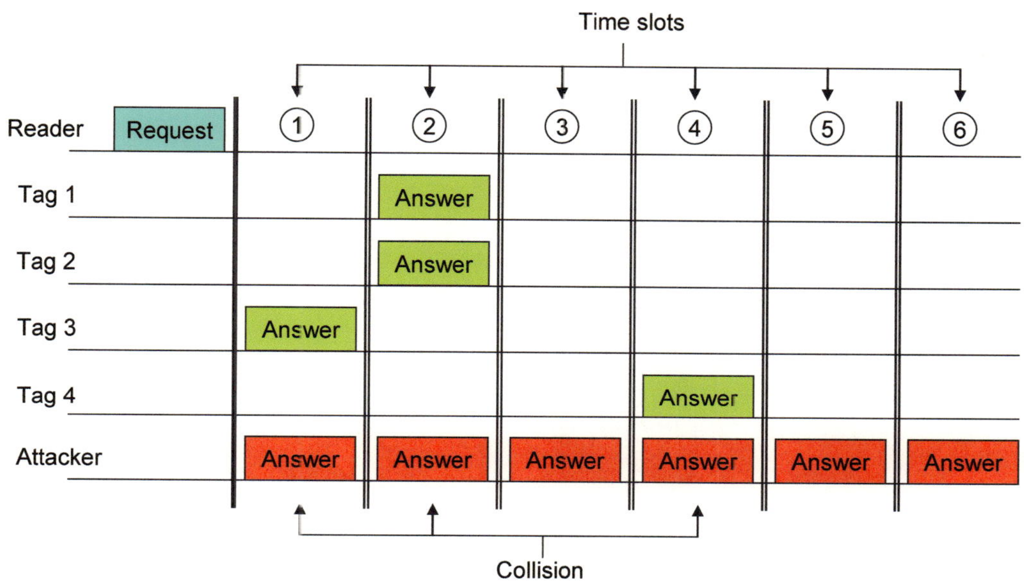

Fig. 17. Blocking protocol for Aloha method

6.2 Reader and card jamming

Jamming is an attack by denial of service which consists in emitting a signal in the same bandwidth as the reader and the card in order to blur the communication between the reader and the card. The only constraint is to flood the reader or tag signal in a higher level noise. The maximal level of emitted magnetic field is defined by ETSI (European Telecommunications Standards Institute) legislations. The ETSI EN300-330 describes a template of magnetic emission at 10 m around 13.56 MHz (ETSI 300-330). According to the following figure, it is illegal to emit more than 42 dBµA/m at 10 m in the 13.56 MHz close range frequency (Figure 18).

As a consequence, any attacker that is able to go over this limitation is sure to create an efficient jamming of an RFID reader. Exceeding the standard value does not necessarily mean that the jamming signal requires a lot of power. If the noisy emission is in the exact bandwidth of the reader signal, only few watts (1 to 2 W) are enough. To blur a tag signal is even easier since its signal is much lower than the reader's.

6.3 Shielding and Faraday cage

Magnetic field can be blocked or dramatically reduced by the process of shielding. It consists in confining an object in a metallic sheet with properties able to stop electromagnetic waves. To prevent the reading of a contactless card by a reader, its owner can insert it in a specific wallet made of metallic sheet (Figure 19). This wallet plays the part

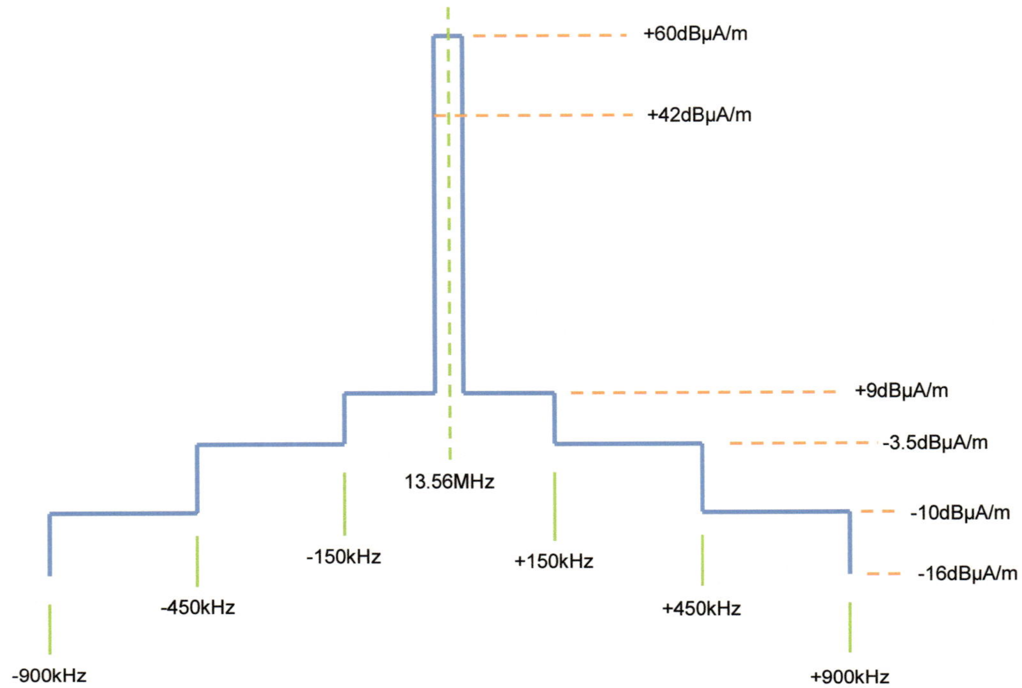

Fig. 18. ETSI EN300-330 13.56 MHz: Maximal magnetic field at 10m from the reader

of a Faraday cage blocking all HF and UHF radio signals of readers. This solution is a counter-measure to avoid the spying of confidential data recorded on cards by a false reader. However, this solution can be used to make denial of service. For example, an attacker can pass through detection system by shielding RFID EAS (Electronic Article Surveillance) to not pay its purchases.

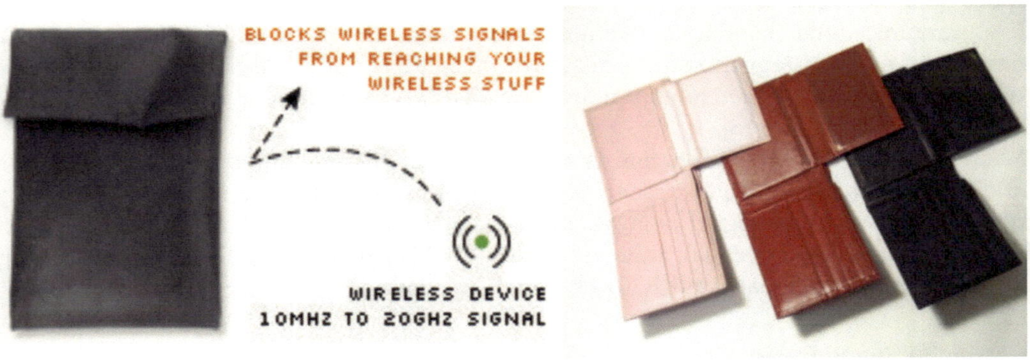

Fig. 19. Such systems block radio communications in the frequency range 10MHz-20GHz (MobileCloak), (DIRFwear)

6.4 Destruction or deactivation

This attack consists in making definitively unusable a contactless card, but it can concern the reader, too. Although this attack threats contactless system availability, it is different from the denial of service attack because once performed, it is irreversible. Destruction is considered as an attack when it is practiced without the holder's consent but it is privacy protection if a card is definitively destroyed with the consent of its holder in order to protect its data from any future attacks. Only physical destructions are treated in this chapter dedicated on physical attacks.

6.4.1 Destruction by electromagnetic field

In the same way that a chip can be damaged when powered by a forbidden voltage, the chip of a transponder can be destructed. In fact, the chip of a card is powered by the magnetic field on the transponder antenna. The chips could not resist to a high level magnetic field. This attack is mainly efficient with inductive loop contactless devices. If the inducted voltage in the loop antenna exceeds a certain value, then the chip can be definitely damaged. This value depends on chip features and on its developers but a 12A/m magnetic field is generally large enough to destruct RFID chip. The strength of this attack is that there is no absolute protection against destruction by electromagnetic field.

Protections, such as Zener diodes or self-heating fuses, can be integrated. However, their protection effect is limited by the available chip area. This attack can be very dangerous in the case of contactless passport or ID card destruction. On the other hand, generating a strong magnetic field requires a large instrumentation and a significant power (BSI, 2004). Then, an attacker is not able to destruct one or many transponders in the same time if they are far away from him.

Fig. 20. Examples of RFID transponders killer (RFIDzapper (EN), 2006), (RFIDzapper)

Building its own RFID transponders destroyer is not a challenge because many websites present schematics and practical implementations. For example, the RFID zapper is a home-made device able to destruct or definitely deactivate close RFID chips. It generates strong electromagnetic pulses by using the electronic of a camera flash and of a power generator. The principle is simple: the generator charges a large value capacitor and the flash is substituted by a simple coil. The discharge of the capacitor generates a strong concentration of energy comparable to an EMP (Electro Magnetic Pulse). The RFID zapper is only an example of devices found on the web but there exists complete devices with tag detector (Figure 20).
The microwave is another way to destruct RFID tags but it can hardly be used by an attacker because this attack needs the access of the cards he want to kill.

6.4.2 Mechanical or chemical destruction

A transponder is usually composed of an antenna and a chip: these two elements can be viewed as two mechanical pieces. The first is just made of a thin copper strip and the chip is fragile and protected by a plastic packaging. It exists many solutions to destruct the transponder. The major part of methods leaves visible signs of damage and needs the access of the card. For example, the inductive antenna is vulnerable; an attacker can easily cut it with scissors. As the antenna is essential to recover power and data, the transponder does not run anymore with a cut antenna. However, an attacker can also destruct a transponder without giving clue on an intentional damage. A solution is to cut the antenna very close to the chip or to hit the chip with a hammer in the case of an unpackaged tag (Instructables, 2008).

7. Substitution, counterfeiting, and replay attack

These three attacks are described in the same part because of they have mainly the same principles. All these attacks need the steal of data from another card. Skimming or eavesdropping attacks allow the dumping of main information recorded in the memory of the attacked chip. In this case, the attacker can store the stolen data in a blank contactless card to have a clone of the attacked card. Write data on a blank tag is not a challenge as anybody can easily buy on the Internet any card from any manufacturer that has a microprocessor which can be easily programmed (DN-Systems, 2007 as cited in Mitrokotsa & al., 2008). Another way is the steal of a contactless card directly on its owner by using violence. In the two cases, the attacker obtains a card with new UID and recorded data. These leads to the three attacks: substitution, counterfeiting and replay attacks.
It is also possible to substitute a RFID tag on an item with another tag of a cheaper item or with a tag that is totally reprogrammed with an aforementioned chip
The replay attack consists to send to a true contactless card or reader, sequences recovered during an eavesdropping or skimming session with a true reader or card. This attack allows the possibility to have transactions without a true card or reader. Grunwald shows its skills at cloning an e-passport, that means a terrorist can pass a frontier with a fake identity (Zetter, 2006).

8. Conclusion

This chapter presents a comprehensive overview of the main attacks on the HF physical layer of contactless technologies. These technologies use radiofrequency waves to transmit data and power between two devices. This communication medium can be exploited by an attacker and then reduce the security and privacy of the card user. The growth of this technology in critical domains such as payments is slowed down because of all the threats which can occur on the card and its user. They do not trust in this technology since it is not secure in terms of confidentiality and privacy of data exchange. Payment smartcards were already a potential source of attacks but the user had a complete control thanks to use of a confidential code and the insertion of the card in the cash dispenser. This weakness can be used in the case of the skimming attack to activate a card as a valid user without the agreement of the card owner. The use of radiofrequency waves disables the management of the communication by the card user. In addition to that, the propagation of the electromagnetic waves in the air enables attacks like eavesdropping. Contactless systems, according to ISO14443-A and B standards, have an operating distance of a dozen of centimetres. However, the experiments and the state of art show that an attacker can eavesdrop a private communication from a reader to a card up to 20 metres. A communication from the card to the reader can be listened at a distance less than 5 metres. Moreover, the use of metallic materials, cables, walls and doors is a great opportunity for attackers as the metal acts as antennas. The major threat is an attack combining the basics of the eavesdropping and skimming attacks. An attacker can create a communication relay between two contactless devices which are not in their communication range. This attack can be used with smartcards but is powered by the use of the air medium. Although the operation distance is close to the dozen of centimetres, an attacker can establish a communication between two devices without distance limit. Basic relay attacks are developed to point out that relay attacks can add delay less than 1µs. The weakness of the air medium is used in a denial of service of the devices to blur the communication by causing fail the anti-collision protocols or by flooding the reader and the card signals. The destruction of the devices or their placement in a faraday cage can also be considered as a denial of service. Most of these attacks can reveal secret information which can introduce attacks like replay or cloning attacks. This overview permits to show the main threats of this technology in terms of security and privacy. It is necessary to develop methods and technologies in order to make these systems more secure and their use more confident.

9. References

BSI (2004). Security Aspects and Prospective Applications of RFID Systems; Federal Office for Information Security, 2004

Carluccio D., Kasper T. & Paar C. (2006). Implementation details of a multipurpose ISO 14443 RFID-tool, Proceedings of RFIDsec'06 Workshop on RFID Security, Graz, Austria July 12-14, 2006

DIFRwear , DIFRwear's RFID Blocking Products, 17.02.11, Available from: http://difrwear.com/

DN-Systems (2007). BBC Reports on Cloning of the new e-passport, 17.02.11, Available from: http://www.dnsystems.de/press/document.2007-01-04.2112016470,

ETSI300-330. Electromagnetic compatibility and Radio spectrum Matters (ERM); Short Range Devices (SRD); Radio equipment in the frequency range 9 kHz to 25 MHz and inductive loop systems in the frequency range 9 kHz to 30 MHz, 17.02.11,

Available from: http://www.etsi.org/website/Technologies/ShortRangeDevices.aspx

Finke T., Kelter H. (2004). Abhörmöglichkeiten der Kommunikation zwischen Lesegerät und Transponder am Beispiel eines ISO14443-Systems, BSI

Finkenzeller K. (2003). RFID Handbook Fundamentals and Applications in Contactless Smart Cards and Identification 2nd Ed., John Wiley & Sons, Ltd,pp. 434 , ISBN: 0-470-84402-7, Munich, Germany, 2003

FOIS(Federal Office for Information Security) (2004). Security Aspects and Prospective Applications of RFID Systems

Guerrieri J. & Novotny D. (2006). HF RFID eavesdropping and jamming test, Electromagnetics division and Electrical Engineering Laboratory, NIST Internal Report 818-7-71

Halvac M. & Rosa T. (2007) A Note on the Relay Attacks on e-passports: The Case of Czech e-passports, IACR Cryptology ePrint Archive, Report 2007/244

Hancke GP. (2005). A Practical Relay Attack on ISO 14443 Proximity Cards, In: Gerhard Hancke homepage, 17.02.11, Available from: http://www.rfidblog.org.uk/research.html

Hancke GP. (2006). Practical attacks on proximity identification systems, Proceedings of S&P'06 IEEE Symposium on Security and Privacy, ISBN 0-7695-2574-1, Oakland, California, USA, May 21-24, 2006

Hancke GP. (2008). Security of Proximity Identification Systems. PhD thesis, University of Cambridge, Cambridge, United Kingdom, February, 2008

Hancke GP. (2008). Eavesdropping Attacks on High-Frequency RFID Tokens, Proceedings of RFIDsec'06 Workshop on RFID Security, Budapest, Hungary, July 9-11, 2008

Hancke GP., Mayes K. & Markantonakis K. (2009). Confidence in Smart Token Proximity: Relay Attacks Revisited, Elsevier Computers & Security, Vol. 28, Issue 7, pp. 615-627

Hancke GP. (2010). Some pratical results and discussion of related industrial and academic work on eavesdropping and skimming attacks, Journal of Computer Security

Hoshida J. (2004), Tests reveal e-passport security flaw, In: EETimes, 17.02.11, Available from: http://www.eetimes.com/showArticle.jhtml?articleID=45400010

Instructables (2008). How to block/kill RFID chips, In: Instructables, 17.02.11, Available from: http://www.instructables.com/id/How-to-blockkill-RFID-chips/step4/How-to-kill-your-RFID-chip/

ISO/IEC14443-2 (2001). Contactless integrated circuit(s) cards -- Proximity cards -- Part 2: Radio frequency power and signal interface, ISO (International Organization for Standardization), Geneva, Switzerland

ISO/IEC14443-3 (2001). Identification cards -- Contactless integrated circuit(s) cards -- Proximity cards -- Part 3: Initialization and anticollision, ISO (International Organization for Standardization), Geneva, Switzerland

ISO/IEC10373-6 (2011). Identification cards -- Test methods -- Part 6: Proximity cards, ISO (International Organization for Standardization), Geneva, Switzerland

ISO/IEC9798-3 (1998). Information technology -- Security techniques -- Entity authentication -- Part 3: Mechanisms using digital signature techniques, ISO (International Organization for Standardization), Geneva, Switzerland

Juels A., Rivest R. & Szydlo M. (2003). The Blocker Tag: Selective Blocking of RFID Tags for Consumer Privacy, Proceedings of CCS 2003 Conference on Computer and Communications Security, pp. 103–111, ACM Press, Washington, USA, October, 2003

Kfir Z. & Wool A. (2005). Picking virtual pockets using relay attacks on contactless smartcard systems, Proceedings of SecureComm 2005 Conference on Security and Privacy for Emerging Areas in Communication Networks , pp. 47-58, ISBN 0-7695-2369-2, Athens, Greece, September 5-9, 2005

Kirschenbaum I. & Wool A. (2006). How to Build a Low-Cost, Extended-Range RFID Skimmer, IACR Cryptology ePrint Archive, Report 2006/054

Lishoy F., Hancke GP., Mayes K. & Markantonakis K. (2010). Practical NFC Peer-to-Peer Relay Attack using Mobile Phones, Proceedings of RFIDSec'10 Workshop on RFID Security, pp. 35--49, ISBN 978-3-642-16821-5, Istanbul, Turkey, June 7-9, 2010

Malherbi-Martins R., Bacquet S., Reverdy J. (2010). Multiple loop against skimming attack. Proceedings of Fifth International Conference on Systems and Networks Communications (ICSNC) , ISBN: 978-1-4244-7789-0, Nice, France, 22-27 Aug. 2010

Mitrokotsa A., Rieback M. R. & Tanenbaum A. S. (2008). Classification of RFID Attacks, IWRT, June 2008

MobileCloak, The off switch for "always on" mobile wireless devices, spychips, toll tags, RFID tags and technologies, 17.02.11, Available from: www.mobilecloak.com

Oren Y. and Wool A. (2009). Relay Attacks on RFID-Based Electronic Voting Systems, IACR ePrint, August 2009

Reinhold P. (1993) Elektrotechnik 1–Felder und einfache Stromkreise, 3rd edn, Springer-Verlag, Berlin/Heidelberg, ISBN 3-540-55753-9, 1993

RFIDzapper . 17.02.11, Available from: https://wiki.c3le.de/wiki/RFID-Zapper

RFIDzapper(EN) (2006). RFID-Zapper(EN), 17.02.11, Available from: http://events.ccc.de/congress/2005/static/r/f/i/RFID-Zapper%28EN%29_77f3.html

HF Antenna Cookbook. Texas Instruments, 17.02.11, Available from: http://www.ti.com/rfid/docs/manuals/appNotes/HFAntennaCookbook.pdf

Tobergte W. & Bienert R. (2007). Eavesdropping and activation distance for ISO/IEC 14443 devices, NXP White Paper, 2007

Verdult R. (2008). Security analysis of RFID tags, Master Thesis, Information Security Group (GSI), Belgium

Youbok L. (1999). Antenna circuit design, AN710, application note, microID 13.56MHz – RFID system design guide, Microchip, 17.02.11, Available from: http://www.microchip.com

Zetter K. (2006). Hackers Clone E-Passports, 17.02.11, Available from: http://www.wired.com/science/discoveries/news/2006/08/71521?currentPage=1

22

Tag Movement Direction Estimation Methods in an RFID Gate System

Yoshinori Oikawa
NEC TOKIN Corporation
Japan

1. Introduction

An RFID system is desired to be introduced in large gate management systems because it can read the ID of a large number of target objects simultaneously in the field of logistics and retail business. Especially, UHF RFID has gathered significant interest since it has the advantage of long distance reading and low cost of tags. Customers using an RFID gate system require several convenient functions. One of them is to know the tag movement direction for the purpose of recognition in warehousing or shipment for inventory management. Moreover it can check for undesirable objects or prevent theft. For this purpose, some sensors are established at the entrance and the exist side of the gate system in an existing system. Therefore the direction of movement of tags is judged by the time difference in the passing time at these sensors. For example, the future store of the Metro group used this gate system for their stock management system of the backyard system [1]. However, in these systems it is necessary to use optional expensive equipment such as several sensors.

In this chapter, an effective tag movement direction detection method is proposed in which an original tag communication system is used as much as possible without using optional equipment.

2. Estimation methods of the RF tag movement direction

It is basically necessary for the judgment of tag movement to obtain two or more time information of an object. For obtaining that information, it is common to use two sensors on both sides of the gate. This method corresponds to the Range-based method, which is a location allocation system (LAS) method using a fixed anchor [2][3][4][5]. A conventional RFID gate system using photoelectric sensors is shown in Fig.1. This gate can detect the movement direction of an RF tag by judging the difference between the two passing times at each sensor. For example, because the RF tag moves from the left side to the right side in the case of Fig.1, sensor 1 detects it in advance of the detection at sensor 2. Here, a new method of applying the Range-free method to RF tag direction detection is proposed.

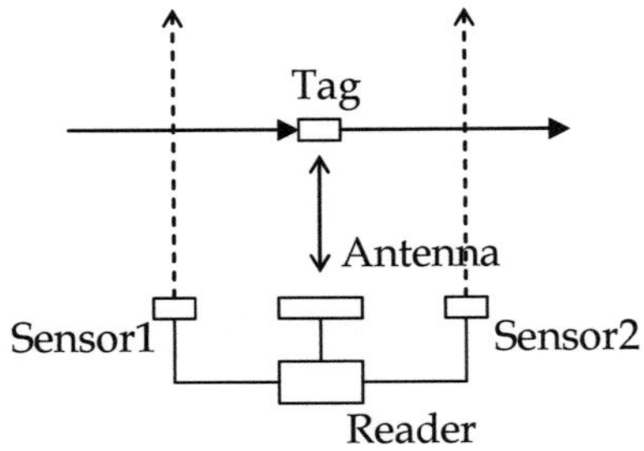

Fig. 1. Conventional RFID gate system

3. Proposed methods

3.1 Basic principle

Detection measures for measuring the time difference are considered for the RF tag and reader antennas. A double antenna method using two antennas is proposed. The configuration of this method is shown in Fig.2. The basic algorithm is that the tag movement direction is estimated by measuring the time difference of two antennas. The merit of this method is that the direction of each tag can be estimated independently. The conventional sensor system can detect only for the bulk in the case of many tags. The proposed method can estimate the movement direction for each tag even if some tags move in the opposite direction toward the other tags simultaneously.

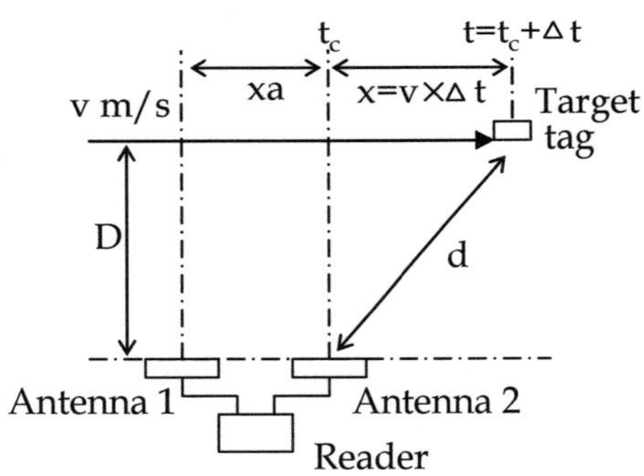

Fig. 2. Double antenna method

3.2 Attributes for estimation

The types of information obtained from a tag is the read count, received power and transmission delay. In this chapter, the former two types of information are studied because they are simpler than the last one. Three methods are considered for judgment of the detection time. They are (1) tag read time, (2) the time over the preset threshold, and (3) total judgment that considers the detection pattern or weighted time. In the case of using the time sequence pattern in the third method above, the processing function is very heavy because of complication of its algorithm. That does not match the philosophy of Range-free. Therefore, in this chapter, a weighted time center method for the third method is proposed. Each method is shown in Table 1.

Method \ Attribute	(a) Read count (n)	(b) Recived power (P_r)
(1) Tag read	$n > 1$	-
(2) Threshold	$n > Th_1$	$P_r > Th_2$
(3) Weighted center	$\sum_i (n_i \cdot t_i) / \sum_i n_i$	$\sum_i (P_{ri} \cdot t_i) / \sum_i P_{ri}$

Table 1. Decision criteria of detection time

3.3 Basic model regarding received power

A basic model of an RFID system is shown in Fig. 3. The received power of a tag (chip) P_{tr} and the received power of a reader P_r are as follows using Friss's formula [6].

$$P_{tr} = P_t + G_{rt} - L_a + G_{tr} \tag{1}$$

$$\begin{aligned} P_r &= P_t + G_{rt} - L_a + G_{tr} - L_m + G_{tt} - L_a + G_{rr} \\ &= P_t + (G_{rt} + G_{rr}) + (G_{tr} + G_{tt}) - 2 \cdot L_a - L_m \end{aligned} \tag{2}$$

$$L_a = 20 \cdot \log\left(\frac{4\pi}{\lambda} \cdot d\right) \tag{3}$$

Here, G_{rt} and G_{rr} is the transmission gain and received gain of the reader antenna, G_{tt} and G_{tr} is the transmission gain and received gain of the tag antenna, L_m is the internal loss of the tag, L_a is the propagation loss in the air, d is the read distance, λ is the wavelength. Generally, the antenna of an RFID system can be used for both transmission and reception. Therefore, let $G_r=G_{rt}=G_{rr}$, $G_t=G_{tt}=G_{tr}$, then eq. (1) and eq. (2) are

$$P_{tr} = P_t + G_r + G_t - L_a \tag{4}$$

$$P_r = P_t + 2 \cdot (G_r + G_t - L_a) - L_m \tag{5}$$

Measurement results of P_r in the case of P_t=1W(30dBm), G_r=6dBiC(circular polarization antenna), G_t=0dBil(linear polarization antenna) are shown in Fig.4. This shows that the results are the same as the calculated values. Since the tag internal loss L_m depends on vendor or input level, the value of the actual used tag chip is applied.

Figure 4 shows that the distance (read range) between the reader and the tag can be approximately estimated by measuring P_r. In eq. (4) and (5), P_r is a maximum when the tag is just in front of the reader antenna. However, P_r decreases as the tag moves into farther from the center of the antenna because of its directional loss. Measurement results and calculated values of P_r vs. the distance x between the center of the reader and tag are shown in Fig.5. From Fig.5, the tag's nearest point (x=0) to the reader can be estimated.

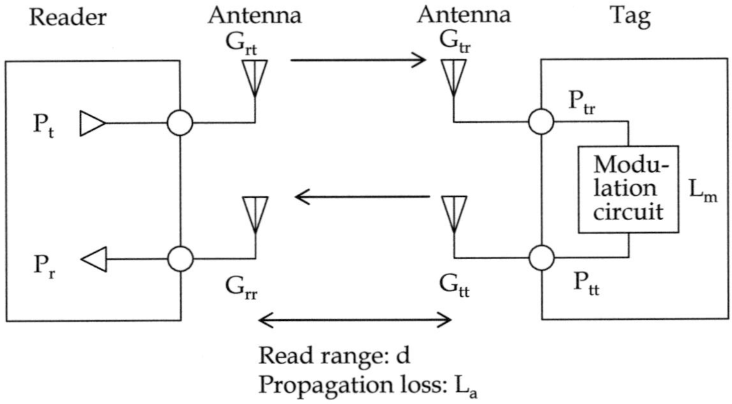

Fig. 3. RFID system model

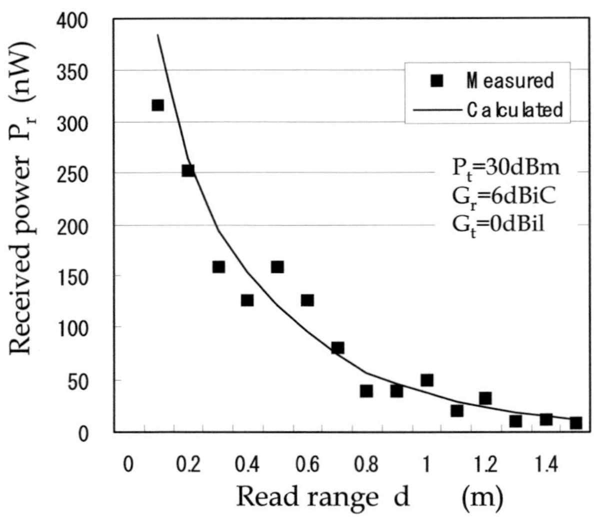

Fig. 4. Read range vs Received power

3.4 Comparison of detection methods

3.4.1 Method 1

In Method 1, the starting time to read a tag is detected as shown in Table 1(1) even if read only one time. In an actual RFID system, because tags are inventoried in advance of reading the tag, the inventory time can also be used. This method is so simple. However, it is hard to increase the decision accuracy since it sometimes happens to inverse the sequence of the read time of the two antennas.

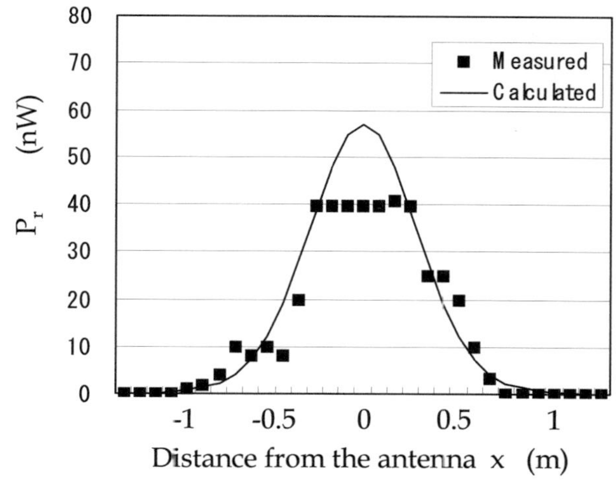

Fig. 5. Measurement result (x vs P_r)

3.4.2 Method 2

Incorrect judgment sometimes occurs due to a passing read for a reflected RF wave in the case of Method 1.

Method 2 uses the threshold of detected values and judges the direction using the time difference between each time when the detected value is over each threshold as shown in Table 1(2). This method is able to increase the accuracy of detection. However, it is sometimes hard to decide the threshold because the read count depends on the speed of movement and the received power depends on the distance between the reader antenna and the tag.

3.4.3 Method 3

Method 3 is proposed for improvement of the two methods, i.e. prevention of tentative read error caused by the influence of reflection or null points. The principal of this method is to estimate the time of the tag's nearest position from the reader antenna. Wilson has proposed the method for localization using the passive tag count percentage [7]. In this approach, tags can be estimated the closest position by detecting the peak point. However, it is difficult to adopt this method as RFID gate system because the variation of detected value reaches up to several tens of meters and is equivalent to the distance between two antennas. Therefore the algorithm we proposed is that each read time is weighted by the read count n or received power P_r, and the tag direction is estimated by the calculated difference between two weighted centers of two antennas. Recently, RFID readers become to have high-performance received power detection function [8]. Therefore, here, this method will be explained using the received power as the tag attribute. Figure 6 shows the judgment procedure of the three methods.

The detailed detection method is explained in Method 3. The received power is a function of time actually because the tag goes through at a speed of v (m/s).

Eq.(5) is shown as eq.(6) from Fig.2 and Fig.5. Δt in Fig.2 is the time deference between the passing time at the front of the reader antenna (t_c) and the present time (t).

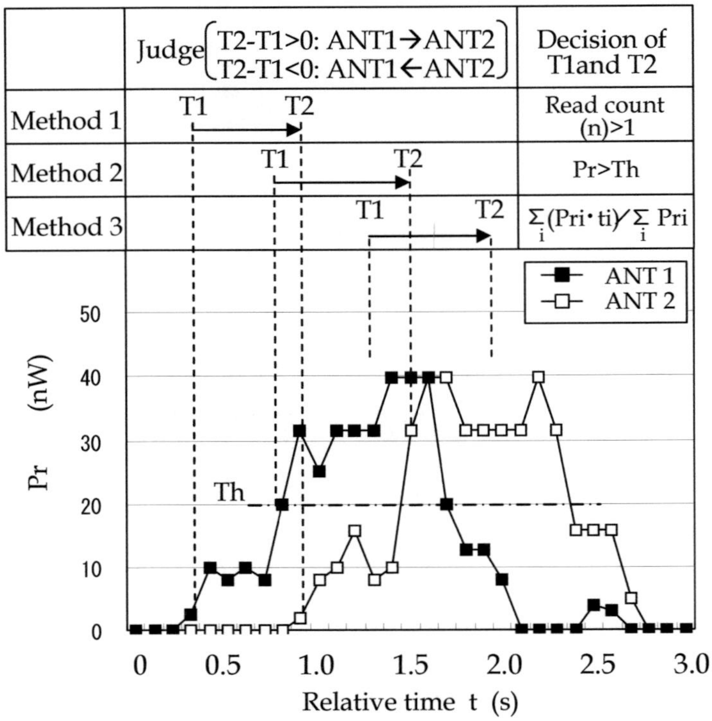

Fig. 6. Three methods in double antenna method

$$P_r(t) = P_t + 2\cdot(G_r(t) + G_t(t) - L_a(t)) - L_m \tag{6}$$

The estimation procedure is as follows.

When the certain time before the reader starts to read tags put t_0, weighted center of read time $t_{w1}(t_k)$ and $t_{w2}(t_k)$ from time t_0 to time t_k of antenna 1 and antenna 2 are

$$t_{w1}(t_k) = \frac{\sum_{i=0}^{k} P_{r1}(t_i)\cdot t_i}{\sum_{i=0}^{k} P_{r1}(t_i)} \tag{7}$$

$$t_{w2}(t_k) = \frac{\sum_{i=0}^{k} P_{r2}(t_i)\cdot t_i}{\sum_{i=0}^{k} P_{r2}(t_i)} \tag{8}$$

where $P_{r1}(t)$ and $P_{r2}(t)$ are the received power of the two antennas at time t.

In the eq.(7) or (8), when $t_{w2}(t_k)-t_{w1}(t_k)>0$, it is judged that the tag moved from antenna 1 to antenna 2, and when $t_{w2}(t_k)-t_{w1}(t_k)<0$, it is judged that the tag moved from antenna 2 to antenna 1. The calculated results in the case of Fig.6 is shown in Fig.7.

When t_{w1} and t_{w2} in the case of stable values after the elapse of a certain period of time put T1 and T2, respectively, the tag direction is finally judged by T2-T1 as shown in Fig.6 .

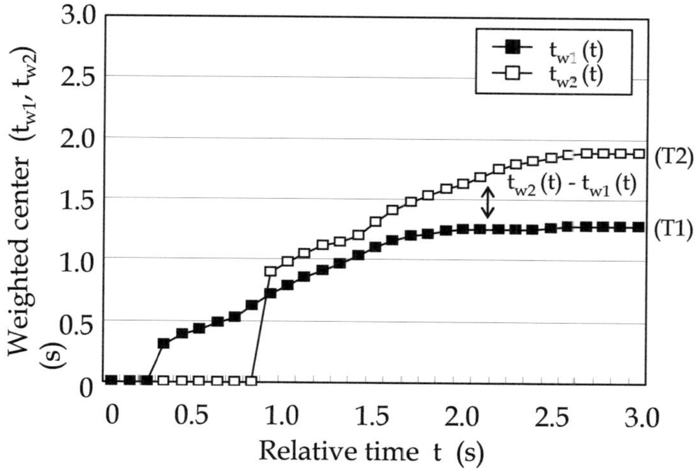

Fig. 7. Shift with time of t_{w1} and t_{w2} in Method 3

Measurement results and the experimental environment using 10 dense tags are shown in Fig.8 and Fig.9. Measurement conditions are shown below.
P_t=30 dBm, G_r=6 dBiC, G_t=0 dBil, D=90 cm, xa=60 cm, v=1 m/s, height of antenna=1.3 m, data rate=80 kbps, Reader: NEC TOKIN (Speedway)
Tags: UPM Raflatac ShortDipole
movement direction: from antenna 1 to antenna 2 (T2-T1>0)
Because the distance between two antennas that are the same type is 60 cm, T2-T1 becomes 0.6 seconds in theory. There are occasional erroneous decisions because of reflection or interference in severe measurement environment, which causes undesirable reading in method 1, and tags placed in the middle (e.g. tag #3, #4, #7 and #8 in Fig.8) are hard to read in method 2.
On the other hand, method 3 is very stable because it is not misjudged, has low deviation and a desirable average. Figure 10 shows the time transition of the difference t_{w2}-t_{w1} in method 3. We can see this method can obtain a stable and correct result (expectant value in the case of Fig.10 is 0.6s) even in the case of misjudgments caused by reflection and interference in the measurement stage.

4. Measurement results in Method 3

4.1 Detection of the tag direction

The detail performance of Method 3 was measured. Figure 11 shows the tag read counts and time difference T2-T1 in the case of two methods.
Though the deviation is wider in the case of a low read count, the judgment result is plus in pattern 1, and minus in pattern 2. Therefore it has enough stability for use as an actual tag direction decision tool. Pattern 3 shows the results in the illegal case assuming turning back in the center of the antenna. In this case, the expectation value is 0. Figure 12 shows the summary of means m and deviation σ of the measurement results of Fig.11.
By the way, an RFID system needs anti-collision technology that prevents no-read situations caused by collision when many tags are read simultaneously. The sequence to read tags is

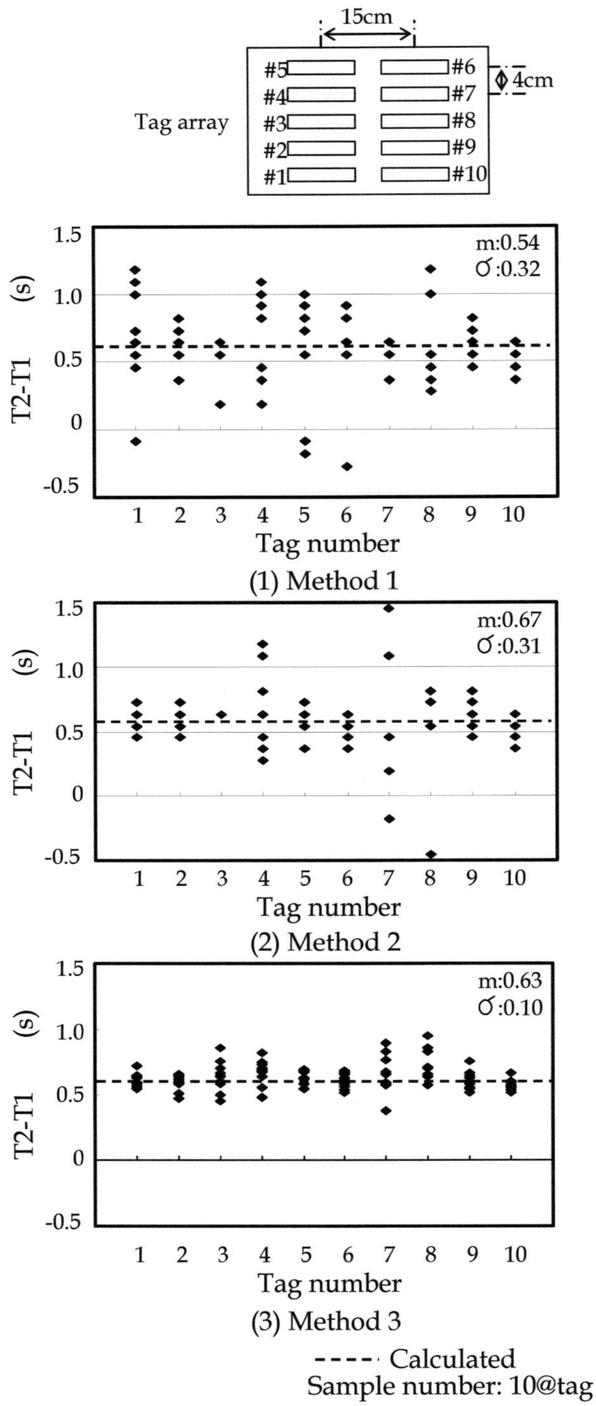

Fig. 8. Different time between two antennas

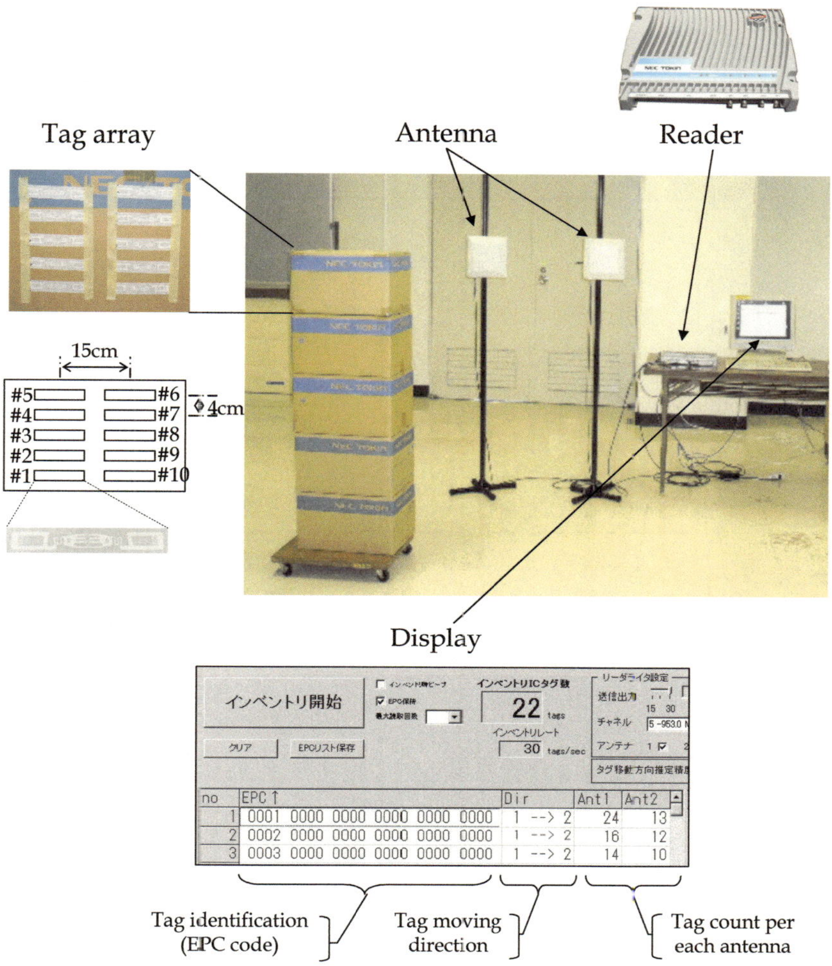

Fig. 9. Photograph of experimental environment

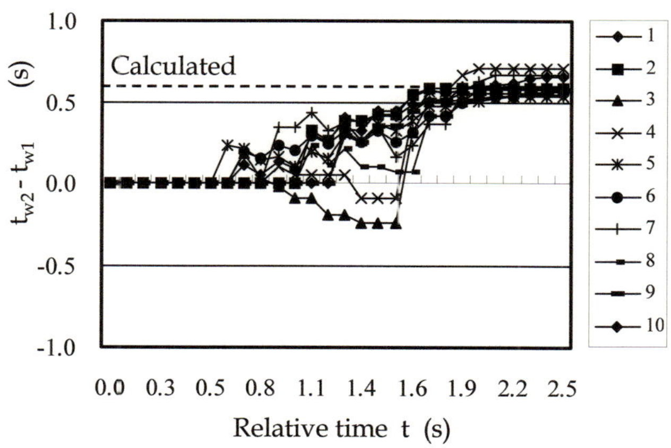

Fig. 10. Relative time vs (t_{w2}-t_{w1}) in method 3

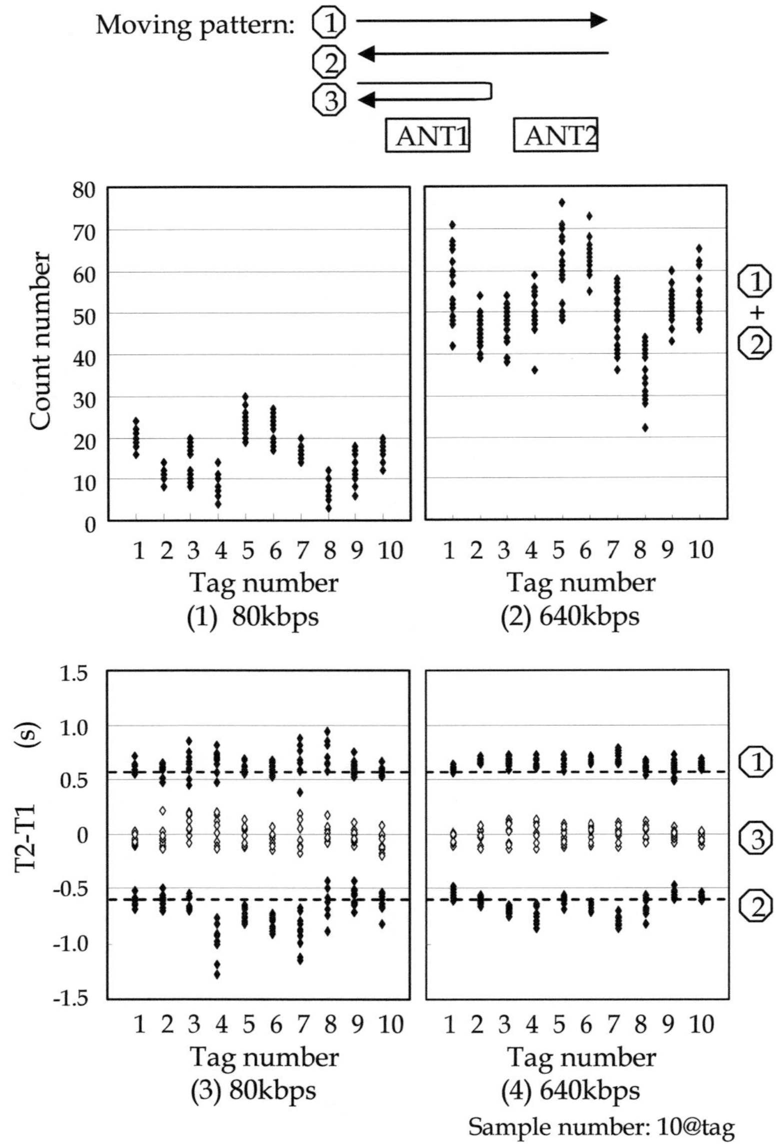

Fig. 11. Relative time vs (t_{w2}-t_{w1}) in method 3

random because a typical anti-collision system is used for using the probabilistic approach [9]. A variation of tag read sequence directly becomes a validation of detection time difference. Therefore, when a weighted center is normally-distributed, the time difference T2-T1 is also independent and identically distributed because of its reproducing property.

From Fig.12, it is assumed that the criteria of detection precisely is 3σ or less, and the tag direction can be judged correctly from the data of pattern 1 and pattern 2. However, a data rate up to round 640kbps is necessary when the difference from abnormal action such as turning needs to be detected (pattern 3 in Fig.12).

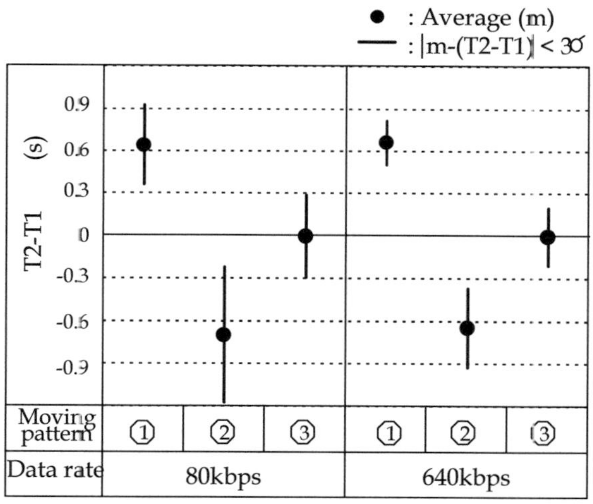

Fig. 12. Measurement results of (T2-T1)

4.2 Estimation of the tag moving speed

The detail performance of Method Moreover, the speed of movement can be also estimated by measuring the time difference T2-T1 because the distance between two antennas is fixed. Figure 13 shows the measurement results of the movement speed.

Variation of measurement results in the case of v=2m/s is larger than in other cases because the precise speed is inversely proportional to the speed of movement of the measurer. Figure 13 shows that this method can estimate not only the tag direction but also the speed of movement. It is very useful to set the threshold Th of movement speed as the decision criteria in order to increase the accuracy. For example, when the threshold Th1 and Th2 are set to -3.5 and 3.5 respectively, it is possible to eliminate abnormal movement such as turning in Fig.13.

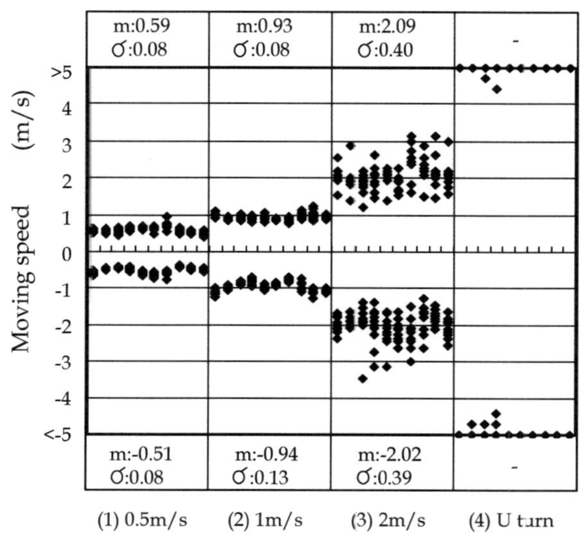

Fig. 13. Measurement results of moving speed

4.3 Effect of the orientation of the tag

Generally, a tag are used a liner polarized dipole antenna in consideration of read range and cost. In this case, the read performance in reader depends on the orientation of the tag. The tag movement detection results of the time difference T2-T1 in three cases is shown in Fig.14.

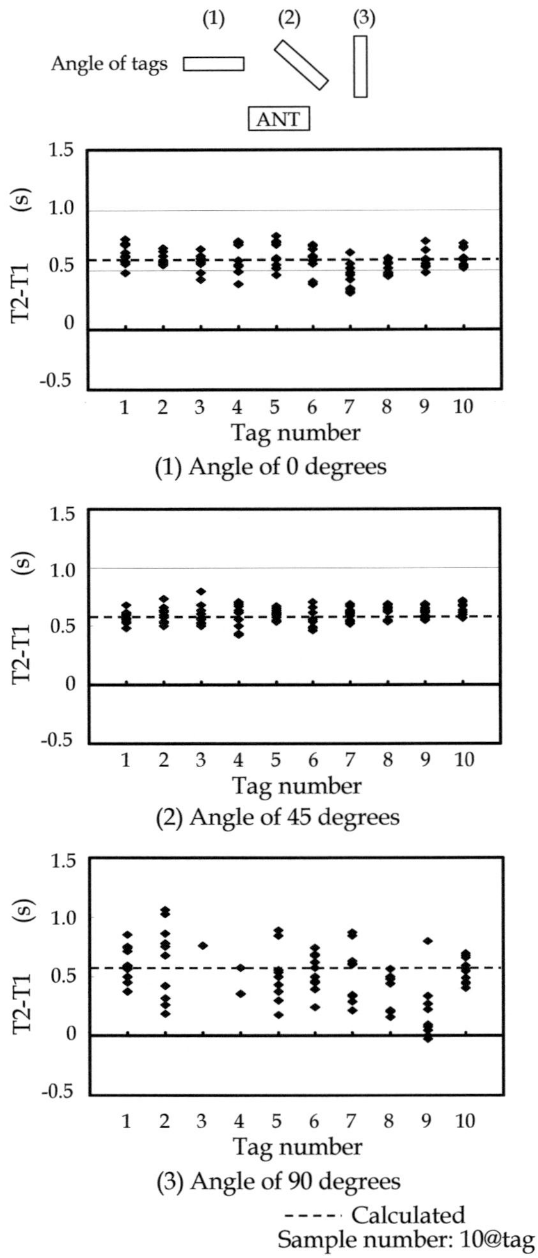

Fig. 14. Different time between two antennas

T2-T1 of the tags that have 90 degrees angle against the reader antenna ((3) in Fig.14) varies widely because they are hard to be read. The percentage of read in this case was 79% and the accuracy among tags to be read was 95%. However, when tags set 45 degrees angle, the movement direction of tags can be detected with as high accuracy as a parallel case ((1) in Fig.14). In other words, it is useful to tilt two antennas of the reader in place of tags.

4.4 Effect of the intersection of the tags

In actual cases, it may happen that two tag groups pass through in the opposite direction individually and simultaneously. The measurement results in that case are shown in Fig.15.

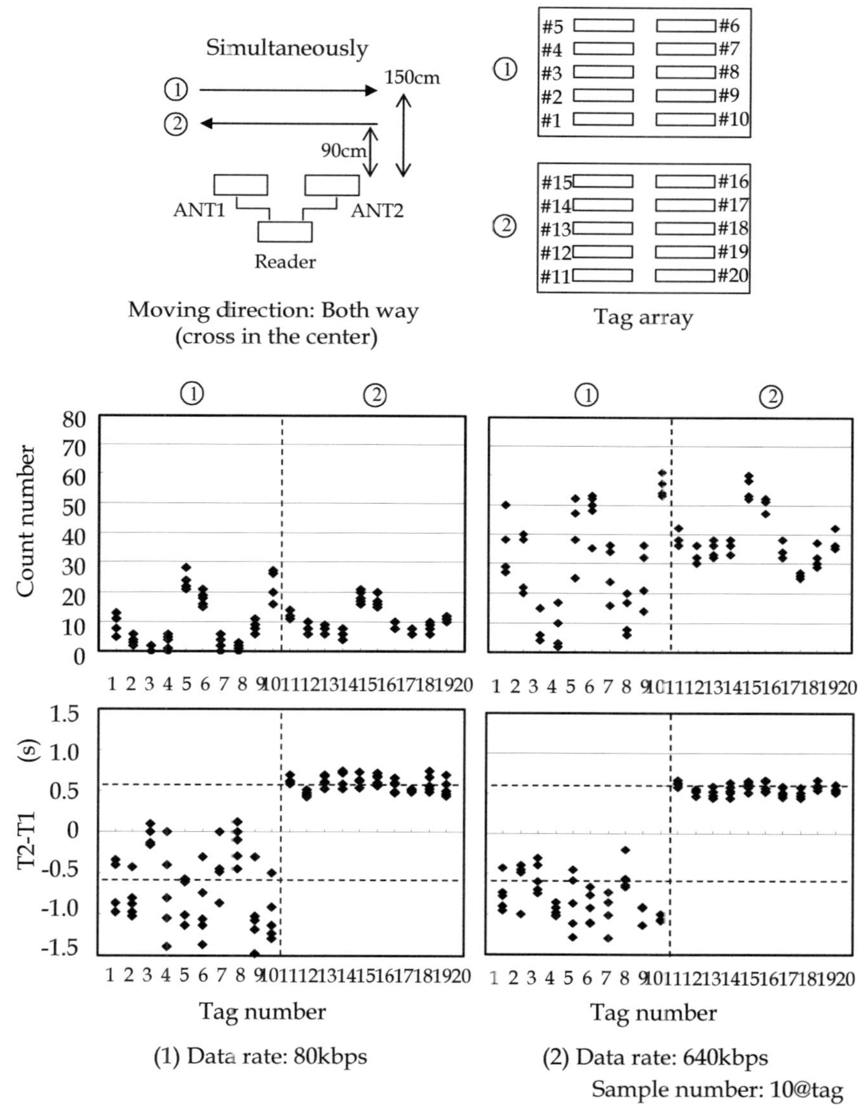

Fig. 15. Measurement results in simultaneous cross moving

One tag group (#1-#10) passed through from antenna 1 to antenna 2, and the other group (#11-#20) passed through in the opposite direction behind the former group. Tag group (#1-#10) has the same characteristics as in Fig.11. However, tag group (#11-#20) is strewn widely because the radio wave is blocked by the other tag group in passing in front of the reader antenna. In the case of 80kbps data rate, 14% of this tag group could not be read and around 5% among all the read tags made an error (that is, the accuracy was about 95%). However, when the data rate is 640kbps, both of the read rate and the accuracy are 100%. Therefore, this method is useful because the tag moving direction can be detected correctly by increasing the data rate even if the most severe case like intersection in front of the antenna.

5. Conclusion

In this chapter, a method for precisely estimating the tag movement direction in an RFID gate system was proposed. This method uses the time difference between two antennas of the reader. This method has the advantage of being able to judge tag direction individually even when there are some tags moving to the reverse direction. Especially, when it uses the proposed algorithm of the weighted center of passing time, the precision of the estimation can be increased. Finally, the feasibility of the method was proved by measurement results.

6. References

[1] "Metro Future store"http://www.rfidjournal.com/article/view/889

[2] H. Ochi, S. Tagashira and S. Fuita, "A localization system for wireless sensor networks," IPSJ SIG Tech. Rep., ARC-160, pp.17 22, Dec.2004

[3] T. He, C. Huang, B. John, A. Stankovic and T. Abdelzaher, "Range-free localization schemes for large scale sensor networks." Mobicom, September 2003, pp.81 93

[4] J. Hightower, G. Borriello and R. Want, "SpotON: an indoor 3D location sensing technology based on RF signal strength" UW CSE Tech. Report #2000-02-02, February 2000

[5] L. Ni, Y. Liu, Y. Lau and A. Patil, "LANDMARC: indoor location sensing using active RFID" Wireless Networks 10, pp.701 710, Kluwer Academic Publishers, Netherlands, 2004

[6] T. Yoshikawa, "Radio engineering B," Tokyo Denki University

[7] P. Wilson, D. Prashanth and H. Aghajan, "Utilizing RFID signaling scheme for localization of stationary objects and speed estimation of mobile objects" International conference on RFID, pp.94 99, March 2007

[8] Y. Oikawa, "UHF IC tag and reader/writer products" NEC Tech. Journal, vol.2, No.4, pp.76 80, December 2007

[9] Y. Kawakita, J. Mitsugi, O. Nakamura and J. Murai, "Acceleration of UHF-band RFID inventory leveraging capture effect." IEICE, vol.J91-B, No.10, pp.1279 1286, October, 2008

23

Third Generation Active RFID from the Locating Applications Perspective

Eugen Coca and Valentin Popa
Faculty of Electrical Engineering and Computer Science
Stefan cel Mare University of Suceava
Romania

1. Introduction

Location systems, both for indoor and outdoor use, are rapidly developing due to the practical need of knowing the position of objects and persons (Harrop, 2008). If for the outdoor world, the GPS system and its variants (DGPS, etc.) is the best possible solution, for indoor use, things are not yet completely solved. Indoor GPS is developing, but in parallel, other projects are running. The vast majority of papers dealing with the subject (Bess, 2009; Chang et al., 2011; Goncalo, 2009; Kathiravan et al., 2009; Khan & Antiwal, 2009; Jeon et al., 2010) present systems based on RF signal measurements. Multiple ways of solving the problem are technically imaginable, starting with those using the signals emitted by the nodes of a common WLAN / Wi–Fi wireless network (Bal et al., 2009; Clulow et al., 2006; Kaemarungsi & Krishnamurthy, 2004; Kushki et al., 2006; Kwon & Song, 2008; Tsui et al., 2010; Yousef & Agrawala, 2005), continuing with RFID systems, WSN networks and finishing with proprietary solutions derived from one of the above, where specialized nodes with one or more coordinators are deployed over the desired locating area (Bahl & Padmanabhan, 2000; Baunach et al. 2007; Chang et al. 2011; Coca et al. 2008; Dai & Su, 2008; Koyuncu & Yang, 2010).

RFID tags are the main factor of progress in identification application development. There are more than 40 year from the first generation (Finkenzeller, 2003), equipped with passive components where the energy is captured from the radio–frequency field generated by the reader, to the third generation where the energy supplied by a battery is used to power a microcontroller and one or several on–board sensors. In terms of price, in 2011 the passive tags may be found at prices as low as 0.05 USD each, whereas the active RFID tags equipped with complex sensors and low–power microcontrollers may cost as much as 100 USD a piece (Harrop, 2008).

From the point of view of RFID tag structure, the changes are obviously influenced by the progress in semiconductors technology. The software for the reader and applications evolved also on the same trend. For the Generation 1 UHF tags, manufacturers provide hardware with their own protocols. Therefore, tags from one specific manufacturer would only work with the RFID reader from the same manufacturer. From the point of view of users, this represents a major limitation and for large-scale implementations, single supplier solutions are not acceptable. Generation 2, the second-generation RFID UHF tags, developed in order to establish a standard for RFID tags, used by the big retailer inventory applications and operating in the ultrahigh frequency (UHF) band (860–960 MHz), offer long range operating distances (at least 8 to 10 meters). A comparison between the differences between Gen 1 and Gen 2 protocols may be found in Table 1 (EPC, 2005).

Four UHF RFID standards exist: Class 0 and Class 1 – from EPCglobal, and 18000–6 Type A and Type B from ISO. ISO's 2006 approval of Gen 2 as an 18000–6C extension opened the way to a single UHF global protocol. Such a protocol will create an open market as well as an open standard, which will force the prices to go down. Even the great efforts made in the direction of unifying the standards, a large RFID market with a strong supply chain and industrial backbone – China, has not accepted either the ISO or EPCglobal standard. Instead, China hopes to develop own standards compatible with Gen–2 tags, their readers being able to communicate with the standard tags (Bijl & Dil., 2010; Roberti, 2005; Razaq et al. 2005).

Gen 2 frequency range is from 860 to 960 MHz and it covers all international frequency spectrums. Tags that comply with EPCglobal's Gen 2 standard operate between these ranges without performance degradation. Gen 1 didn't do well in Europe due to European radio frequency spectrum allocation didn't leave enough open bandwidth for US radio frequencies, but Gen 2 offers Europe's required frequency range of 865 to 868 MHz, while fulfilling US frequency sub band of 902 to 928 MHz. The ISO 18000–6C extension makes Gen 2 a real flexible international standard (Jong & Bijl, 2010; Roberti, 2005; Razaq et al. 2005).

Description	**Gen 1**	**Gen 2**
Acceptance level	Not a global standard Global standard after an amendment in	ISO–UHF 18000–6 standard
Arbitration	Deterministic binary tree for Class 0 and deterministic slotted for Class 1	Probabilistic slotted
Anticollision/tag-sorting algorithm	Binary tree algorithm with persistent state/wake states	Q algorithm, which is a variant of the slotted aloha protocol
Air interface	Pulse width modulation (PWM) for Class 0 and Class 1	Pulse interval encoding (PIE–ASK), Miller, FM0
Data rate	40/80 Kbits for Class 0 and 70/140 bits for Class 1	40 to 640 Kbits
Distance	Less than 10 meters	Less than 10 meters
Frequency range	850-930 MHz	860 to 960 MHz
Security password	8 and 24–bit passwords, respectively, for Class 1 and Class 0	32 bits
Data write verification	No	Yes
Write speed (for 96–bit electronic product code)	Three tags per second	Minimum five tags per second

Table 1. Main differences between the Gen 1 and Gen 2 protocols

The working frequency is a key design issue in RFID locating systems. The ability for signals to propagate within crowded environments is dependent on the signal wavelength. Within warehouses, truck yards, office buildings, and other industrial or commercial facilities, the ability for an RFID system to operate in and around obstructions is critical (Han et al., 2008; Hsu et al., 2009; Jeon et al., 2010; Kiang et al., 2009). These obstructions are often made of metal, such as vehicles and metal racks, requiring signals to propagate around rather than through them. Signals propagate around obstructions by means of diffraction, and the level of diffraction is dependent on the size of the object over the signal wavelength ratio. Diffraction occurs when the wavelength approaches the size of the object. For example, at 433 MHz the wavelength is approximately a meter, enabling signals to diffract around vehicles, containers, and other large obstructions.
Regarding the frequencies used by active tags systems regulations, a summary is presented in Table 2:

Band	303 MHz	315 MHz	418 MHz	433 MHz	868 MHz	915 MHz	2400 MHz
Working frequency band	302–305 MHz	314.7–315 MHz 42 dBuA/m @10m	418.95–418.975 MHz 10 mW ERP	433.050–434.790 MHz 10mW ERP 10%	868–868.6 MHz 25mW ERP 1%	902–928 MHz	2400–2483.5 MHz
USA	x	x	x	x		x	x
Canada	x	x	x	x		x	x
UK				x	x		x
France				x	x		x
Germany				x	x		x
Netherlands				x	x		x
Singapore		x		x	x	x	
Taiwan	x	x	x	x		x	x
China		x				x	x
Australia				x		x	x

Table 2. Summary of global frequency regulations for the most common Active RFID bands

At 2.4 GHz, the wavelength is approximately 10 centimeters and diffraction is very limited with these obstructions, creating blind spots and areas of limited or even no coverage. Frequencies above 2 GHz present significant challenges for operation in crowded environments and are therefore not recommended for most RFID applications.
One may notice only 433 MHz and 2400 MHz working frequencies bands systems are allowed in almost all countries. Even both frequency bands overlap with the ISM bands, these are the most accepted in the RFID world. Despite in the 2400 MHz band there are many wireless systems (Wi-Fi, Bluetooth, ZigBee, etc.) making the frequency spectrum very crowded, producers continue to develop new systems and communication protocols working in this free band, design simplicity, small dimensions and low power consumption being solid arguments for continuing the researches.

2. RFID systems in localization applications

Real Time Locating Systems (RTLS) help users to locate and track objects in real-time. This could be done in many ways, along the time different technologies being developed around the idea. The RTLS term was introduced in 1988 to describe a technology that provided the Automatic Identification capabilities of active RFID, but added the ability to see the physical location of the tagged object.

From the locating perspective, the RFID has a long history. Conventional active RFID tags are the first used for real time locating applications. Starting with Gen 2 tags, the EPCglobal Class−1 Generation−2 UHF RFID Protocol for Communications standardized the active tags working in 860 − 960 MHz frequency band (EPC, 2005). The included battery helps them to initiate a signal, giving longer range compared to passive tags. This type of tags are mostly known for the end users as locking the cars as over two billion dollars were spent on car clicker systems to date (Harrop, 2008). Millions of other tags were also deployed in postal services monitoring applications, and supplies or assets management. Localization systems were developed for this type of active tags, an example being the RFID radar (RFID−Radar, 2005). RFID radar is a mixed localization system, based on both ToA − Time of Arrival and AoA − Angle of Arrival methods (Coca & Popa, 2007). It uses a system based on one emitting and two receiving antennas. The working principle, based on a tag-talks-first protocol (Coca et al., 2008), is as follows: when a transponder enters the area covered by the emitting antenna, it will send its ID and memory content. Two dedicated antennas receive the signal transmitted by the transponder. Based on the time difference between the two received signals and the range information, it computes the angle and the distance. The system uses a central frequency of 870.00 MHz with a bandwidth of 10 kHz. The performances in term of localization precision are modest, even the location information is available for tags placed up to 40−45 meters in front of the antennas (Popa et al., 2010).

The second generation of active RFID tags is present in the true Real Time Locating Systems (RTLS) used today for continuous monitoring applications. The active tag includes a battery used also to supply the on-board sensors and a low power microprocessor, improving the capability to store the measured data over a significant period. When using many readers, distances over several hundred meters are usually obtainable. Even these systems were initially designed for assets tracking, localization applications were position and speed information were added as a plus. In terms of location precision, it is strongly influenced by the reflections on obstacles and moving objects positioned between the reader and the tags.

The third generation of active RFID tags overlaps with the well-known Wireless Sensor Networks (WSN) or Ubiquitous Sensor Networks (USN) (Bess, 2009; Harrop, 2008). The most important characteristic of the tags is they communicate one with each other and in the same time with the central node. A central node, named also the coordinator or the gateway, pays the role of the RFID reader from a classic system. Even the maximum distance between the two nodes is limited to 10 to 30 meters (mainly due to maximum power emission regulatory restrictions but also due to attenuation, reflections and interference with other systems), the networks could be easily extended over hundreds of meters or even more, based on the inter−nodes communications capabilities.

3. Third generation RFID system in localization applications

Wi-Fi technologies, developed for delivering wireless communications between mobile terminals, are also used in locating applications by processing the identification data from multiple Access Points (AP) and the Received Signal Strength Indication (RSSI) information.

The signal strengths of received signals from at least three access points are used to determine the location of the object being tracked. To increase accuracy, more sophisticated methods use RF fingerprint maps that are based on calibrations of the strength of Wi-Fi signals at various points in a predefined area. Applications using Wi-Fi combined with Time Difference of Arrival (TDOA) techniques were also developed.

In an RSSI system, the distance between a tag and a reader is computer by converting the value of the signal strength at the reader into a distance measurement, based on the known signal output power at the tag and on a particular path loss model.

Wi-Fi location technique has some advantages over other systems:

- It uses the existing infrastructure;
- Position information is available both at the coordinator and at each node, information that could be shared with neighbor nodes.

Some major disadvantages of these systems include:

- Signal power measurements are affected by fixed and mobile objects, thus generating random measuring errors, even a power map was created for the specified measuring area;
- Network traffic congestions affect the system availability and the results;
- Power consumption is higher compared to RFID or WSN solutions.

To be effective, RSSI requires a dense deployment of Access Points, which adds considerably to the systems cost. The key problem related to RSSI based systems is that an adequate path loss model must be found for both non-line-of-sight and non-stationary environments. In practice, the estimated distances are not quite precise. RSSI locating systems may also be disqualified from security applications as an attacker can easily alter the strength of received signals by amplifying or attenuating it, or by other methods distorting the signal strength received from one more Access Points used as fixed references.

The disadvantages above made the Wi-Fi locating system not to develop as rapid as other technologies did, and positioning system solutions based on it are not widely spread in the real world.

RFID locating implementations were investigated and test setups are already used in real world applications both for indoor or outdoor locating services, even this technology was created as a bare code replacement. RFID systems were initially developed with the need of data storage in mind, and other aspects were not taken into consideration. Many efforts were done in order to modify RFID systems and make them suitable for indoor locating applications. A proprietary system derived from a RFID system (RFID Radar, 2005) is a good example for outdoor and indoor location, as only a small quantity of information is transmitted, the processing power being used for position estimation. One of the major disadvantages of such systems is the user is unable to modify the application or to write his own code due to copyright restriction. Communication protocol details are not always completely disclosed, so creating new system configurations could be a difficult task. In addition, the high power level used by the system makes it unsuitable for indoor location application or for populated areas (Coca et al., 2008).

The third generation RFID systems have the characteristics of a network of wireless sensors, the nodes being the tags. There are even no notable differences between the active RFID tags and WSN nodes, as both are powered from external energy sources, contain sensors and small data processing capabilities. This is the reason the research was focused on the WSN networks for using them in applications where standard active RFID systems were unable to deliver the required performance levels.

Wireless Sensors Networks contains nodes with one or more sensors connected with a RF transceiver. When multiple WNS nodes are deployed over an area, signals transmitted by them could easily used for location purposes. The performances in the cited literature (Buta et al., 2010; Halgamuge et al., 2009; Kim & Yang, 2008; Jeong & Nof, 2008; Kuang et al., 2008; Lanzisera et al. 2004; Mao et al., 2007; Miorandi et al., 2007, Ota & Wright, 2006) are low enough to justify future investigations. For stationary environments, especially for indoor situations, location of objects is a relatively easy task. When moving objects have to be located and eventually traced, the WSN is a challenging solution. The typical structure of a third generation RFID locating network is shown is Figure 1. The message transmitted from one node to the gateway contains the node IDentification data along with the physical coordinates.

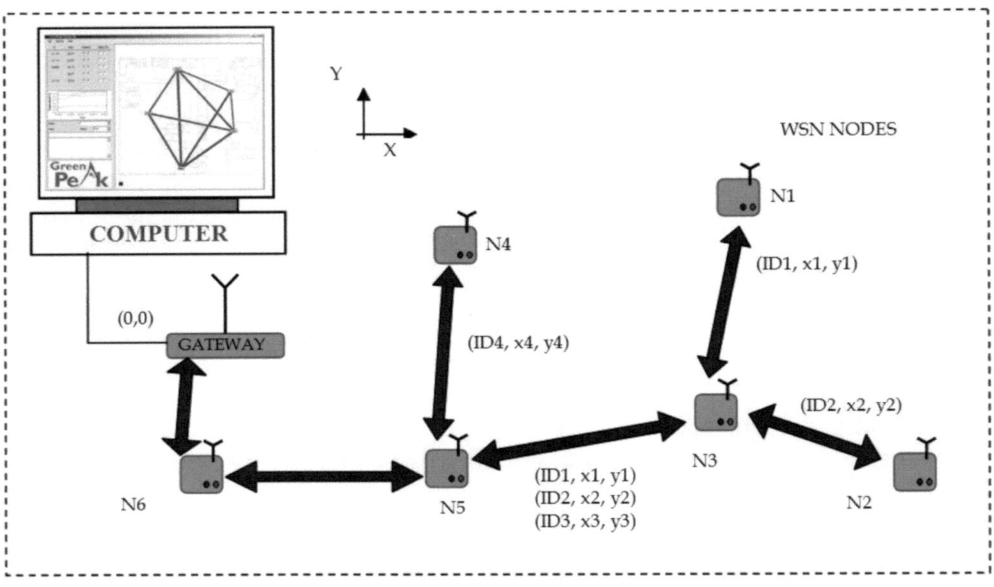

Fig. 1. Typical third generation RFID locating network overview

The position information may be obtained from on-board sensors (like GPSs, accelerometers, etc.) or may be computed from the information received from nearby nodes (RSSI is the most common information computed in order to obtain position information). The key of this architecture is the communication protocol that allows the information to be transmitted from one node to the gateway through any available path. This way in the event a node is not available, the information is routed through the healthy nodes.

4. Experimental results

4.1 Test system characteristics

For performance evaluation, we used a Wireless Sensor Network development system from Green Peak Technologies (GreenPeak, 2010). The system consists of a coordinator node (known as Gateway) and nine nodes, operating in the ISM band (2.4 GHz), with 16 channels and 250 kbps data rate and is certified to meet EN 300 440 (Europe), FCC CFR47 Part 15 (US) and ARIB STD–T66 (Japan) standards. The node architecture is presented in Figure 2:

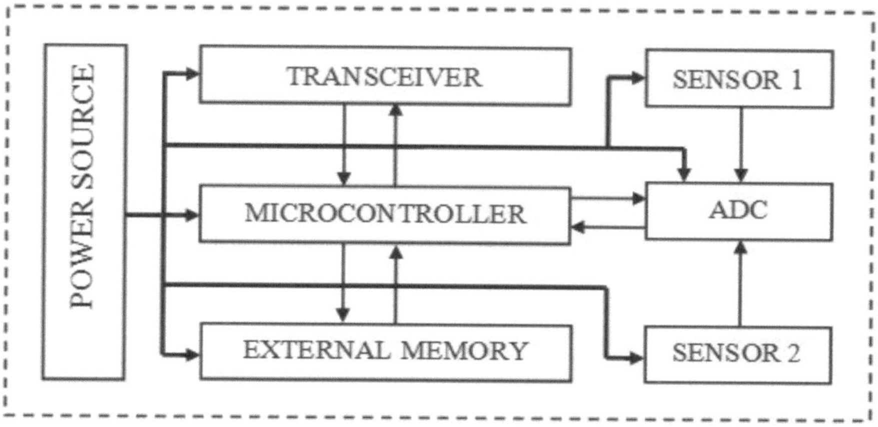

Fig. 2. WSN node architecture

The node is built around an Atmel AVR 1281 microcontroller and powered by 3 AAA batteries (Figure 3). On the board, there are temperature and humidity sensors, analog and digital inputs. In complex applications, the node may be upgraded to support a more powerful processor and multiple inputs.

In terms of operating distance, the typical values declared by the producer vary from 40–100 meters indoor, to 160–400 meters outdoor and up to 1000 meters outdoor in light–of–sight view. In the presence of blocking objects, shorter ranges are expected. The gateway is equipped with a RISC processing unit and a RF module, very similar with the one on the node.

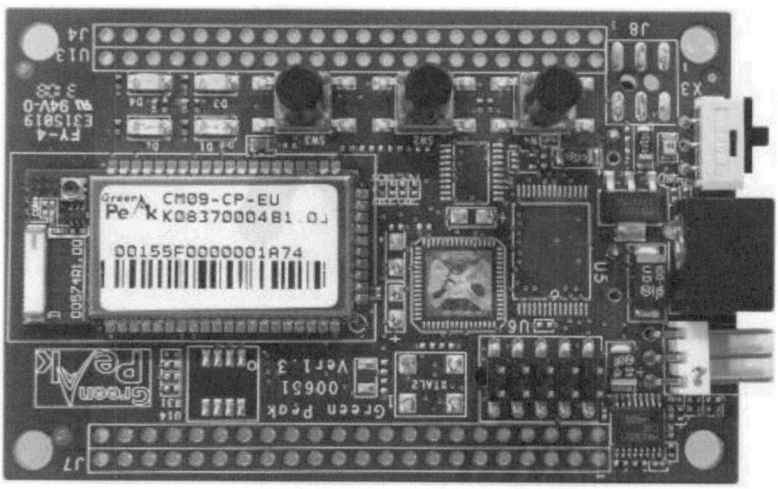

Fig. 3. WSN sensor node with integrated temperature sensor (Power provided by 3 AAA bateries placed underneath)

The WSN Gateway has a wireless communication module connected to its interface board (Figure 4), allowing TCP/IP, USB or RS232 serial communication with the external world (the processing software installed on a standard PC). The main characteristics of the communication stack are:

- Mesh network: messages travel from source node to destination node through intermediate nodes thereby multiplying range as a function of number of hops. The multi-hop feature does not require any application intervention.
- Self-forming: mesh network forms automatically, without any application intervention
- Self healing: when individual links fail the mesh network reestablishes a reliable route autonomously
- Security: data transfer through message encryption (AES 128 bit)
- Support for mobile nodes: Nodes can physically move through the network without requiring network re-association
- Support for ultra low power end devices: Reduced functionality devices can operate for years without replacing batteries
- Support for network visualization: network topology can be visualized using the optional JadeMonitor PC software component
- Robust against interference: able to operate in the presence of other wireless devices such as Wi-Fi, Bluetooth and others
- Scalability: the network can scale up to 100s of nodes without reconfiguration

Fig. 4. Gateway/Coordinator node built around a RISC microcontroller

From the communication protocol side, we have to choose between 2 types of network stacks, namely PeakNetZ and PeakNet LPR. The API to both stacks is almost identical. The different properties are given in Figure 5.

In applications where the nodes have access to mains power instead of batteries and some devices operate on batteries, the PeakNet™ Z is the best solution. The network consists of Full Functionality Devices (FFD), Reduced Functionality Devices (RFD) and one or more

Network Coordinators. All FFDs automatically become part of the wireless mesh networks and take active part of routing messages.
Sensors may be connected to these nodes. The RFDs nodes interface to sensors and actuators and connect wirelessly to a nearby FFD. As they are set in a sleeping–state most of the time, they consume very little power. The RFD will not actively route messages for other devices. The Network is self–healing and self–forming and is managed by the coordinator node(GreenPeak, 2010).
When all nodes are battery–powered, PeakNet™ Low Power™ (LPR) is the most convenient solution. PeakNet LPR does not require always–on, mains–powered devices. All devices are in low–power state and still form a mesh and route messages through the network. The low–power routing meshing capability is obtained by occasionally waking up the low power nodes along a synchronized scheme.

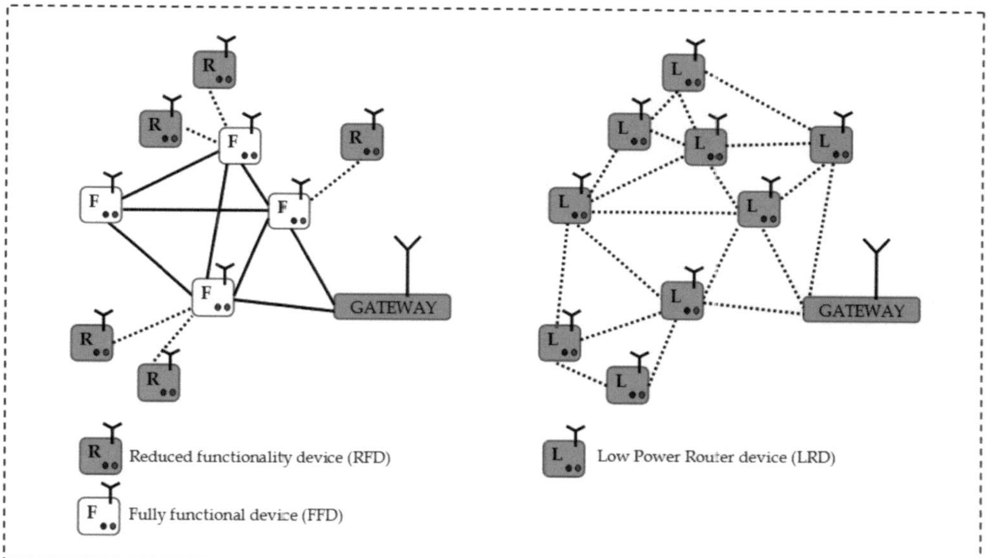

Fig. 5. The 2 types of network stacks PEAKNET™ Z (left) and PEAKNET™ LPR (right) (GreenPeak, 2010)

Hence, devices can pass messages through the network and in the same time conserve the battery power. Devices can be woken up according to a pre–defined schedule or when an external event occurs, or on a combination of both.
When powered up, the nodes automatically associate to the coordinator node. This coordinator also functions as a serial gateway: it allows the user to access the remote nodes in the network from a PC connected to the coordinator module.
All the software necessary for the network to work is embedded in the coordinator node. This means that the network can run stand–alone, without attaching a PC to the gateway/coordinator module.
The development software offered by the producer has an interface showing each node relative to other nodes positions. A map of the installations location permits to calibrate the distances and to display the real positions of all nodes having the coordinator node as a reference. For each node, the software displays the information read from the sensors and

from the inputs (Figure 6). Graphs showing the history of values recorded from the sensors are also available. Outputs may be activated remotely, making the system also useful in controlling remote processes where mobility is necessary.

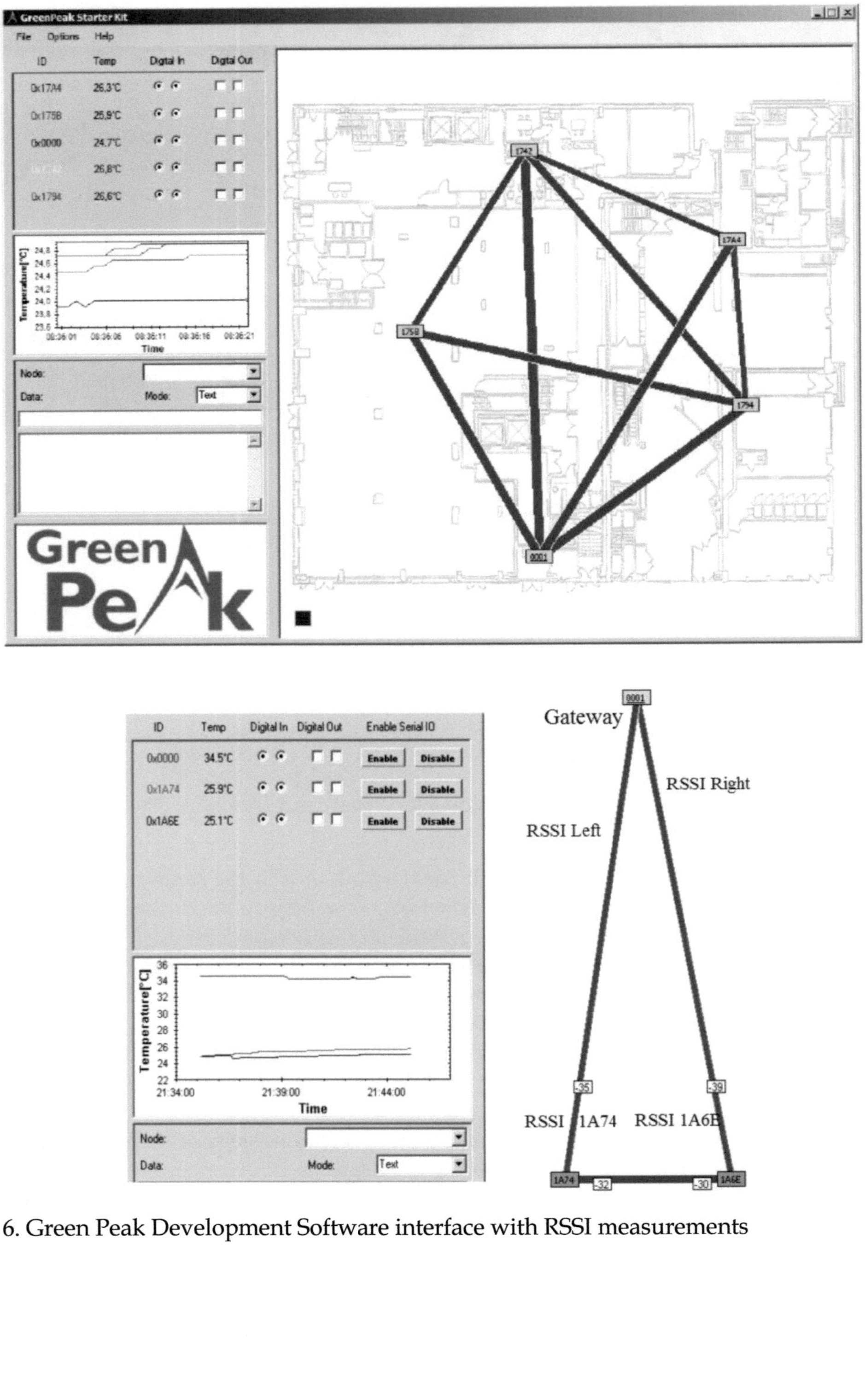

Fig. 6. Green Peak Development Software interface with RSSI measurements

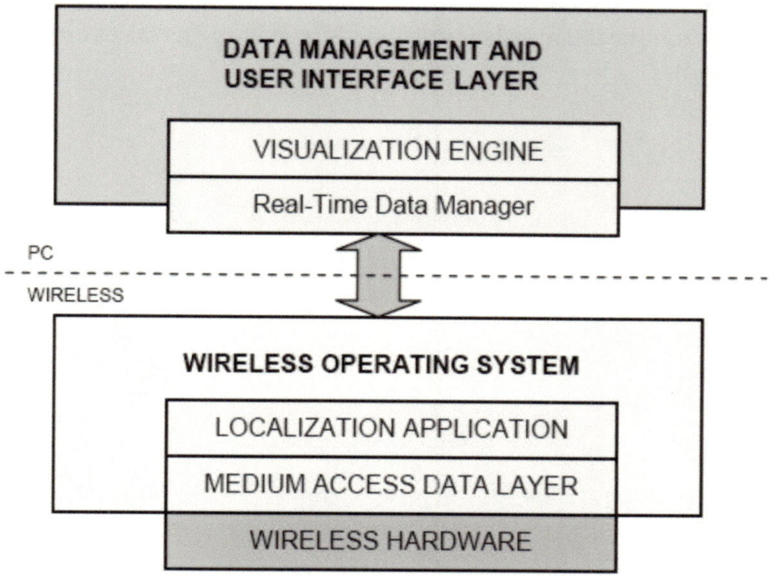

Fig. 7. Development software architecture

4.2 Test configuration

We used a dedicated LAN interface for connecting the gateway (Fig. 6) to the computer. As we already mentioned in the introduction, in WSN networks, a sensor node (Fig. 7) can have different roles, like network coordinator, router node (Full Functionality Device – FFD) and end device (Reduced Functionality Device – RFD, as described in IEEE 802.15.4 standard). The user can select the role of a WSN node, by modifying the software installed on-board.

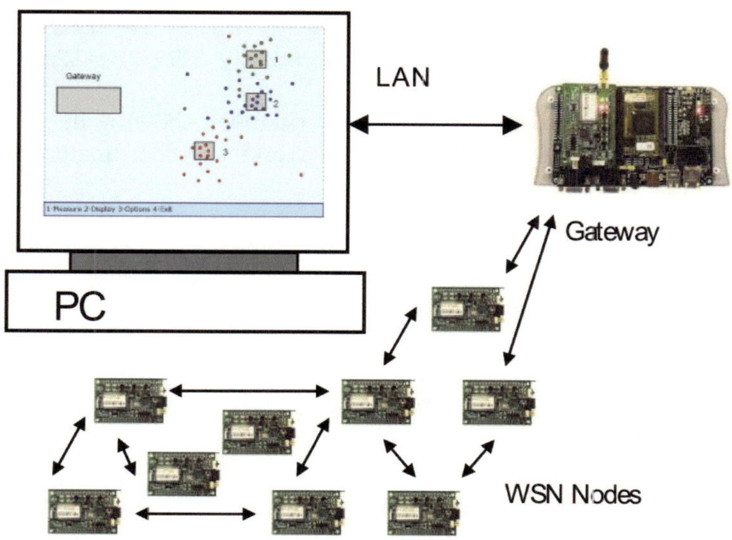

Fig. 6. WSN system overview

In our experiment, the WSN nodes were configured in FFD mode, in order to eliminate the effects of wake-up routine delay (when the node is in standby mode in order to reduce the power consumption).

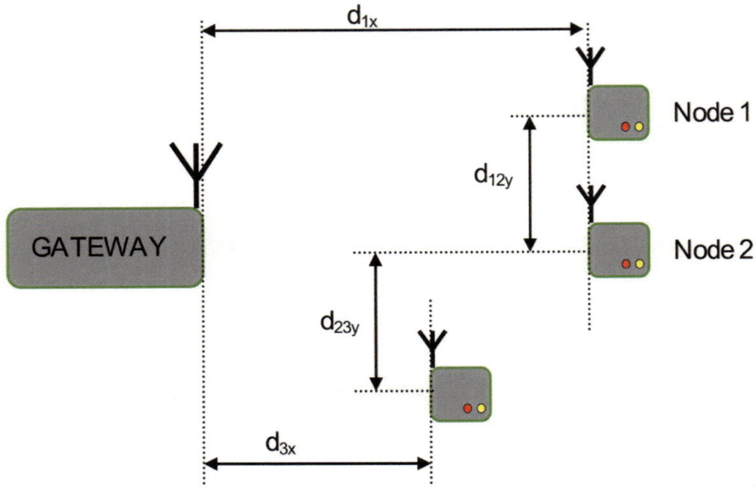

Fig. 7. Test setup with the Gateway and three WSN nodes (d_{1x}=2.5m, d_{3x}=2.2m, d_{12y}=0.4m, d_{23y}=0.8m)

The control software installed on the PC communicates with the gateway and process the RSSI information transmitted by the nodes. RSSI is a relative value (between 0 and the RSSI maximum), and a conversion routine transforms it in distance.

Regarding the physical positioning of the nodes and the gateway, we used for the tests the same configuration, both for the laboratory office room and for the anechoic chamber measurements. In Fig. 7 one may see the arrangement of the coordination node (the gateway) and the WSN nodes. The nodes and the gateway where placed 1 m above the ground level. As shown in Fig. 7, the distances were d_{1x}=2.5m, d_{3x}=2.2m, d_{12y}=0.4m and d_{23y}=0.8m.

In Figure 8 there is a photo taken in the anechoic chamber, showing the whole setup: the three nodes at 3 meters in front of the antenna and the coordinator node behind it (presented in a detailed photo in Figure 9).

For the measurements inside the laboratory, the nodes were positioned on the same relative distances between them and the gateway node, the same as in the semi anechoic chamber test. Wood furniture, other equipments emissions and moving humans are the perturbing elements present in this setup.

4.3 Experimental measurement results

The software on the PC was developed using the information provided in the SDK kit. For connectivity between the PC and the gateway, the LAN option was the single choice, as we had to extract the data from the semi anechoic chamber without using any metal cables from outside to the inside of the room. Ethernet cooper to optical fiber converter were used for this task. For the tests in the laboratory, both the serial RS−232 and the Ethernet interfaces may be used.

Fig. 8. Semi–anechoic chamber setup with threee WSN nodes and the Gateway node

Fig. 9. Detail with the Gateway node positioned behind the receiving antenna

The RSSI signal received by the Gateway from the three sensors nodes and the similar information transmitted between the nodes were recorded, in order to compute the distances. The developed software calculates the distances d_{1x}, d_{3x}, d_{12y}, d_{23y}, save them in a local file for future processing, and displays on a picture the positions of the nodes, considering known the gateway position.
For every setup, we made a set of 30 measurements, one at each 10 seconds.

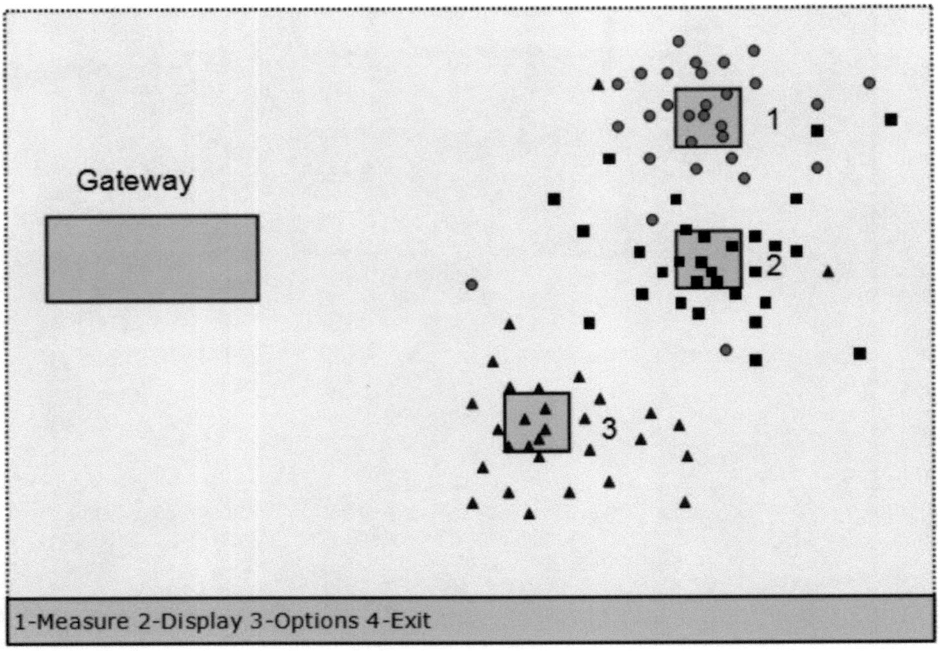

Fig. 10. Graphic display of the distances computes using the RSSI information from the laboratory

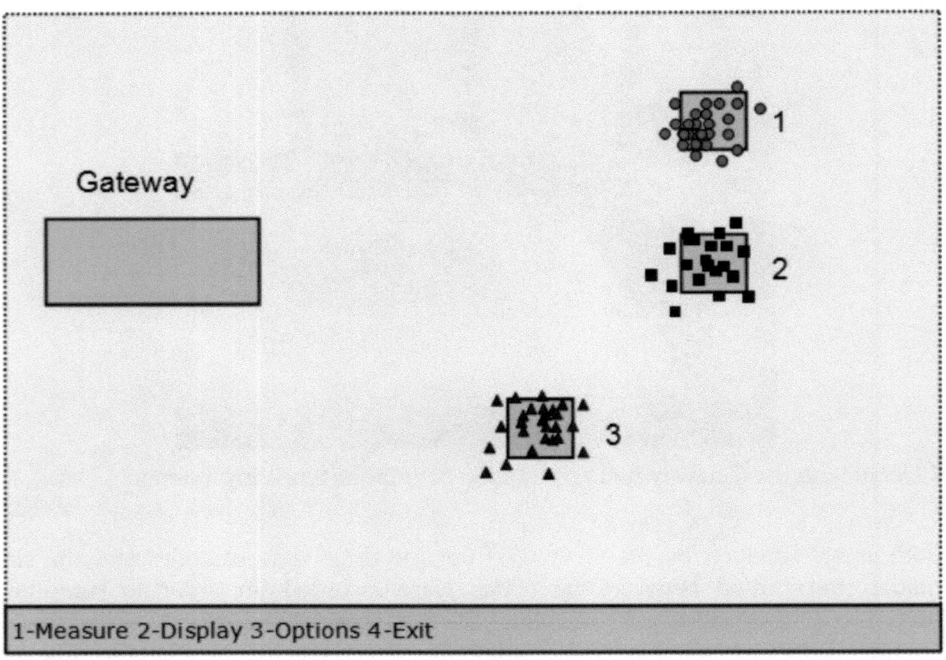

Fig. 11. Graphic display of the distances computes using the RSSI information from the anechoic chamber setup

For the first set of measurements, we used a standard laboratory room with furniture, chairs and moving humans. In addition, there were nearby emissions from two WLAN access points and other personal communication devices (mobile phones, PDAs, laptops, etc.).
The graphical representation of the positions of the three nodes in Fig. 10 shows us a great dispersion of the computed values.
For the second set of measurements, we used the same setup in terms of gateway and nodes positions and distances, but the equipments were positioned in the semi anechoic chamber, with virtually zero emissions from the outside world (noise floor at less than 120 dBm) and no furniture or humans present. The link between the computer and the gateway was made by using a pair of fiber optic to Ethernet converters. The graphical representation of the positions of the nodes is presented in Fig. 11.

Laboratory Room	**Distances**			
	d1x	**d3x**	**d12y**	**d23y**
Real distance (m)	2.50	2.20	0.4	0.8
Average value (m)	2.24	2.19	0.38	0.82
Max/Min value (m)	2.80/1.95	2.65/1.85	0.65/0.10	1.25/0.45
Standard deviation	0.39	0.29	0.24	0.26

Table 3. Results from the laboratory room measurements

Numerical results for both situations are summarized in Table 3 and Table 4. The results from the laboratory room setup show a great dispersion of the values for all distances. Despite this, the average values calculated for the distances between the nodes are quite good, with very small errors, while instantaneous ones may lead to wrong conclusions (Fig. 11). For larger distances, the standard deviation is greater, indicating the reflections on the walls and objects, and the presence of electromagnetic field emitting devices have a big influence on the results.

Anechoic Chamber	**Distances**			
	d1x	**d3x**	**d12y**	**d23y**
Real distance (m)	2.50	2.20	0.4	0.8
Average value (m)	2.42	2.20	0.39	0.81
Max/Min value (m)	2.65/2.20	2.45/1.95	0.60/0.15	1.05/0.55
Standard deviation	0.09	0.14	0.12	0.07

Table 4. Results from the anechoic chamber measurements

The influence of external electromagnetic fields from wireless devices operating in the 2.4 GHz band could not be neglected, and the results from the open area measurements are relevant in this direction.
The results obtained in the anechoic chamber are much better, the average values being closer to the real distances between the nodes. In addition, the standard deviations are smaller, meaning one single measurement have a better chance to be near the real value than in the previous case.

4.4 Electromagnetic field measurements

In order to estimate the emission level of a single WSN node, we measured it in an isolated environment. The measurements have been done in a 3m TDK semi anechoic chamber using

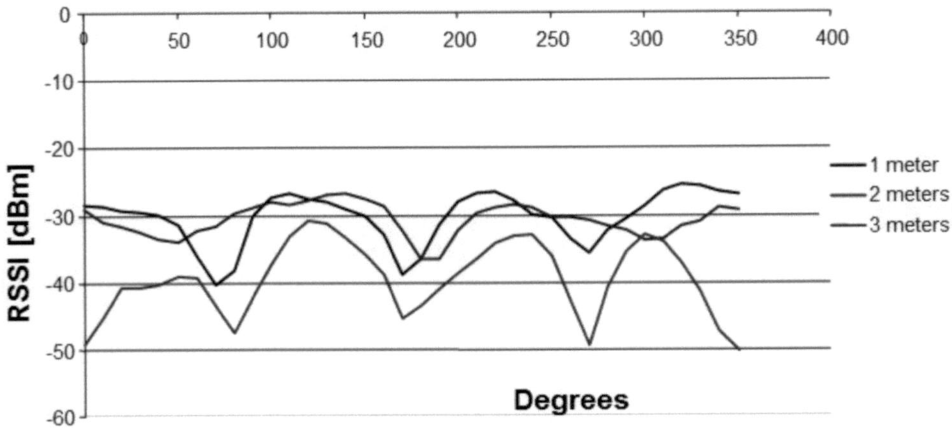

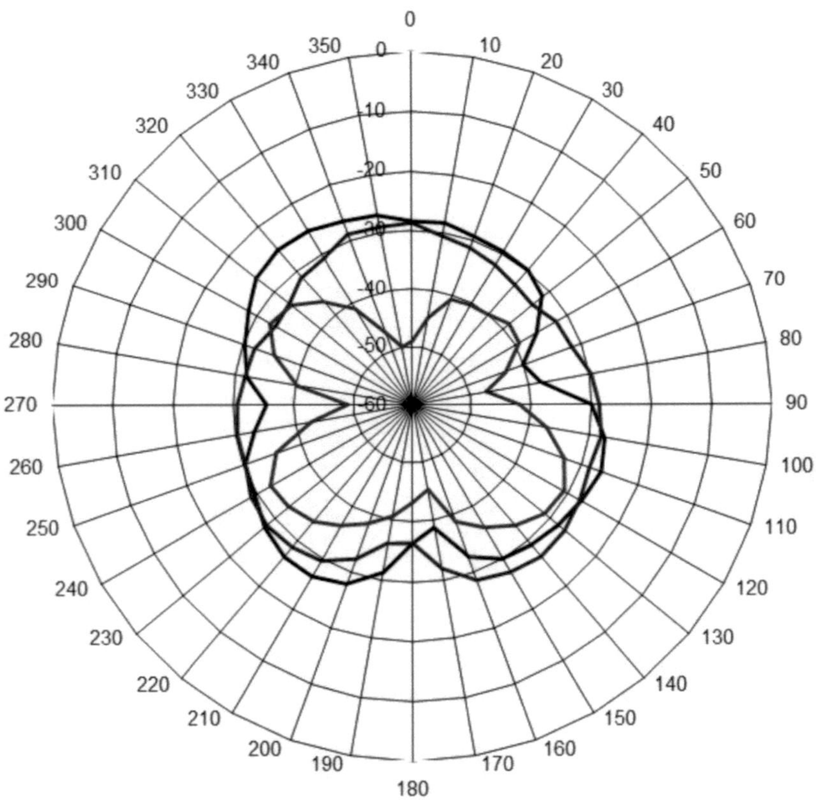

Fig. 12. WSN antenna radiation pattern at 1 meter, 2 meters and 3 meters distance away of the gateway

a Rohde & Schwarz - ESU 26 EMI Test Receiver, calibrated antennas and cables. The turntable and the antenna mast were operated by using an in-house made software program. The international standard specifying the emissions level for SRD–RFID

equipments is EN 55022 (CISPR 22) - "Information technology equipment - Radio disturbance characteristics - Limits and methods of measurements", while EN 300−220 - "Electromagnetic compatibility and Radio spectrum Matters (ERM); Short Range Devices (SRD)" is used for the operating performances and functional characteristics evaluation.
A standard configuration was used for the tests, as the equipment to be measured (EUT – Equipment Under Test) was positioned on a turn table at 0.8 meter above the ground and at 3 meters distance from the antenna tip. The gateway was positioned behind the receiving antenna system at 0.8 meter height. During the measurements, the antenna moved from 1 m to 4 m height and the EUT rotated 360 degrees, to find out the maximum emission level in the 30 to 3000 MHz band (more than the 1000 MHz limit specified in the standards, in the final scan procedure the operating frequencies being excluded from the measurement interval). In accord to the standards mentioned above, the readings were made continuously, one measure per second, using quasi-peak and peak detectors for the pre-scan and the final scan measurements, respectively. Even the standards do not specify a limit for the radiated emissions for frequencies over 1000 MHz we recorded those levels.
The maximum power level recorded for one measured node was around −30 dBm (with a minimum of −55 dBm) in the working frequency band, no other emissions being detected.
If there are multiple nodes in the same indoor environment, the field strength increases, but due to discontinuous emissions of nodes, the average field will remain much lower compared to the field generated by the continuous emission of an IEEE 802.11 b/g access point, for example.
The electromagnetic pollution will increase in the future due to extensive use of 2.4 GHz ISM band devices, including all types of portable computers, mobile phones, wireless gadgets, locating RFID systems contributing also to this increase but with a small quota.

5. Conclusions

Radio signals based indoor location systems is a hot topic. Even many papers deals with this subject, and some solutions were tested, currently we have no mature commercial implementations. Based on Wi-Fi, RFID, WSN, ZigBee or proprietary solutions, locating systems working principles implies the measurement of radio signals of information transmission using radio signals. Due to propagation issues in real working conditions, the practical demonstrated performances are far enough from theoretical calculated or simulation results. In indoor environments, the presence of different objects in rooms may cause multiple propagation paths, dynamic position changing objects or human presence may influence the measurement precision.
An evaluation of a WSN system was made by using it in a distance measurement and position estimation application. The obtained results, from measuring the distances in two different situations, were compared: in real life conditions (in a laboratory room with furniture and moving humans inside) and in a shielded room (completely isolated from the outside world electromagnetic fields and without interfering objects or humans). A set of 30 measurements for all distances were done, at 10 seconds time interval, in both situations.
From the results obtained in the two cases, one may conclude the average values for all distances are good enough in both cases, but the dispersion is greater in real life conditions. In mission critical applications where the position of an object must be known in real time, the WSN positioning solution could not be recommended. On the contrary, in applications where the position of an object have to be known, but the time is not critical, this solution

could be implemented with success, the price of a node being the single restrictive factor for large deployment areas.

Problems related to human safety will also emphasize due to high level of electromagnetic field intensity levels generated by all the wireless devices, not only in the free bands but also in regulated frequency bands. Continuous exposure to low levels of electromagnetic fields in domestic and industrial areas is a hot debate theme among the specialists and a definitive and scientific demonstrated conclusion is not yes available for the public.

Despite the significant research work in the area, there are still many difficult problems in indoor wireless sensors localization. In terms of positioning precision, different software algorithms may be used in order to process the measurement data and estimate the position of the nodes with only a small set of results. If we add a RF map and use path loss models adapted to particular application, the results may justify a rapid adoption of this technology in the real world applications.

6. References

Bahl, P., Padmanabhan, V., (2000). "RADAR: An In-Building RF-Based User Location and Tracking System," Proc. IEEE INFOCOM, vol. 2, pp. 775-784

Bal, M., Liu, M., Shen, W., Ghenniwa, H., (2009). "Localization in cooperative Wireless Sensor Networks: A review", 13th International Conference on Computer Supported Cooperative Work in Design, Santiago, Chile, April 22-24, pp. 438-443

Baunach, M., Kolla, R., Muhlberger, C., (2007). "Beyond Theory: Development of a Real World Localization Application as Low Power WSN," lcn, pp.872-884, 32nd IEEE Conference on Local Computer Networks (LCN 2007)

Bess, C., (2009). Third Generation RFID and the expanding Edge of the Enterprise, The HP Blog Hub, 27 Feb 2009

Bijl, M., Dil, B., (2010). Ambient 3000 Series White Paper - Localization, Ambient Systems, 2010

Buta, G., Coca, E., Graur, A., (2010). "Path Loss Exponent Influence on Distance Estimation between Wireless Sensor Nodes," Advances in Electrical and Computer Engineering, vol. 10, no. 1, pp. 110-115, 2010. [Online]. Available: http://dx.doi.org/10.4316/AECE.2010.01020

Chang, J. M., Huang, Yo-., Liu, S., (2011). "Real-Time Location Systems and RFID," IT Professional, pp. 12-13, March/April, 2011

Clulow, J., Hancke, G. P., Kuhn, M. G., Moore, T., (2006). "So Near and Yet So Far: Distance-Bounding Attacks in Wireless Networks", Computer Laboratory, University of Cambridge

Coca, E., Popa, V. (2007). "Experimental Results and EMC Considerations on RFID Location Systems", Proceedings of the 1st International RFID Eurasia Conference, 4-6 September 2007, Istanbul, Turkey, pp. 279-283, ISBN 978-975-01566-0-1, Digital Object Identifier 10.1109/RFIDEURASIA.2007.4368138

Coca, E., Popa, V., Gaitan, V.G., Turcu, C.O., Turcu, Cr., (2008). "Speed Measurement of a Moving Object by using a RFID Location System and Active Transponders", Electronics and Electrical Engineering (Elektronika ir Elektrotechnika), Kaunas

University of Technology, Lithuania, No. 8(88), 2008, ISSN 1392-1215, pp. 63-66

Dai, H., Su, D., (2008). "Indoor Location System Using RFID and Ultrasonic Sensors," Proc. 8th International Symposium Antennas on Propagation and EM Theory, IEEE Press, 2008, pp. 1179-1181

Finkenzeller, K., (2003). RFID Handbook : Fundamentals and Applications in Contactless Smart Cards and Identification: Wiley, 2003

Goncalo, G., Helena, S., (2009). "Indoor Location System Using ZigBee Technology," sensorcomm, pp.152-157, 2009 Third International Conference on Sensor Technologies and Applications, 2009

Halgamuge, M. N., Chan, T.-K., Mendis, P., (2009). "Experiences of Deploying an Indoor Building Sensor Network," sensorcomm, pp.378-381, 2009 Third International Conference on Sensor Technologies and Applications, 2009

Kaemarungsi, K., Krishnamurthy, P., (2004). "Properties of Indoor Received Signal Strength For WLAN Location Fingerprinting," Proc. First Ann. Int'l Conf. Mobile and Ubiquitous Systems: Networking and Services (MOBIQUITOUS), pp. 14-23, 2004

Kathiravan, K., Pradeep, P., Ronak, G., Roshan, S. S., (2009). "Modeling Location Monitoring System Using Directional Antennas," Computer Modeling and Simulation, UKSIM European Symposium on, pp. 488-493, 2009 Third UKSim European Symposium on Computer Modeling and Simulation, 2009

Khan, M. A., Antiwal, V. K., (2009). "Location Estimation Technique using Extended 3-D LANDMARC Algorithm for Passive RFID Tag," Proc. International Advance Computing Conference, IEEE Press, 2009, pp. 249-253

Kim, H.-J., Yang, J., (2008). "The Practical System Architecture for the Wireless Sensor Networks," MUE, pp.547-551, 2008 International Conference on Multimedia and Ubiquitous Engineering (MUE 2008)

Koyuncu, H., Yang, S. H., (2010). "A Survey of Indoor Positioning and Object Locating Systems", IJCSNS International Journal of Computer Science and Network Security, Vol. 10, No. 5, pp. 121-128, 2010

Kuang, X. H., Shao, H. H., Feng, R., (2008). "A New Distributed Localization Scheme for Wireless Sensor Networks," Acta Automatica Sinica, 34(3), 344-348, 2008

Kushki, A., Plataniotis, K., Venetsanopoulos, A. N., (2006). "Location Tracking in Wireless Local Area Networks with Adaptive Radio Maps," Proc. IEEE Int'l Conf. Acoustics, Speech, and Signal Processing (ICASSP), vol. 5, pp. 741-744, 2006

Kwon, O. H., Song, H. J., (2008). "Localization through Map Stitching in Wireless Sensor Networks," IEEE Trans. on Parallel and Distributed Systems, 19(1), 93-105, 2008

Han, X.-L., Zhao, W.-D., Ji, J., (2008). "Indoor Location Algorithm Based on RFID Technology and Its Improvement," Computer Engineering, vol. 34, Nov. 2008, pp. 225-270

Harrop, P., (2008). Third-generation active RFID bursts onto the scene, Retail Technology Review, 14 Oct 2008

Hsu, P.-W., Lin, T. H., Chan,g H. H., Chen, Y. T., Yen, C. Y., Tseng, Y. J., Chang, C. T., Chiu, H. W., Hsiao, C. H., Chen, P. C., Lin, L. C., Yuan, H. S., Chu, W. C., (2009). "Practicability Study on the Improvement of the Indoor Location Tracking Accuracy with Active RFID," Communications and Mobile Computing, International Conference on, pp. 165-169, 2009 WRI International Conference on Communications and Mobile Computing, 2009

Huang, Y., Lui, Z., Ling, G., (2008). "An Improved Bayesian-based RFID Indoor Location Algorithm," Proc. International Conference on Computer Science and Software Engineering, IEEE Press, 2008, pp. 511-514

Jeon, S., Choi, M., Kim, G., Hong, B., (2010). "Localization of Pallets Based on Passive RFID Tags," Information Technology: New Generations, Third International Conference on, pp. 834-839, 2010 Seventh International Conference on Information Technology, 2010

Jiang, X., Liu, Y., Wang, X., (2009). "An Enhanced Approach of Indoor Location Sensing Using Active RFID," Information Engineering, International Conference on, pp. 169-172, 2009 WASE International Conference on Information Engineering, 2009

Jeong, W., Nof, S. Y., (2008). "Performance evaluation of wireless sensor network protocols for industrial applications," Journal of Intelligent Manufacturing, vol.19, pp.335-345, 2008

Jong E., Bijl, M., (2010). Ambient 3000 Series White Paper - Technology Overview, Ambient Systems, 2010

Lanzisera, S., Lin, D., Pister, K., (2004). "RF Time of Flight Ranging for Wireless Sensor Network Localization," 4th Workshop on Intelligent Solutions in Embedded Systems (WISES), June 2006

Liu, M. L. Y., "LANDMARC: Indoor location sensing using active RFID," Wireless Network, vol.10, Jun. 2004, pp. 701-710

Li, Y., Wang, Z., Song, Y.Q., (2006). "Wireless Sensor Network Design For Wildfire Monitoring," Proc. of The Sixth World Congress on Intelligent Control and Automation, WCICA, Vol.1, pp. 109-113, Dallan, 2006

Mao, G., Fidan, B., and Anderson B. D. O., (2007). "Wireless Sensor Network Localization Techniques," The International Journal of Computer and Telecommunications Networking, vol. 51, pp. 2529-2553, 2007

Miorandi, D., Uhlemann, E., Vitturi, S., Willig, A., (2007). "Guest Editorial Special Section on Wireless Technologies in Factory and Industrial Automation—Part II," Industrial Informatics, IEEE Transactions on , vol.3, no.3, pp.189-190, Aug. 2007

Nikitin, P. V., Martinez, R., Ramamurthy, S., Leland ,H., Spiess, G., Rao, K. V. S., (2010). "Phase Based Spatial Identification of UHF RFID Tags," in Proc. IEEE International Conference on RFID, 2010

Ota, N., Wright, P., "Trends in wireless sensor networks for Manufacturing," Int. Journal of Manufacturing Research, Vol. 1, No. 1, 2006

Popa, V., Coca, E., Dimian, M., (2010). "Applications of RFID Systems - Localization and Speed Measurement", Radio Frequency Identification Fundamentals and Applications Bringing Research to Practice, Cristina Turcu (Ed.), ISBN:

978-953-7619-73-2, InTech, Available from: http://www.intechopen.com/articles/show/title/applications-of-rfid-systems-localization-and-speed-measurement

Razaq, A., Luk, W. T., Shum, K. M., Cheng, L. M., Yung, K. N., (2008). "Second-Generation RFID," IEEE Security and Privacy, pp. 21-27, July/August, 2008

Roberti, M., "Understanding the EPC Gen 2 Protocol," RFID J. Special Report, 28 Mar. 2005

Tsui, A. W. T., Lin, W.-C., Chen, W.-J., Huang, P., Chu, H.-H., (2010). "Accuracy Performance Analysis between War Driving and War Walking in Metropolitan WiFi Localization," IEEE Transactions on Mobile Computing, 28 Jun. 2010. IEEE computer Society Digital Library. IEEE Computer Society. [Online]. Available: http://doi.ieeecomputersociety.org/10.1109/TMC.2010.121

Wada, T., Uchitomi N., Ota Y., Hori, T., Mutsuura K., Okada H., (2009). "A Novel Localization Scheme for Passive RFID Tags Communication Range Recognition (CRR)," in Proc. IEEE International Conference on RFID, 2009

Wang, Q., Yan, C., Liu, F., (2009). "Knowledge integration based operation mode for workshop manufacturing system," Computer Integrated Manufacturing Systems, vol. 15, Apr. 2009, pp. 698-704

Youssef, M., Agrawala, A., (2005). "The Horus WLAN Location Determination System," Proc. Third Int'l Conf. Mobile Systems, Applications, and Services, pp. 205-218, 2005

Yihua, H., Zongyuan, L., Guojun, L., (2008). "An Improved Bayesian-Based RFID Indoor Location Algorithm," Computer Science and Software Engineering, International Conference on, pp. 511-514, 2008 International Conference on Computer Science and Software Engineering, 2008

Zongwei, L., Chan, T., Li, J. S., (2005)."A Lightweight Mutual Authentication Protocol for RFID Networks," Proc. IEEE Int'l Conf. e-Business Eng., IEEE CS Press, 2005, pp. 620-625, 2005

***, Green Peak WSN Development Tool, Green Peak Technologies [Online]. Available from: http://www.greenpeak.com

***, EN-55022:2007 Information technology equipment. Radio disturbance characteristics. Limits and methods of Measurement

***, EPC Radio-Frequency Identity Protocols Class-1 Generation-2 UHF RFID, EPCglobal, Jan. 2005

***, ISO/IEC 24730-1:2006 Information technology -- Real-time locating systems (RTLS) -- Part 1: Application program interface (API)

***, ISO/IEC 24730-2:2006 Information technology -- Real-time locating systems (RTLS) -- Part 2: 2,4 GHz air interface protocol

***, ISO/IEC 24730-5:2010 Information technology -- Real-time locating systems (RTLS) -- Part 5: Chirp spread spectrum (CSS) at 2.4 GHz air interface

***, ISO/IEC 18000-6:2004/FPDAM 1, Amendment 1, extension with type C and update of type A,

ISO/IEG Available from:
www.iso.org/iso/iso_catalogue/catalogue_tc/catalogue_detail.htm?csnumber=43923

***, ISO/IEC FDIS 18000−6:2003(E, Information Technology Automatic Identification. and Data Capture Techniques, ISO/IEC, JTC 1/SC 31/WG4, Nov. 2003

***, RFID−Radar − Brochure on Development model of RFID−Radar, Trolley Scan(Pty) Ltd, 2005. [Online]. Available: http://www.rfid−radar.com

24

Optimization of RFID Platforms: A Cross-Layer Approach

Ramiro Sámano-Robles and Atílio Gameiro
Instituto de Telecomunicações, Campus Universitário, Aveiro
Portugal

1. Introduction

RFID (Radio Frequency Identification) is a technology that uses radio frequency signals for purposes of identification and tracking of objects, humans or animals. Since it allows automated identification and potential new features such as sensing of environmental parameters, RFID is gaining preference over legacy identification technologies. RFID is also being implemented in future mobile terminals, thereby paving the way for new ubiquitous applications. RFID is thus expected to enable the concept of the Internet-Of-Things by closing the gap between the worlds of computer networks and physical objects (Darianian & Michael (2008)).

As any emerging application, RFID at the item level is facing several obstacles towards massive consumer adoption. These obstacles include: high implementation costs, standards in early stages of adoption, privacy and security threats, low consumer acceptance levels, and reading reliability issues (Jahner et al. (2008)). Dissemination activities have been organized worldwide with the aim of improving end-user knowledge of RFID technology and thus boost both acceptance levels and standard adoption. Furthermore, several improvements on RFID technology have been recently proposed in order to increase reading reliability levels (e.g., Sabesan et al. (2009)), reduce privacy/security threats (e.g., Park et al. (2006)), and lower implementation costs (e.g., Subramanian et al. (2005)).

Despite these advances in RFID technology, optimization of algorithms across different layers, commonly known as cross-layer design, has been scarcely explored in RFID systems. Cross-layer design has been proved crucial in the evolution of conventional wireless networks towards broadband solutions (Srivastaya & Montani (2005)). In the RFID arena, however, only a few solutions using context-aware mechanisms have been shown to significantly improve reading reliability levels (e.g., Ahmed et al. (2007)) and security/privacy features (e.g., Kriplean et al. (2007)). In addition, recent studies suggest that RFID systems would obtain great benefits from using information across different layers (Samano & Gameiro (2009)). Therefore, there is a big potential in using advanced cross-layer design techniques in order to improve existing platforms and propose future algorithms for RFID applications.

Cross-layer design is expected to make most of its impact upon the two lower layers of RFID platforms: medium access control (MAC) and physical layers (PHY)(Samano & Gameiro (2008)). In particular, mobile RFID systems raise new interesting issues that can be appropriately tackled by using cross-layer methodologies. For example, in networks with large numbers of mobile readers, where reader collisions may constantly occur, resolution algorithms with joint power and scheduling control will be required. Furthermore, in mobile terminals with embedded reader functionalities cross-layer optimization can be used to adapt low level reader protocols to bandwidth- and resource-constrained environments. Therefore, cross-layer design will also lead to a better optimization and cost reduction of RFID platforms. The specific objectives of this chapter are: 1) to provide an overview of reading reliability impairments that affect RFID and that need to be tackled by cross-layer solutions (Section 3); 2) to review existing trends and current issues in the design of RFID systems, particularly focusing on identifying algorithms suitable for cross-layer optimization (Sections 2 and 4); 3) to propose a framework for cross-layer optimization and complexity impact analysis that will help in the design and optimization RFID platforms (Section 5); and 4) to propose a set of examples of cross-layer optimization algorithms for RFID (Section 5).

2. RFID system architecture

A typical RFID system consists of tags, readers and back-end processing servers (Chandramouli et al. (2005)). Tags have the only function of responding to readers' requests. Conversely, readers are in charge of responding to requests from application layers, as well as requesting, collecting and processing tag information. Finally, back-end processing servers are in charge of high level information management and application level execution. In mobile RFID systems, additional components might be required to provide networking connectivity and mobility features. A general architecture for cross-layer optimization of RFID platforms showing the potential functionalities of each element is displayed in Figure 1. An optional mobile-proxy entity is used in this figure to provide mobility to a reader platform. For example, a mobile terminal acting as proxy can be used to control nearby readers via Bluetooth and also to relay their data to a remote controller using a 3G data connection.

As observed in Figure 1, some of the functionalities of an RFID platform can be hosted by more than one entity. Therefore, it is possible to reduce the complexity of those parts of the network that are limited in processing capacity, and push functionalities towards less critical elements. For example, in centralized architectures most of the operations are performed by a central controller while readers perform only tag processing operations. By contrast, in decentralized architectures readers host most of the processing and middleware functionalities and only report the results to external application layers (Floerkemeier & Sarma (2008)). In a mobile RFID scenario, functionalities can also be hosted by mobile terminals (e.g., the NFC -near field communication- system). These different architectures affect in different ways the interfaces and protocols used for the communication between network entities. This impact is mainly in terms of signaling and monitoring mechanisms which in turn affect the required processing complexity and channel bandwidth. Since these two resources are limited in certain RFID deployments, cross-layer optimization of protocols under bandwidth- and resource-constrained environments will be required. Before addressing this optimization it is first necessary to analyze the impairments to be modeled, to review issues of current RFID solutions, and select potential algorithms that are good candidates for performance and complexity optimization.

3. Reading reliability impairments

The act of reading/writing the information of a tag via a wireless connection, particularly in passive RFID systems, is prone to impairments that may considerably degrade its reliability. Reading reliability is regarded in this document as the ability of an RFID system to maintain

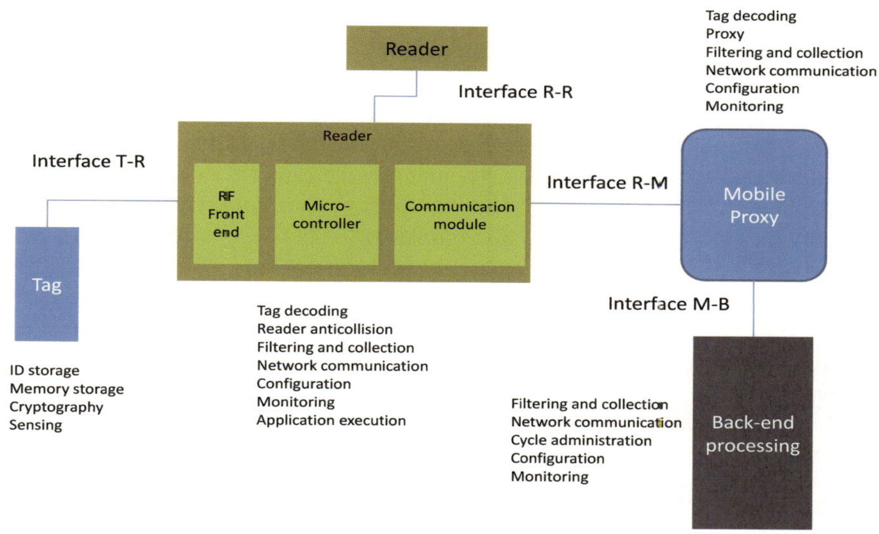

Fig. 1. Reference RFID system architecture

some performance metrics such as correct number of tag readings, reading range, false positive readings, false negative readings, etc. within certain boundaries.

3.1 Physical layer impairments

3.1.1 Propagation channels

Perhaps the most evident impairment in wireless communications is the one of attenuation or path-loss (Sklar (1997)). Signals propagate in different directions distributing the initial power over larger surfaces as waves travel. The free space loss model considers that wave-fronts travel in concentric spheres so the power loss is proportional to the area of such spheres (path loss exponent 2). In RFID systems at low frequencies (e.g., high frequency -HF- bands), where tags use induction coupling to activate their chip, free space loss is a slightly inaccurate assumption as high-order exponent terms tend to appear in induction fields. By contrast, in RFID systems working in the UHF (ultra-high-frequency) band, where tags use backscattering load modulation, free space models fit better as tags are usually located in the far-field of analysis. Other effects such as non-line-of-sight (NLOS) might modify the path loss exponent experienced by some applications. In ultra-wideband (UWB) RFID systems appropriate path loss modeling still has to be accurately studied.

Wireless systems are also prone to the effects of fast fading. Fast fading refers to the fluctuations of the received signal due to random scatterers of small size causing the signal to arrive at the destination with destructive superposition (Sklar (1997)). It is called fast because channel fluctuations occur at a relative high speed with respect to the transmission rate. Since range of RFID systems is relatively short, fast fading is considered only in certain scenarios in combination with line-of-sight components (e.g., Floerkemeier & Sarma (2009)). Furthermore, Doppler effects due to fast moving tags/readers are not expected to cause major impairments except perhaps in applications such as toll payment systems in highways.

RFID systems can also be affected by shadowing, which arises when large obstacles "shadow" the received signal. Shadowing causes variations on the signal that change at a relative slow speed with respect to transmission rates (Sklar (1997)). In RFID, shadowing can affect supply chain applications where large objects may block the line of sight between readers and tags. Shadowing modeling, however, needs to be studied in more detail in RFID settings.

Another source of impairment is multi-path propagation. Multi-path propagation results from signals traveling through different paths that experience random delays within the order of a symbol duration. Multi-path propagation causes inter-symbol interference at the receiver, which can only be overcome by means of complex equalization (Proakis (1997)). Since RFID tags cannot, in general, host advanced equalization schemes multi-path propagation usually has a negative effect in reading reliability. Multi-path will be mainly considered at high frequencies (UHF bands) where its effects are more evident than at lower frequencies.

The problem of interference can also reduce reliability figures of RFID systems. Interference is caused by signals of other devices being transmitted at the same time and in the same frequency band of the desired signal. In RFID systems, interference can be caused by other readers or by electronic devices operating nearby. Therefore, methodologies are needed to mitigate the effects of interference (e.g., Kim et al. (2009)). The work in (Cheng & Prabhu (2009)) presents a detailed report of EMI (Electro-Magnetic Interference) measurement of an industrial floor environment with machines that interfere with RFID systems. It was observed that reliability levels were reduced up to 40% for typical RFID deployments, thus concluding that design of RFID systems must consider the effects of local EMI sources.

NLOS environments also affect RFID signal reception. However, existing approaches focus on simple models with free space loss and Rice channels (e.g., Floerkemeier & Sarma (2009)) without making clear distinction between line-of-sight (LOS) and NLOS conditions. Other studies have been carried out to tune RFID parameters according to particular application and environmental conditions (e.g., Hariharan & Bukkatapatman (2009)). More accurate propagation models, such as those used in conventional wireless systems, are still required in RFID systems. For example, multi-slope propagation models for LOS-to-NLOS transitions have been extensively analyzed in (WINNER (2007)) for typical wireless systems. Indoor propagation models such as the well known multi-wall floor (MWF) propagation model in (COST 231 (2006)), which includes the loss of waves traveling through different materials, could also be proposed in RFID supply chain settings with pallets and boxes.

3.1.2 Impairments due to technical issues

Impairments on reading reliability also arise due to imperfections of RFID technology. Several issues currently affect tag, readers and middleware designs. At the tag side electromagnetic decoupling, inappropriate material for tag construction, inefficient power utilization and high chip activation thresholds may reduce performances of reliability and reading range. At the reader side, low sensitivity and inefficient isolation between the down-link and up-link chains can be mentioned as the main sources of impairments (Wang et al. (2007)).

3.1.3 Metallic environments and other effects

Metallic plates reflect electromagnetic waves, thereby increasing the number of multi-path components in indoor environments and causing further fading phenomena (Wagner et al. (2007)). When tags are attached to a metallic surface the antenna port may suffer from grounding, which affects the signals received by the tag (Qing & Chen (2007)). In addition, metallic obstacles may also affect the operational frequency of the tags or they can simply shield the tags from reader's signals.

The authors in (Qing & Chen (2007)) presented the analysis of proximity effects of metallic environments on the properties of HF tag antennas. Resonant frequency of the antenna was found to be shifted in the presence of metallic surroundings, thereby reducing its efficiency. The magnitude of these effects was found dependant on the size of the metallic plates, distance to the metallic plate, and tag orientation. Thus RFID systems can be tuned according to the particular metallic environment. A similar work has been presented in (Wagner et al. (2007)). Three main effects were analyzed: reflections, shielding, and de-tuning of the tag at different distances from a metallic plate. Guidelines to the design of RFID systems to reduce the effects of metallic environments were further provided. For example, a dielectric material between the tag and the metallic plate was proposed to avoid tag grounding.

Reading reliability can also be affected by the relative orientation of tags, material absorption, the influence of other tags (mutual impedance), and the bending of the tag when attached to irregularly-shaped objects. RFID tags are commonly designed as flat antennas. However, tagged objects often have irregular shapes so tags have to be deformed to fit the shape of the object, thus reducing the effectiveness of RF power conversion. The authors in (Siden et al. (2001)) have calculated the performance loss of a dipole UHF antenna under different angles of bending. While the work in (Siden et al. (2001)) used theoretical analysis based on the method of moments (MoM) and the finite element method (FEM), the authors in (Leung & Lan (2007)) have proposed a new definition of effective antenna area to predict the performance of loop antennas for inductive coupling RFID tags over curvilinear surfaces.

In some RFID applications electromagnetic interactions between neighbor tags may also arise. The authors in (Chen et al. (2009)) have analyzed electromagnetic interaction between stacked NFC tags and they have concluded that considerable losses are obtained only in some regions of the space. The authors in (Lu et al. (2009)) have reached similar conclusions using both mutual impedance and radar cross-section (RCS) calculations.

3.2 Medium access control layer impairments

3.2.1 Tag-to-tag collision problem description

In RFID, readers broadcast a signal that can be received by a group of tags. Several tags inside this group may simultaneously respond to the same request causing the potential loss of information. A collision resolution algorithm is thus required. These algorithms rely on retransmission of the information by the involved tags. This retransmission process requires extra power and transmission resources, which further reduces reading reliability. Therefore, resolution algorithms that reduce the number of retransmissions of each tag and ensure the reliable reading of all the contending tags are potentially good candidates for RFID applications (Samano & Gameiro (2008)).

3.2.2 Reader collision problem

RFID tags may receive signals from one or more readers at the same time. When two readers transmit with enough power to interfere with each other, then the tag is not able to decode the information from any of the readers (Birari & Iyer (2005)). This is known as the multiple-reader-to-tag collision problem. Several schemes have been proposed in the literature including solutions with power control or scheduling. Another type of interference is called reader-to-reader, in which the signal received by a reader from a tag can be degraded by the signal from another active reader nearby (Birari & Iyer (2005)).

3.3 Upper-layer impairments

3.3.1 Security and privacy issues

The possibility of malicious users tracking consumer shopping habits in retailers or scanning personal information from tagged passports represent examples of privacy issues of RFID (Juels (2006)). An eavesdropper reader located at even hundreds of meters can be listening to the transmissions of another reader and deduce tag-related information (Xiao et al. (2006)). Another common example is an unauthorized reader requesting information from tags. Since tags usually have limited processing capabilities, complex authentication and encryption mechanisms cannot be employed. Conversely, tags might also contain malicious code that can be used to pose security threats to middleware applications. The area of security/privacy issues of RFID has attracted loads of attention in recent years (see Juels (2006)).

3.3.2 Middleware and networking issues

Middleware platforms have to be designed to deal with the particularities of RFID systems. Impairments may arise when RFID specific procedures fail. The main functionality of an RFID middleware platform is that of filtering and aggregating RFID raw data to cope with incorrect tag readings due to the low reliability of physical layer interfaces (Floerkemeier et al. (2007)). Therefore, when middleware procedures fail reliability can be seriously compromised. Similarly, incorrect forwarding and routing of the information, particularly in mobile RFID, cannot only cause reliability problems but also privacy and security issues (e.g., Park et al. (2006)). The design of an appropriate middleware and networking architecture to ensure reliability as well as security and privacy features is crucial in RFID systems.

4. Algorithms to improve reading reliability

4.1 Physical layer schemes

4.1.1 Signal processing schemes

Due to recent advances in wireless communications, a wide set of tools generated in this framework can be used to improve the PHY layer of RFID systems. Among these tools, signal processing algorithms exploiting the concept of diversity stand as promising options. Diversity refers to the ability of transmitting/receiving the information via two or more independent sources that when correctly combined help to improve the correct reception of the information. Diversity sources may span frequency, code, time, or space domains. Space diversity can be achieved by means of multiple antennas at the transmitter, at the receiver or at both ends. Space diversity can also be achieved via relaying, where the signal is received by relay nodes that forward the signal towards the destination. For example, a tag antenna with two ports that can be used to implement a receive diversity algorithm has been presented in (Nikitin (2007)). Another example is the work in (Quiling (2007)) where the authors propose spread spectrum techniques for RFID to achieve diversity in the code domain. However, since the processing capabilities of passive tags are limited, diversity mechanisms will be more efficient at the reader side. Multiple antennas can be used to implement maximum ratio combining (MRC), successive interference cancelation (SIC), parallel interference cancelation (PIC) and multiuser detection (MUD) schemes. The authors in (Angerers et al. (2009)) have tested an MRC receiver at the reader side that is used to increase diversity and thus reliability. Beam-forming or smart antennas with fixed or adaptive beams can also be used to improve reliability of the reading process. In addition, smart antennas can be used to direct the radiated energy towards a desired area while suppressing signals radiated towards insecure zones with potential eavesdropper readers. For example, the authors in (Chia et al. (2009)) have designed a multi-band (900 MHz and 2.4 GHz) integrated circuit which is suited for electronic beam steering. The beam steering design allowed improving the performance of a reader in the 900 MHz band. Another smart antenna system for RFID readers has been reported in (Kamadar et al. (2008)) where the authors proved the benefits of this type of technology by improving RFID reading rates. Another type of antenna deployment for RFID is the one called distributed antenna system (DAS). DAS systems have been used in RFID in (Sabesan et al. (2009)), where an increase of 10dB on the received tag signals as compared to a switched multi-antenna system was reported. Unlike conventional approaches with co-located antennas, in DAS the antennas are spaced by long distances and are interconnected to a controller via a coaxial or optical link, thereby achieving large diversity gains (Choi & Andrews (2007))

Channel coding can also be used to improve reliability of RFID. Since tags have limited capabilities, aggressive channel coding is more feasible in uplink rather than in the down-link. However, only those coding schemes with simple encoding rules such as FEC (Forward Error Correct) codes can be potentially implemented in tags.

Additional signal processing capabilities have an impact on the complexity of reader and tags. Therefore, it is necessary to estimate such complexity for an appropriate technical-economical evaluation. Complexity of multiuser detection schemes can be expressed in terms of the number of users (K) and the number of stages (P). In comparison with multiuser detection schemes, whose complexity orders are in the range from K to K^3, PIC and SIC have complexity orders of PK and K, respectively, with acceptable performance results (Andrews (2007)).

Summarizing, in the down-link the most attractive schemes were beam-forming (smart-antennas) and DAS in terms of performance and backwards compatibility. Other solutions such as polarization diversity, Alamouti space-time coding, spread spectrum, and forward error codes (FEC) are also attractive but depend on changes in tag designs. The down-link is the most critical in RFID since tag sensitivity is the main limitation. By contrast, the uplink can be enhanced by several techniques such as multiuser detection, interference cancelation, maximum ratio combining, and also smart and distributed antennas. Distributed antennas and interference cancelation schemes are also promising schemes in terms of low hardware complexity.

4.1.2 Antenna and integrated circuit design

In general, there are three main types of passive tags: chip-based tags using induction coupling at low frequencies, chip-based tags using backscattering at high frequencies, and chip-less tags based on SAW (surface acoustic waves) filters. While the main limitation of chip-based tags is the power threshold required to activate the chip, SAW-based tags are based on a continuous piezoelectric effect that allows operation under any power level. The only limitation of these tags is thus given by the reader's sensitivity, which is generally better than chip-based tag's sensitivity. Therefore SAW tags have better reading ranges than passive tags (Hartman & Clairborne (2007)). Their main disadvantage is their inability to have cryptographic features or memory registers to write information.

At low frequencies tags are relatively small with respect to the operational wavelength. Thus, antennas should be designed to operate in the induction field of the interrogator. Induction-based passive tags store the energy radiated by the interrogator by means of a capacitor and use it to activate a chip that will transmit a signal back to the interrogator carrying the ID of the tag using load-based modulation (Weinstein (2005)). Design of these induction-based tags is focused on the efficiency of the coil antenna (e.g., Leung & Lan

(2007); Nummela (2007)). Reliability levels of inductive RFID systems can be improved by an appropriate electromagnetic design of antenna and modulation circuits.
At high frequencies sizes of tags become comparable to the operational wavelength. Thus antenna design should consider far field analysis. Antennas at high frequencies are designed by using either aperture or linear antenna theory assisted by the method of moments (MoM) or the finite element method (FEM) (Balanis (2000); Siden et al. (2001)). An increase on the electric aperture or length of the antenna is translated into increased gain. Improved gain of the antenna is directly related to longer reading rages and higher reliability levels. However, size of the antennas is also limited by the size of the tag. Thus another way of increasing the gain of the antenna without increasing its size is by improving its efficiency. The work in (Rautio (2010)) uses advanced electromagnetic tools in the analysis of RFID tags. Impedance analysis of RFID tags can also be found in (Qing et al. (2009)), where the authors have proposed a methodology to matching impedances of UHF RFID tags with the underlying circuits thereby obtaining enhanced reading ranges. Other antenna designs for UHF tags can be found in (Chen et al. (2009); Chen (2009); Gao et al. (2009); Guo et al. (2006); Leung & Lan (2007); Nikitin (2007); Pillai et al. (2007)). The effects of antenna properties on the reliability and reading range of RFID systems at high frequencies have been addressed in (Tang et al. (2009)). The authors have performed an analysis of the effects of antenna properties (gain, radar cross-section, half power beam-width, etc.) on the reading reliability of RFID systems.
On chip antenna technology (OCA) allows building antennae together with application chips considerably reducing size and production costs. For example, the authors in (Guo et al. (2006)) present an OCA design of a UHF inductive coupling tag with 1mm reading distance for access control applications. Dielectric materials used for antennas have also reduced their cost, thereby allowing reduction of price of passive tags. However, materials such as paper, which is common in consumer goods, have been found to decrease tag performances.
Regarding metallic environments the results reported in (Qing & Chen (2007)) suggest that RFID systems and antennas can be designed according to the constraints of particular metallic environments. Another work on this subject was carried out in (Wagner et al. (2007)), where the authors study the effects of metal on the final performance of RFID systems and propose the use of a dielectric material to avoid grounding of the antenna port. Other approaches to avoid the effects of metal using antenna design can be found in (Chen et al. (2009); Chen (2009); Gao et al. (2009)). The authors in (Gao et al. (2009)) have also designed an antenna with a dielectric substrate that avoids the antenna port to be grounded. A different approach is followed in (Chen et al. (2009)), where the authors design the antenna directly over a metallic plate using an I-shaped hole or feed port, thereby avoiding metal grounding. Since the size of antennas for metallic objects can result quite large, the authors in (Chen (2009)) have proposed a method to reduce the size of this type of antenna by introducing a conducting line that increases the inductance of the antenna without the need of increasing its size.
Since UHF tags are limited by the activation threshold of the chip and by the efficiency of the energy harvesting mechanism, lowering power consumption is crucial in improving reliability (Hartman & Clairborne (2007)). Reduction of power consumption can be achieved in different ways. For example, reducing voltage and reducing clock rate of the tag have been proposed in (Wang et al. (2007)) and references there in. Power consumption of the analog devices in the tags has also been discussed, particularly of local oscillators and voltage regulators. Mechanisms developed in other papers are claimed to provide further reduction in power consumption. The design of efficient voltage rectifiers with efficiencies as high as 37% is proposed as another way to further reduce power consumption. A low power tag design

has been reported in (Pillai et al. (2007)) where the authors describe an ultra low power UHF and microwave tag that can be used as active or passive tag. The tag design allows achieving ranges of more than 24 m at 900 MHz and 3.5 m at 2.4GHz.
At the reader side, the main challenge to improve reading reliability is to avoid the problem of carrier leakage (Wang et al. (2007)), which is due to the continuous transmission of carrier waves from the reader to the tags while the reader is overhearing tag responses. Carrier leakage can be reduced by means of efficient isolators or dynamic interference cancelation schemes. Reader improvement schemes may also include low power consumption designs, improvement in receiver sensitivity, antenna design (which may include smart and distributed antennas) and improved algorithms for reader collision.

4.2 MAC-layer schemes

4.2.1 Tag anti-collision algorithms

Tag anti-collision algorithms in RFID have been limited to ALOHA and binary tree schemes. ALOHA protocols are the simplest of all: they consist of allowing users to transmit at free will, and in case of collision each user enters into a back-off random retransmission state (Abramson (1970)). The implementation simplicity of ALOHA algorithms comes at the expense of a low channel utilization and stability problems. The non-slotted version of ALOHA only reaches 18% (e-2) of channel utilization, while the slotted version only reaches 36%(e-1) (Bertsekas & Gallager (1992)). In addition, ALOHA without appropriate retransmission strategy has been proved unstable. Thus, tags need either to adapt their retransmission schemes according to traffic load, or use a fixed retransmission scheme at the expense of losing stability and reduce even more reading reliability. ALOHA schemes can be also improved by optimizing the retransmission strategy with context information. For example, if large numbers of tags are expected in one of the readers, the retransmission strategy can be adapted accordingly to reduce collisions during back-off periods.
In RFID a modified ALOHA protocol called Framed-ALOHA has been implemented to allocate different tags in consecutive frames thereby avoiding tags being detected in consecutive slots (see Burdet (2004)). A further improvement on Framed-ALOHA has been presented in (Liu (2009)) where frames of different sizes are used in order to reduce the effects of idle and non-successful slots. The work is also an evolution of TEM techniques (tag estimation method) that are used to improve performance of RFID MAC algorithms. ALOHA protocols are usually improved by using carrier-sense or resource reservation approaches (Bertsekas & Gallager (1992)). However, these schemes are unfeasible if we desire to keep tags as simple as possible. By contrast we have the area of splitting tree protocols (see citeCapetanakis79a,bertsekas92). Unlike ALOHA, these algorithms have the ability of being stable under favorable channel conditions. In these algorithms tags are allowed to transmit at free will too, but once a collision has been detected they split into two or more groups by means of a binary/m-ary decision. Tags in one group are allowed to retransmit in the next slot while the others remain silent. The procedure is repeated until all the contending tags are decoded free of collisions. Despite their good stability properties tree algorithms may suffer from delay as compared to ALOHA and they also reach limited channel utilization. Binary algorithms reach at most 34% of channel utilization, while the well known FCFS (First Come First Served) algorithm has been proved to reach 48% of channel utilization (Bertsekas & Gallager (1992)). Tree algorithms are also prone to eavesdropper readers that listen to the feedback broadcast by the reader. Since the reader transmits at higher power levels than tags,

the information can be overheard by eavesdropper readers at long distances. Thus security mechanisms for the feedback of tree algorithms have been proposed (Xiao et al. (2006)).

Further improvements can be achieved by using dynamic tree algorithms (e.g., Capetanakis-a (1979)). The ability of these schemes is to act as ALOHA protocols at light traffic loads thereby achieving low delay figures, and act as TDMA protocols at high traffic loads thus reducing the number of collisions. For example, the adaptive binary splitting (ABS) protocol proposed in (Myung et al. (2006)) exploits the information collected from previous collision periods to avoid starting with sets of multiple tags in the next contention period. This scheme allows reducing access delay and outperforms previous binary splitting algorithms for RFID applications. A similar approach has been used by the authors in (Yan & Zhu (2009)) where they enhance the performance of a binary tree algorithm by estimating the tag population of the next time slot and thus adapt the variables of the tree algorithm accordingly. Another example of population estimation for RFID can be found in (Xue et al. (2009)) where a fuzzy logic algorithm is used to group tags and improve collision resolution algorithms.

4.2.2 Reader collision resolution algorithms

Reader collision resolution algorithms can be broadly classified here as scheduling-based or coverage-based (power-control), and also as centralized or decentralized, depending on whether a central server schedules the different readers or they autonomously decide when to transmit.

RFID standards have defined schemes for reader collision resolution. For example, early versions of EPC (Electronic Product Code) RFID standards considered a simple frequency division multiple access (FDMA) scheme for reader collision avoidance. By contrast, ETSI (European Telecommunication Standards Institute) standards have used an ALOHA-based reader anti-collision algorithm with carrier sense features, also known as the Listen-before-talk (LBT) algorithm. However, ALOHA efficiency is not as high as required in RFID applications, while carrier-sense features are prone to hidden/expose terminal problems and also suffer from complexity issues at the reader side in order to implement the sensing mechanism (Birari & Iyer (2005)). A proposal for a medium access technique for RFID readers is provided in (Quan et al. (2008)) and is referred to as Slotted-LBT (S-LBT). Based on carrier-sensing or LBT, this algorithm makes use of several channels, and in case the selected channel is sensed as busy, a new channel is considered for the transmission. Slotted LBT does not require any control from the middleware, but the readers must implement a reader-to-reader communication protocol in order to synchronize themselves.

The scheme called Colorwave implements a distributed time division multiple access protocol where readers select at random a particular slot or color to transmit. If a collision occurs the reader is able to detect it and to retransmit in other color/slot while informing its neighbors of such a change (Waldrop et al. (2003)). Unfortunately this type of solution relies on collision detection schemes at the readers and also requires environments with relatively low numbers of readers. Furthermore, collision detection and stabilization mechanisms require feedback from tags, which is not yet supported by current commercial technologies.

Another reader anti-collision algorithm, referred to as Pulse (Birari & Iyer (2005)), has been proposed for mobile RFID reader scenarios. Pulse uses two non-interfering channels, one for control and another one for data transfer. The control channel is used to inform neighbor readers of possible transmissions and thus avoid collisions. Power of the control channel is adjusted to make sure other readers hear the beacon signals. However, collisions between pulses may still arise. Pulse has been proved effective against the collisions among readers,

even in mobility scenarios. The disadvantage of Pulse is the deployment of additional channels that need to be decoded by readers. A similar approach to Pulse has been followed in (Eom et al. (2009)) where a control server is in charge of organizing a semi-decentralized resource allocation algorithm. Each reader follows the commands transmitted by the server and also transmits a beacon to identify collisions with neighbor readers. The algorithm reduces the large overhead required by other solutions. In (Hsu et al. (2009)), an improved version of Pulse has been proposed. The transmission range of the control channel is dynamically adjusted based on the density of the neighbor readers estimated by each device. A learning algorithm called HiQ has been proposed in (Junius (2003)) where dynamic solutions to the reader collision problem are obtained by learning the collision patterns of the readers and by effectively assigning frequencies over time. HiQ relies on a centralized server called Q-server that runs the learning algorithm and that assigns resources in order to minimize collisions. Another approach to solve both reader and tag collisions is presented in (Kim et al. (2009)) where the authors have presented a master-slave algorithm for both readers and tags with different frequency hopping sequences. The algorithm reduces both reader and tag collisions at the expense of complexity to switch to different frequency hopping sequences. Two approaches can be found in coverage-based algorithms: those that reduce the overlapping area between neighbor readers (e.g., Kim et al. (2009)) and that also aim at reducing the interference from multiple readers to tags, and those that monitor the interference between readers and adapt the transmit power of each one of them accordingly (e.g., Cha et al. (2007)). The work in (Kim et al. (2009)) has addressed two problems: a homogenous case where all readers have the same computing power and a heterogeneous case where readers are allowed to have different computing powers. The algorithm assumes a centralized server where the LLCR algorithm (low energy localized cluster for RFID) is run with the information retrieved from every reader (position and energy state). The algorithms are divided into two phases: one for initial phase control and another one for iterative policy. Two optimization rules were used: non-linear programming (NLP) and vector computation (VC). The algorithm has shown good results in reducing overlapping areas between readers thereby reducing the problem of multiple-readers-to-tag collision. A slightly different approach is followed by (Cha et al. (2007)) where the proposed scheme aims at reducing the interference from reader-to-reader. The authors present a novel distributed and adaptive power control algorithm followed by a selective back-off algorithm.
The complexity of collision algorithms can be determined by the number of operations per unit of time or per reading rate. ALOHA schemes are the simplest and the complexity increases as additional functionalities such as carrier-sensing, tag estimation and control channels are implemented. Distributed algorithms and context aware improvements also require additional feedback channels to be supported by readers.

5. Cross-layer algorithms

During the last century wire-line communication systems experienced considerable success and technological development. Part of this success was due to the concept of layered architecture design which allowed distribution of simplified tasks between semi-isolated layers and consequently manufacturer inter-operability. Wireless systems during the 80s and 90s were designed as extensions of their wireline counterparts, thereby reusing layered methodologies. Over the last few years, however, layered models have shown several drawbacks in achieving the data rates required by modern wireless applications (Dimic et al. (2004)). Reliable communication through wireless channels has been found to require, inherently, design across different layers, which has been coined cross-layer design.

Cross-layer design solutions can be generally classified as follows (Srivastaya & Montani (2005)): downward information flow, where information from an upper layer is used to tune the parameters of a lower layer; upward information flow, where the parameter exchange is in the opposite direction; back-and-forth, where the information flow is in both directions; design coupling, where one of the layers is fixed and another one is redesigned to cope with the fixed layer; vertical calibration, where parameters across different layers are simultaneously tuned; and merging of adjacent layers where two or more adjacent layers are completely jointly optimized.

Cross-layer solutions can adopt either only slight or tight interaction rules between layers. Tight cross-layer design can also be translated into a loss of architectural rules, which in the long term affects manufacturer inter-operability and increases the signaling bandwidth required for interaction between layers. Thus, cross-layer design must be accompanied by a careful evaluation of signaling loads and impact on architectural principles.

Typical examples of downward information flow are schedulers based on application layer priorities. In upward information flow we find channel-aware schedulers and transport adaptation schemes for wireless networks. Back-and-forth algorithms can be exemplified by schedulers with power control, while vertical calibration can be observed in solutions with error correction capabilities across different layers. Finally, the case of merging of adjacent layers represents the most attractive solution with examples given by joint design of scheduling, power control and link adaptation, as well as random access protocols jointly assisted by source separation and retransmission control.

Cross-layer design has been recognized as a key factor in achieving the stringent data rates required by future wireless networks. Therefore, wireless standards have adopted cross-layer design not only as a potential option but as mandatory for new schemes such as MIMO (multiple-input multiple-output) and distributed antenna systems. In the context of RFID, only context aware solutions and some multiple access protocols with tag estimation methods can be considered as early examples of cross-layer design. However, given the results of these few examples and the literature on cross-layer design it is envisioned a lot of potential improvement in RFID schemes by using this new paradigm, particularly at medium access control and physical layers.

5.1 MAC/PHY cross-layer design

Perhaps the best example of cross-layer design in wireless networks is the joint analysis of PHY and MAC layers. The physical layer is in charge of transmitting raw bits of information across a communication channel. It also defines modulation parameters, signal amplitudes, and mechanical and electrical specifications for reliable transmission of information. On the other hand, the MAC layer is in charge of scheduling the initially uncoordinated transmissions of a group of terminals who share the same medium, thus being in charge of avoiding or resolving the possible conflictive interactions between them.

Traditionally, MAC protocols were designed by considering the PHY layer as a "black box" with a behavior that was assumed to remain constant over long periods of time (as in a wire-line channel). However, the random phenomena that govern wireless environments (such as fading and multi-path transmission) create completely different conditions and thus other assumptions must be considered (Shakkotari et al. (2003)). Furthermore, the last two decades have witnessed the revolution of digital communications and the advent of faster

and more reliable signal processors. This has made possible the implementation of complex signal processing techniques to cope more efficiently with harsh propagation conditions. The consequences of improved physical layer operations and the random behavior of wireless channels have not been appropriately modeled by conventional protocols at the MAC layer.
The first works that can be considered as cross-layer were the studies of the influence of wireless channels on ALOHA protocols (e.g., Abramson (1970; 1977)). Further investigations of throughput and stability of ALOHA under the power capture effect have been reported since then (e.g., Zorzi & Rao (1994)). The power capture effect allows the correct decoding of a packet if its power is much larger than the combined power of all the other contending packets. The power capture effect was used in channel aware stabilization schemes of ALOHA showing the direct relation between the maximum stable throughput (MST) and the roll-off parameter of the channel (Zorzi & Rao (1994)). Since most of the tag anti-collision algorithms in RFID (for tag and reader collision) are based on the ALOHA system, all these results can be potentially used to further optimize the operation of current solutions.
Another relevant work in MAC/PHY cross-layer design was presented in (Ghez et al. (1988)). The authors analyzed the stability properties of ALOHA with multi-packet reception under symmetrical and infinite user scenarios. The novelty brought by this approach was a stochastic multi-packet reception matrix that represents in an accurate way the impairments of wireless channels and signal processing schemes with multiple antenna diversity. A further improvement was presented in (Naware et al. (2005)), where the authors extended the model to the asymmetrical user scenario and proposed a stochastic reception model based on conditional reception probabilities. The relevance of these works for RFID systems is that random access protocols with multiple antennas can be used to improve tag reading rates in the uplink. Readers can implement modified ALOHA protocols with multi-packet reception and considerably reduce tag collisions. Thus, the tools developed in these works can be used directly in the analysis of advanced cross-layer features for RFID including the signal processing schemes and impairments discussed in previous subsections.
A different approach to achieve diversity in multiple access protocols was presented in (Tsatsanis et al. (2000)), where packet collisions are resolved by means of protocol-induced retransmissions. In NDMA, a MIMO system is created by collecting consecutive packet retransmissions. The packets are then recovered using conventional multiuser detection schemes. NDMA has been proposed for RFID applications in (Samano & Gameiro (2008)). NDMA is particularly attractive for RFID applications since it allows using signal processing tools to combine several tag readings received at different times.

5.1.1 Context-aware solutions

Context aware solutions are employed in RFID applications to enhance security/privacy features and to improve reading reliability levels. Security/privacy enhanced features are based on the concept that some tags will follow a given trajectory inside a business process or factory. Therefore, tags will be read with higher probability by some readers rather than others. Middleware applications can easily detect unauthorized attempts to read a tag by a reader which is not supposed to do that, and vice versa to detect unauthorized tags that attempt sending information from unauthorized location. Therefore correlation between tags and physical locations has been found useful in improving security and privacy features.
A similar approach can be used to improve reading reliability figures. For example, tags that move across a supply chain follow known paths and locations. Therefore, their movements can be predicted with certain accuracy. Whenever a false negative occurs, the middleware can

perform a modified decision based on previous outcomes to infer that the tag is in the vicinity of a reader with high probability and that perhaps a reading error caused the tag not being detected. Outcomes from different readers can be stored to provide a historical record to infer the real trajectory of the tag and thus eliminate both false negative and false positive readings. For example, a middleware approach to security is given by the authors in (Du et al. (2009)) where they use an access control scheme as a security layer of a reconfigurable middleware platform. The middleware platform is especially designed to provide security in ubiquitous environments. Security issues are also tackled by the security-enhanced RFID middleware platform proposed in (Song & Kim (2006)). This platform deploys a novel context aware access control service. The access control scheme prevents unauthorized users from having access to consolidated data provided by the middleware server.

Physical access control policies for captured RFID data has been addressed by the work in (Kriplean et al. (2007)), where a visibility metric is used to control access to data captured by authorized readers. Another work is given by the data cleaning model used by the authors in (Song et al. (2009)). The authors propose a virtual spatial granularity concept and a Bayesian estimation algorithm to cope with false positives and false negatives. The virtual spatial granularity concept exploits the fact that tags across a supply chain follow similar movements and spatial locations. The algorithm classifies tags according to their spatial movements and thus improves their probability of correct detection by estimating their next movement.

Another approach to improve reliability in RFID systems is given in (Ahmed et al. (2007)). The authors have proposed a middleware architecture called RF^2ID which is based on the concept of context aware design assisted by virtual reader and path abstraction models. Additionally, their design is oriented to organize queries in an efficient manner and provide high levels of reliability and scalability. The concept consists of creating virtual readers, which consider the unreliable nature of each interrogator, and virtual paths, which serve as a higher level abstraction that can identify and follow a tag moving across the environment. A virtual path can cope with false negative and false positive reads of a tag moving across different virtual readers. Another work on data cleaning models is reported in (Peng et al. (2009)), where the authors propose a P2P (Peer to Peer) collaborative model. The model exploits redundancy information that is exchanged between the different nodes across the path of a tag.

5.2 Cross-layer framework for optimization

Consider a set $\mathcal{R}$ of R readers $\mathcal{R} = \{r_1, r_2, \ldots, r_R\}$ and a set $\mathcal{T}$ of J tags $\mathcal{T} = \{t_1, t_2, \ldots, t_J\}$. We consider that a subset of tags $\mathcal{T}_A \subset \mathcal{T}$ appears in the vicinity of the area under analysis with probability $\Pr\{\mathcal{T}_A\}$. For context-aware purposes we further define the conditional probability of inter-tag arrival as $\Pr\{t_j \in \mathcal{T}_A | t_l \in \mathcal{T}_A\}$ for tag-to-tag correlation, $\Pr\{t_j \in \mathcal{T}_A | \mathcal{S} \in \mathcal{T}_A\}$ for correlation between a single tag t_j and a group $\mathcal{S}$ of tags, and $\Pr\{\mathcal{U} \in \mathcal{T}_A | \mathcal{S} \in \mathcal{T}_A\}$ for correlation between two groups of tags ($\mathcal{U}$ and $\mathcal{S}$). The transmit power level of reader r_k will be denoted by $P_k^{(r)}$ and the subset of scheduled readers can be denoted by $\mathcal{R}_t \subset \mathcal{R}$. The probability of transmission of reader r_k will be denoted by $p_k^{(r)}$. Additionally, the transmit power level of tag t_j will de denoted by $P_j^{(t)}$, the set of activated tags will be $\mathcal{T}_P$ and the subset of tags that transmit their ID once they have been activated, also called contending tags, will be denoted by $\mathcal{T}_t$.

Now consider that the instantaneous channel between reader r_k and tag t_j is given by $h_{k,j}^{(rt)}$ for the main multi-path component and $g_{k,j}^{(rt)}$ for the combined effect of additional multi-path components. Similarly, the channel experienced between reader r_k and reader r_m is given by $h_{k,m}^{(rr)}$ for the main component and $g_{k,m}^{(rr)}$ for additional multipath components. Finally, the channel experienced between tag t_n and tag t_j is given by $h_{n,j}^{(tt)}$ for the dominant multipath component and $g_{n,j}^{(tt)}$ for the combined effect of the remaining multi-path components. All channels may include both fast- and slow-fading distributions, as well as path loss and radiation patterns as the result of using, for example, smart antennas or beamforming algorithms. The signal-to-interference-plus-noise ratio (SINR) experienced by tag t_j due to a transmission from reader r_k will be denoted by $\gamma_{k,j}^{(rt)}$ and can be mathematically expressed as follows:

$$\gamma_{k,j}^{(rt)} = \frac{P_k^{(r)} |h_{k,j}^{(rt)}|^2}{I_{k,j}^{(s)} + I_{k,j}^{(r)} + I_{t,j} + \sigma_{v,j}^2}, \qquad r_k \in \mathcal{R}_t \tag{1}$$

where $I_{k,j}^{(s)} = P_k^{(r)} |g_{k,j}^{(rt)}|^2$ is the inter-symbol interference due to multi-path distortion, $I_{k,j}^{(r)} = \sum_{m \in \mathcal{R}_t, m \neq k} P_m^{(r)} (|h_{m,j}^{(rt)}|^2 + |g_{m,j}^{(rt)}|^2)$ is the interference created by other active readers, $I_{t,j} = \sum_{n \in \mathcal{T}_t, n \neq j} P_n^{(t)} (|h_{j,n}^{(tt)}|^2 + |g_{j,n}^{(tt)}|^2)$ is the interference created by other tags, and $\sigma_{v,j}^2$ is the noise component. The SINR expression in eq.(1) can be also modified to represent multiple antenna schemes or other diversity mechanism by considering the contributions from different diversity sources.

If the SINR experienced by tag t_j is above the tag sensitivity threshold $\widehat{\gamma}_j^{(t)}$, then the tag is considered as active. The probability of tag t_j being activated is given by $\Pr\{t_j \in \mathcal{T}_P\} = \Pr\{\max_k \gamma_{k,j}^{(rt)} > \widehat{\gamma}_j^{(t)}\}$. In the strict sense the set of active tags should be a subset of the set of available tags, i.e., $\mathcal{T}_P \subset \mathcal{T}_A$. However, in some cases another tag which does not belong to the set of targeted tags $t_n \notin \mathcal{T}_A$ can also be activated by mistake, i.e.,$t_n \in \mathcal{T}_P$, thus being considered as a potential false positive. Tags are also considered to start a random transmission process once they have been activated, which will prevent collisions with other actives tags. This random transmission control will be characterized as a Bernoulli process with parameter p_j.

Now consider the backscattering factor function $\beta_j(\gamma_{k,j}^{(rt)})$ and the transmission power of tag t_j which can be calculated as $P_j^t = \beta_j(\gamma_{k_{opt},j}^{(rt)}) P_{k_{opt}}^{(r)} |h_{k_{opt},j}^{(rt)}|^2$. The term $r_{k_{opt}}$ (where $k_{opt} = \arg\max_k \gamma_{k,j}^{(rt)}$) denotes the reader that has activated the tag. Thus, the SINR of the backscattered signal from tag t_j upon reader r_k can be written as:

$$\gamma_{j,k}^{(tr)} = \frac{P_j^t |h_{j,k}^{(tr)}|^2}{I_{j,k}^{(s)} + I_{r,k} + I_{j,k}^{(t)} + P_k^{(r)} \eta_k + \sigma_{v,k}^2}, \qquad t_j \in \mathcal{T}_t \tag{2}$$

where $I_{j,k}^{(s)} = P_j^{(t)} |g_{j,k}^{(tr)}|^2$ is the inter-symbol interference due to multi-path distortion, $I_{r,k} = \sum_{m \neq k} P_m^{(r)} (|h_{m,k}^{(tr)}|^2 + |g_{m,k}^{(tr)}|^2)$ is the interference created by other active readers, $I_{j,k}^{(t)} = \sum_{n \neq j} P_n^{(t)} (|h_{n,k}^{(tr)}|^2 + |g_{n,k}^{(tr)}|^2)$ is the interference created by other active tags and $\sigma_{v,k}^2$ is the noise component at the reader side. Interference cancelation schemes or multiple access protocols based on diversity can help in reducing the interference terms in the denominator, thus improving the SINR received at the reader side. Furthermore, the backscattering function works as an abstraction model of all tag physical layer schemes. New electromagnetic antenna or chip designs with reduced power consumption can be easily abstracted into this function. Let us now consider that tag t_j can be detected by reader r_k if the received SINR is above a threshold denoted by $\widehat{\gamma}_k^{(r)}$. The set of detected tags by reader r_k will be denoted by $\mathcal{T}_D^{(k)}$, hence the probability of tag t_j being in $\mathcal{T}_D^{(k)}$ will be given by $\Pr\{t_j \in \mathcal{T}_D^{(k)} | t_j \in \mathcal{T}_P\} = \Pr\{\gamma_{j,k}^{(tr)} > \widehat{\gamma}_k^{(r)}\}$. For context aware purposes we can also consider correlation between different readers (spatial correlation) $\Pr\{t_j \in \mathcal{T}_D^{(k)} | t_j \in \mathcal{T}_D^{(m)}\}$, which can be further extended along the time domain as $\Pr\{t_j \in \mathcal{T}_D^{(k)}(\Delta) | t_j \in \mathcal{T}_D^{(m)}(\Delta + \delta)\}$.

5.2.1 Optimization

The parameters to be optimized are the set of scheduled readers $\mathcal{R}_t \subset \mathcal{R}$ (or the vector of transmission probabilities $\mathbf{p}^{(r)} = [p_1^{(r)}, \dots p_R^{(r)}]^T$), the vector of transmit powers $\mathbf{P}^{(r)}$ whose elements are the transmit power levels $P_m^{(r)}$ of $r_m \in \mathcal{R}_t$, and the transmission probabilities of the active tags $\mathbf{p}^{(t)}$ whose elements are the transmission probabilities $p_j^{(t)}$ of $t_n \in \mathcal{T}_P$. The main target of the optimization will be the maximization of the number of correctly detected tags per reader ($|\mathcal{T}_D^{(k)} \cap \mathcal{T}_A|$ where $|\cdot|$ is the cardinality operator) and optionally the minimization of the number of false positives readings ($|\mathcal{T}_D^{(k)} \cap \overline{\mathcal{T}}_A|$, where $\overline{(\cdot)}$ denotes the complement set operator). There are several ways to express the optimization problem. A straightforward option can be optimizing the summation of all the correctly detected tags per reader as follows:

$$\{\mathbf{P}^{(r)}, \mathbf{p}^{(t)}, \mathcal{R}_t\}_{opt} = \arg \max_{\{\mathbf{P}^{(r)}, \mathbf{p}^{(t)}, \mathcal{R}_t\}} \sum_{r_k \in \mathcal{R}} |\mathcal{T}_D^{(k)} \cap \mathcal{T}_A| \tag{3}$$

However, this type of optimization, which is similar to a sum-rate optimization problem, leads to unfairness by giving too much weight to readers with good conditions. To counteract this problem it is possible to use a transmit power constraint as follows:

$$\{\mathbf{P}^{(r)}, \mathbf{p}^{(t)}, \mathcal{R}_t\}_{opt} = \arg \max_{\{\mathbf{P}^{(r)}, \mathbf{p}^{(t)}, \mathcal{R}_t\}} \sum_{r_k \in \mathcal{R}} |\mathcal{T}_D^{(k)} \cap \mathcal{T}_A| \quad \text{s.t.} \quad \mathbf{P}^{(r)} < \mathbf{P}_0^{(r)} \tag{4}$$

Or by optimizing one individual reader subject to the throughput of all the other readers being constant:

$$\{\mathbf{P}^{(r)}, \mathbf{p}^{(t)}, \mathcal{R}_t\}_{opt} = \arg \max_{\{\mathbf{P}^{(r)}, \mathbf{p}^{(t)}, \mathcal{R}_t\}} |\mathcal{T}_D^{(k)} \cap \mathcal{T}_A| \quad \text{s.t.} \quad \mathbf{P}^{(r)} < \mathbf{P}_0^{(r)}, \quad |\mathcal{T}_D^{(m)} \cap \mathcal{T}_A| = \theta_m, m \neq k \tag{5}$$

This particular optimization can be modified to cope complexity-constrained environments. Defining a complexity measure of the reader as a function of the tag reading rate, i.e. $C_k = f_{ck}(|\mathcal{T}_D^{(m)} \cap \mathcal{T}_A|)$, then the expression $|\mathcal{T}_D^{(m)} \cap \mathcal{T}_A| = \theta_m$ or $|\mathcal{T}_D^{(m)} \cap \mathcal{T}_A| < \theta_m$, represents a complexity constraint. Another approach is to minimize the total power of the readers subject to a constant level of successful tag readings per reader:

$$\{\mathbf{P}^{(r)}, \mathbf{p}^{(t)}, \mathcal{R}_t\}_{opt} = \arg \min_{\{\mathbf{P}^{(r)}, \mathbf{p}^{(t)}, \mathcal{R}_t\}} \sum_k P_k^{(r)} \quad \text{s.t.} \quad |\mathcal{T}_D^{(m)} \cap \mathcal{T}_A| = \theta_m, \tag{6}$$

The above optimization problems assume perfect knowledge of channels between readers and tags, which is an unrealistic assumption. However, the optimization problem can be modified to use average channel values instead of instantaneous values. These average channel values can be defined over a given optimization area for each reader.

5.2.2 Reader and tag ALOHA protocols: joint optimization

Consider a symmetrical system where all devices of the same kind (readers or tags) are statistically equivalent and with fixed transmit power. Slotted ALOHA protocol will be used as contention mechanism both in the reader and tag sides including incorrect detection and activation probabilities. Two main assumptions will be used: one in which readers and tags do not interfere with each other except for the powering-up process, and another one in which they have close interaction.

Scenario without reader-tag interference. In this subsection we consider that the activation process of tags from readers and tag transmissions toward readers do not interfere with each other. The probability of a group of u tags being activated, denoted here by p_u, assuming ALOHA operation will be given by the probability that only one reader transmits in a time-slot and that its signal strength is high enough to power-up the tag, which occurs with probability $P_{dt} = \Pr\{\gamma_{rt} > \widehat{\gamma}_t\}$. Therefore p_u can be written as:

$$p_u = \binom{J}{u} P_{dt}^u (1 - P_{dt})^{J-u} R p_r (1 - p_r)^{R-1}. \tag{7}$$

The tag throughput (T) of all the readers can thus be expressed as the modified formula of ALOHA for each possible number of active tags u:

$$T = \sum_u u p_u (1 - P_{dr})^R p_t (1 - p_t)^{u-1}, \tag{8}$$

where P_{dr} is the probability that a single tag transmission is correctly detected by any of the readers, and which can be written as $P_{dr} = \Pr\{\gamma_{tr} > \widehat{\gamma}_r\}$. Results for a scenario with 15 tags and 5 readers with power-up probability $P_{dt} = 0.7$ and probability of detection at the reader of $P_{dr} = 0.95$ are displayed in Fig. 2a. It can be observed that optimum probabilities of the reader and tag anti-collision components are independent (no need of joint optimization).

Scenario with full reader-tag interference. Consider now that activation of tags from readers and tag transmissions toward readers interfere with each other. The state of the system is defined as the number of powered-up tags. The transition probability between state m and state n is given by

$$p_{mn} = \begin{cases} \binom{m}{m-n} p_t^{m-n} (1 - p_t)^n & n < m \\ \binom{J-m}{n-m} P_{dt}^{n-m} (1 - P_{dt})^{J-n} R p_r (1 - p_r)^{R-1} (1 - p_t)^m & n > m \\ (1 - p_t)^m & m = n = J \\ (1 - R P_{dt} p_r (1 - p_r)^{R-1})(1 - p_t)^m & m = n, n \neq J \end{cases} \tag{9}$$

The transition probabilities define a Markov chain that can be solved using standard tools. Throughput can be finally assessed using

$$T = \sum_u u p_u (1 - (1 - P_{dr})^R) p_t (1 - p_t)^{u-1} (1 - p_r)^R. \tag{10}$$

Results for the same scenario as in the previous example are displayed in Fig. 2b. It can be observed that probabilities of the reader and tag anti-collision components are now dependent on each other. This means that, unlike the previous example, joint optimization of reader and tag algorithms is now justified.

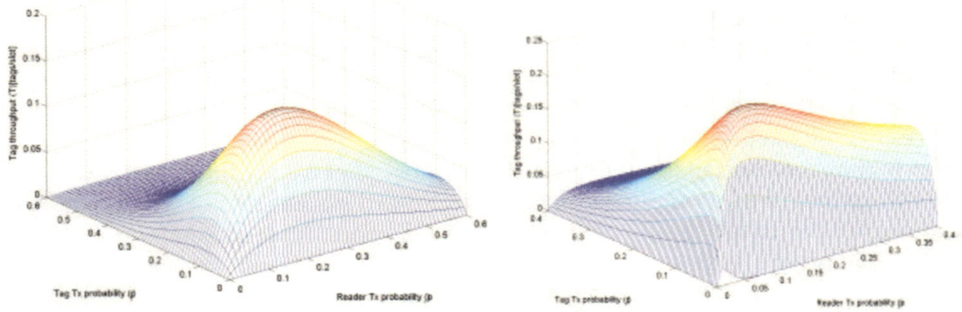

(a) ALOHA model 1 (b) ALOHA model 2

Fig. 2. Joint evaluation of ALOHA reader and tag anti-collision algorithms optimization

5.2.3 ALOHA tag protocol with imperfect tag detection

We now address an asymmetrical ALOHA tag protocol with incorrect tag detection and false alarm. We define the probability of correct detection of tag t_j as $P_{D,j}$ and the false alarm as $P_{F,j}$. The throughput of ALOHA can thus be expressed as the probability that only one tag transmits provided it has been correctly detected and that the remaining tags either do not transmit or do not experience a false alarm:

$$T_j = p_j P_{D,j} \prod_{n=1;n\neq j} (1-p_n)(1-P_{F,n}) \tag{11}$$

Note that this expression is only for tag t_j and it belongs to the observations made by only one reader. Since probabilities of detection and false alarm are related to each other via the activation threshold of the tag, the joint optimization problem for calculation of the throughput region can thus be written as follows:

$$\{\mathbf{p}, \mathbf{P}_F\}_{opt} = \arg \max_{\{\mathbf{p},\mathbf{P}_F\}} T_j, \quad T_m = \theta_m, m \neq j \tag{12}$$

where $\mathbf{p} = [p_1, \ldots, p_J]^T$ and $\mathbf{P}_F = [P_{F,1}, \ldots, P_{F,J}]^T$. The optimization problem in the previous expression can be solved using the technique of Lagrange multipliers as follows:

$$\partial T_n / \partial p_j - \sum_{l\neq j} \nu_l \partial T_l / \partial p_j = 0 \qquad \partial T_n / \partial P_{F,j} - \sum_{l\neq j} \partial T_l / \partial P_{F,j} = 0 \tag{13}$$

where ν_l is the Lagrange multiplier associated to the l-th throughput constraint. Solving this system of simultaneous equations for different throughput constraints results in the derivation of the boundaries of the throughput region. An alternate approach consists of obtaining from the previous expression $J-1$ expressions that are independent from the

Lagrange multipliers and then solve the problem for the remaining variables and equations. For each one of these $J-1$ expressions, it is necessary to select J out of the $2J$ equations in the previous expression. Since each one of the selected combinations involves a different selection of variables either from $\mathbf{p}$ or $\mathbf{P}_F$, we will denote the m-th combination by the dummy vector variable $\mathbf{x}^m = [x_1^m, \ldots, x_J^m]^T$, where x_j^m denotes any of the probabilities of transmission or false alarm from the m-th combination. The derivation of the desired expression for the m-th combination is equivalent to solving the following equation:

$$det(\mathbf{Ja}^m) = 0$$

where $det(\cdot)$ denotes the determinant operator and $\mathbf{Ja}^m$ is the Jacobian matrix with elements given by $Ja_{k,j}^m = \frac{\partial T_k}{\partial x_j^m}$. Some simplifications of the expressions yield the following equations for the optimum transmission and detection probabilities, respectively:

$$\sum_{j=1}^{J} p_j = 1 \qquad \sum_{j=1}^{J} \left\{ \frac{P_{D,j}/(1-P_{F,j})}{\partial P_{D,j}/\partial P_{F,j} - P_{D,j}/(1-P_{F,j})} \right\} = 1. \tag{14}$$

The term $\frac{\partial P_{D,j}}{\partial P_{F,j}}$ describes the operational curve of the tag detector and it depends on the adopted reader collision algorithm, interference and noise. ALOHA performances using different SNR values with Rayleigh fading channels are displayed in Figure 3(a).

5.2.4 Retransmission diversity multiple access

The performance analysis and optimization of NDMA for RFID is similar to the ALOHA protocol in the previous subsection. The throughput of NDMA can be expressed as follows(see Samano et al. (2009) and references therein):

$$T_j = (1/L) p_j P_{D,j} \prod_{n=1, n\neq j}^{J} P_{U,n} \tag{15}$$

Where $P_{U,j} = p_j P_{D,j} + (1-p_j)(1-P_{F,j})$, $L = \sum_{j=1}^{J} P_{A,j} + \prod_{j=1}^{J}(1-P_{A,j})$ and $P_{A,j} = p_j P_{D,j} + (1-p_j)P_{F,j}$. The optimization, using the previous subsection, gives(Samano et al. (2009)):

$$\sum_{j=1}^{J} \left\{ \frac{(\partial P_{U,j}/\partial x_j^m)(L/P_{U,j}) - (\partial L/\partial x_j^m)}{(\partial P_{U,j}/\partial x_j^m)(1/P_{U,j}) - \partial(p_j P_{D,j})/\partial x_j^m (1/p_j P_{D,j})} \right\} = L \tag{16}$$

Figure 3(b) shows the benefits of the NDMA protocol as compared to other multiple access protocols, thus being amongst the most attractive for RFID solutions (Samano et al. (2009)).

5.2.5 Two-user ALOHA protocol with context aware analysis

Let us now analyze an ALOHA protocol in the case where we have knowledge of the joint tag-arrival distribution, which is called here context information. Using this joint distribution we can infer the presence of another tag that due to imperfect channel conditions was not correctly detected. For convenience we analyze a system with only two tags. The probability space for tag arrival is given by $\Pr\{1\}$, which indicates the arrival of tag 1 only; $\Pr\{2\}$, which indicates the arrival of tag 2 only; $\Pr\{1,2\}$ which indicates the joint arrival of tags 1 and 2; and $\Pr\{\emptyset\}$ which indicates the probability that none of the tags are available. To denote the

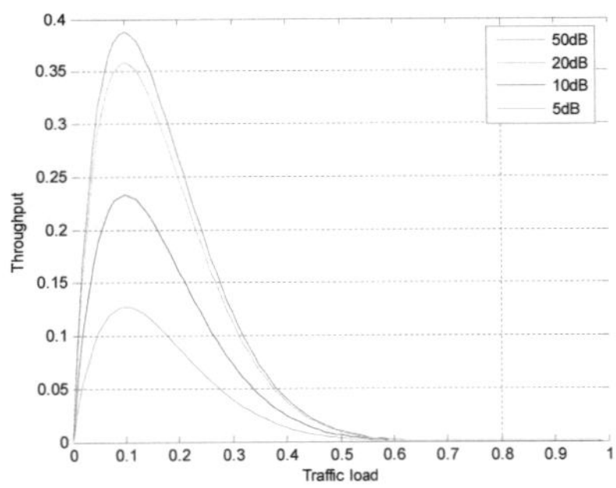

(a) ALOHA with imperfect tag activation

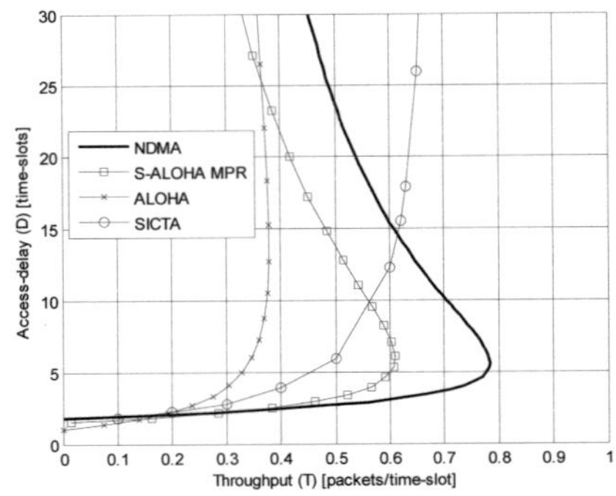

(b) NDMA vs. other multiple access protocols

Fig. 3. Performance of MAC algorithms with cross-layer optimization

imperfect detection of tags we will use P_{D1} and P_{D2} as the correct detection probabilities of tag 1 and 2, respectively, and P_{F1} and P_{F2} as the false detection probabilities of tag 1 and 2, respectively. The throughput of user one can be written as follows:

$$T_1 = \Pr\{1\}P_{D1}\overline{P}_{F2} + \Pr\{1,2\}P_{D1}\overline{P}_{D2} + x(\Pr\{1,2\}P_{D2}\overline{P}_{D1}), \tag{17}$$

where $\overline{(\cdot)} = 1 - (\cdot)$ and the parameter x is used to regulate the average probability of false positives. Note that the throughput with perfect tag detection is simply given by $T_1^0 = \Pr\{1\}$. Now the false positives of tag 1 are given by

$$F_1 = \Pr\{\emptyset\}P_{F1}\overline{P}_{F2} + \Pr\{2\}P_{F1}\overline{P}_{D2} + x(\Pr\{\emptyset\}P_{F2}\overline{P}_{F1} + \Pr\{2\}P_{D2}\overline{P}_{F1}) \tag{18}$$

The previous expressions indicate that we can increase the throughput of correct tag detections as much as we want by using the parameter x, which is an indication of how many correct tag detections we can infer of tag 1 from the observations of tag 2 being detected. Let us now illustrate an example with the following values: $\Pr\{1\} = 0.2$, $\Pr\{2\} = 0.2$, $\Pr\{1,2\} = 0.2$, $\Pr\{\emptyset\} = 0.4$, $P_{D1} = 0.3$,$P_{D2} = 0.3$,$P_{F1} = 0.4$,and $P_{F2} = 0.4$, which can be found in Figure 4(a) for several values of x. Note that the throughput of correctly detected tags lies below the throughput with perfect detection and the number of false positives. If we now use $\Pr\{1,2\} = 0.3$, $\Pr\{\emptyset\} = 0.3$ we obtain the graph in Figure 4(b). Similarly, we study the system for $\Pr\{1,2\} = 0.5$, $\Pr\{\emptyset\} = 0.1$ and $\Pr\{1,2\} = 0.6$, $\Pr\{\emptyset\} = 0$ in Figure 4(c) and Figure 4(d), respectively. Note that the higher the joint arrival probability the higher the throughput of correctly detected tags and the lower the number of false positives. Note that the throughput can surpass the one with perfect detection of tags. Therefore, context aware detection can improve the number correctly detected tags and false positives.

5.2.6 Complexity optimization

Consider the reader complexity C and the occupied bandwidth B as functions of the tag traffic λ: $C = f_c(\lambda)$ and $B = f_b(\lambda)$. As discussed in Section 2 reduction in complexity can be translated into an increase of traffic due to extra signaling procedures. On the contrary, an increase of signaling traffic is also translated into an increase of complexity to handle remote commands. The optimization problem can be thus be tackled in two different ways: to optimize complexity subject to bandwidth constraints ($\min C, \text{s.t.} B < B_0$) or to optimize

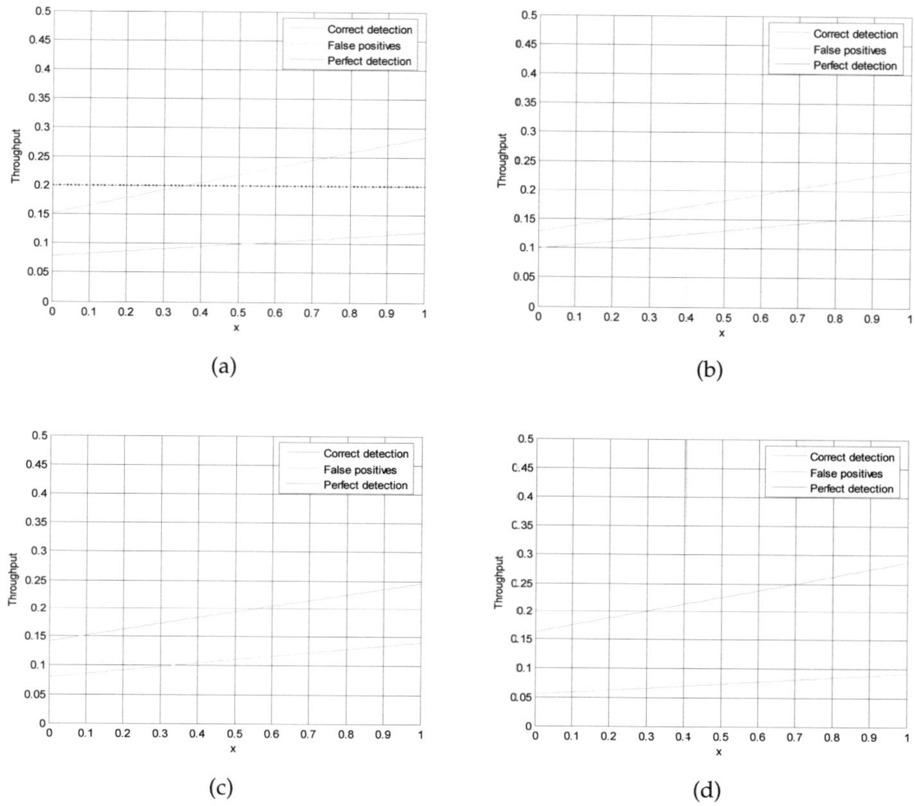

Fig. 4. Performance of ALOHA with context aware analysis

bandwidth subject to complexity constraints ($\min B, \text{s.t.} C < C_0$). Since complexity and bandwidth functions are difficult to express in analytical form, these optimization problems could be solved by exhaustive search in tables mapping hardware complexity figures and bandwidth requirements. Performance figures can also be included using the cross-layer framework proposed in this subsection.

6. Conclusions

Cross-layer design has been identified in this chapter as an attractive tool in the optimization of RFID platforms. After a thorough investigation of different impairments that affect RFID, as well as a review of existing algorithms and technologies across different layers of RFID architectures, it has been found that cross-layer methodologies can provide additional useful gains, particularly at the MAC and PHY layers. A general framework for MAC/PHY cross-layer optimization has been proposed for the design of RFID platforms. The framework can be potentially used for a wide variety of PHY and MAC algorithms, thereby paving the way for interesting research topics. Particular examples of MAC/PHY optimized algorithms with imperfect tag detection, reader collision, retransmission diversity and context aware mechanisms have shown the benefits of joint optimization of algorithms across different layers and its importance for future RFID applications. Future work includes to adopt security/privacy parameters and different signal processing schemes with multiple antenna diversity, beam-forming, and smart-antennas in the cross-layer framework proposed in this chapter.

7. References

Abramson N. (1979). The ALOHA system-Another alternative for computer communications, *Proc. Fall Joint Computer Conf., AFIPS,* Vol 44, Montvale, N.J., 1970, pp 281-285.

Abramson N.(1977).The throughput of packet broadcasting channels, *IEEE Transactions on Communications*, Vol. COM-25,No. 1, January 1977, pp. 117-128.

Ahmed, N.; Kumar, R.; French R.S. & Ramachandran, U.; (2007). RF2ID: A Reliable Middleware Framework for RFID Deployment. *IEEE International Parallel and Distributed processing Symposium*, March 2007, pp. 1-10.

Andrews J.G (2007). Interference cancellation for cellular systems: A contemporary overview. *IEEE Wireless communications Magazine* Vol. 12, No. 2, April 2007, pp. 19-29.

Angerers C.; Langwieser R.; Maier G. & Rupp M. (2009). Maximal Ratio Combining Receivers for Dual Antenna RFID Readers, *IEEE MTT-S International Microwave Workshop on Wireless Sensing, Local Positioning, and RFID* (IMWS 2009 - Croatia), pp. 1-4.

Balanis A. (2000). *Antenna Theory*, Ed. Wiley, 3rd edition, 2005.

Bertsekas D. & Gallager R. (1992). Data Networks.Prentice Hall 1992.

Birari S.M. & Iyer S. (2005). Mitigating the Reader Collision Problem in RFID Networks with Mobile Readers, *13th IEEE International Conference on Networks*, 2005, pp. 463-468

Burdet L.A. (2004) RFID Multiple Access Methods, *IEEE Antennas and wireless propagation letters,* ETH Zürich, Summer semester 2004, Seminar Smart Environments.

Capetanakis J. (1979). Generalized TDMA: the multi-accessing tree protocol, *IEEE Transactions on Communications,* Vol. 27, No. 10, October 1979, pp. 1476-84.

Cha K.; Jagannathan S. & Pommerenke D. (2007). Adaptive power control with hardware implementation for wireless sensor and RFID networks, *IEEE Systems Journal,* Vol. 1, No. 2, December 2007, pp. 145-159.

Chandramouli, R., Grance T.; Kuhn R. & Landau, S. (2005). Security Standards for the RFID Market. *IEEE Security and Privacy,* Vol 3, No. 6, pp. 85-89.

Chen X.; Lu F. & Ye T. (2006) Mutual Coupling of Stacked UHF RFID Antennas in NFC Applications. *IEEE Antennas and Propagation Society International Symposium*, 2009, pp. 1-4.

Chen S.L.; Lin K.H. & Mittra R. (2009). A novel triple-feed I-shaped slot antenna for an RFID tag designed for metallic objects, *IEEE Antennas and Propagation International Symposium*,2009, pp. 1-4.

Chen S.L. (2009) A Miniature RFID Tag Antenna Design for Metallic Objects Application, *IEEE Antennas and Wireless propagation Letters* 2009, pp. 1043-1045.

Cheng, C. & Prabhu V. (2009). Experimental Investigation of EMI on RFID in Manufacturing Facilities. *5th IEEE Conference on Automation Science and Engineering,* Bangalore, 2009, pp. 241-245.

Chia M.et al. (2007). Electronic Beam-Steering IC for Multimode and Multiband RFID. *IEEE Transactions on Microwave Theory and techniques* Vol. 57, No. 5, Part 2, May 2009, pp. 1310-1319.

Choi W. & Andrews J. (2007). Downlink performance and capacity of distributed antenna systems in a multi-cell environment, *IEEE Transactions on Wireless Communications,* Vol. 6, No. 1, January 2007, pp.69-73.

COST 231 Final report, COST (COperation européenne dans le domaine de la recherche Scientifique et Technique).

Darianian, M.& Michael, M.P. (2008). Smart home Mobile RFID-based Internet-of-Things systems and services. *International Conference on Advanced Computer Theory and Engineering ICACTE,* 2008, pp.116-120.

Dimic G.; Sidiropoulos N.D & Zhang R. (2004). Medium Access Control-Physical Layer Design, *IEEE Signal Processing Magazine,* Vol. 21, No. 5, September 2004, pp. 40-50.

Du B.; Ju S. & Wang D. (2009). Access Control for OSGi-Based Reconfigurable RFID Middleware, *ICCIT 09, fourth International Conference on Computer Sciences and Convergence Information Technology,* 24-26 Nov. 2009, pp. 1010 - 1014.

Eom J.B.;Yim S.B. & Lee T.J. (2009). An Efficient Reader Anticollision Algorithm in Dense RFID Networks With Mobile RFID Readers. *IEEE Transactions on Industrial Electronics,* Vol. 56, No. 7, July 2009, pp. 2326-2336.

Floerkemeier C.; Roduner C. & Lampe M. (2007). RFID Application Development with the Accada Middleware Platform *IEEE Systems Journal* Vol. 1, No. 2, December 2007, pp 82-92.

Floerkemeier, C. & Sarma, S.; (2008). An Overview of RFID system interfaces and reader protocols. *IEEE International Conference on RFID,* 2008, pp. 232-240.

Floerkemeier, C. & Sarma S. (2009). RFIDSim-A Physical and Logical Layer Simulation Engine for Passive RFID. *IEEE Transactions on Automation Science and Engineering,* 2009, pp. 33-43.

Gao B.; Heung C.H., Yuen M. & Murch D. (2009). Low Cost Passive UHF RFID Packaging with Electromagnetic Band Gap (EBG) Substrate for Metal Objects, *Proceedings of the 57th Electronic Components and Technology Conference (ECTC),* 2007, pp. 974-978.

Ghez S.; Verdu S. & Schwartz S. (1988).Stability properties of slotted Aloha with multipacket reception capability, *IEEE Transactions on Automated Control,* Vol. 33, No. 7, July 1988, pp. 640-649.

Guo L.H.; Popov A.P.;Li H.Y.;Wang Y.H; Bliznetsov V.;Lo G.Q.; Balasubramanian N. & Kwong D. (2006). A Small OCA on a 1 0.5-mm2 2.45-GHz RFID Tag-Design and Integration Based on a CMOS-Compatible Manufacturing Technology, *IEEE Electron Device letters*Vol. 27, No. 2, February 2006. pp. 26-28.

Hariharan, S. & Bukkatapatman S. (2009). Misplaced Item Search in a Warehouse using an RFID-based Partially Observable Markov Decision Process (POMDP) Model, *5th IEEE Conference on Automation Science and Engineering* Bangalore, India, August 22-25, 2009, pp. 443-448.

Hartman C.S.& Clairborne L.T. (2007). Fundamental limitations on reading range of passive IC-based RFID and SAW-based RFID*IEEE International Conference on RFID* Gaylord Texan Resort, Grapevine, TX, USA March 26-28, 2007,pp. 41-48.

Hsu C.H.; Chen S.C.; Yu C.H. & Park J.H. (2009). Alleviating reader collision problem in mobile RFID networks, *Personal and Ubiquitous Computing,* Vol. 13, No. 7 October 2009, Springer-Verlag.

Jahner, S.; Leimeister J.M.; Knebel U. & Krcmar, H. (2008). A Cross-Cultural comparison of perceived strategic importance of RFID for CIOs in Germany and Italy. *Proceedings of the 41st Annual Hawaii International Conference on System Sciences,* 2008, pp. 405.

Juels, A. (2006). RFID security and privacy: A research survey. *IEEE Journal on Selected Areas in Communications*, Vol. 24, No. 2, February -2006, pp.381-394.

Junius K.H. (2003). Solving the reader collision problem with a hierarchical q-learning algorithm. Master's thesis, Massachusetts Institute of Technology, February 2003.

Kamadar N.C.; Roy S.M. & Ikram M.S. (2008). Development of Smart Antenna for RFID reader*2008 IEEE International Conference on RFID*, Las Vegas, Nevada, April 16-17, 2008, pp. 65-73.

Kim, S.Y.; Lee J.K. (2009). A study on control method to reduce collisions and interferences between multiple RFID readers and RFID tag. *2009 International Conference on New Trends in Information and Service Science NISS*, 2009, pp.339-343.

Kriplean, T.; Kumar, R.; French R.S. & Ramachandran, U.; (2007). Physical Access control for captured RFID Data. *IEEE Pervasive Computing*, Vol. 6, No. 4, 2007, pp.48-55.

Leung S.Y. & Lan D.C. (2007). Performance of printed polymer-based RFID antenna on curvilinear surface *IEEE Transactions on Electronic Packaging Manufacturing* Vol. 30, No. 3, July 2007, pp. 200-205.

Liu D. (2009). ALOHA algorithm considering the slot duration difference in RFID system, *2009 IEEE International Conference on RFID*, 2009, pp. 56-63.

Lu F.; Chen X. & Ye T. (2009) Performance Analysis of Stacked RFID Tags. *IEEE International Conference on RFID*,2009, 330-337.

Myug J. & Srivastava J. (2006). Adaptive Binary Splitting for Efficient RFID Anti-collision, *IEEE Communication Letters*,Vol. 10, No. 3, March 2006, pp. 144-146.

Naware V.; Mergen G. & Tong L. (2005). Stability and delay of finite-user slotted ALOHA with multipacket reception, *IEEE Transactionns on Information Theory*, Vol. 51, No. 7, July 2005, pp. 2636-2656.

Nikitin P.V. & Rao K.V.S. (2007) Performance of RFID tags with multiple RF ports, *IEEE International Symposium of the Antennas and Propagation Society*, 2007,pp. 5459-5462.

Nummela J. (2007). 13,56 MHz RFID Antenna for Cell Phone Integrated Reader, *IEEE International Symposium of the Antennas and Propagation Society*,2007, pp. 1088-1091.

Park, N.; Lee, H.; Kim, H. & Won, D. (2006). A Security and privacy enhanced protection scheme for secure 900MHz UHF RFID reader on mobile phone. *IEEE 10th International Symposium on Consumer Electronics ISCE*, 2005, pp.1-5.

Park N.; Lee J.; Kim H.; Chung K. & Sohn S. (2009). A Layered Approach to Design of Light-Weight Middleware Systems form Mobile RFID Security (SMRM : Secure Mobile RFID Middleware System) *IEEE Workshop on Computational Intelligence in Vehicles and Vehicular Systems (CIVVS)* March 2009, pp 51-57.

Peng X.; Ji Z.; Luo Z.; Wong E.C. & Tan C.J. (2008). A P2P Collaborative RFID Data Cleaning Model, *International Conference on Grid and Pervasive Computer Work.*, 25-28 May 2008, pp. 304 - 309.

Pillai V.; Heinrich H.;Dieska D.;Nikitin P.V. & Rao K.V. (2007). An Ultra-Low-Power Long Range Battery/Passive RFID Tag for UHF and Microwave Bands With a Current Consumption of 700 nA at 1.5 V, *IEEE Transactions on circuits and systems: I: Regular papers*,Vol. 54, No. 7, July 2007, pp. 1500-1512.

Proakis, J. (1997). *Digital Communications*, McGraw-Hill.

Qing X. & Chen Z. (2007). Proximity effects of metallic environments on high frequency RFID Reader antenna: study and applications *IEEE Transactions on Antennas and Propagation* Vol. 55, No. 11, November 2007, pp. 3105-3111.

Qing X.; Goh C.K. & Cheng Z.N. (2009). Impedance Characterization of RFID Tag Antennas and Application in Tag Co-Design. *IEEE Transactions on Microwave Theory and Techniques*,vol. 57, No. 5, May 2009, pp. 1268-1274.

Quan C.H.; Choi J.C.; Choi G.Y. and Lee C.W.(2008). The Slotted-LBT: A RFID Reader Medium Access Scheme in Dense Reader Environments, *IEEE International Conference on RFID*, The Venetian, Las Vegas, Nevada, USA, April 16-17, 2008,pp. 207-214.

Quiling Z.; Chun Z.; Zhongqi L.; Jingchao W.; Fule L. & Zhihua W. (2007) A Robust Radio Frequency Identification System Enhanced with Spread Spectrum Technique, *International Symposium on Circuits and Systems ISCAS*, 2009, pp. 37-40.

Rautio J. (2010). RFID Design Using EM Analysis. *2010 Long Island Systems Applications and Technology Conference LISAT* 2010, pp. 1-6.

Sabesan, S. ; Crisp, M. ; Penty, R.V. & White, I.H. (2008). Demonstration of improved passive UHF RFID coverage using optically-fed distributed multi-antenna system. *IEEE International Conference on RFID*, 2009, pp. 217 - 224.

Samano-Robles; R. & Gameiro, A. (2009). Collision resolution algorithms for RFID applications. *Asia-Pacific Microwave Conference*, 2008, pp.1-4.

Samano-Robles; R. & Gameiro, A. (2009). Integration of RFID readers into wireless mobile telecommunications networks. *First International Conference on wireless commun., vehicular tech., information theory, and aerospace & electronics system technology, Wireless VITAE*, 2009, pp.327-331.

Samano-Robles; R. Ghogho M. & McLernon, D.C. (2009). Wireless Networks with Retransmission Diversity and Carrier-Sense Multiple Access. *IEEE Transactions on Signal processing*, Vol. 57, No. 9, pp. 3722-3726.

Shakkotari S.; Rappaport T.S. & Karisson P.C. (2003). Cross-layer design for wireless networks, *IEEE Communications Magazine*, Vol. 41, No. 10, October 2003, pp. 74-80.

Siden J.; Jonsson T. & Wang G. (2001) Performance degradation of RFID system due to the distortion in RFID antenna, *11th International Conference on Microwave and Telecommunication Technology*, 2001, pp. 371-373.

Sklar, B. (1997). Rayleigh fading channels in mobile digital communication systems Part 1: characterization. *IEEE Communications Magazine*, Vol. 6, No. 4, 1997, pp.48-55.

Song J. & Kim H. (2006). The RFID Middleware System supporting Context-Aware Access Control Service, *The 8th International conference on Advanced Communication Technology (ICACT2006)*. Vol. 1, 20-22 February 2006, pp. - 866.

Song B.; Qin P.; Wang H.; Xuan W. & Yun G. (2006). bSpace: A Data Cleaning Approach for RFID Data Streams Based on Virtual Spatial Granularity,*2009 fifth International Conference on Hybrid Intelligent Systems*, Vol. 3, 12-14 Aug. 2009, pp. 252 - 256.

Srivastaya, V. & Montani, M.; (2005). Cross-layer design: a survey and the road ahead. *IEEE Communications Magazine*, Vol. 43, No. 12, December 2005, pp.112-119.

Subramanian, V.; Chang, P.C.; Lee, J.B.; Molesa, S.E. & Volkman, S.K.; (2008). Printed organic transistors for ultra-low-cost RFID applications. *IEEE Transactions on Components and Packaging Technologies*, Vol. 28, No. 4, 2005, pp.742-747.

Tang Z.; He Y.; Hou Z.; Li B. & Tang Z. (2009). The Effects of Antenna Properties on read Distance in Passive backscatter RFID Systems, *International Conference on Networks Security, Wireless Communications and Trusted Computing*,Vol. 1, 2009, pp. 120-123.

Tsatsanis M.K.; Zhang R. & Banerjee S. (2000). Network-Assisted Diversity for Random Access Wireless Networks, *IEEE Transactions on Signal Processing*, Vol. 48, No. 3, March 2000, pp. 702-711.

Wagner J.; Fischer R. & Günther W.A. (2007). The influence of metal environment on the performance of UHF smart labels in theory, experimental series and practice. *First Annual RFID Eurasia conference* 2007, pp. 1-6.

Waldrop N.; Engels D.W.& Sarma E. (2003). An anticollision algorithm for the reader collision problem, *IEEE Internationl Conference On Communications (ICC '03)*, Ottawa, Canada, 2003, pp. 1206-1210.

Wang Z.; Sun X.; Zhang C.& Li Y.(2007). Issues in integrated circuit design for UHF RFID, *IEEE International Workshop on RF Integration Technologies,* Dec. 9-11, 2007, Singapore, pp. 322-328.

Weinstein R. (2005). RFID: A Technical overview and its application to the enterprise,*IT professional*, Vol. 7, No. 3, 2005, pp. 27-33.

WINNER deliverable D1.1.2 (2007). Final link level and system level channel models. Available at www.ist-winner.org

Xiao Y.; Shen X.; Sun B.; and Cai L. (2006). Security and privacy in RFID and Applications in Telemedicine, *IEEE Communications Magazine,* Vol. 44, No. 4, April 2006, pp. 64-72.

Xue Y.; Sun H. & Zhu Z. (2009). RFID Dynamic Grouping Anti-collision Algorithm Based on FCM, *International Joint Conference on Bioinformatics, Sysstems Biol. and Intelligent Computing,*2009,pp. 619-622.

Yan X. & Zhu G. (2009). An Enhanced Query Tree Protocol for RFID Tag Collision Resolution with Progressive Population Estimation, *International Conference on Mobile Adhoc and Sensor Systems (MASS)*, 2009, pp. 935-940.

Zorzi M. & Rao R. (1994).Capture and retransmission control in mobile radio, *IEEE Journal on Selected Areas of communications*, Vol. 12, No. 8, October 1994, pp. 1289-1298.